Agroforestry for Climate Resilience and Rural Livelihood

OTHER RELATED BOOKS

Title	Author	ISBN
• A Manual for Dryland Afforestation and Management	*G. Singh*	9788172339784
• A Text Book of Environmental Science	*Vidya Thakur*	9788172337568
• Agroforestry: Theory and Practices	*A.J. Raj*	9788172338664
• Applied Microbiology (Agriculture, Environmental, Food and Industrial Microbiology)	*S.M. Reddy*	9789386237903
• Biodiversity Conservation	*M.S.B. Kumar*	9788172334893
• Biodiversity, Traditional Knowledge and • Intellectual Property Rights	*S.R. Reddy*	9788172339692
• Forest Fires and their Control	*E.S. Artsybashev*	9789383692590
• Forest Mensuration: A Handbook for Practitioners	*G.J. Hamilton*	9789388399012
• Forest Nursery: How to Raise and Manage	*S.S. Sagwal*	9789388043991
• Forest Tree Seeds: Handbook	*S.S. Sagwal*	9789388043663
• Forest Tribology and Anthropology	*V.M. Mhaiske*	9788172339661
• Forestry Extension Making it Work	*Wazeka*	9789388399128
• Forestry Extension Methods	*D. Sim*	9789388399029
• Forestry: Principles and Applications	*A.J. Raj*	9788172338107
• Forestry Research Extension: Challenges & Strategies	*C.S. Dange*	9788172335960
• Forestry Technologies: A Complete Value Chain Approach	*K.T. Parthiban*	9789386102607
• Industrial Agroforestry: Perspectives and Prospectives	*K.T. Parthiban*	9788172339050
• Introduction to Forestry & Agroforestry	*K.T. Parthiban*	9789388172677
• Nursery and Plantation Practices in Forestry	*Vinod Kumar*	9788172337155
• Plantation and Agroforestry Pulpwood Value Chain Approach	*K.T. Parthiban*	9789386347985
• Sacred Groves of Rajasthan	*G. Singh*	9789386102881
• Shushk Kshetra Varnikaran Evam Van Prabandhan: Takniki Evam Kaaryavidhiya	*G. Singh*	9789386237941
• Silvipasture in India: Present Perspectives and Challenges Ahead	*D.R. Palsaniya*	9788172336868
• Traning Manual for Applied Agroforestry Practices	*Michael Gold*	9789388399104
• Tree Mortality Assesment and Mitigation	*J.D.S. Negi*	9788172339876
• Tropical Forest Ecosystems Structure and Function	*V.P. Singh*	9788172333706
• Vegetational Wealth of Aravalli Ranges	*A.M. Otaghvari*	9788172339265
• Wildlife Ecology and Conservation	*M. Balakrishnan*	9788172339746
• Wood Handbook: Wood as an Engineering Material	*Usda*	9789388399111

For complete details visit www.scientificpub.com

Agroforestry for Climate Resilience and Rural Livelihood

EDITORS

Inder Dev
Asha Ram
Naresh Kumar
Ramesh Singh
Dhiraj Kumar
A.R. Uthappa
A.K. Handa
O.P. Chaturvedi

Published by
SCIENTIFIC PUBLISHERS (INDIA)
5 A, New Pali Road, P.O. Box 91
Jodhpur 342 001 (INDIA)
E-mail: info@scientificpub.com
Website: http://www.scientificpub.com

ISBN: 978-93-87309-06-3 (Hardbound)

ISBN: 978-93-87991-92-7 (Ebook)

Printed in India

त्रिलोचन महापात्र, पीएच.डी.
एफ एन ए, एफ एन ए एस सी, एफ एन ए ए एस
सचिव एवं महानिदेशक

TRILOCHAN MOHAPATRA, Ph.D.
FNA, FNASc, FNAAS
SECRETARY & DIRECTOR GENERAL

भारत सरकार
कृषि अनुसंधान और शिक्षा विभाग एवं
भारतीय कृषि अनुसंधान परिषद
कृषि एवं किसान कल्याण मंत्रालय, कृषि भवन, नई दिल्ली 110 001

GOVERNMENT OF INDIA
DEPARTMENT OF AGRICULTURAL RESEARCH & EDUCATION
AND
INDIAN COUNCIL OF AGRICULTURAL RESEARCH
MINISTRY OF AGRICULTURE AND FARMERS WELFARE
KRISHI BHAVAN, NEW DELHI 110 001
Tel.: 23382629; 23386711 Fax: 91-11-23384773
E-mail: dg.icar@nic.in

FOREWORD

Agroforestry is a traditional land use system that balances productivity *vis-à-vis* ecological services, thus discussed globally for its relevance in mitigation of climate change impacts. Agroforestry meets almost half of the demand of fuelwood, two-thirds of small timber, 70–80% wood of plywood, 60% raw material for paper pulp and 9–11% of the green fodder requirements, in addition to various environmental benefits. Adoption of agroforestry on large scale thus offers a great potential and appears to be the most desirable strategy for maintaining social, economic and ecological sustainability in the country. At present, 25.32 million ha is under agroforestry, which is about 8.32% of total geographic area of the country. The Task Force on Greening India (Planning Commission, 2001) had further recommended an additional 28 million ha area to be brought under plantation through agroforestry to achieve one-third of the total geographical area of the country under forest cover. To this effect, India is the only South Asian country to have enacted 'National Agroforestry Policy' in the year 2014.

I am sure that this book "**Agroforestry for Climate Resilience and Rural Livelihood**" is an excellent reference for those seeking knowledge on importance of agroforestry providing livelihood and ecosystem services enabling climate resilience. This volume covering novel information on agroforestry will be of immense use to policy makers, planners, scientists, academicians, and students working in the field of agroforestry.

(T. MOHAPATRA)

Dated the 8th January, 2018
New Delhi

Preface

Developing countries mostly rely on agriculture for rural livelihood and food security. Nevertheless, these countries have maximum vulnerability for agricultural production system due to high dependency on rainfall and climate change. On the other hand, land degradation exacerbates the productivity decline. In present scenario, the goal of sustainable development could only be achieved through diversification of existing cropping systems by promoting the integration of trees in different land use systems. Infact, agroforestry is one of the prime options for attaining the target of 33% of forest cover. Apart from nutritional security, agroforestry also provide other benefits like climate change mitigation and adaptation, land tenure security, increased farm income, restoration and maintenance of above and below ground carbon storage capacity, restoration and maintenance of biodiversity, watershed hydrology and soil conservation. Majority of farmers in our country belong to marginal and small category, who struggle every day to meet out their livelihood and to perform the farming practices under the threat of aberrant weather conditions. Thus, it is high time for them to adopt the agroforestry practices to meet out their livelihood requirements and also to sustain productivity with economic security and minimum environmental degradation. This compiled book entitled "Agroforestry for Climate Resilience and Rural Livelihood" would definitely be helpful for policy makers, planners, academicians, students and scientists to suggest the technologies and strategies to the farmers for enhancing their productivity, economic stability, meeting nutritional security under the changing climatic scenario. The voluminous compilation will act as a boost for farmers to adopt agroforestry system in their pursuit for better environmental management and resilience against the climate change. We have segregated the contents of our book under two theme areas, climate resilience and rural livelihood.

We are very thankful to the authors for their contributions in the form of chapters dealing with different aspects of agroforestry, climate change and nutritional security. We are really very pleased to bring this publication by compiling the information received through various experts working in the field of agroforestry and brought number of case studies from different quarters of the country and the issues related to present scenario. The views of the readers on this edition will be highly appreciated for improvement and guidance in future editions.

Editors:

Inder Dev, Asha Ram, Naresh Kumar, Ramesh Singh
Dhiraj Kumar, AR Uthappa , AK Handa, OP Chaturvedi

BRIEF BIODATA OF DR. INDER DEV

Dr. Inder Dev was born on 20thNovember, 1968 at village Tharass in Kullu district of Himachal Pradesh. He has devoted about 21 years of his career in research, extension and management in silvipastoral management, grassland management of temperate, alpine and cold desert region of the Himalaya; watershed management and agroforestry in semi-arid CSK Himachal Pradesh Krishi Vishwavidayala, Palampur, Himachal Pradesh and Ph. D (Agronomy) from IARI, New Delhi. He also obtained an International Diploma in Professional Capacity Building Programme in Agricultural Research for Development from International Centre for Development Oriented Research in Agriculture, Wageningen, with field study in Ghana. He has published 35 research publications in refereed journals.

At present Dr Inder Dev is working as Principal Scientist (Agronomy) at ICAR-Central Agroforestry Research Institute, Jhansi, Uttar Pradesh, India, with a major focus on Agroforestry based conservation agriculture and watershed management.

BRIEF BIODATA OF DR ASHA RAM

Dr. Asha Ram was born on 15th August, 1986 at village Inderwara in Pali district of Rajasthan. He has a PhD in Agronomy with emphasis on soil fertility management. At CAFRI, he is working with major focus on improving productivity of wasteland through silvipasture management with soil and moisture conservation practices, Agroforestry based Integrated farming system, tree-crop interactions study in agroforestry systems and agroforestry based conservation agriculture. In short span of his scientific career, he has published more than 25 research papers, reports and articles, 6 book chapters, 16 popular articles and edited 19 books, technical bulletins and extension folders. Presently he is working as a scientist (Agronomy) at the Indian Council of Agricultural Research (ICAR) - Central Agroforestry Research Institute (CAFRI), Jhansi, Uttar Pradesh, India.

BRIEF BIODATA OF DR. NARESH KUMAR

Dr. Naresh Kumarwas born on 11th October, 1975 in Bilaspur District of Himachal Pradesh. He did his B.Sc. (Forestry), M.Sc. (Forestry) and Ph. D. (Agroforestry) from Dr. Y.S. Parmar University of Horticulture and Forestry, Solan, Himachal Pradesh. He has been engaged in teaching and research for more than 12 years and have more than 40 publications in the form of research papers, book, book chapters, technical bulletins, extension folders etc. Earlier worked as Assistant Professor in College of Horticulture and Forestry, Central Agricultural University, Pasighat, Arunachal Pradesh. Presently,he is working as Sr. Scientist (Agroforestry) in ICAR - Central Agroforestry Research Institute (CAFRI), Jhansi, Uttar Pradesh, India. He is associated with different research projects focusing on short rotation tree species, improvement of wastelands through silvipastoral system, agroforestry based farming system, conservation agriculture etc.

BRIEF BIODATA OF DR. RAMESH SINGH

Dr. Ramesh Singh was born on 1st July, 1970 at village Ori in Mirzapur district of UP. He has devoted 20 years of his career in research and extension in the field of watershed management and hydrology of agroforestry systems. He did his B. Tech (Agri Engg.) and Ph. D (Irrigation & Drainage Engineering)) from GBPUA&T, Pantnagar, Uttarakhandand M.E. (IWM) from CTAE, Udaipur, Rajasthan.He is recipient of Gold Medal for post-graduation. He has published about 45 research publications and conducted national and international trainings on NRM. He has evaluated more than 50 IWMP projects. He has been awarded by number of awards including Vasant Rao Naik Award for Research Applications in Dryland Farming System 2010 by ICAR, New Delhi.

At present, Dr. Ramesh Singh is working as PS (SWC Engg.) at ICAR-Central Agroforestry Research Institute, Jhansi with a major focus on management of natural resources through agroforestry led watershed interventions.

BRIEF BIODATA OF DR. DHIRAJ KUMAR

Dr. Dhiraj Kumar was born on 19th March, 1987 at Hazaribagh district of Jharkhand. He did his B.Sc. Agriculture (Hons.) from UBKV, Pundibari, Cooch Behar (West Bengal) and M.Sc. as well as Ph.D. in Soil Science and Agricultural Chemistry from Indian Agricultural Research Institute (IARI), New Delhi. He has already contributed nine research papers, six book chapters, three technical bulletins, research reports and popular articles published in journals of repute.His major area of interest is soil carbon and nitrogen dynamics studies. Presently, he is involved in studies on tree-crop interaction, soil biological parameters under conservation agriculture based agroforestry system, soil quality assessment in agroforestry based watershed interventions etc. At present Dr. Dhiraj Kumar is working as Scientist (Soil Science) at ICAR-Central Agroforestry Research Institute, Jhansi, Uttar Pradesh, India, with a major focus on fine roots and litter dynamics studies under agroforestry system.

BRIEF BIODATA OF UTHAPPPA A R

Uthappa A R was born on 20 May 1987 at village Ponnampet in Kodagu district of Karnataka. He did his B.Sc. Forestry from UAS, Bangalore, Karnataka and M.Sc. (Agriculture) in Agroforestry from GBPUAT, Pantnagar, Uttarakhand.

He has published more than 20 research papers, popular articles and book chapters in different refereed journals and books. He has also authored two books. At present he is working as a Scientist (Agroforestry) at ICAR-Central Agroforestry Research Institute, Jhansi, Uttar Pradesh, India, with a major focus on Tree Improvement of fodder trees and Agroforestry based conservation agriculture.

BRIEF BIODATA OF DR. A.K. HANDA

Dr. A.K. Handa was born on 21.08.1967. Dr A K Handa is B.Sc., M.Sc. and Ph.D. in Forestry from Dr Y S Parmar University of Horticulture and Forestry, Solan, Himachal Pradesh. He has an experience of more than 23 years of research and development in the field of Agroforestry / Forestry. He has been Principal Investigator of different research projects including externally funded projects. He has more than 80 research and technical publications to his name published in International and National journals. He was been awarded with Dr K G Tejwani Award for Agroforestry 2013. He is currently working as Principal Scientist (Agroforestry) at ICAR- Central Agroforestry Research Institute, Jhansi, Uttar Pradesh, India.

CURRICULUM VITAE OF DR. O.P. Chaturvedi

Dr. O.P. Chturvedi was born on 1st January, 1956 at Barwan, Uttar Pradesh. He did his Ph.D. from Kumaun University, Nainital. He has an experience of more than 36 years of research and teaching in the field of Agroforestry / Forestry. He is specialized in Forestry; agroforestry; agrisilvicultural systems; agrihorticultural systems; silvipastoral systems; quality plant production; growth and production modeling of trees; wasteland reclamation and development; watershed management and development on ecosystem basis. He was been awarded with number of prestigious awards such as Forestry Teacher Award U.K. 1988, Overseas Technical Trainers Award U.K., 1989, K G Tejwani Award for Agroforestry 2008 and NAAS Fellow Award 2008. Dr. O.P. Chaturvedi was Ex. Director of ICAR- Central Agroforestry Research Institute, Jhansi, Uttar Pradesh, India.

List of Contributors

A Balasubramanian	Forest College and Research Institute, Mettupalayam, Coimbatore, TamilNadu, India
A Pattanayak	ICAR- Vivekananda Parvatiya Krishi Anusandhan Sansthan, Almora, India
Anand Kumar Singh	ICAR-Central Agroforestry Research Institute, Jhansi, India
A. G. Yewale	Uttarakhand University of Horticulture and Forestry, Tehri Garhwal, Uttarakhand, India
A.R. Uthappa	ICAR-Central Agroforestry Research Institute, Jhansi, India
Abhishek M Tripathi	Global Change Research Institute AS CR, Brno, Czech Republic
AK Handa	ICAR-Central Agroforestry Research Institute, Jhansi, India
AK Singh Parihar	N.D. University of Agriculture & Technology, Kumarganj, Faizabad
Anil Kumar	Navsari Agricultural University, Navsari, Gujarat, India
Anil Kumar	ICAR-Central Agroforestry Research Institute, Jhansi, India
Ankur Jha	Krishi Vigyan Kendra, Auraiya, Uttar Pradesh, India
AR Manju hashini	Forest College and Research Institute, Mettupalayam, Coimbatore, TamilNadu, India
Asha Ram	ICAR-Central Agroforestry Research Institute, Jhansi, India
Ashish Yadav	ICAR-National Organic Farming Research Institute, Sikkim, India
Atul Shrivastava	Jawaharlal Nehru Krishi Vishwa Vidyalaya, Jabalpur, Madhya Pradesh, India
B Palanikumaran	Tamil Nadu Agricultural University, Tamil Nadu, India
Babli Joshi	SCVB Govt. Collee, Palampjur (H.P), India.
Badre Alam	ICAR-Central Agroforestry Research Institute, Jhansi, India
BH Kittur	University of Agricultural Sciences, Dharwad, India
Bimlendra Kumari	CCS Haryana Agricultural University, Hisar, India
BM Kumar	Nalanda University, Rajgir, District Nalanda, Bihar, India
BM Pandey	ICAR- Vivekananda Parvatiya Krishi Anusandhan Sansthan, Almora, India
BP Mishra	College of Horticulture and Forestry, CAU Pasighat, Arunachal Pradesh, India
BS Kasana	Rajmata Vijayaraje Scindia Krishi Vishwa Vidyalaya, Gwalior, India
C Cinthia fernandaz	Tamil Nadu Agricultural University, Tamil Nadu, India

C Valli	Tamil Nadu Veterinary and Animal Sciences University, Kattupakkam, Tamil Nadu, India.
C.Bandeswaran	Tamil Nadu Veterinary and Animal Sciences University, Kattupakkam, Tamil Nadu, India.
CB Pandey	ICAR-Central Arid Zone Research Institute, Jodhpur, India
Chaman Lal	Dr. Y. S. Parmar UHF, Nauni, Himachal Pradesh, India
CP Rahangdale	Jawaharlal Nehru Krishi Vishwa Vidyalaya, Jabalpur, Madhya Pradesh, India
D.R. Palsaniya	ICAR-Indian Grassland and Fodder Research Institute, Jhansi, India
D.S.Chauhan	Hemvati Nandan Bahuguna Garhwal University, Srinagar, Garhwal, Uttarakhand, India
DB Patel	Navsari Agricultural University, Navsari, Gujarat, India
Dhiraj Kumar	ICAR-Central Agroforestry Research Institute, Jhansi, India
Dinesh Kumar	Institute of Agricultural Sciences, Banaras Hindu University, Varanasi, India
DSN Raju	Srikiran Biotech, Telangana, India
G S Shamet	Dr. Y.S. Parmar University of Horticulture and Forestry, Nauni, Solan, India
H Suvera	Navsari Agricultural University, Navsari, Gujarat, India
H. Mehta	ICAR-Indian Institute of Soil and Water Conservation, Dehradun, India
HS Mishra	G. B. Pant University of Agriculture & Technology, Pantnagar, Uttarakhand, India
Inder Dev	ICAR-Central Agroforestry Research Institute, Jhansi, India
JC Tewari	ICAR-Central Arid Zone Research Institute, Jodhpur, India
JK Bisht	ICAR- Vivekananda Parvatiya Krishi Anusandhan Sansthan, Almora, India
JR Jat	Sardarkrushinagar Dantiwada Agricultural University, Sardarkrushinagar, Gujarat, India
JS Guleria	Chaudhary Sarwan Kumar Himachal Pradesh Krishi Visvavidyalaya, Palampur, India
Jyotimala Sahu	Rajmata Vijayaraje Scindia Krishi Vishwa Vidyalaya, Gwalior, India
K B Sridhar	ICAR-Central Agroforestry Research Institute, Jhansi, India
K Srinivas	Srikiran Biotech, Telangana, India
K Sudhakara	Kerala Agricultural University, Vellanikkara, Thrissur, Kerala, India
K. Alagusundaram	NRM Division, Indian Council of Agricultural Research, New Delhi, India
K. Patle	Ministry of Agriculture and Farmers Welfare, Govt. Of India
Kavita Satyawali	Govind Ballabh Pant University of Agriculture and Technology, Pantnagar, India

Kaushal K Garg	International Crop Research Institute for Semi-Arid Tropics, Patancheru, Hyderabad, India
KK Vikrant	Hemvati Nandan Bahuguna Garhwal University, Srinagar, Garhwal, Uttarakhand, India
KN Patel	Sardarkrushinagar Dantiwada Agricultural University, Sardarkrushinagar, Gujarat, India
KS Bangarwa	Chaudhary Charan Singh Haryana Agriculture University Hisar, Haryana, India
KS Verma	Dr. Y.S Parmar University of Horticulture and Forestry, Solan, H.P.
KT Parthiban	Tamil Nadu Agricultural University, Tamil Nadu, India
Lal Chand	ICAR-Central Agroforestry Research Institute, Jhansi, India
LD Koshta	Jawaharlal Nehru Krishi Vishwa Vidyalaya, Jabalpur, Madhya Pradesh, India
Madhulika Srivastava	ICAR-Central Agroforestry Research Institute, Jhansi, India
Mahendra Singh	ICAR-Central Agroforestry Research Institute, Jhansi, India
Manisha Thakur	Dr. Y.S. Parmar University of Horticulture and Forestry, Nauni, Solan, India
Matber Singh	ICAR-National Organic Farming Research Institute, Sikkim, India
Mayank Chaturvedi	ICAR-Central Agroforestry Research Institute, Jhansi, India
Michal V Marek	Global Change Research Institute AS CR, Brno, Czech Republic
ML Sahu	Jawaharlal Nehru Krishi Vishwa Vidyalaya, Jabalpur, Madhya Pradesh, India
MS Pathania	Chaudhary Sarwan Kumar Himachal Pradesh Krishi Visvavidyalaya, Palampur, India
M Ramachandran	Tamil Nadu Veterinary and Animal Sciences University, Kattupakkam, Kancheepuram District, Tamil Nadu, India
Naresh Kumar	ICAR-Central Agroforestry Research Institute, Jhansi, India
Neelam Bisen	Institute of Agricultural Sciences, B.H.U. Varanasi. U.P., India
Neema Bisht	Govind Ballabh Pant University of Agriculture and Technology, Pantnagar, India
Neeraj	N. D. University of Agriculture and Technology, Kumarganj, Faizabad, India
NS Thakur	Navsari Agricultural University, Navsari, Gujarat, India
O.P. Chaturvedi	ICAR-Central Agroforestry Research Institute, Jhansi, India
OP Rao	N. D. University of Agriculture and Technology, Kumarganj, Faizabad, India
OP Yadav	ICAR-Central Arid Zone Research Institute, Jodhpur, India
P.K. Ghosh	ICAR – Indian Grassland and Fodder Research Institute, Jhansi, India
P.K. Mishra	ICAR-Indian Institute of Soil and Water Conservation, Dehradun, India

PK Sarkar	ICAR Research Complex, Patna, India
Poornima Malviya	Jawaharlal Nehru Krishi Vishwa Vidyalaya, Jabalpur, Madhya Pradesh, India
PR Renganayaki	Tamil Nadu Agricultural University, Tamil Nadu, India
Pradyuman Singh	N. D. University of Agriculture and Technology, Kumarganj, Faizabad, India
R. K. Sahu	Jawaharlal Nehru Krishi Vishwa Vidyalaya, Jabalpur, Madhya Pradesh, India
R. M. Singh	Institute of Agricultural Sciences, Banaras Hindu University, Varanasi, India
R.H.Rizvi	ICAR-Central Agroforestry Research Institute, Jhansi, India
Radek Pokorny	Global Change Research Institute AS CR, Brno, Czech Republic
Rahul Dongre	Jawaharlal Nehru Krishi Vishwa Vidyalaya, Jabalpur, Madhya Pradesh, India
Rahul Singh	Dr. Y. S. Parmar UHF, Nauni, Himachal Pradesh, India
Rahul Singh	Punjab University Patiala, Punjab, India
Rajesh Kaushal	ICAR-Indian Institute of Soil and Water Conservation, Dehradun, India
Ram Newaj	ICAR-Central Agroforestry Research Institute, Jhansi, India
Ramesh Singh	ICAR-Central Agroforestry Research Institute, Jhansi, India
Raveena Negi	Dr. Y.S. Parmar University of Horticulture and Forestry, Nauni, Solan, India
RC Dhiman	ITC-PSPD, Uttarakhand, India
RK Avasthe	ICAR-National Organic Farming Research Institute, Sikkim, India
RK Tewari	ICAR-Central Agroforestry Research Institute, Jhansi, India
RN Sharma	Rajmata Vijayaraje Scindia Krishi Vishwa Vidyalaya, Gwalior, India
RP Dwivedi	ICAR-Central Agroforestry Research Institute, Jhansi, India
RP Yadav	ICAR- Vivekananda Parvatiya Krishi Anusandhan Sansthan, Almora, India
RR Shakhela	Sardarkrushinagar Dantiwada Agricultural University, Sardarkrushinagar, Gujarat, India
S Gunasekaran	Tamil Nadu Veterinary and Animal Sciences University, Kattupakkam, Tamil Nadu, India.
S Radhakrishnan	Forest College and Research Institute, Mettupalayam, Coimbatore, TamilNadu, India
S Sarvade	Jawaharlal Nehru Krishi Vishwa Vidyalaya, Jabalpur, Madhya Pradesh, India
S. Bhaskar	NRM Division, Indian Council of Agricultural Research, New Delhi, India
S.B. Chavan	ICAR-Central Agroforestry Research Institute, Jhansi, India

Salil Tewari	GB Pant University of Agriculture & Technology, Pantnagar, Uttarakhand, India
Sanjeev Kumar	College of Horticulture and Forestry, CAU Pasighat, Arunachal Pradesh, India
SB Agrawal	Jawaharlal Nehru Krishi Vishwa Vidhyalaya, Jabalpur, India
Seema Maikhuri	Govind Ballabh Pant University of Agriculture and Technology, Pantnagar, India
SK Chauhan	Punjab Agricultural University, Ludhiana, India
SK Dhyani	NRM Division, Indian Council of Agricultural Research, New Delhi, India
SK Jha	Navsari Agricultural University, Navsari, Gujarat, India
Shishir Kumar Singh	Bundelkhand University, Jhansi, India
SK Verma	N.D. University of Agriculture & Technology, Kumarganj, Faizabad, India
Sourav Gupta	Rajmata Vijayaraje Scindia Krishi Vishwa Vidyalaya, Gwalior, India
SPS Tanwar	ICAR-Central Arid Zone Research Institute, Jodhpur, India
Subrata Sharma	Jawaharlal Nehru Krishi Vishwa Vidyalaya, Jabalpur, Madhya Pradesh, India
Sudhir Kumar	ICAR-Central Agroforestry Research Institute, Jhansi, India
Sumit Chaturvedi	G. B. Pant University of Agriculture & Technology, Pantnagar, Uttarakhand, India
Sushil Yadav	Rajmata Vijayaraje Scindia Krishi Vishwa Vidyalaya, Gwalior, India
T. Thomas	National Institute of Hydrology, Ganga Plains South Regional Centre, WALMI CAMPUS, Bhopal, India
Tarun Kumar	CCS Haryana Agricultural University, Hisar, India
TK Kunhamu	Kerala Agricultural University, Vellanikkara, Thrissur, Kerala, India
US Thakur	Rajmata Vijayaraje Scindia Krishi Vishwa Vidyalaya, Gwalior, India
Usha	RLB Central Agricultural University, Jhansi, India
Varun Attri	Dr. Y. S. Parmar UHF, Nauni, Himachal Pradesh, India
VB Upadhyay	Jawaharlal Nehru Krishi Vishwa Vidyalaya, Jabalpur, Madhya Pradesh, India
Veeresh Kumar	ICAR-Central Agroforestry Research Institute, Jhansi, India
Vikas Kumar	Kerala Agricultural University, Kerala, India
Vikas Kumar	ICAR-Indian Grassland and Fodder Research Institute, Jhansi, India
Vishal Johar	CCS Haryana Agricultural University, Hisar, India

Contents

CHAPTER 1

Role of Agroforestry in Current Scenario

Inder Dev*, Asha Ram, S Bhaskar[1] and OP Chaturvedi

[1]ICAR-Central Agroforestry Research Institute, Jhansi (U.P.), India

Natural Resource Management Division, ICAR, New Delhi, India

Corresponding author's Email: drinderdev@gmail.com

ABSTRACT

Agroforestry has traditionally been practiced in India as a way of life and livelihood since ancient time. Agroforestry meets almost half of the demand of fuelwood, 65 per cent (small timber), 70–80 per cent (wood for plywood), 60 per cent (raw material for paper pulp) and 9–11 per cent (green fodder requirement of livestock), in addition to its environmental benefits. Agroforestry in developing nations attracted considerable attraction for addressing multiple problems and delivering a range of economic, environmental and socio-economic benefits especially after the Kyoto Protocol to the United Nations Framework Convention on Climate Change (UNFCCC). Integration of trees with agricultural crops has the potential to improve local economy through stable income, diversification of landuse and rural skills, improved food, fuel and fodder security besides improving the environment. Agroforestry systems store carbon in the soils and woody biomass, and these also reduce greenhouse gas emissions from soils. India targets a 33-35 per cent reduction of emission intensity of its Gross Domestic Products (GDP) by 2030 from 2005 levels. India has announced the landmark National Agroforestry Policy 2014, which will help to bringing agroforestry in forefront in country. National Agroforestry Policy (NAP) also recommends the sound database for Agroforestry in India and setup the institutional mechanism for promoting agroforestry research at national level. The Prime Minister of India has also emphasized on several occasions about the importance of tree plantation on farm boundary.

Keywords: Agroforestry, Climate change, C sequestration, Ecosystem services, National Agroforestry Policy.

INTRODUCTION

India has critical imbalanced natural resource base with about 18 per cent human, 15 per cent livestock population of the world, which is supported on 2.4 per cent geographical area, 1.5 per cent forest and pasture lands and 4.2 per cent water resources. There are serious concerns on the performance of agriculture sector in the country, because the agriculture sector growth has not been higher in the post-reform period as compared to that of 1980s. This sector showed lower than 2 per cent per annum growth during the last decade (Dev 2008) because of plateauing of productivity of major food crops. The Indian agriculture is highly dependent on variations of weather and climate. At present farmers are leaving agriculture because of less remunerative and high risk due to weather vagaries. Though agriculture in India reflects series of adaptation with time, but presently climate change influenced the overall agriculture negatively (Mishra et al 2011, Ram et al 2016). Hence, a serious threat is being predicted for increasing food grain production and meeting the food, fibre, fuel and fodder requirements for the increasing population. Agroforestry has traditionally been practiced in India as a way of life and livelihood since ancient time. Increasing area under agroforestry in the country can help to addresses some of major challenges arising due to climate change (CAFRI Vision 2015, 2020, Dhyani et al 2016). Projection of decreased area under agriculture and increased demand for food grain and fuel (2 times), fodder (1.5 times) and timber production (3 times) is further increasing the importance of agroforestry (Dhyani and Handa 2014). Agroforestry meets almost half of the demand of fuelwood, 65 per cent (small timber), 70–80 per cent (wood for plywood), 60 per cent (raw material for paper pulp) and 9–11 per cent (green fodder requirement of livestock), in addition to its environmental benefits (NRCAF 2013). Agroforestry provide various ecosystem services *viz.* food, fuelwood, fodder, timber, poles etc. (provisioning), hydrological benefits and modifications of micro-climatic (regulatory), nutrient cycling and agro-biodiversity conservation (supporting service), recreation and aesthetics (cultural service), besides securing the livelihood of its practitioners. The country has also been in the forefront since organized agroforestry research started worldwide. It developed robust agroforestry science, innovations and practices that are attracting global interest (CAFRI Vision 2015). Agroforestry in developing nations attracted considerable attraction for addressing multiple problems and delivering a range of economic, environmental and socio-economic benefits especially after the Kyoto Protocol to the United Nations Framework Convention on Climate Change (UNFCCC). Agroforestry can play effective role in lowering the vulnerability, enhancing resilience of agriculture and buffering households against climate extremes (Dhyani 2014).

At present in India, agroforestry is practiced between 17.45 million ha (Rizvi et al 2014) to 25.32 million ha (Dhyani et al 2013). Rizvi et al (2014) estimated the area under agroforestry through using the geo-spatial technologies and ranked Uttar Pradesh (1.86 m ha), Maharashtra (1.61 m ha) and Rajasthan (1.55 m ha) as first, second and third, respectively. However, there are still many challenges that impede the growth of agroforestry in India. These include a lack of uniform policies and regulations relating to felling and transporting farm-grown timber and other products in different states. Agroforestry ventures are not covered under agricultural insurance schemes or entitled to marketing support, and cannot access the soft bank loans available for crop farming. India has announced the landmark National Agroforestry Policy 2014 that will mainstream the growing of trees on farms to meet a wide range of developmental and environmental goals (Dhyani 2014, Dhyani et al 2016). In terms of specific empirical evidence, the area under forest will not change much in future in view of the developmental needs of the country, though it has increased by 8.24

per cent from 63.83 million ha to 69.09 million ha in last forty years. In next 40 years it is expected that there will be substantial increase in area under tree cover i.e. Tree Outside Forest and a large portion of it will be devoted to agroforestry. This is possible as Task Force on Greening India (Planning Commission 2001) projected that additional 28 million ha area can be brought under plantation through agroforestry to meet the national goal of increasing forest cover on one-third of the total geographical area. For this purpose, 10 million ha of irrigated but problematic lands and another 18 million ha of rainfed lands have been earmarked for agroforestry development. A sizeable area will also be contributed from barren and uncultivable land, permanent pastures and other grazing land, cultivable waste land and fallow land (CAFRI Vision 2050, 2015). Furthermore, agroforestry is likely to play important role in the National Mission on Greening India, NM on Sustainable Agriculture, NM on Energy and NM for Sustaining the Himalayan Ecosystem among the Prime minister's eight National Missions (NM) under National Action Plan on Climate Change.

HISTORY OF AGROFORESTRY IN INDIA

Agroforestry has traditionally been important and a way of livelihood in India since Mesolithic period (10,000 to 4,000 before present). Some of vedic literature like Brhat Samhita also describe relationship between irrigation tank and trees (Kumar and Sikka 2014). Practice of pasturing in forest area by humans is one of the oldest landuse practices in India. Civilization was started in forests and most of *Rishis* lived in harmony with the nature. There were harmonized combination of agriculture, pasture, trees, birds and animals near to rishi's ashram. Tree worshiping is quite common after Harappan and Mohanjodaro civilization. The cultivation of various fruits trees like banana, pomegranate, datepalm, aonla, ber, coconut and many agricultural crops and rearing livestock can be traced in chalcolithic period of civilization. Ancient literature of Vedas and other scriptures has mentioned religious importance of many trees like *P. cineraria, Ficus religiosa, Butea monosperma* etc. Importance of the trees during *Rigveda* period could be seen in the hymn (Bhatla et al 1984):

'May plants, the water and the sky
Preserve us and woods and mountains
With their trees for tresses' (*Rigveda* V. 41.11)

AGROFORESTRY FOR IMPROVING RURAL LIVELIHOOD AND EMPLOYMENT OPPORTUNITIES

The ultimate aim of agroforestry is to improve and sustain the livelihood of poor marginal and small farmers in era of climate change. Integration of trees with agricultural crops has the potential to improve local economy through stable income, diversification of landuse and rural skills, improved food, fuel and fodder security besides improving the environment (Smith 2010). Agroforestry has the high biomass production potential with diverse products resulting in higher financial benefits compared to monoculture of crops. Product diversification reduces the risk associated with farming of single crop and gives the year round income. There are different uses of tree products (timber, fuel, fodder and bioenergy) which can reduce the inputs and enhance the eco-efficiency of agroforestry systems (AFS). Furthermore, agroforestry landuse is more labour intensive as compared to monoculture of crops, hence different component of AFS diversifies the skills base of the labour force (Yamada and Gholz 2002, Campos et al 2008). Besides creating the employment opportunities, agroforestry reduces the pressure on natural forests by providing various products to the rural households.

In Haridwar and Yamunanagar district of North India, poplar based agrisilviculture shown has shown a B:Cration of 3.85 with net return of ₹ 1.97 lakh per hr. (Singh et al 2016). *Casuarina equisitifolia* has emerged an important fast growing tree species in Coastal states of India and reported highly suitable for agroforestry. Saravanan and Vijayaraghavan (2014) reported ₹ 89,000 net annual return per ha with groundnut + casuarina based agroforestry system. *Melia azedarach* multipurpose tree species has widely used for insecticidal and medicinal properties. It is commonly planted along the bund / canal in irrigated area and its foliage is used for goat feeding. *Melia azedarach* planted with soybean crop gave 19 per cent higher net return than the sole soybean (Al-Rubae 2009, Patil et al 2009). Similar types of the agroforestry systems are enhancing the livelihood of the farmers especially in rainfed conditions. Agroforestry systems have the good potential for lifting the socioeconomic status of farmers and for contributing to the overall development of the region. This can be seen in terms of increased income and in the creating employment opportunities (Dhyani et al 2003). It is estimated that 943 million person-days employment can be further generated annually from 25.4 million ha (Table 1). Dhyani et al (2005) had estimated the only in Himalayan region agroforestry has the potential to generate employment upto 5.76 million person days per year. The overall ratio of rate of return to investment in different agroforestry systems is estimated to be 117 per cent per year (Basu 2014).

Table 1: *Employment generation potential of agroforestry in India*

Agroforestry system	Area (million ha)	Additional employment (Person-day ha^{-1} yr^{-1})	Total annual employment (million person-days)
Silviculture	1.8	30	53.3
Agrisilviculture (irrigated)	2.3	40	91.3
Agrisilviculture (rainfed)	1.3	30	38.0
Agrihorticulture (irrigated)	1.5	50	76.1
Agrihorticulture (rainfed)	0.5	40	20.3
Silvipasture	5.6	30	167.4
Tree-borne oil seeds	12.4	40	497.1
Total	25.4	—	943.4

Source: NRCAF (2007)

AGROFORESTRY: SMART PRACTICE FOR CLIMATE CHANGE MITIGATION AND ADAPTATION

Developing countries like India will be attected more due to due to negative impacts of climate change especially of agriculture. Global conferences like COPs are not sufficiently effective to stop the increase of atmospheric greenhouse gases (GHG) concentrations. Mitigation efforts will therefore only provide a partial softening of the effects of climate change (Verchot et al 2007). Agroforestry provides unique opportunity to combine the twin objectives of climate change adaptation and mitigation. Agroforestry systems store carbon in the soils and woody biomass, and these also reduce greenhouse gas emissions from soils (Dhyani et al 2016). Recent studies under various agroforestry systems in diverse ecological conditions emphasized that agroforestry systems increase and store carbon stocks in above ground biomass and in soil and also has an important role in increasing livelihood security and reducing vulnerability to climate change.

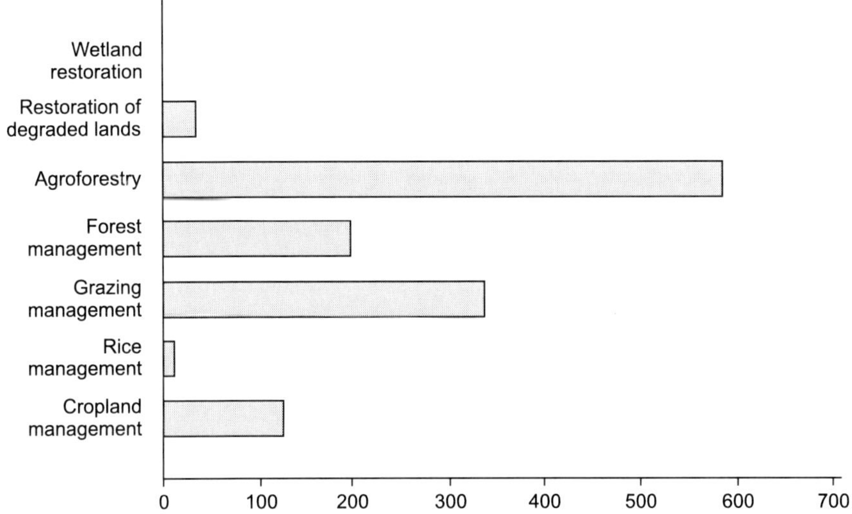

Figure 1: *Carbon sequestration potential of different land use and management options (adapted from IPCC 2000)*

Out of all landuse analyzed, agroforestry has a great potential for sequestration of carbon from the atmosphere (Fig. 1). India recently submitted voluntary pledges for action on mitigating climate change in the form of the Intended Nationally Determined Contributions (INDCs) to the United Nations Framework Convention on Climate Change (UNFCCC). India targets a 33-35 per cent reduction of emission intensity of its GDP by 2030 from 2005 levels. The target will be achieved through increasing area under trees through afforestation, agroforestry and increase in share of renewable source of energy like biofuel, bioenergy, solar energy etc.

AGROFORESTRY FOR ECOSYSTEM SERVICES

After the green revolution in 1960s, intensification and expansion of agriculture with dwarf varieties of rice and wheat changed India's status from begging bowl to the self-sufficient in food grain production (Ram et al 2014). High and imbalanced use of resources in modern agriculture leads to degradation of ecosystem services provided by agriculture (Fagerholm et al 2016). However, several alternate land-use systems such as agroforestry can safeguard and enhance the various ecosystem services (O'Farrell and Anderson 2010, Dhyani et al 2016). Agroforestry has long been seen as a better option for adaptation and mitigation of challenges arises due to global climate change (Nair and Garrity 2012, Dhyani et al 2016). Different studies have shown the importance of agroforestry in conserving the biodiversity, sustainable use of natural resources and providing various ecosystem services (Alam et al 2014).The Millennium Ecosystem Assessment (2005) has categorized the ecosystem services into provisioning service, regulating service, supporting service, and cultural service (Table 2).

Table 2: *Categories of ecosystem services provided by agroforestry*

Ecosystem services	Definition
Provisioning services	Goods, such as food or freshwater, that ecosystems provide and humans consume or use, fuel-wood, fodder, timber, poles etc.
Regulatory	Services such as flood reduction and water purification, other hydrological benefits, micro-climatic modifications provided by agroforestry systems.
Cultural	Intangible benefits, such as aesthetic enjoyment or religious inspiration and recreation.
Supporting	Basic processes and functions, such as soil formation and nutrient cycling and biodiversity conservation that are critical to the provision of the first three types of ecosystem services.

Ecosystem services and benefits provided by the agroforestry systems are not confined to in-situ, but occur over a wide range of spatial and temporal scale (Jose 2009). Agroforestry systems have a high potential for carbon sequestration and thus could constitute a carbon sink, helping to reduce adverse global climate change. It is reported that, agroforestry in India could sequester 0.25 (Kaur et al 2002) to 19.14 Mg C ha^{-1} yr^{-1} (Nath and Das 2012). Trees reduce erosion by binding the soil through roots. Trees also act as great air purifier by removing carbon dioxide and other pollutants from atmosphere. Trees provide shade and wind breaks to keep us cooler in summer and warmer in winter. Trees are an integral and indispensable part of the farming systems; they are deliberately retained on farmlands to support agriculture and are either planted or preserved to provide various amenities of life.

AGROFORESTRY POLICY FOR CHANGING THE SCENARIO OF AGROFORESTRY IN INDIA

Agroforestry is playing vital role in maintaining the natural resources and increasing overall productivity with minimizing risk against the vagaries of weather. Although agroforestry is practiced in India since ancient time, but as of now agroforestry not gained proper importance due to dependency on multi-institutions and multi-disciplinary approach (Dhyani 2014). In India agriculture is a state subject and forestry is placed under concurrent list, but Ministry of Agriculture and Farmers' Welfare has the mandate for research in Agroforestry. State government of many states imposed the restriction on felling and transportation of many tree species (as perSection-41 of India Forest Act) which are locally grown in natural forests as well as on farmers land. Besides lack of institutional finances, absence of adequate infrastructure for wood produce, lack of price discovery mechanism, lack of market information and poor extension of agroforestry technologies are major hurdle for disseminating agroforestry on farmer's field. To overcome these problems, India has announced the landmark National Agroforestry Policy 2014 (http://www.indiaenvironmentportal.org.in/content/389156/national-agroforestry-policy-2014/). This policy will help to bringing agroforestry in forefront in country. In this policy initially 20 multipurpose tree species which are mainly grown under agriforesty systems and having national importance have been identified for exemption from harvesting, transportation and marketing. National Agroforestry Policy (NAP) also recommends the sound database for Agroforestry in India and setup the institutional mechanism for promoting agroforestry research at national level.

After the policy announcement, the ICAR's National Research Centre for Agroforestry (NRCAF) was upgraded to the level of a central institute on 1st December, 2014. The ICAR-Central Agroforestry Research Institute (CAFRI) has revised its mandate, resources and staffing pattern. The NAP, 2014 also recommends the insurance and credit facility for the agroforestry practitioner. Credit facility and insurance to small and marginal land holder will boost the area under agroforestry. Tamil Nadu being the first state in India started the tree insurance by United India Insurance, Chennai. The insurance agency is ensuring the insurance against natural calamities (flood and cyclone), wild animal damage, and pest and disease for the species (Chavan et al 2015). In coming days more states are likely to come out with their own state agroforestry policy or rule which will definitely promote the agroforestry in the country. The Prime Minister of India has also emphasized on several occasions about tree plantation on farm boundary.

WAY FORWARD

The National Agroforestry Policy 2014 has pointed the way forward to promoting agroforestry among various stakeholders. Unfavorable legislation must be amended and regulations related to agroforestry, forestry and agriculture should be simplified. Government should include agroforestry in all policies related to climate change and management and conservation of natural resources, funding agroforestry related infrastructure, education and research. There should be proper coordination between various stakeholders of agroforestry like researchers, extension workers, industries, farmers etc. India is a land of variability in terms of climate; soil etc., hence location specific and economically sound agroforestry systems need to be developed. Agroforestry land use has great potential in providing the ecosystem services; hence these services should be quantified with best possible accuracy.

REFERENCES

Alam M, Olivier A, Paquette A, Dupras J, Reveret JP, and Messier C 2014 A general framework for the quantification and valuation of ecosystem services of tree-based intercropping systems, *Agroforestry Systems* **88** 679–691 DOI 10.1007/s10457-014-9681-x

Al-Rubae AY 2009 The potential uses of *Melia Azedarach* L. as pesticidal and medicinal plant, review, *American-Eurasian Journal of Sustainable Agriculture* **3**(2) 185-194

Asha Ram, Kumar D, Singh N, and Anand A 2014 Effect of sulphur on growth, productivity and economics of aerobic rice (*Oryza sativa*), *Indian Journal of Agronomy* **59**(3) 404-409

Basu JP 2014 Agroforestry, climate change mitigation and livelihood security in India, *New Zealand Journal of Forestry Science* **44** (Suppl 1): S11

Bhatla N, Mukherjee T and Singh S 1984 Plants: Traditional worshiping. *Indian Journal of History of Science,* **19**(1): 37-42

CAFRI Vision 2050 Jhansi: CAFRI, 2015.

Campos P, et al 2008 Economics of Multiple Use Cork Oak Woodlands: Two Case Studies of Agroforestry Systems, in *Agroforestry in Europe: Current Status and Future Prospects*, A. Rigueiro-Rodríguez, et al., (eds), Springer

Chavan SB, Keerthika A, Dhyani SK, Handa AK, Ram Newaj, and Rajarajan K 2015 National Agroforestry Policy in India: a low hanging fruit, *Current Science* **108** (10) 1826-1834

Dev SM 2008 Challenges for Revival of Indian Agriculture, in *1st Prof. Dayanatha Jha Memorial Lecture, 2 May 2008:* © 2008 National Centre for Agricultural Economics and Policy Research.

Dhyani SK 2014 National Agroforestry Policy 2014 and the need for area estimation under agroforestry, *Current Science*, **107**(1) 9-10

Dhyani SK and Handa AK 2014 Agroforestry in India: Current scenario, *Indian Farming* **63**(11) 6–8

Dhyani SK, Asha Ram, and Dev I 2016 Potential of agroforestry systems in carbon sequestration in India, *Indian Journal of Agricultural Sciences* **86** (9) 1103–12

Dhyani SK, Handa AK, and Uma 2013 Area under agroforestry in India: An assessment for present status and future perspective, *Indian Journal of Agroforestry* **15**(1) 1-11

Dhyani SK, Sharda VN, and Sharma AR 2003 Agroforestry for water resources conservation: issues, challenges and strategies, in *Agroforestry: Potentials and Opportunities* (eds) Pathak PS and Newaj, Ram Jodhpur, India: Agribios

Dhyani SK, Sharda VN, and Sharma AR 2005 Agroforestry for sustainable management of soil, water and environmental quality: Looking back to think ahead, *Range Management and Agroforestry* **26**(1) 71-83

Fagerholma N, Torralbaa M, Burgessc PJ, and Tobias Plieningera 2016 A systematic map of ecosystem services assessments around European agroforestry, *Ecological Indicators* **62** 47–65

IPCC 2000 Land-use, land-use change and forestry, Special report of the intergovernmental panel on climate change. Cambridge University Press, UK, p 375

Jose *S* 2009 Agroforestry for *Ecosystem Services* and Environmental Benefits: An Overview. *Agroforestry Systems*, **76**, 1-10

Kaur B, Gupta SR, and Singh G 2002 Carbon storage and nitrogen cycling in silvopastoral systems on a sodic soil in northwestern India, *Agroforestry Systems* **54** 21–9

Kumar BM and Sikka AK 2014 Agroforestry in South Asia: Glimpses from Vedic to present times, *Indian Farming* **63**(11): 2-5.

Millennium Ecosystem Assessment 2005 Ecosystems and human wellbeing-volume 4-multiscale assessments: findings of the sub-global assessments working group, Island Press, Washington DC, 388p

Mishra M, Panda GR and Padhi BK 2011 Sustaining Indian Agriculture in the Era of Climate Change in *Genetics, Biofuels and Local Farming Systems, Sustainable Agriculture Reviews* (ed.) E. Lichtfouse 7, DOI 10.1007/978-94-007-1521-9 8

Nair PK and Garrity D 2012 Agroforestry—the future of global land use, *Advances in Agroforestry* **9**, Springer, New York doi 10.1007/978-94-007-4676-3_16

Nath AJ and Das AK 2012 Carbon pool and sequestration potential of village bamboos in agroforestry systems in Northeast India, *Tropical Ecology* **53(3)** 287–93

NRCAF 2007 Vision 2025. National Research Centre for Agroforestry, Jhansi

NRCAF 2013 Vision 2050, National Research Centre for Agroforestry, Jhansi

O'Farrell PJ and Anderson PML 2010 Sustainable multifunctional landscapes: are views to implementation, *Current Opin*ion in *Environmental Sustain*ability **2** 59–65 http://dx.doi.org/10.1016/ j.cosust.2010.02.005.

Patil SJ, Mutanal SM, and Patil HY 2009 *Melia azedarach* based agroforestry system in transitional tract of Karnataka, *Karnataka Journal of Agriculture Sciences* **25**(4) 460-462

Planning Commission 2001 Report of the Task Force on Greening India for Livelihood Security and Sustainable Development, Planning Commission, Govt. of India, New Delhi, 231 p.

Ram A, Dev I, Kumar D, Uthappa AR, Tewari RK, Singh R, Sridhar KB, Singh M, Srivastava M, Kumar V and Chaturvedi OP 2016 Effect of tillage and residue management practices on blackgram and greengram under bael (*Aegle marmelos* L.) based agroforestry system, Indian Journal of Agroforestry **18**(1) 90-95.

Rizvi RH, Dhyani SK, Newaj R, Karmakar PS, and Saxena A 2014 Mapping agroforestry area in India through remote sensing and preliminary estimates, *Indian Farming* **63**(11) 62-64

Saravanan S and Vijayaraghavan A 2014 *Casuarina equisitifolia* based agroforestry system for higher economic returns for the farming communities in Tamilnadu, India, Fifth International Casuarina workshop, Mamallapuram, India, 03-07 February, 2014 p.61

Singh R, Singh C, Gulati A, and Kujur S 2016 Current status of poplar based agroforestry for economic development: A case study of Haridwar and Yamunanagar Districts, *Indian Forester* **142**(50) 487-492

Smith J 2010 Agroforestry: reconciling production with protection of the environment. Organic Research Centre, Elm Farm, Hamstead Marshall, Newbury, p 24

Verchot LV, Noordwijk MV, Kandji S, Tomich T, Ong C Albrecht A, Mackensen J Bantilan C, Anupama KV, and Palm C 2007 Climate change: linking adaptation and mitigation through agroforestry, *Mitigation and Adaptation Strategies for Global Change* DOI 10.1007/s11027-007-9105-6

Yamada M and Gholz HL 2002 An evaluation of agroforestry systems as a rural development option for the Brazilian Amazon, *Agroforestry Systems* **55** 81-87

CHAPTER 2

Livelihood Opportunities Through Agroforestry Interventions

K Alagusundaram, SK Dhyani* and S Bhaskar

Natural Resource Management Division, ICAR, KAB II, PUSA, New Delhi, India

**Corresponding author's E-mail: shivkdhyani@gmail.com*

ABSTRACT

Planting trees in agricultural landscapes not only generates income in the form of timber and non-timber produce, but also provides livelihood security. Agroforestry has an important role in reducing vulnerability, increasing resilience of farming systems and buffering households against climate related risks. Agroforestry involves a complete chain of activities from seedling/sapling production to their transportation to planting sites, field preparation, layout and planting, maintenance, pruning trees, harvesting, transportation, marketing, and processing of tree produce in the wood based industry. It thus generates a huge employment in rural areas where the employment opportunities are meagre due to less development activities in such remote locations. In addition, there are many options which can be integrated with agroforestry to ensure livelihood security, such as apiculture, sericulture, lac cultivation, gum and resin, medicinal, aromatic plants and many more. However, it requires appropriate marketing interventions and capacity-building programmes for post-harvest technology with value addition in these areas to provide better economic returns and to check the migration of rural youth towards urban areas. The present paper aims at potential and role of agroforestry in providing livelihood opportunities.

Keywords: Agroforestry interventions, Livelihood security, Post-harvest and value addition

INTRODUCTION

Agroforestry is practised by millions of farmers worldwide, and has been a feature of agriculture for millennia. It encompasses a wide range of trees that are grown on farms and in rural landscapes, and includes the generation of science-based tree enterprise opportunities that can be important in the future. Zomer et al (2009) conducted a global assessment of tree cover on agricultural land and found that 48 per cent of all agricultural land had at least 10% tree cover, which indicates that the earlier World Bank 2004 report estimation of 1.2 billion rural people practising agroforestry on their farms and in their communities on lower side. Nair et al (2009) estimated the total area under agroforestry was 2.35 billion ha. At present agroforestry in India meets almost half of the demand of fuel wood, two thirds of the small timber, 70-80% wood for plywood, 60% raw material for paper pulp and 9-11% of the green fodder requirement of livestock, besides meeting the subsistence needs of households for food, fruit, fibre, medicine etc. However, the biomass productivity per unit area is less than 2 t $ha^{-1}yr^{-1}$. Agroforestry practices have demonstrated that this could be safely increased to 10 t $ha^{-1}yr^{-1}$ by carefully selecting tree-crop combinations. Area under forest is degrading due to demographic pressure and infrastructure growth needs while agricultural area is almost stable. In India, nearly 120.72 million ha land or 37% of the total geographical area is under one or other forms of soil degradation (e.g. water erosion: 93 million ha (m ha), wind erosion: 11 m ha, salt affected soils: 6.74 m ha and 16.53 m ha of open forest area) (ICAR, 2010). Up to 2010-11, 56.96 m ha has been treated under various watershed development programmes, however, a sizeable area is yet to be treated. Agroforestry can play a vital role in such endeavours by meeting the diverse needs of people through trees, crops and livestock in addition to the multiple produce viz. food, fruit, fodder, fuel, fertilizer and fibre from limited land resources. Trees are integral part to our traditional farming systems, for the innumerable benefits that they provide. However, over time, with shrinking land holdings, annual crops replaced trees for various reasons. Trees complement farming in terms of manure, fodder and fuel needs of the farmer. They form the backbone for practicing integrated farming systems which is necessary for self reliant and sustainable agriculture. Agroforestry systems due to diverse options and products provide opportunities for employment generation in rural areas. Increased supply of wood in the market has triggered a substantial increase in the number of small-scale industries dealing with wood and wood based products in the near past.

DIVERSITY OF AGROFORESTRY SYSTEMS

The Diagnosis and Design (D&D) surveys of the existing agroforestry systems were conducted to identify suitable species and traditional agroforestry practices. The various improved agroforestry practices and systems such as home gardens, block plantation, energy plantation, shelterbelts and improvement or alternative to shifting cultivation are some of the specialised agroforestry systems developed. Dhyani et al (2009) have reported the common twenty agroforestry systems being practised in different agro-ecological regions of India and these are, 1. Agri-silviculture; 2. Boundary plantation; 3. Block plantation; 4. Energy plantation (trees + crops during initial years); 5. Alley cropping (hedges + crops); 6. Agri-horticulture (fruit trees + crops); 7. Agri-silvi-horticulture (trees + fruit trees + crops); 8. Agri-silvi-pasture (trees + crops + pasture or animals); 9. Silvi-olericulture (tree + vegetables); 10. Horti-pasture (fruit trees + pasture or animals); 11. Horti-olericulture (fruit tree + vegetables); 12. Silvi-pasture (trees + pasture/ animals); 13. Forage forestry (forage trees + pasture); 14. Shelter-belts (trees + crops); 15. Wind-breaks (trees + crops); 16. Live fence (shrubs and under- trees on boundary); 17. Silvi or Horti-sericulture (trees or fruit trees + sericulture); 18. Horti-apiculture (fruit trees + honeybee); 19. Aqua-forestry (trees + fishes); and 20. Homestead (multiple combinations of trees, fruit trees, vegetable etc). Agronomic practices for raising the promising multipurpose tree species (MPTS) in association with annual crops have been developed and standardised.

The All India Co-ordinated Research Programme on Agroforestry (AICRP-AF) and ICAR-Central Agroforestry Research Institute, Jhansi has developed 35 agroforestry models for different agro-climatic zones of the country (Chaturvedi et al, 2016). In general, *Morus* and *Grewia*-based system for the western Himalayas, alder-based for the North Eastern Hill (NEH) region, poplar-based for the Indo-Gangetic region, aonla and khejri-based for the semi-arid and arid regions, teak-based for the tropical region and *Gmelina* and *Acacia*-based system for Humid and Sub humid regions are commonly recommended. Home-gardens in Kerala and other coastal states promote food security and diversity and provide basic needs of food, fuel-wood, fodder, plant-derived medicines, and cash income from their small holdings. In addition, agroforestry practices have been intertwined with the various programmes/ schemes like watershed development, rehabilitation of problem soils, treatments of degraded and other wastelands etc. Deliberate growing of trees on field bunds (risers) and in agricultural fields as scattered trees, and the practice to utilise the open interspaces in the newly-planted orchards and forests for cultivating field crops are also widespread in the sub-continent. It is estimated that currently agroforestry (outside the forest) has a greater number of trees than the "State" forests. The prominent agroforestry systems from different region of the country are enumerated and presented in Table 1.

Table 1: *Prominent agroforestry systems in different regions of India*

Agro-climatic Zone	**Agroforestry systems**	**Tree Component**	**Crop/Grass**
Western Himalayas	Silvipasture (RF)	*Grewia optiva*	*Setaria* spp.
		Morus alba	*Setaria* spp.
	Agrihorticulture	*Malus pumila*	Millets, Wheat
	Agrihorticulture	*Prunus persica*	Maize, Soyabean
Eastern Himalayas	Agrisilviculture	*Anthocephalus cadamba*	Rice
	Agrihorticulture	*Alnus nepalensis*	Large Cardamom/ Coffee
	Silvipasture	Bamboos, *Parkia roxburghii, Morus alba*	–
	Silvipasture	*Bauhinia variegata, Ficus spp., Morus alba*	Napier
Lower Gangetic Plains	Agrisilviculture (Irri)	*Eucalyptus spp., Albizia lebbeck*	Rice
	Agrihorticulture (Irri)	Mango/Banana/Litchi	Wheat, Paddy, Maize
	Silvipasture	*Morus alba, Albizia lebbeck*	*Dicanthium, Pennisetum*

Middle Gangetic Plains	Agrisilviculture (Irri)	*Populus deltoides*	Sugarcane-Wheat
	Agrisilviculture (Irri)	*Eucalyptus spp.*	Rice-Wheat
	Agrisilviculture	*Dalbergia sissoo*	Sesamum
	Agrihorticulture (Irri)	Mango/Citrus	Rice-Wheat
	Silvipasture	*Albizia lebbeck*	*Chrysopogan, Dicanthium*
Trans Gangetic Plains	Agrihorticulture (Irri)	*Emblica officinalis*	Black gram/Green gram
	Agrisilviculture	*Azadirachta indica*	Black gram- Wheat/ Mustard
	Silvipasture	*Bauhinia variegata, Albizia lebbeck*	*Cenchrus, Pennisetum*
Upper Gangetic Plains	Agrisilviculture (Irri)	*Populus deltoides*	Wheat, Bajra fodder
	Agrisilviculture (Irri)	*Eucalyptus spp.*	Rice-Wheat
	Silvipasture	*Bauhinia variegata, Albizia lebbeck*	*Chrysopogon, Poa*
Eastern Plateau & Hills	Agrisilviculture	*Gmelina arborea*	Rice, Linseed
	Agrisilviculture	*Acacia nilotica*	Rice
	Silvipasture	*Acacia mangium, A. nilotica, bamboos*	-
	Silvipasture	*Leucaena leucocephala*	*Chrysopogon, Pennisetum, Dicanthium*
Central Plateau & Hills	Agrihorticulture (Irri)	*Psidium guajava*	Bengal Gram/Groundnut
	Agrihorticulture (RF)	*Emblica officinalis*	Black gram/Green gram
	Agrisilviculture	*Acacia nilotica / Leucaena leucocephala /Azadirachta indica / Albizia lebbeck*	Soyabean, Black gram-Mustard/Wheat
	Silvipasture (RF- and degraded lands)	*Albizia amara, Leucaena leucocephala, Dichrostycus cinerea*	*Chrysopogon, Stylosanthes hamata, S. scabra*
	TBOs (RF)	*Jatropha curcas*	-
Wastern Plateau & Hills	Agrihortisilviculture (Irri)	*Tectona grandis, Achrus zapot*	Rice, Maize
	Agrihorticulture	*Areca catechu*	Black pepper, Cardamom
	Silviculture	*Prosopis juliflora, Ailanthus excels*	-
	Silvipasture	*Acacia mangium, Albizia amara*	*Cenchrus*

Southern Plateau & Hills	Agrisilviculture (RF)	*Eucalyptus, Casuarina equisetifolia, Ailanthus excelsa*	Cotton, Groundnut
	Agrisilviculture (Irri)	*Eucalyptus tereticornis, Melia dubia*	Chilli
	Silviculture (RF)	*Leucaena leucocephala, Acacia leucopholea*	-
		Eucalyptus	-
	Agrihorticulture	*Tamarindus indica*	Chilli
	TBOs	*Pongamia pinnata*	-
East Coast Plains & Hills	Agrisilviculture (RF)	*Ailanthus excelsa, Acacia leucophloea*	Cow pea
	Silviculture	*Casuarina equisetifolia, Leucaena leucocephala*	-
	TBOs	*Pongamia pinnata*	-
	Silvipasture	*Artocarpus spp.*	*Chrysopogon, Napier, Cenchrus*
West Coast Plains & Hills	Agrisilviculture (RF)	*Acacia auriculiformis*	Black Pepper
	Agrihorticulture (RF)	*Artocarpus heterophyllus*	Black Pepper
	Agrisilviculture (RF)	*Acacia auriculiformis*	Rice
	Agrihorticulture	*Cocos nucifera/Areca catechu*	Rice
	Agrisilviculture	*Casurina equisetiofolia*	Rice
	Silvipasture	*Hardwickia binnata, Albizia lebbeck*	*Cenchrus*
Gujarat Coast plains & Hills	Agrisilviculture	*Azadirachta indica, Ailanthus excelsa*	Cow pea, Green gram
	Silviculture	*Prosopis juliflora, Acacia nilotica*	-
	Silvipasture	*Leucaena leucocephala*	*Cenchrus, Setaria*
Western Dry Region	Agrisilviculture	*Prosopis cineraria, Tecomella indica, Acacia nilotica, Azadirachta indica*	Pearl millet
	TBOs	*Jatropha curcas*	-
	Silvipasture	*Albizia lebbeck, Hardwickia binnata*	*Cenchrus*
All Islands	Agrihorticulture	*Cocos nucifera*	Rice
	Silvipasture	*Bauhinia* spp, *Erythrina indica, Leucaena leucocephala*	*Cenchrus, Pennisetum*

Irri: irrigated; RF; rainfed, TBOs: tree borne oil seed (Source: Dhyani et al 2009)

AGROFORESTRY FOR FOOD AND NUTRITIONAL SECURITY

The country's food production has increased five-fold since independence, but recent improvements in food supply have been insufficient to fulfil the nutritional needs of the average person owing to the increasing population. As per estimates of economic survey in 2015, the wheat and rice yield was 90.76 and 102.54 million tonne respectively, which has the greatest increase in yields, while production of coarse grains (40.22 m t) and pulses (17.38 m t) has changed only marginally. Increasing the production and consumption of protein-dense foods, pulses in particular, will be necessary if the country is to meet its protein needs. Agroforestry with appropriate tree-crop/ legume combination is one option to achieve it. Agroforestry systems provide the desired diversification to increase the food security of the country and act as a shield against poor production during drought and other stress conditions. Agroforestry also provides nutritional security because of diverse production systems which include fruit, vegetables, oilseed crops, medicinal and aromatic plants in addition to normal food crops grown by the farmers. Vegetable crops yield a higher return on an average than the common field crops. Crops like peas and cowpeas can be grown successfully during winter and summer months respectively under trees and these crops can also fix the atmospheric nitrogen and improve the fertility status of the soil along with providing additional income from the production of vegetable crops. Agroforestry has the immense potential to increase and sustain the food production per unit area in systems like the *Prosopis cineraia*-based agroforestry system of the arid region of India.

AGROFORESTRY FOR LIVELIHOOD SECURITY

Agroforestry systems have been developed for different agro-ecological regions of the country. In addition to providing higher returns, these also improve livelihood security as a cover against crop failure due to climatic aberrations, particularly in arid and semi-arid regions. Some of the important activities which can be integrated with agroforestry for livelihood security include lac cultivation, sericulture, apiculture and gum production. Conventional tree hosts of lac like ber (*Ziziphus mauritiana*) and *Palas* (*Butea monosperma*) can easily be integrated in bunds of cropped fields. Other quick-growing hosts like *Flemingia semialata* hold potential for integration in farming system models for lac production. Another option is to integrate sericulture as one of the four hosts of the silk worm viz., Mulberry (*Bombyx mori*), *Tasar* (*Antheraea myllitte*), *Eri* (*Phelosamia ricini*) and *Muga* (*Antheraea assamensis*) are cultivated in India. These silk worm insect can feed on leaves of different agroforestry tree species such as *Terminalia tomentosa* (*asan*), *T. arjuna* (*arjun*) and *Shorea robusta* (*sal*) for the *Tasar* insect; the commonly used *Morus alba*, *M. indica*, *M. serrata* and *M. latifolia* for the mulberry insect; can be included in the agroforestry models. As per study of Sridhar et al (2015) the lac cultivation on *Butea monosperma*-based agroforestry is quite helpful for livelihood security in drought-prone regions of Central India and a single tree of *Butea monosperma* can produce 1.5-2.5 kg and gives an annual income of ₹ 700/- to ₹ 800/- (10-12 US$).

An apiculture-based agroforestry system is another option which will not only provide additional income to the farmers, but will also result in other benefits to the agricultural production due to the role of honeybees as pollinators. A number of agroforestry tree species have been identified which can act as host species for honey bees and there is ample scope for including such trees in different agroforestry models to sustain round-the-year honey production. Gums and resins form another important group of non-wood forest products. India is traditionally the largest producer of guar gum and karaya gum. Gum-yielding trees species such as *Acacia senegal* (gum Arabic), *Acacia nilotica*, *Butea monosperma* (Bengal kino gum) and *Boswellia serrata* (yielding *salai guggul*) are potential species for agroforestry models and need to be promoted on a large scale to provide sustained income and livelihood security to resource poor rural masses. Venkatesh et al (2011) reported that a single tribal family can collect up to 65 kg of *Butea* gum and earn a maximum amount of ₹ 2628 (40-42 US$) in a season.

The most spectacular evidence in context of livelihood improvement potential of agroforestry in arid region came through *Acacia senegal* based localized traditional agroforestry system in parts of Barmer and Jodhpur districts (15 villages). *A. senegal* is a source of gum Arabic, which has very high commercial value. In addition to harvesting crop grain for food and crop straw for fodder, the farmers harvested substantial quantity of gum Arabic through small intervention using gum inducer provided by ICAR-Central Arid Zone Research Institute, Jodhpur.

During the year 2008-09, 12,000 trees were treated through ICAR-CAZRI gum inducer which resulted in exudation of 5.4 t of gum Arabic and farmers earned ₹ 27 Lakhs from the sale of gum. Subsequently in the years 2009-10 and 2010-11; 21,000 and 22,600 trees were treated, resulting in production of 10.5 t and 7.6 t gum, respectively. In the process farmers earned 52 lakhs in the years 2009-10 and ₹ 38 lakhs in the year 2010-11. In this way, the farmers of 15 villages earned ₹ 117 Lakhs from the sale of gum arabic with the average modest sale price ₹ 500 kg^{-1} in local market. This technology has been so successful that farmers have made their own cooperatives for purchasing of gum inducer and sale of produce in the market.

EMPLOYMENT GENERATION AND RURAL DEVELOPMENT

Agroforestry generates a huge employment in rural areas where the employment opportunities are meagre due to less development activities. In such remote locations. Locals, especially women, find part-time employment in many of these activities and is a source of income generation to them. In a study an employment generation potential of 943 million person days annually from agroforestry area with improved agroforestry systems was estimated, while Dhyani and Sharda (2005) indicated the potential of agroforestry for rural development and employment generation to the tune of 5.763 million human days yr^{-1} from Indian Himalayas alone.

Commercial agroforestry involves a complete chain of activities from seedling or sapling production to their transportation to planting sites, field preparation, layout and planting, maintenance, pruning trees, harvesting, transportation, marketing, and processing of tree produce in the wood based industry. Most of these activities are labour intensive. Some of them are performed by the farmers themselves, many others are carried out by outsourced labour at market driven costs. A poplar tree generates

5 person days employment in its production, harvesting, transportation and usage in the wood based industry. Poplar based agroforestry was estimated to generate 100 million man-days per annum when there was planting and harvesting of around 20 million poplar saplings per annum. Since its planting number has now gone up to 30 million saplings per annum now, a direct employment of 150 million person days equivalent to persons is generated each year. About 796 agroforestry based units at Yamunanagar, Haryana are reported to engage 4075 persons per day. In addition, 671 industries subsidiary to agroforestry are providing employment to 6068 persons per day. It is estimated that around 2000 persons are daily engaged in felling of trees in command area of Yamunanagar market. West Coast Paper Mills estimated employment generation of 760 persons (http://westcoastpaper.com/index.php?q=node/45) and harvesting to 430 by ITC in Eucalyptus plantations. Overall, it was estimated that wood based industries of Yamunanagar generate about 35 million man-days work yielding ₹ 3,000 million every year. Indian pulp and paper industry being the 15th largest industry in the world is now largely based on commercial agroforestry having a bearing on the socio economic development as it provides employment to nearly 1.5 million people. ITC established 140,989 ha plantations provide an estimated 63 million person days of employment from tasks such as nursery, planting, logging and maintenance operations. Taking an average person days of 800/ha, a total of 4000 million person days are generated from the commercial agroforests in the country (*pers. comm*. RC Dhiman, WIMCO).

Growing trees as cash crops has prompted numerous growers to venture into growing planting stock and plantations as commercial activity on own and leased land. A classical model of poplar nursery production adopted by WIMCO, the main promoter of poplar culture in North India is to grow its nurseries for one-two years in a location and shift it to another site during the following year. The activity has helped a great deal in developing trained human resources on the left over sites where growing of poplar nurseries is picked up by the locals. Besides, numerous employees separated from the company undertook poplar nursery production to supply the same to the farming community. As a result this section of self-employed nursery growers grew more poplar nurseries than any of other single stake holders in the country. A similar trend is happening in clonal Eucalyptus plant production in South India where around 100 million Eucalyptus plants are produced by the small nursery growers.

AGROFORESTRY FOR FODDER PRODUCTION

Trees and shrubs often contribute substantial amounts of leaf fodder in arid and semi-arid and hill regions during the lean period, through lopping/pruning of trees, popularly known as top feed. The leaf fodder yield depends on species, initial age, lopping intensity and interval as well as agro-climatic conditions. Fodder from trees is provided by leaf twigs and pods. This forage is usually rich in proteins, vitamins and minerals like calcium. They are however in general low in phosphorus and crude fibre. Top feed plays an important role in human food security through the function as an animal feed resource. The importance of top feeds increases with the severity of drought and progression of the drought season. Other uses such as for live fencing, are complementary, as they encourage cultivation of the species and increase the availability of feed. The top feeds are considered very important in vegetation stabilisation and sustained productivity of rangelands. They also play an important role as windbreaks and by providing shade for the grazing animal. The most important top feed species are *Prosopis cineraria, Albizzia lebbeck. Acacia* spp., *Leucaena leucocephala, Dalbergia sissoo, Ailanthus excelsa, Azadirachta indica, Acacia leucophoela* for the arid and semi –arid regions and *Grewia optiva, Morus alba, Celtis australis* for the hilly regions. Tree fodder provides enough nutrients and can serve very well as a green fodder supplement. In addition, leaves are also conserved in the form of hay and silage to supplement feed during shortages.

AGROFORESTRY FOR BIO–FUEL PRODUCTION

A large part of India's population, mostly in rural areas, does not have access to conventional source of energy. Biofuels are gaining worldwide acceptance as a solution for problems of environmental degradation, energy security, restricting imports, rural employment and improving the agricultural economy. Trees with potential for biofuel are *Jatropha curcas, Pongamia pinnata, Simarouba, Azadirachta indica, Madhuca* spp., etc. In agroforestry these can be intercropped with annual crops such as cowpea, sesame, sunflower, french bean, black gram, green gram and groundnut. The promotion of oil crops also provides a poverty alleviation option in rural areas as vacant, waste and marginal land can be used for these trees, providing annual produce and income. The increased green cover will also benefit the environment. These oils are also CO_2 neutral, which would mitigate the Green House Effect. The major source of biofuels in India is non-edible oil seeds *i.e.,* Tree Borne Oilseeds (TBOs). The requirement of TBOs in the country as per the National Oilseeds and Vegetable Oils Development Board, Gurgaon (India) is more than 5 million tons. However, only 0.8-1.0 million tons are being collected from different species such as Neem (*Azadirachta indica*), Karanj (*Pongamia pinnata*), Mahua (*Madhuca longifolia*), Jatropha (*Jatropha curcas*), Kusum (*Schleichera oleosa),* Pilu *(Salvadora oleoides),* Bhikal (*Prinsepia utilis),* Undi (*Calophyllum inophyllum*), *Simarouba glauca,* Sal (*Shorea robusta*) and Jojoba (*Symondsia chinensis*) (Dhyani et al, 2015). Some of the TBO species can be grown in wasteland under relative unfavourable environmental conditions and such species needs to be promoted. They meet the dual objectives of rehabilitating wastelands and providing an energy substitute. However, there is still a need for research on genetic improvement for increasing seed and oil yield.

AGROFORESTRY FOR BIO-ENERGY

In rural areas 70-80% energy comes through biomass from trees and shrubs. Due to the agroforestry initiatives, large amounts of wood are now being produced from outside the conventional forestlands. Small landholdings and marginal farmers, through short rotation forestry and agroforestry practices are now providing the bulk of country's domestically produced wood products. *Prosopis juliflora* is the major source of fuel for the boilers of the power generation plants in Andhra Pradesh. The fuel wood potential of indigenous (*Acacia nilotica, Azadirachta indica, Casuarina equisetifolia, Dalbergia sissoo, Prosopis cineraria* and *Ziziphus mauritiana*) and exotic (*Acacia auriculiformis, A. tortilis, Eucalyptus camaldulensis* and *E. tereticornis*) trees revealed that calorific values ranges from 18.7 to 20.8 MJ DM $kg^{-1}$for indigenous tree species and 17.3 to19.3 MJ DM kg^{-1} for exotics. Species such as *Casuarina equisetifolia, Prosopis juliflora, Leucaena leucocephala* and *Calliandra calothyrsus* have become prominent due to their potential for providing wood energy at the highest efficiency, shorter rotation and also their high adaptability to diverse habitats and climates. Agroforestry plantations on community land and live fences on farm boundaries have immense potential in this regards. There is also an option for biomass based power plants where electricity generation can use crop/plant residue is another option.

Chauhan et al (2015) conducted extensive study on socio-economic status of fuelwood utilization in Bendelkhand region of India and concluded that plantation of locally available multipurpose tree species can be promoted in the region in the wasteland, community and private land. Moreover, the promotion to use biomass as one of the source of energy (through agroforestry, for example) in the semi arid Bundelkhand region, not only improved the livelihood condition but will also bring better ecological health in this area. In continuation of this Chavan et al (2016) carried out fuelwood analysis of farmer preferred tree species and recommended *Acacia catechu, Acacia senegal, Anogeissus pendula* and *Prosopis juliflora* as a best fuelwood tree species for agroforestry plantation in Bundelkhand region.

ECONOMICS OF AGROFORESTRY SYSTEMS IN INDIA

There are a number of studies from different parts of the country suggesting that agroforestry is more profitable to farmers than agriculture or forestry for a particular area of land. Ram Newaj and Rai (2007) analyzed 13 years aonla-based agroforestry system in marginal lands under rainfed conditions and found a B: C ratio of 3.28 (on discounted rate it was 2.61) which indicated its profitability. Similarly, there are a number of studies indicating profitability of the systems. An analysis indicated the economic viability with IPR ranging from 25 to 68 and B:C ratio of 1.01 to 4.17 for 24 agroforestry systems from different agro-climatic regions of the country (Planning Commission, 2001). Agroforestry models adopted by the farmers in upper Gangetic region and especially in Haryana, Punjab and Western Uttar Pradesh are highly lucrative, therefore, attracting farmers in a big way. In these areas poplar planted on agricultural fields and field boundaries are harvested at 6 to 8 years rotation, and the average economic return of poplar based agroforestry systems is very high compared to that of sole agriculture crop. *Kinnow* based horti-agricultural system is highly remunerative. In general, average gross income of ₹ 50,000 to ₹ 1,00,000 acre^{-1} is obtained from *Kinnow* plantation after 6-7 years of plantation depending upon tree density and management. Some of the progressive farmers even earn more than ₹ 1.00 lakh from an acre. The net return from the agricultural crops after 4 years of plantation is considerably less because of reduced yields. Another examples of eucalyptus from Haryana state that the economic feasibility of *Eucalyptus* hybrid both with and without agriculture under varying spacing viz. 2.5×2.5 m, 3×1.5 m, 4×2 m and 6×1 m was worked out by Dhillon and Bangarwa (2016). The results of the study indicate that cultivation of eucalyptus at wider spacing (6×1 m) with rotation of 8 years in combination with agricultural crops on farm lands has given maximum NPV and B/C ratio of 2.28. The influence of eucalyptus trees grown as a single row on one side of the field in East-West direction on the growth and yield of wheat was studied in Haryana state. Poplar-based agroforestry has been very successful in Northern States, but we do not have a similar agroforestry model for other parts of the country except *Eucalyptus* and *Casuarina*. It was because of the success of these trees in agroforestry, credit support was provided by banking institutions. Hence, there is tremendous scope for promoting viable agroforestry models for different agro-ecological regions of the country through quadripartite arrangement among industries, banks, research institutes and farmers. Research Institutes under Indian Council of Forestry Research and Education and Indian Council for Agricultural research have developed a number of agroforestry models but these are mostly restricted to research farms. These models need to be taken to cultivators' field for improvement of farmer's livelihood.

CONCLUSION

The nurturing trees in fields or homestead gardens are way of life in Indian culture. They are integral to our traditional farming systems, for the innumerable direct and indirect benefits that they provide. These agroforestry system act as a source of tree based products like gum, resins, floss, fibers, fodder, brooms, fruits etc, for the subsistence of farmers and provide immense employment opportunities. Therefore, agroforestry acts as backbone of marginal farmers for self-reliant and sustainable agriculture. Achieving food security and environmental security is prime benefit of agroforestry systems. The varied agroforestry systems in India also form a strong feasible option against climate change. However, all of these approaches trying to reduce the ecological footprint are focused on farm scale, and not at the larger landscape level which is need of the hour.

REFERENCES

Chaturvedi OP, Sikka AK, Handa AK, and Bajpai, CK 2016 Agroforestry technologies for different agro-climatic zones of the country, Jhansi: *ICAR-CAFRI* Publication, 196 pages

Chauhan D, Chavan SB, Keerthika A, Jha A, Dwivedi, RP, Ram Newaj and Kumar A 2015 Socio-economic fuelwood survey in Bundelkhand region of Central India in *Energy Research and Environmental Management: an innovative approach* (eds) GC Mishra and SK Yadav) New Delhi: Krishi Sankriti Publications,124-129

Chavan SB, Chauhan DS, Keerthika A, Uthappa AR, Jha Ankur and Ram Newaj 2016 Fuelwood characteristics of selected tree species from Bundelkhand region of Central India. *Ecology Environment and Conservation* **22** 95-100

Dhillon RS and Bangarwa, KS 2016 Agroforestry practices in Haryana. Department of Forestry, CCSHAU, Hissar, Haryana: 35

Dhyani SK and Sharda VN 2005 Agroforestry systems as rural development options for the Indian Himalayas, Journal *of Tree Sciences* **24**(1) 1-19

Dhyani SK, Ram Newaj, and Sharma AR 2009 Agroforestry: its relation with agronomy, challenges and opportunities. *Indian J. Agronomy,* **54**(3) 249-266

Dhyani SK, Vimala Devi S, Handa AK, Gupta VK, Kumar RV, Ahlawat SP, Kumar, Sudhir, Datta A, and Singh V 2015. A decade of research on tree borne oilseeds to strengthen National Biofuel Programme: A review, *Indian Journal of Agroforestry*, **17**(1) 1-12

ICAR (Indian Council of Agricultural Research) 2010 Degraded and wastelands of India: status and spatial distribution *Indian Council of Agricultural Research, New Delhi*, 158 pages

Nair PKR, Kumar BM, and Nair VD 2009 Agroforestry as a strategy for carbon sequestration, *Journal of Soil Science and Plant Nutrition* **172** 10–23

Planning Commission 2001 Report of the Task Force on Greening India for Livelihood Security and Sustainable Development. Government of India, New Delhi, 254 pages

Ram Newaj and Rai P 2005 Aonla-based agroforestry system: A source of higher income under rainfed conditions, *Indian Farming,* **55** 24-27

Sridhar KB, Dhyani, SK, Kumar S, Dwivedi RP, Singh M, Venkatesh A, Monobrullah Goshal S, Dev I, Tewari RK, Singh R, Chavan S, Uthappa AR, Singh R and Tripathi VD 2015 A tree with a purpose: *Butea monosperma* (Lam.) (Improving livelihood of disadvantaged rural people of central India) *In* the Proceedings of the *XIV World Forestry Congress, Durban, South Africa*, 7-11 September 2015

Venkatesh A, Dwivedi RP, Tewari RK, Dhyani, SK, Singh R, Palsaniya DR, Yadav RS, Singh R and Kareemulla K 2011 Livelihood opportunities in drought prone Bundelkhand: lessons learnt from Garhkundar-Dabur Watershed, *Indian Journal Dryland Agricultural Research and Development*, **26** 72-82

Zomer RJ, Trabucco A, Coe R, and Place F 2009 Trees on farm: Analysis of global extent and geographical patterns of agroforestry, ICRAF Working Paper No. 89. Nairobi, Kenya, ICRAF: 63

CHAPTER 3

Agroforestry for Sustainable Land Use

R Kaushal*, H Mehta and PK Mishra

ICAR-Indian Institute of Soil and Water Conservation 218 Kaulagarh Road, Dehradun (Uttrakhand), India

** Corresponding author's Email: kaushalrajesh1@rediffmail.com*

ABSTRACT

Natural resources are vital for meeting the food, livelihood and environmental security. It is now widely accepted that future of food, livelihood and environmental security depends upon the attention paid to conservation, sustainable development and management of natural resources. There is urgent need to devise/refine land use systems and management practices which can enhance the productivity besides conserving the resources on sustainable basis. Agroforestry in this context offers affordable alternative for providing significant benefits. It has potential to provide provisioning services, regulating services, cultural services and supporting services. Agroforestry measures are simple, cost effective and help in relieving pressure on traditional cultivated lands and forests. Joint efforts of research in last 3 decades have resulted in development of different agroforestry models which are suitable for production and resource conservation on arable and non-arable lands. The paper highlights the importance of agroforestry in sustainable production through increasing production, conserving natural resources, providing environmental and livelihood security.

Keywords: Agroforestry, Environmental challenges, Resource conservation, Soil health

INTRODUCTION

Increasing demographic pressure, is leading to overexploitation of the natural resources which are important for food, livelihood and environmental security. Owing to population pressure, per capita availability of cultivated land has declined from 0.53 ha in 1950 to 0.15 ha in 2000 and was expected to further reduced to 0.12 ha by 2015. The consumption of fuel-wood and fodder from forests in the country is largely unsustainable. Nearly 86 million tonnes of fuel wood is being removed from the forests and plantations every year in excess of what they are capable of producing on sustained basis (Planning commission, 2003). Soil erosion has become one of the major environmental hazards. About half of the country's geographical area is affected by different form of soil erosion and land degradation. About 5334 m tones of soil is lost (16 t ha^{-1}) due to wind and water erosion annually which is highest in black soils (24-112 t ha^{-1}) followed by Shiwaliks (80 t ha^{-1}) and shifting cultivation (27-40 t ha^{-1}) (Mandal et al. 2008). The problem is further being aggravated by increase in concentration of greenhouse gases which is leading to climate change, a major threat to food production. Corrective measures are therefore immediately required to check damage to the resource base and environment. In this context, agroforestry (AF) is emerging as sustainable land use system for addressing wide range of environment and livelihood issues.

Agroforestry provides a land-management system that increases production and ecological stability, as well as supporting sustainable development. Agroforestry (AF) has potential to provide different services such as provisioning services (food, fuel, fibre, fresh water and genetic resources), regulating services (climate regulation, water maintenance and erosion control), cultural services (spiritual enrichment, reflection, recreation and aesthetic values) and supporting services (production of oxygen and soil formation). Agroforestry meets almost half of the demand of fuel wood, two thirds of that for small timber, 70-80%wood for plywood, 60% raw material for paper pulp and 9-11% of the green fodder requirement of livestock, besides meeting the subsistence needs of households for food, fruit, fibre, medicine etc (Handa et al 2016). In developing countries, approximately 1.2 billion people (20% of the world's population) depend directly on agroforestry products and services (Leakey and Sanchez, 1997).

The statistics of the area under agroforestry is not very clear. Almost 10% of the world's agricultural area is under agroforestry (Zomer et al. 2007), whereas Nair et al. (2009) estimated the total area under agroforestry is 2.35 billion ha. Trees outside forests in India is estimated between 14,224 million (Ravindranath and Hall, 1995) and 24,602 million (Prasad et al. 2000) which is spread over an equivalent area of 17 million ha (GOI, 1999) supplying 49% of the 201 million tonnes of fuel wood and 48% of the 64 million m^3 of timber consumed annually by the country (Rai and Chakrabarti, 2001). Dhyani et al. (2009) reported 25.72 million ha under various types of tree plantations, which include agroforestry, social forestry, farm forestry and others.

Realizing importance of agroforestry, various AF practices have been included in the various national developmental programmes/schemes like Flood Control/Management Programmes, Multipurpose River Valley Projects, Agriculture Development Programmes, Integrated Rural Development Programmes (IRDP), National Watershed Development Programme for Rainfed Areas (NWDPRA), Forestry Development Scheme, Drought Prone Area Development Programme (DPAP) and Desert Development Programme (DDP). The task force on Greening India has also identified a potential of agroforestry on 10 million ha in irrigated lands and another 18 million ha in rainfed areas (Planning Commission 2001). Agroforestry can help in sustainable development through increasing production, conserving natural resources, providing environmental and livelihood security. The various role of agroforestry are described as under:

AGROFORESTRY FOR SUSTAINABLE LAND USE

Agroforestry systems primarily involve multiple land use systems for maximizing productivity per unit area and time. Perennial and annual components in agroforestry help in enhancing the productivity on sustainable basis. Joint efforts of research in last 3 decades have resulted in development of different agroforestry models which are suitable for arable and non-arable lands. Some of the important agroforestry systems used for sustainable production are as under:

AGRISILVICULTURE SYSTEM FOR ARABLE LAND

Popular, Eucalyptus, Subabul, Gmelina, Ailanthus based agroforestry systems have been adopted by farmers on large scale. They are grown as block plantation or on bunds with agriculture crops (agrisilviculture system) for diversification of farm income. Numerous industries are promoting these agroforestry systems for meeting daily industrial requirement and providing employment opportunities. Poplar (*Populus deltoides*) based agroforestry is common in north India. All *rabi* and *kharif* crops can be grown during the first three years except paddy. Third year onwards wheat, cabbage, chilly, tomato, barley, coriander, turmeric, ginger, strawberry, oats, *barseem*, mustard etc. are raised throughout the rotation. Poplar can provide yield of 150-200 m^3 ha^{-1} in block plantation and 12-20 m^3 ha^{-1} in boundary plantation at 8 years rotation. The above ground biomass production in poplar ranged from 63 t ha^{-1} (3-year) to 105.4 t ha^{-1} (10- years) in bock plantation and 7.87 to 49.51 t ha^{-1} in boundary plantation (Lodhiyal et al. 1995; Rizvi et al. 2011; Kanime et al.2007).

Eucalyptus based agroforestry is common in both northern and southern part of India. These plantation are capable of providing productivity of 20 to 25 m^3 ha^{-1} year^{-1} under un-irrigated conditions and 50 m^3 ha^{-1}year^{-1} under good care (Lal 2004). Subabul *(Leuceana leucocephala*) based agroforestry is being practiced for pulpwood and animal husbandry in south India. It is grown at spacing of 1x1 m (for pulpwood), 2.5 x 0.75 m (with intercrops) and 1.25x1.25m (fodder production in animal husbandry) which can provide yield of 160 t ha^{-1} (Pathak et al 2006). In regions of high fertility and rainfall in southern and eastern part of India, homestead practice is commonly followed by small farmers for maximizing the returns from land. In this multi-storeyed cropping, perennial crops such as coconut, arecanut, rubber and pepper are intercropped with seasonal and annual crops like tapiaco, banana, pulses and vegetables which provide return on sustainable basis.

HORTICULTURE BASED SYSTEM FOR ARABLE AND MARGINAL LAND

Fruit based agroforestry systems have assumed greater importance due to better economic returns on sustainable basis. Studies conducted in the past have suggested that on marginal rainfed and dryland area, these agroforestry systems have great potential. In sub-montane area in Dehradun, peach based agroforestry system has been recommended along with vegetables. On marginal lands, intercropping of guar, cowpea or pearl millet with peach; turmeric with papaya; *Chrysopogon* gorda or napier grass with *aonla* or *ber* has been identified as an economically-viable (Sharda and Venktaswaralu, 2007). For marginal (class II) rainfed lands, Arora and Mohan (1986) reported that agri-horticulture system of kinnow-turmeric in V-shaped micro-catchment with *Morus alba* on field bunds produced 4.34 t ha^{-1} of *kinnow* fruits, 1.11 t ha^{-1} of turmeric from interspaces and 2.24 t ha^{-1} of canes (for basket making) along with 2.16 t ha^{-1} of wood and 0.69 t ha^{-1} of mulberry leaves for sericulture. Samra et al. (1995) reported that in Relmajra watershed, plantation of *kinnow* at 4 m x 4 m spacing and *bhabar* grass at 50 cm x 50 cm after minor levelling of land provided early returns to the farmers and was highly profitable. Dhyani, et al (1995) reported that under silvi-horticulture system with *Acacia auriculiformis* and *Leucaena leucocephala*, pineapple produced yield of 11.7 t ha^{-1} and 13.4 t ha^{-1}, respectively.

SILVIPASTURE SYSTEM FOR NON-ARABLE LAND

On degraded lands, fodder and fuelwood production could be increased several- folds by adopting silvipasture systems (Chaturvedi et al 2014). In Bundelkhand region, Jhansi Rai et al. (1999, 2001) reported enhanced productivity through *Acacia nilotica, Albizia lebbek, A. procera*, and *Hardwickia binata* based silvopastoral systems. Sharma et al (1994) reported that biomass productivity of arid lands can be increased two to three times by replacing natural grass cover with *Cenchrus ciliaris* + *Z. nummularia* and *Grewia tanax*. Prajapati et al (1993) reported 1.5 to 2.0 t ha^{-1} yr^{-1} grass yield and 4.2 t ha^{-1} yr^{-1} from trees in a 14 year old *Acacia nilotica* and *Acacia tortalis* + *Cenchrus ciliaris* silvipastoral system in Yamuna ravines. Silvipastoral system comprising *Alnus nepalensis*, pineapple and forage crops like *Panicum maximum* or *Setaria sphacelata* coupled with *Stylosanthes guyanensis* in 1:1 ratio was found to sustainable in soils having 30-60% slope. Forage yield of 13.5 t ha^{-1} was obtained from the combination of Stylo and Setaria. In addition to this, 2.3 t ha^{-1} litter from *Alnus nepalensis* and 4000 fruits ha^{-1} from pineapple were obtained. This system also restored the fertility of soils (Chauhan et al. 1993). In Doon valley, *Albizia lebbek* (siris), *Grewia optiva* (bhimal), Bauhinia *purpurea* (kachnar) and *Leucaena leucocephala* (subabul) were found promising when grown with *Chrysopogon fulvus* (gorda) and *Eulaliopsis binata* (bhabar). Silvipasture systems can provide mean annual production of 8-10 t ha^{-1} which comprised4.5-5.0 t ha^{-1} of fodder from grasses, 1.5-2.0 t ha^{-1} of leaf fodder from tree loppings and 2.0-2.5 t ha^{-1} of fuelwood from the lopped branches (Raizada and Singh 2010).

On degraded ravine land in Gujarat, bamboo (*Dendrocalamus strictus*) and Anjan grass (*Cenchrus ciliaris)* based silvipasture system was found very effective for enhancing productivity. The system is capable of yielding 10 t ha^{-1} $year^{-1}$grass during initial years. After 7 years of plantation, 1000 bamboo poles ha^{-1} $year^{-1}$ can be harvested giving net present value of benefits worth Rs. 30075 per ha (at 2008 prices) over a production period of 20 years with benefit cost ratio of 1.85: 1. Organizations including Anand-based Foundation for Ecological Security (FES), Gujarat State Watershed Management Agency (GSWMA), Gujarat State Land Development Corporation (GSLDC), Forest, and Agricultural departments and other users agencies have taken up this technology in nearly 1000 hectare of community and waste lands in Mahi river stretch in Gujarat for stabilizing the degraded ravine lands and improving livelihood of primary stake holders through reclamation and productive utilization of ravine lands(http://www.icar.org.in/node/5101).

BIO-ENERGY PRODUCTION

With inception of Clean Development Mechanism (CDM), energy plantations are gaining importance. Research on bio- fuel started under All India Coordinated Research Project (AICRP) on agroforestry revealed that bio-energy plantation is economical for degraded lands. Species like *Acacia nilotica, Azadirachta indica, Casuarina equisetifolia, Dalbergia sissoo*, *P. cineraria* and *Ziziphus mauritiana* and exotic species like *Acacia auriculiformis*, *A. tortilis*, *E. camaldulensis* and *E. tereticornis* have huge potential to be used for energy production due to higher energy efficiency, shorter rotation and high adaptability to diverse habitats and climates (Pathak, 2002). *Eucalyptus* provided 11.9 t ha^{-1} (3 years old plantation) to 196.5 t ha^{-1} (9 year old plantations) biomass when grown extensively as wood lots. On degraded lands, *Casuarina equistifolia provided* 120 tonnes of firewood at 7 to 10 years rotation (Planning Commission 2003). On old river bed lands in the NW Himalayas, densified plantations of *Grewia optiva*, *Dalbergia sissoo* and *Bahunia variegata* raised at 2x2m (close), 3x2m (medium) and 4x2m (wide) spacing were found very effective for firewood and fodder production (Raizada et al. 2015).

Planning Commission, (2004) reported that *Jatropha* yielded about 2 kgs of seed $tree^{-1}$. If planted in hedges, the reported productivity of *Jatropha* varied from 0.8 kg- 1.0 kg of seed m^{-1} of live fence, which is equivalent to seed production of between 2.25 t ha^{-1} and 5 t ha^{-1}, depending upon whether the soils are poor or average for plantations and between 2.5 t ha^{-1} $year^{-1}$ and 3.5 t ha^{-1} $year^{-1}$for hedges.

AGROFORESTRY FOR SOIL AND WATER CONSERVATION

Water erosion leads to huge soil loss, nutrient loss, pollutes water and contributes to land degradation. The importance of tree in controlling soil erosion is widely accepted. In agroforestry, tree canopy check soil erosion mainly by intercepting rainfall thereby reducing the impact of raindrops and decreasing its erosive capacity (Andreu et al. 1998; Kaushal et al. 2017). Litter helps in producing water stable aggregates thereby reducing surface run off volume. Stem and roots trap sediments, stop their down slope movement and facilitates water infiltration. Deep tree roots stabilize slopes and help prevent shallow landslides. In long term, vegetation influences the fluxes of water and sediment by increasing the soil aggregate stability and cohesion as well as by improving soil water infiltration (Zuazo and Pleguezuelo 2008). Plantation crop combinations, multilayer tree gardens and home gardens, hedgerow (alley) cropping, and windbreaks and shelterbelts are the common agroforestry practices that are used commonly for erosion control (Young, 1997). McDonald et al. (2002) reported that agroforestry caused a seven-fold decrease in surface runoff and 21-fold decrease in soil erosion than the control soils. Lal (1989) observed that established hedgerows of *Leucaena* at 2-m spacing were very effective in reducing water runoff and controlling erosion. A mean runoff, in the order of 17.0, 4.9, 3.3, and 2.4 percent was reported from maize for plow-till, *Leucaena* - 4m, *Leucaena* - 2m, *Gliricidia*- 2m treatments, respectively. Soil erosion of 4.3, 0.57, 0.10, and 0.60 t ha^{-1} was observed in the same order of treatments.

Results of research on soil losses under agroforestry systems in India are relatively very few. Some of the agroforestry technologies have been evaluated on the experimental scale by ICAR-Indian Institute of Soil and water Conservation (IISWC) (Formerly CSWCRTI), Dehradun, University of Horticulture and Forestry, Solan and ICAR Research Complex at Barapani. In Shiwalik region, Grewal (1993) reported that the soil loss and runoff under *Eucalyptus tereticornis* + *Eulaliopsis binata* reduced to 0.07 t ha^{-1} and 0.05 per cent in comparison to 5.65 t ha^{-1} and 23.0 per cent under cultivated fallow and 2.69 t ha^{-1} and 20.50 per cent under *Sesamum indicum* - *Brassica campestris* systems, respectively. In Doon valley, Narain et al. (1998) reported that contour tree-rows or *Leucaena* hedges reduced the runoff and soil loss by 40% and 48%, respectively, over the maize plot, reducing soil loss to about 12.5 t ha^{-1}. In Shiwalik region, Yadav et al (2005) reported that the effectiveness of different intercrops in reducing soil loss was in the order of *Chrysopogon* > napier > pigeon pea with 81, 56 and 25 per cent reduction over sole aonla. Gupta et al (2006) compared poplar based agroforestry system with non-agroforestry system and found the dispersion ratio reduced, whereas, the water stable aggregates (> 0.5 mm) increased under a poplar based agroforestry system irrespective of the soil type as compared to a non-agroforestry system of cultivation. In North-east region (NER), Saha et al (2012) reported that hedgerows reduced soil loss by 94% and run-off by 78%. Further, use of twigs and tender stem of hedge plants for mulch helped in conserving 83% of the soil and 42% of rainfall. In degraded ravine land of Gujarat, bamboo (*Dendrocalamus strictus*) and Anjan grass (*Cenchrus ciliaris)* based silvipasture system was found to absorbs more than 80% of rainfall which was utilized by the plant or percolated deep to recharge the ground water. Soil loss under the system was reduced to less than 1 t ha^{-1} $year^{-1}$ than about 20 t ha^{-1} $year^{-1}$ from degraded ravines prior to plantation

Bioengineering measures comprising of tree, grasses and mechanical measures were very effective in resource conservation in mine spoils at Shastardhara in Dehradun (Juyal et al 2007). These measures reduced equivalent slope in the treated hilly terrain from 38 % to about 19 % over a period of 14 years. Runoff and soil loss also decreased considerably over the time with mechanical and biological measures. Satapathy (2005) reported that mixed land use systems with appropriate soil conservation measures, namely, bench terraces, contour trenches were effective in retaining 90-100% annual rainfall and simulated the effects of natural forest in NER. The watersheds having continuous stream flow characteristics generated base flow to the extent of 70-90% of its total water yields. The watershed treated with *jhum* (shifting) cultivation yielded the highest peak runoff while the one left undisturbed with natural vegetation gave the minimum peak runoff.

MAINTENANCE OF SOIL HEALTH

Agroforestry helps in maintaining the soil physico-chemical properties. Trees act as a nutrient pump by taking up nutrients from deeper layers of the soil. Decay of litter and fine roots helps in addition of organic matter to the soil (Kaushal et al 2012). Tree roots act as safety net by trapping and recycling nutrients which would otherwise be lost by leaching. Improvements in soil are also reflected through lower bulk density (Belsky et al 1993), lower surface resistance to penetration (Campbell et al 1994), increased porosity, higher rain water infiltration and greater aggregate stability (Kaushal et al 2016). Tree canopy shade also alters soil conditions to promote microbial activity and the rate of soil mineralization (Martius et al 2004). Growing of nitrogen fixing hedge species on the field can contribute as much as 358 kg nitrogen (N) ha^{-1}, 28 kg phosphorus (P) ha^{-1} , 232 kg potassium (K) ha^{-1}, 144 kg calcium (Ca) ha^{-1}, and 60 kg magnesium (Mg) ha^{-1} (Palm 1995). The changes/improvements in soil are however, species-specific and dependent on size and age of the trees, and site conditions. Such improvement in soil properties has a direct bearing on long-term sustainability, productivity and soil quality (Saha et al 2012).

Ameliorative effect of agroforestry has also been reported for problematic soils which are affected by salinization, ravines, gullies, and other water and wind erosion hazards.

Different agroforestry technologies have been successfully demonstrated for rehabilitation of bouldry lands, salt-affected soils, gullied and ravine lands, mine spoil areas, degraded village community lands, areas along road sides, canals and railway tracks (Dagar 2012). In waterlogged condition, trees in agroforestry pump excess water out of the soil profile (Dhillon et al. 2009; 2010). In water logged areas near canals planting cloned *Eucalyptus* on bunds (of one meter height) on farmers field (in two lines in space of 1m x 1m) each strip separating 66 m (growing crops between two strips), proved very useful, which not only controlled rise in water table but also helped in revenue generation. The average draw down of groundwater table underneath the plantation was 1.01 m below the average (4.15 m) natural ground water table level (Jeet Ram 2004).

In mine spoils planting of slips of *Eulaliopsis binata* was ideal on the degraded, steep mine spoil areas. Planting of leguminous species, such as *Leucaena leucocephala* and *Peuraria hirsuta* provided foliage rich in N which served as fodder, organic manure, mulch etc (Juyal et al. 2007). Species, such as *Thysonoleana maxima*, *Saccharum munja*, *Pennisetum purpureum*, *Eulaliopsis binata*, *Ipomoea carnea* and *Vitex negundo* performed well with geotextiles conservation measures. The vegetal cover increased from 10 % to > 90 % due to reforestation over a period of 20 years. On an average, after 18 years, pH of the spoil came down from 8.1 to 7.4, organic carbon increased from 0.13 to 0.45 %, whereas $CaCO_3$decreased from 54.6 % to 29.5 % and bulk density from 1.63 Mg m^{-3} to 1.45 Mg m^{-3}.

ENVIRONMENTAL CHALLENGES

CLIMATE CHANGE AND CARBON SEQUESTRATION

Carbon dioxide is the main focus of global climate changes studies, because it accounts for about 70-80% of the greenhouse gases. Agroforestry can pool more CO_2 from the air and can play an important role in reducing vulnerability, increasing resilience of farming systems and buffering households against climate related risks. Tree based system provides a long-term strategy to increase the amount of carbon while still allowing for growing of food crops. Average sequestration potential of agroforestry has been estimated to be 25 t C ha^{-1} over 96 million ha of land in India (Sathaye and Ravindranath 1998). Sequestering carbon through tree based system is now being considered as an attractive economic opportunity for carbon trading (Verma et al. 2014; Kaushal et al. 2016). Of all the land uses analyzed in the Land-Use, Land-Use Change and Forestry report of the IPCC, agroforestry offered the highest potential for carbon sequestration in non-Annex I countries Improved agroforestry systems reduce the vulnerability of small-scale farmers and help them to adapt to changing conditions often meet the conditions for an eligible afforestation / reforestation (A/R) activity in the Clean Development Mechanism (CDM). Table 1 provides the role of agroforestry in climate change mitigation and adaption (Schoreneberger et al. 2012).

Table 1: *Measures of climate change adaptation and mitigation through agroforestry*

Climate change activity	Major CC Functions	Agroforestry role
Mitigation		
Activities that reduces GHGs in the atmosphere or enhance the storage of GHGs stored in ecosystems	Sequester carbon	• Accumulate C in woody biomass and in soil
	Reduce GHG emission	• Conserve existing C stock available in the forest providing alternate source of fuel, fodder and timber through AFS • Enhance forage quality, thereby reducing CH_4 • Reduce fossil fuel consumption: • Reduce equipment runs in areas with trees • Reduce farmstead heating and cooling • Reduce CO_2 emission by C sink • Reduce N_2O emissions: • By greater nutrient uptake through tress • By reducing N fertilizer consumption in tree systems
Adaptation		
Action to reduce or eliminate the negative effects of CC or take advantage of the positive effects	Reduce threats and enhance resilience	• Amelioration of microclimate i. to reduce impact of extreme weather events on crop production ii. to maintain quality & quantity of forage to reduce livestock stress • Provide greater habitat diversity to support organisms (e.g. native pollinators & useful insects) • Provide greater structural and functional diversity to maintain and protect natural resources • Create diversified production opportunities to reduce risk under aberrant weather
	Allow species to mitigate to more favorable conditions	• Provide habitat corridors for species migration

Source: Schoreneberger et al. 2012

MICROCLIMATIC AMELIORATION

Trees on farm bring about favourable changes in the microclimatic conditions by influencing radiation flux, air temperature, wind speed, saturation deficit of understorey crops all of which will have a significant impact on modifying the rate and duration of photosynthesis and subsequent plant growth, transpiration, and soil water use (Monteith et al 1991). The low radiation load and wind speed under canopies reduced water demand by crops and soil evaporation, which are particularly important during drought periods. Leaves by transpiration also control the atmospheric humidity. The effect of solar radiation at day and night times increase surface temperature considerably and affects the crop during critical periods such as flowering and seed maturing. Dela Cruz and Luna (1992) reported 2.4-5.9 °C lower temperature under *Acacia auriculiformis* plantation compared to grassland during day time. Steffan-Dewenter et al (2007) reported that removal of shade trees increased soil surface temperature by about 4 °C and reduced relative air humidity at 2 m above ground by about 12%. Under *Acacia tortilis,* Mann and Muthana (1984) reported that mean daily maximum soil temperature beneath these trees were 10-16 °C lower in top soil zone (10-15cm) and 4-5 °C upto 30cm depth than those recorded in the open indicating better soil thermal regimes under the plantations for intercropping. Hazra and Patil (1986) observed that air temperature under the tree cover was lower (36.8-37.8 °C) as against non-tree open cover (38.4 °C). The leaf temperature of grasses under different tree covers was 31.5 °C- 32.2 °C whereas, it was 33.8 °C in open. Relative humidity plays an important role in growing agricultural crop beneath the trees. Ramakrishna and Sashtri (1977) observed that *Acacia tortilis* (7 years old plantation) intercropped with guar crop at Jodhpur showed 7 per cent higher mean daily relative humidity values beneath the tree canopy during the active cropping season.

ECONOMIC AND LIVELIHOOD SECURITY

Agroforestry plays an important role in income and employment generation. Agroforestry has a potential to increase employment by up to 943 million person days annually from the 25.4 m ha of agroforestry. In the Indian Himalaya alone there is an estimated potential increase of 5.763 million person days each year (Dhyani et al. 2005). Chaturvedi and Khan (2009) reported that silvi-pastoral system on an average cycle of 10 years can generate 120 man days ha^{-1} employment. In Tamil Nadu, charcoal making from P. *juliflora* is providing employment to the tune of 6.34 million man days and 7.03 million women days per annum.

Studies from different parts of the country revealed that agroforestry is more profitable to farmers than agriculture or forestry for a particular area of land. Agroforestry models adopted by farmers in Haryana, Punjab, Western Uttar Pradesh, Andhra Pradesh are highly lucrative and provide immense opportunity of livelihood to the farmers. Rizvi et al. (2011) reported that benefit: cost ratio of poplar based agroforestry systems from 2.35-3.73 indicating them lucrative returns. Gandhi and Joshi (2002) reported an earning of Rs 0.4 million from the inter-cultivation of strawberry per annum under poplar. Dhillon and Bangarwa (2016) reported that cultivation of eucalyptus at wider spacing (6×1 m) with rotation of 8 years in combination with agricultural crops on farm lands provided maximum NPV and B/C ratio of 2.28. Ram Newaj and Rai (2005) reported that aonla-based agroforestry system gave B:C ratio of 3.28 in marginal lands under rainfed conditions. *Grewia optiva* + hybrid napier system was also found to be economically viable with B:C ratio of 1.6 and pay back period of 8 years when calculated for a 15 year cycle at 12 % discount rate (Singh and Dadhwal 2012). Subabul based agroforestry system provided net returns of Rs 50,000 ha^{-1} at spacing of 1x1 m, Rs 72000 ha^{-1} when intercropped and Rs 1,68,632 ha^{-1} when grown for animal husbandry after three years (Pathak et al. 2006).

Horticulture based system have also been found to improve the economy and livelihood of the farmers. In Bihar, litchi based agri-horti system has been found to generate employment opportunities of 130-140 man days ha^{-1}. In Dehradun, combinations of peach as an over-storey and blackgram-toria as under-storey yielded a net profit of ₹ 10,000 $ha^{-1}yr^{-1}$ and had a B:C ratio of 1.65 (Rathore and Raizada 2010). Evaluation of mango based agri-horticulture systems indicated that cowpea-toria sequence was quite remunerative with gross income of ₹ 16,850 ha^{-1} (Rathore et al. 2012). In Meghalaya, the most promising agroforestry practices were arecanut + black pepper + pineapple (₹ 42,750 ha^{-1}) followed by arecanut + black pepper (₹ 36,500 ha^{-1}) and pineapple + mandarin (₹ 29,000 ha^{-1}). Rathore et al. (2012) reported that Mango based agri-horticulture system can give net benefit between ₹ 12582 to 16849 ha^{-1} yr^{-1}. Turmeric crop can be taken after 9-10 years of mango plantation as it grows well in shade and its yield is not affected much by mango canopy. After 10 years, net income from mango + turmeric and mango + colocasia was ₹ 29808 and ₹ 21308 ha^{-1} yr^{-1}, respectively. The B:C ratio of some of the important agroforestry systems in the country are compiled and given in Table 2 (Planning Commission 2001).

Table 2: *Agroforestry systems with respective B: C ratios in different agro-climatic zones*

Zone	Tree	Crops	Years	B:C ratio
Eastern Himalayas	*Anthocephalus kadamba*	Paddy	8	1.6
Western Himalayas	*Prunus persica*	Maize, Soybean	15	1.21
Lower Gangetic	*Eucalyptus hybrid*	Paddy, wheat	10	2.33
Middle Gangetic	*Dalbergia sissoo*	Sesamum	20	1.15
Upper Gangetic	*Populus deltoides*	Paddy, wheat	7	3.31
Trans Gangetic	*Acacia nilotica*	Pearl Millet	20	2.61
Eastern plateau	*Hardwickia binata*	Pennisetum	15	2.13
Central plateau	*Emblica officinalis*	Ground nut	8	1.01
Western plateau	*Albizia amara*	Cenchrus	16	1.56
Southern plateau	*Tamarindus indica*	Chilli	10	2.03
West Coast and Ghat	*Casuarina equisetifolia*	Paddy	7	1.39
Gujarat plains	*Dalbergia sissoo*	caster	20	2.14
Western dry	*Prosopis cineraria*	Pearl millet	20	1.65
Island	*Cocos nucifera*	Paddy	6	1.17

Source: Planning Commission (2001)

REFERENCES

Andreu V, Rubio JL and Cerni R 1998 Effects of Mediterranean shrub cover on water erosion, *Journal of Soil Water Conservation* **53**112-120

Arora YK and Mohan SC 1986 Agri-horti systems for watershed management, *Indian Journal of Soil Conservation* **14**100-104

Belsky AJ, Mwonga SM and Duxbury JM 1993 Effects of widely spaced trees and livestock grazing on understory environments in tropical savannas, Agroforestry *Systems* **24** 1–20

Campbell BM ,Frost P, KingJA ,Mawanza M and Mhlanga L 1994 The influence of trees on soil fertility on two contrasting semi-arid soil types at Matopos Zimbabwe, *Agroforestry Systems* **28** 159–172

Chaturvedi OP, Kaushal R, Tomar JMS, Prandiyal AK, Panwar P 2014 Agroforestry for wasteland rehabilitation: mined, ravine, and degraded watershed areas in *Agroforestry Systems in India: Livelihood Security & Ecosystem Services* (eds) J C Dagar, AK Singh and A Arunachalam New Delhi Springer, 233-271pages

Chaturvedi OP, Venkatesh A, Dwivedi RP, Singh R and Dhyani SK Satish 2009 Agroforestry research perspectives and potential, in *Agroforestry Natural Resource Sustainability, Livelihood and Climate Moderation* Delhi Serial Publishing House, 27-42 pages

Chauhan DS, Dhyani SK and Desai AR 1993 Productivity potential of grasses in association with *Alnus nepalensis* and pine apple under silvi-horti-pastoral system of agroforestry in Meghalaya, *Indian Journal of Dryland Agriculture Research & Development* 860-64

Dagar JC 2012 Utilization of degraded lands/habitats and poor quality water for livelihood security and mitigating climate change, *Indian Journal of Agroforestry* **14** (1)1-16.

Dela Cruz LU and Luna AC 1992 Effect of *Acacia auriculiformis*and*Gmelina arborea* on soil and microclimate of degraded grassland in Nueva Ecuja, Phillippines. *Paper presented at an International Symposium on Rehabilitation of Degraded Grasslands* held during Sept 15-20, 1992 in Tsukubu, Japan

Dhillon WS, Chauhan SK and Singh N 2009 Physiology and yield of turmeric under poplar canopy, *Asia-Pacific Agroforestry News* **35** 5-6

Dhillon WS, Srinidh HV, Chauhan SK, Singh C, Singh N andJabeen N 2010 Micro-envoriment and physiology of turmeric cultivated under poplar tree canopy, *Indian Journal of Agroforestry* **12**(2) 23-37

Dhillon RS and Bangarwa KS 2016 Agroforestry practices in Haryana. Department of Forestry, CCSHAU, Hissar, Haryana, 35

Dhyani SK, Kareemulla K and Handa A 2009 Agroforestry potential and scope for development across agro-climatic zones in India, *Indian Journal of Agroforestry* **32** 181–190

Dhyani SK, Chauhan DS, Singh BP and Prasad RN 1995 Tree effect on yield and quality of pineapple (kew) in a silvi-horti systems of agroforestry in NEH region of India, *Range Management & Agroforestry* **16**(2) 89-93

Dhyani SK, Sharda VN and Samra JS 2005 Agroforestry for sustainable management for soil, water and environment quality: looking back to think ahead, *Range Management& Agroforestry* **26**, 71-83

Gandhi JN and Joshi BC 2002 Poplar based agroforestry and planting stock availability, *in Tree Crop Interface* (eds) SK Chauhan R Chauhan and SS Gill, Ludhiana: Department of Forestry and Natural Resources, PAU 157-161pages

GOI 1999 National Forestry Action Programme Vols. 1/2, New Delhi: Ministry of Environment and Forests Government of India

Grewal SS 1993 Agroforestry systems for soil and water conservation in Shivalik, in *Agroforestry in 2000 AD for Semi-arid and Arid Tropics*, NRCAF, Jhansi, India 82-85 pages

Gupta N, Kukal SS and Singh P 2006 Soil erodibility in relation to poplar based Agroforestry system in north western India, *International Journal of Agriculture and Biology* **8**(6) 859-861

Handa AK, Toky OP, Dhyani SK and Chavan SB 2016 Innovative agroforestry for livelihood security in India, *World Agriculture* 7-16

Hazra CR and Patil BD 1986 Forage production under silvipastoral system, light and temperature relation, *Indian Journal of Range Management* **7**(1) 33-36

Jeet Ram 2004 Agroforestry for the reclamation of water logged areas, 92-93. In Proceedings of National Workshop on Agroforestry organised by Forest Department, Haryana, Chandigarh

Juyal GP, Katiyar VS, Dhadwal KS, Joshie P and Arya RK 2007 Mined area rehabilitation in Himalayas-Shastradhara experience. Bulletin No.T-53/ D-34, Central Soil and Water Conservation Research and Training Institute, Dehradun

Kanime N, Kaushal R, Tewari SK, Raverkar KP, Chaturvedi S and Chaturvedi OP 2013 Biomass production and carbon sequestration in different tree based systems of Western Himalayan Tarai Region, *Forest Trees and Livelihood* **22** (1) 38-50

Kaushal R, Kumar Ambrish, Alam NM, Mandal D, Jayaparkash J, Tomar JMS, Patra S, Gupta AK, Mehta, H Panwar P, Chaturvedi OP and Mishra PK 2017 Effect of different canopy management practices on rainfall partitioning in *Morus alba, Ecological Engineering* 102 374 – 380

Kaushal R, V Subbulakshmi, JMS Tomar, NM Alam, J Jayparksh, H Mehta and Chaturvedi OP 2016 Predictive models for biomass and carbon stock estimation in male bamboo (*Dendrocalamus strictus* L.) in Doon valley, India, *Acta Ecologica Sinica* **36** (6) 469-476

Kaushal R, Verma A, Mehta H, Mandal D, Tomar JMS, Jana C, Rajkumar, Jayparkash J and Chaturvedi OP 2016 Soil quality under *Grewia optiva* based agroforestry systems in western sub-Himalaya, *Range Management and Agroforestry* **37** (1) 50-55

Kaushal R, Verma KS, Chaturvedi OP, Alam NM 2012 Leaf litter decomposition and nutrient dynamics in four multipurpose tree species, *Range Management and Agroforestry* **33** (1) 20-27

Lal R 1989 Agroforestry systems and soil surface management of a tropical alfisol II: Water runoff, soil erosion, and nutrient loss, *Agroforestry Systems* **8** (2) 97–111

Lal P 2004 Integrated development of farm forestry plantation and wood based industries, *Indian Forester* **130** (1) 71–78

Leakey RRB and Sanchez PA 1997 How many people use agroforestry products? *Agroforestry Today* **9** (3) 4–5

Lodhiyal LS, Singh RP and Singh SP 1995 Structure and function of an age series of poplar plantation in central Himalaya I Dry matter dynamics, *Annals of Botany* **76** 191–199

Mandal D, Sharda VN and Tiwari A 2008 *Soil loss tolerance limits in Planning Conservation Measures in Different Districts of India.* CSWCRTI, Dehradun, India

Mann HS and Muthana KD 1984 Arid Zone Forestry, ICAR - CAZRI Monograph No. 23, Jodhpur

Martius C, Hofer H, Garcia MVB, Rombke J, Forster B and Hanagarth W 2004 Microclimate in agroforestry systems in central Amazonia does canopy closure matter to soil organisms? *Agroforestry Systems* **60** 291-304

McDonald MA, Healey JR and Stevens PA 2002 The effects of secondary forest clearance and subsequent land use on erosion losses and soil properties in the Blue mountains of Jamaica. *Agriculture Ecosystem and Environment* **92** 1–9

Monteith JL, Ong CK and Corlett JE 1991 Microclimatic interactions in agroforestry systems, *Forest Ecology and Management* **45** 31– 44

Nair PKR, Kumar BM and Nair VD 2009 Agroforestry as a strategy for carbon sequestration, *Journal of Soil Science and Plant Nutrition* **172** 10–23

Narain P, Singh RK, Sindhwal NS and Joshie P 1998 Agroforestry for soil and water conservation in the western Himalayan Valley Region of India 1. Runoff, soil and nutrient losses, *Agroforestry Systems* **39** (2) 175-189

Pathak PS, Suresh G, KhanTA and Gupta SK 2006 *Subabul in the Rural Economy of Andhra Pradesh-A Case Study.* Indian Grassland and Fodder Research Institute. Jhansi, UP India

Planning Commission 2001 Reportof the Task Force on Greening India for Livelihood Security and Sustainable Development. Planning Commission, New Delhi, Government of India

Planning Commission 2004 Report of the committee on development of Biofuels. Planning Commission, New Delhi, Government of India 200-222

Prajapati MC, Nambiar KTN, Puri DN, Singh JP and Malhotra BM 1993 Fuel and fodder production in Yamuna ravines at Agra, *Indian Journal of Soil Conservation* **21**(3) 8 13

Prasad R, Pandey DN, Kotwal PC 2000 Trees Outside Forests in India: A National Assessment. Indian Institute of Forest Management, Bhopal, India

Rai P, Solanki KR and Rao GR 1999 Silvipasture research in India – a review, *Indian Journal of Agroforestry* **1** 107–120

Rai P, Yadav RS, Solanki, KR Rao GR and Singh R 2001 Growth and pruned biomass production of multipurpose tree species in silvipastoral system on degraded lands in semiarid region of Uttar Pradesh, India, *Forest Tree and Livelihood* **11** 347-364

Rai SN and Chakrabarti SK 2001 Demand and supply of fuel wood and timber in India, *Indian Forester* **127** 263–279

Raizada A and Singh C 2010 Silvipastoral systems for wasteland utilization in foothills of the western Himalayas. Technology Brochure. CSWCRTI, Dehardun

Raizada A, Ghosh BN, Jayaprakash J and Singh C 2014 Biomass production, carbon sequestration and water transmission properties as influenced by densified plantations raised on old river bed lands in the north West Himalayas, *Indian Journal of Soil Conservation* **42** (1) 85-97

Ram Newaj and Rai P 2005 Aonla-based agroforestry system: A source of higher income under rainfed conditions, *Indian Farming* **55** 24-27

Ramkrishna, YS and Sastri AS 1977 Microclimate under *Acacia tortilis* plantation, Annual Progress Report, CAZRI, Jodhpur. 69-70

Rathore AC, Saroj PL, Sharma NK, Shrimali SS, Jayaprakash J, Chaturvedi OP and Dadhwal KS 2012 Mango based agri- horticultural system for degraded lands of north western Himalaya. Technology Brouchere CSWCRTI, Dehradun

Rathore, AC and Raizada A 2010 Peach based agri-horticultural practices for utilization of marginal lands. Technology Brochure. CSWCRTI, Dehardun

Ravindranath, NH and Hall DO 1995 *Biomass, Energy and Environment: A Developing Country Perspective from India,* New York: Oxford University Press

Rizvi RH, Dhyani SK, Yadav RS and Singh R, 2011 Biomass production and carbon stock of popular agroforestry systems in Yamunanagar and Saharanpur districts of north western India, *Current Science* **100** 736-742

Saha SR, Chaudhary RS and Somasundaram J 2012 Soil health management under hill agro-ecosystem of north east India, Applied *and Environmental Soil Science* 1-9

Samra JS, Bansal RC, Sikka AK, Mittal SP and Agnihotri Y 1995 Resource conservation through watershed management in Shiwalik foothills – Relmajra Bulletin No. T-28/C-7, CSWCRTI, Research Centre, Chandigarh

Satapathy KK 2005 Runoff production on hill slopes under different land use systems, *In Agroforestry in North East India Opportunities and Challenges* (eds) BP Bhatt and K M Bujarbaruah ICAR Research Complex for NEH Region, Umiam, Meghalaya 451–459

Sathaye JA and Ravindranath NH 1998 Climate change mitigation in the energy and forestry sectors of developing countries, *Annual Review of Energy Environment* **23** 387-437

Schoreneberger M, Bendrub G, Gooijer H, Soolanayakanaha IR, Tom Sauer, Brandle J, Zhan X and Den Cuer 2012 Branching out agroforestry as a climate change mitigation and adaptation for agriculture, *Journal of Soil and Water Conservation* **67**(5) 128-136

Sharda VN and Venkateswarlu B 2007 Crop diversification and alternate land use systems in watershed Management, 111-128. *In: Best-bet Options for Integrated Watershed Management. Proceedings of the Comprehensive Assessment of Watershed Programs in India,* ICRISAT, Patancheru, Andhra Pradesh, India

Sharma SK, Singh RS, Tiwari JC and Burman U 1994 Silvipastoral studies in arid and semiarid degraded lands of western Rajasthan, 749–754, in *Agroforestry Systems for Degraded Lands* (eds) P Singh, PS Pathak and M Roy. Oxford & IBH, New Delhi.

Singh C and Dadhwal KS 2012 *Grewia optiva* + Hybrid napier based silvi-pastoral system for degraded lands in north-west Himalayas. Technology Brochure. CSWCRTI, Dehardun.

Steffan-Dewenter I, Kessler M, Barkmann J, Bos M, Buchori D, Erasmi S, Faust H, Gerold G, Glenk K, Gradstein RS, Guhardja E, Harteveld M, Hertel D, Hohn P, Kappas M, Kohler S, Leuschner C, Maertens M, Marggraf R, Migge-Kleian S, Mogea J, Pitopang R, Schaefer M, Schwarze S, Sporn GS, Steingrebe A, Tjitrosoedirdjo SS, Tjitrosoemito S, Twele A, Weber R, Woltmann L, Zeller M and Tscharntke T 2007 Tradeoffs between income, biodiversity, and ecosystem functioning during tropical rainforest conversion and agroforestry intensification, PNAS **104** 4973–4978

Verma A, Kaushal R, Alam NM, Mehta H, Chaturvedi OP, Mandal D, Tomar JMS, Rathore AC and Singh C 2014 Predictive models for biomass and carbon stocks estimation in *Grewia optiva* on degraded lands in western Himalaya, *Agroforestry Sysytems* **88** 895–905

Yadav RP, Ram Prasad, Aggarwal RK, Agnihotri Y and Samra JS 2005 Aonla (*Embilica officinalis*) based silvi-pastoral systems for soil and water conservation in degraded lands of Shivalik foothills, *Indian Journal of Soil Conservation* **33**(3) 207 – 212

Young A 1997 *Agroforestry for Soil Management,* Wallingford, UK C.A.B. International

Zomer RJ, Bossio DA, Trabucco A, Yuanjie L, Gupta DC and Singh VP 2007 *Trees and Water Smallholder Agroforestry on Irrigated Lands in Northern India.* Colombo, Sri Lanka International Water Management Institute. 47 (IWMI Research Report 122)

Zuazo VHD and Pleguezuelo CRR 2008 Soil erosion and runoff prevention by plant covers. A review, *Agronomy for Sustainable Development* **28** 65-86

CHAPTER 4

Augmenting Grasslands and Agroforestry for Sustaining Forage Production

P. K. Ghosh and D. R. Palsaniya*

ICAR – Indian Grassland and Fodder Research Institute, Jhansi (UP), India

** Corresponding authors Email: ghosh_pk2006@yahoo.com*

ABSTRACT

Livestock sector play an important role in Indian economy through its contribution in GDP and employment. Rapidly developing peri-urban livestock farming, emerging fodder markets, vibrant cooperative dairy sector, livestock product based super markets and increasing demand of livestock based food are indicators of fast changing economic scenario in livestock sector. However, the low productivity of Indian livestock is a serious concern which largely due to breed and feed related problems. Grasslands and agroforestry are the two important niches of fodder supply in India. Most of the grazing animals derive their fodder from grasslands and pastures. Similarly, agroforestry also significantly contribute fodder to animals especially small ruminants. However, the productivity and carrying capacity of our grazing lands have deteriorated considerably due to heavy grazing pressure and lack of proper management. The whopping demand and supply gap of fodder can effectively be bridged by rehabilitation and management of our grasslands and agroforestry systems. This chapter provides information on current fodder and livestock scenario, major issues and strategies required for sustainable forage production with special focus on grasslands and agroforestry.

Keywords: Agroforestry, Ecosystem, Forage production, Grasslands

INTRODUCTION

The total livestock population in India is 512 million numbers in 2012 (19th Livestock Census 2012, www.dahd.nic.in/). The livestock sector alone contributes nearly 25.6% of value of output at current prices of total value of output in agriculture and allied sector. The contribution of livestock sector in total GDP is nearly 4.11% at current prices. India is the largest producer (146 million tonnes) and consumer of milk in the world. Livestock sector which recorded a growth in value of output about 4.8% per annum in 11th Five Year Plan has excellent potential for higher growth. Global trend in animal production indicates a rapid and massive increase in the consumption of livestock products. Urbanization has brought a marked shift in the lifestyle of people in feeding habits towards milk products, meat and eggs with resultant increase in demand of livestock products (Ghosh and Palsaniya 2014b). Rapidly developing peri-urban livestock farming, emerging fodder markets, vibrant cooperative dairy sector and livestock product based super markets are indicators of fast changing economic scenario in livestock sector (Ghosh et al 2016; Ghosh and Palsaniya 2014a and Ghosh et al 2013). Strong farm-gate prices and rising demand for value-added products due to increasing consumer income are stimulating increased milk production in India. Growing private investment in dairy processing facilities is providing further impetus. The animal husbandry and livestock sectors are critical for the rural economy, especially the small and marginal farmers (Ghosh et al 2013). They not only contribute to their income but also their best insurance against any natural calamity. Animal products are also playing an important role in their nutritional security. Marginal, small and semi-medium operational holdings farmers (area less than 4 ha) own about 87.7% of the livestock. Quite often livestock is the only source of cash income for subsistence farms and also serves as insurance in the event of crop failure. Hence, development of livestock sector would be more inclusive.

ISSUES

At present, India faces a critical imbalance in its natural resource base with about 18% (human) and 15% (livestock) population of the world being supported only by 2.4% (geographical area), 1.5% (forest and pasture lands) and 4.2% water resources (Palsaniya et al 2012a). The animal adapted to various agro-climatic regions are low producers due to their genetic potential and inadequate and poor quality of feeds and fodder supply. Many of the animal health problems are further compounded due to poor fodder availability. Presently, the country faces a net deficit of 35.6% green fodder, 10.95% dry crop residues and 44% concentrate feed ingredients (IGFRI Vision, 2050). Moreover, in case of forages, the regional and seasonal deficiencies are more important than the national deficiencies, as it is not economical to transport the forages over long distances. Farmers maintain large herds of animals to compensate for the low productivity, which adds to the pressure on fodder and other natural resources (Palsaniya et al 2008, Palsaniya et al 2009 and Palsaniya et al 2010a).

Table 1. *Demand and supply estimates of dry and green forages (million tonnes)*

Year	Demand		Supply		Net Deficit		% Deficit	
	Dry	Green	Dry	Green	Dry	Green	Dry	Green
2010	508.9	816.8	453.2	525.5	55.72	291.3	10.95	35.66
2020	530.5	851.3	467.6	590.4	62.85	260.9	11.85	30.65
2030	568.1	911.6	500.0	687.4	68.07	224.2	11.98	24.59
2040	594.9	954.8	524.4	761.7	70.57	193.0	11.86	20.22
2050	631.0	1012.7	547.7	826.0	83.27	186.6	13.20	18.43

Source: IGFRI Vision: 2050

The three major sources of fodder supply in India are crop residues, cultivated fodder and fodder from common property resources like forests, permanent pastures and grazing lands. Nearly 2/3rd of total cost in livestock production is due to feed and fodder. Hence, any attempt towards enhancing availability of feeds and fodder, and economizing the feed cost would result in better remuneration to livestock farmers. The area under cultivated fodder is only 8.4 million ha, and has been static during last two decades. The scope for further increase seems to be very low due to demographic pressure for food crops. The recent crop diversification where commercial crops replacing the traditional cereal crops especially the coarse cereals, is likely to have an impact on the availability of crop residues for animal production (Ghosh and Palsaniya, 2014a). Similarly, the productivity of some prominent cultivated forages is highly variable. Among the *kharif* forages, sorghum, maize, cowpea, Napier x bajra hybrid and Guinea grass have wide amplitude. However, during *rabi*, the choice is limited to oat, lucerne and berseem. Emphasis has to be made on area-specific new crops that can break the yield barriers and meet the challenges of fodder deficits.

Genetic resource enhancement in forage crops remained confined to cultivated crops over many decades. Improvement programmes on range grasses and range legumes were not given proper attention which is a major constraint in the development of grassland and pastures. Twenty prioritized crops are still to be embraced by crop improvement programmes with a due focus on agronomic requirements. There is also need of a separate National Livestock Policy with greater emphasis on feed and fodder. Mission mode approach with a multi- pronged strategy policy and research interventions are required to take care of all aspects of seed production technology, quality, seed standards, certification, distribution and marketing (Ghosh et al 2016).

The other major issue in fodder production is to reduce the cost of production with international quality standards and in eco-friendly environment. Climate change and its impact on sustainable forage production is another important issue in recent times (Sunil Kumar et al 2013, Palsaniya et al 2012b and Ghosh and Palsaniya, 2014b). The warming trend in India over the past 100 years (1901 to 2007) was observed to be 0.51 °C with accelerated warming of 0.21 °C per every 10 years since 1970. The impact of global warming is seen in the form of increase in intensity and frequency of hydro-meteorological disasters like droughts, cyclones and floods etc. at many locations (IPCC 2007). The forage production per unit area is a consequence of the interactions between genotypes and environment. Farm level adaptation options can be made in tillage practices, planting and harvest dates, crop rotations, selection of suitable crops and crop varieties for cultivation, water consumption for irrigation, fertilizer management and application of additional nitrogen can reduce the impacts of climate change on forage crops (Palsaniya et al 2012b, Ghosh and Palsaniya, 2014b and Sunil Kumar et al 2013). Moreover, adaptation of forage species particularly perennial species and trees (agroforestry) has the ability to produce more forage yield under changing climate and reduces the ill-effect of climate change through carbon sequestration (Dhyani et al 2010; Rai and Palsaniya 2015 and Palsaniya et al 2011b).

GRASSLANDS AND AGROFORESTRY: KEY AREAS FOR FODDER

Grassland is one of the largest (30% of the earth's ice-free land surface) land use system on planet earth and plays significant role in the livelihood of millions of livestock farmers particularly in developing countries and Sub Saharan Africa. However, the present status of grasslands is very poor and there is urgent need for a holistic approach for this sector. Major grasslands of the world and India are listed in table 2.

Table 2: *Major grasslands of the world and India*

Location	Grassland name
Africa	Savannas, Velds, Western Zambezian grasslands
Australia	Downs, Rangelands, Cantebury grassland (Newzealand)
Asia	Steppes, Trans fly and Terai-Duar savanna and grasslands
North America	Prairies, Western gulf coastal grasslands
South America	Pampas, Campos, Llanos, Cerrados, Gran chaco
India	*Sehima-Dichanthium, Dichanthium-Cenchrus-Lasiurus, Phragmites- Saccharum-Imperata, Themeda-Arundinella* and Temperate Alpine

GRASSLAND PRODUCTION SCENARIO

Our studies on grass communities including their succession and trends and the level of productivity have shown that *Sehima-Dicanthium* cover is highly productive at its climax stage, whereas *Themeda-Arundinella* yields are the lowest (table 3). But the early seral stages of *Phragmites-Saccharum-Imperata* and *Dicanthium-Cenchrus-Lasiurus* are comparatively highly productive. Most of these communities are in their last stage of retrogression, and thus represent poor condition grasslands. Production potentials of these grazing lands can be doubled through improved management practices and protection from uncontrolled grazing and farming. Their actual harvestable biomass is between 0.2 and 3.5 t/ha as against the potentials of 4-6 t/ha (Table 3).

POTENTIAL THREATS TO GRASSLANDS

Population explosion, overgrazing, biodiversity loss, competition between livestock and wildlife, encroachment, land degradation and clearing of land for crops and lack of suitable policy support and government incentives are the main threats to grasslands. About 16% of tropical grasslands have been converted for agriculture or urban development. Desertification is also a significant threat. The climate change and its resultant effects may also affect grasslands and agroforestry. Social conflicts may also be there between pastoral (especially migratory herders) and crop production systems. There is also a severe imbalance in stocking rate and carrying capacity of arid and semi-arid grasslands in India (Fig. 1).

Table 3: *Retrogressive seral stages in sub-climax tropical-sub-tropical grazing lands along with levels of their production (Grassland Survey of India, Indian Council of Agricultural Research 1954-62)*

Grass cover (ecosystems)	Harvestable biomass	
	Actual production ($g/m^2/yr$)	Potential production ($g/m^2/yr$)
(Sub mountain region)		
1. *Themeda -Arundinella*		
Themeda- Arundinella	220	400
protection ↑ ↓grazing, burning		
Arundinella- Chrysopogon	220	300
protection ↑ ↓ grazing	200	
Heteropogon- Bothriochloa		350
protection ↑ ↓ grazing	50	
Cynodon		100
(Sub-tropics – High humid)		
2. *Phragmites- Saccharum-Imperata*	500	500
Phragmites- Saccharum-Imperata		
protection ↓ ↓burning, cutting ↓ grazing	450	750
Desmostachya- Imperata- Sclerostachya		
protection ↑ ↓burning, cutting	200	300
Desmostachya - Imperata		
↑ ↓	100	150
Sporobolus-Paspalum-Chrysopogon		
(Dry and Arid tracts)		
3. *Dichanthium- Cenchrus- Lasiurus*	3	500
Dichanthium- Cenchrus- Lasiurus	330	
protection ↑ ↓*grazing*		500
Cenchrus - Lasiurus	250	
protection ↑ ↓grazing		200
Sporobolus- Cynodon - Eleusine	100	
protection ↑ ↓grazing		150
Chloris-Cenchrus-Biflorus		
Compact soil Loose soil	100	
Central and Southern Semi arid tracts		
4. *Sehima-Dichanthium*		
(Buring-Eroding slope, gravelly soil)		
Sehima- Dichanthium	350	600
↑ ↓ (Increasing soil moisture)		
Themeda-Cymbopogon-Iseilema		
↑ ↓ ↓ grazing		
Chrysopogon- Bothriochloa	250	500
protection ↑ ↓ grazing		
Heteropogon-Eremopogon	150	260
protection ↑ ↓ grazing		
Aristida-Eregrostis-Melanocenchrus	20	40

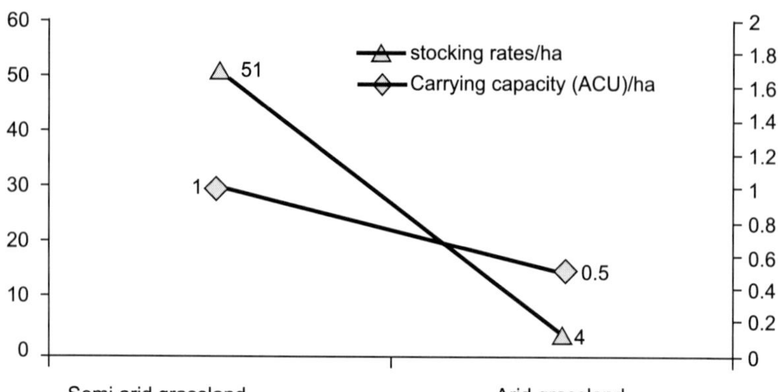

Figure 1: *Stocking rate and carrying capacity of arid and semi-arid grasslands in India*

AUGMENTING GRASSLANDS ECOSYSTEMS AND PRODUCTION

Grassland scientists are only successful if they have a hard data base, which describes the particular grassland ecosystems in which they are involved. This requires physical information about the interrelations of plant growth, senescence, consumption, persistence, nutritive value and nutrient flows with environmental factors, farm inputs and animal performance using modern tools and techniques. We can establish patterns of plant behaviour or affinities between plant attributes and environmental characters, using rigorous mathematical techniques. It is usually impossible to predict the outcome of an interaction since we are unable to weight the component factors by thought alone; we have to measure plant behaviour and devise a successful grazing strategy which is tested by field experiment.

REVITALIZING GRASSLAND ECOSYSTEMS

Palsaniya et al (2011a) recognized protection of area, cleaning of bushes, adequate soil and water conservation practices, reseeding, inclusion of legumes, a complete rest from grazing during the establishment period and other general management practices as a variety of mechanisms of grassland regeneration and rehabilitation. This will allow the vegetation to recover and natural species to re-establish from the seed bank in the soil or from spread of plants by vegetative means. The community participation in rehabilitation of degraded grasslands is an important step in enhancing the success rate. Grassland systems based on local knowledge and traditions and in partnership with local people on local problems had more success (Herlocker 1999 and Palsaniya et al 2011a).

MANAGEMENT OF GRASSLANDS

The management of grasslands aims for a desirable composition that gives both high production and maintenance of the resources. The desirable composition varies both temporally and spatially and, given the range of requirements, a mix of species is almost always needed. Productive legumes are important both to provide nitrogen for associate grasses and for their high herbage quality, but legume decline is a major issue. In sown tropical pastures both legume dominance and lack of legume persistence are problems. In the *Stylosanthes* grasslands of semi-arid tropics, stylo dominance (particularly *Stylosanthes scabra*) is causing some concern. In stylo-grass pastures, animal preference for grass is very strong during the early wet season (Gardener 1980) leading to heavy selective grazing of the grasses, which are susceptible at this stage of their growth (Hodgkinson et al 1989 and Mott et al 1992). On the other hand, in the drier subtropical grasslands, legume persistence remains a problem, and the success of legume-based pastures has been more variable. Similarly, weeds also continue to be a problem. A number of measures have been recommended to reduce the stylo dominance - fire, pasture spelling, sowing grazing-tolerant grasses, and raising soil phosphorus levels (McIvor et al 1998).

BREEDING GRASS-LEGUME IDEOTYPES

Though intensive researches in range grasses and legumes breeding augmented with allied disciplines resulted in development of many varieties with a high genetic yield potential, but the demand for future are expected to increase several folds. The problem emanates due to frequent occurrence of polyploidy, self-incompatibility and apomixis resulting into various degree of sterility, poor seed setting and difficulty in desirable and controlled hybridization in many tropical range grasses and legumes species, which complicate their genetic improvement. Grassland production depends on herbage yield, its seasonal distribution, nutritive value, seed production and persistence. The tropical pasture species are very poor in these attributes. There has been constant effort for genetic improvement for pasture grasses and legumes on these fronts, but to a very little success. Plant breeders using the primary gene pool of various individual range grasses and legumes should further reorient breeding efforts and develop the varieties for higher biomass production and fodder quality. The varieties should be free from diseases and pests, photosynthetically efficient and abiotic stress tolerant and reduce need of fertilizer etc. The new techniques provided by biotechnology are relatively fast, resource efficient and highly specific. It may offer opportunities to increase sustainability and profitability vis-a-vis competitiveness.

GRAZING MANAGEMENT

There has been widespread interest in grazing systems, with considerable debate over the merits of systems based on short grazing periods and long rest periods (e.g. short-duration grazing, time-control grazing, cell grazing) compared with continuous grazing. A number of producers have enthusiastically adopted these systems and report large benefits, both for financial performance and for sustainability of resources (McCosker 1994 and McArthur 1998), although other producers have not had positive results (Waugh 1997). In contrast to the positive benefits reported from these grazing systems, reviews of grazing systems research conclude that continuous grazing is no worse than rotational grazing, and may be better for animal production considered reasons for this divergence of views, and concluded that differences were related to uniformity of utilization - this was constant, even in small experimental paddocks, irrespective of grazing system, so the experiments showed no benefit for rotational grazing (Norton 1998). However, utilization is often not uniform in large commercial paddocks, and more even utilization is achieved when they are subdivided for rotational systems, leading to production benefits.

LEGUME EFFECTS

Legumes differ markedly from grasses, cereals and other non-legume crops because much of the nitrogen they require is produced through fixation of atmospheric nitrogen by bacteria in nodules on their roots and their inclusion is essential for the regeneration of nutrient-deficient soils and for providing needed protein, minerals, and vitamins to humans and livestock (Palsaniya et al 2014). Legumes supplemented with grasses significantly increase overall animal productivity. Adding the residue from legume plants into livestock forage can increase the digestibility and overall quality of cereal crop residues. Legumes have long been recognized and valued as "soil building" crops. Growing legumes improves soil quality through their beneficial effects on soil biological, chemical and physical conditions. Studies on intercropping of various pasture legumes (*Clitoria ternatea, Stylosanthes hamata, Centrosema pubescens, Atylosia scarabaeoides, Macroptelium atropurpureum, Phaseolus lathyroides, Vigna luteola, Glycine javanica*) in *Chrysopogon fulvus* under rainfed condition conducted at IGFRI, Jhansi revealed that *Atylosia scarabaeoides, Centrosema pubescens* and *Stylosanthes hamata* enriched the nitrogen (74–90 kg ha^{-1}) status of soil and approximately doubled the biomass production (10 to 11.6 t ha^{-1}) as well as seed yield (78–81 kg ha^{-1}) of *Chrysopogon fulvus* (Dwivedi et al 1988).

POST-HARVEST MANAGEMENT

Post-harvest management of surplus fodder should be promoted as a mitigating strategy for abating regular phenomenon of seasonal and regional deficit of forage and during natural calamities like drought and flood. Bailing and enrichment of grass residues and other leguminous crop residues for proper storage, balanced feeding with green fodder and minimizing wastage and storage loss. Use of leaf meals from leguminous fodder and crop residues from pulses will substitute costly concentrate (stylo, grass pea, subabool, gliricidia etc.). Technology has been developed to conserve surplus green fodder for use during the lean period either as hay or silage. Manually operated hay baler having a baling capacity of 2.5 tonnes ha^{-1}day^{-1} has been developed. The weight of bale is 25 kg and density is 150-165 kg m^{-1}. The densifying machine produces high density (350-400 kg m^{-1}) bales of chaffed stovers, grasses and tree leaves for economic storage, handing and transport. Complete feed block rations in the form of blocks for grazing calves, heifers and milking animals have been standardized. Effect of drying or processing of leaf meal showed the reduction of anti-nutritional factor mimosine from leucaena leaf. Final CP content of stored leucaena leaf meal for one year storage was 19.87% as against initial value of 22.01%. Mimosine content during storage reduced from 2.90 to 2.53%.

POLICY SUPPORT

The grassland development is a more complex issue than food and commercial crops. Due to multiplicity of range grasses and legumes grown in different season and region, surplus and deficit in different regions, non-commercial nature of crops and production of forage with minimal inputs from degraded and marginal lands has led to huge gap in fodder availability and requirement. Some of the prominent aspects related to policy to provide favorable environment for accelerated grassland development are – collection of database of fodder production and utilization from grasslands, investment in grassland development, credit facility, support price for forage and marketing of seed, non-diversion of edible crop residues to other use like packaging, policy on grazing and common property resources and legal protection of grasslands.

AGROFORESTRY FOR FODDER

Agroforestry is widely known as a sustainable form of land use for augmentation of biomass production from agricultural systems (Nair 2001; Kang and Wilson, 1987), natural resource conservation (Palsaniya et al 2011b and Palsaniya et al 2012a) problems soil management (Singh et al 1998 and Palsaniya et al 2011b) and recently for mitigation of climate change by way of sequestering carbon in both standing biomass and soil (Ajit et al 2013 and Palsaniya et al 2011a). The integrated approach associated with agroforestry makes it the most self-sustaining and ecologically sound land management system. In agroforestry, fodder trees can be accommodated mainly in following ways:

- Agrisilvicultural system
- Silvipastoral system
- Hortipasture system
- Boundary plantation

Among above fodder based agroforestry systems, silvipasture and hortipasture are the important for sustainable fodder production for livestock. These systems enhance the overall biomass productivity and profitability of the degraded lands. During initial few years, there is no adverse effect of tree canopy on the productivity of the under storey pasture and in later years, the yield reduction is compensated by the forage yield and other benefits from the tree component and thus, the overall yield is, in general, higher than the yields obtained from pasture or tree alone (Rai and Palsaniya, 2015 and Palsaniya et al 2011a). Forage production under silvipastoral system is directly influenced by its proper initial establishment, subsequent management, edaphic and climatic conditions, tree density, canopy development etc. Trees and shrubs in the silvipastoral system often contribute substantial amount of top feed in the form of leaves and pods especially in arid and semiarid region during lean periods through lopping/pruning of trees. The productivity of silvipasures is much more than the natural and improved pastures (Fig. 2). The potentials of various regions for silvipastoral and hortipastoral system in India is given in table 4.

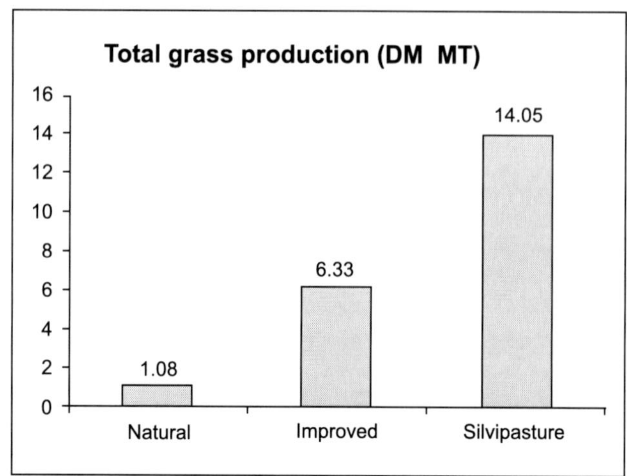

Figure 2: *Productivity of natural, improved and silvipastoral systems.*

Table 4: *Potentials of various regions for silvipastoral and hortipastoral systems in India*

Region	Accents and components	Results
Himalayas	Pastures in alpine area and intensive silvipasture and hortipasture in temperate areas (i) *Alnus nepalensis-Stylosanthes guyensis+Panicum maximum* (ii) *Ficus hookerii+Thysanolaena maxima* (iii) *Dalbergia sissoo+Chrysopogon fulvus* (iv) *Acacia catechu+Eulaliopsis binata* (v) *L. leucocephala+Hybrid napier* (vi) *Bauhinia purpurea+ E. binata* (vii) *Albizia lebbek+ E. binata*	• 5-13.5 t $ha^{-1}year^{-1}$ dry forage yield • 8-10 kg $tree^{-1}$ dry matter yield • 64-71 t ha^{-1} fuel wood and 5.2-5.5 t ha^{-1} $year^{-1}$ fodder grass in a short rotation of 4-5 years
Indo-Gangetic plains	Tree-grass combinations as silvipasture and hortipasture *A. nilotica/Madhuca latifolia/Leucaena leucocephala/ Gmelina arborea/Mangifera indica/Pithecellobium dulce+ Phragmites australis/ Saccharum, spontaneum/ S. arundinaceum/Imperata cylindrica* combinations	• 8-12.3 t ha^{-1} $year^{-1}$ biomass production
Arid and semi-arid	Silvipasture and hortipasture and grazing lands *Prosopis cineraria/Zizyphus nummularia/Acacia nilotica/Ailanthus excelsa/A. leucophloa/A. procer/L. leucocephala+Dichanthium/Cenchrus/Lasiurus/Panicum/ Heteropogon/Chrysopogon*	• *A. excelsa- C. setigerus+ Vigna anguiculata* produced dry biomass of 10.0 t ha^{-1} (3.93 t forage + 3.13 t leaf fodder + 2.91 t fuel wood). • 10.2 t/ha DM from *L. leucocephala- C. ciliaris+C. ternatea*

Humid and sub-humid	Quick growing leguminous trees in crop rotations, plantations and silvipastures *L. leucocephala/Gliricidia/Sesbania/Ficus hookeri*+fodder grasses like *Pennisetum pedicellatum* and *Stylo*	• Have high forage yield potential
Coastal plains	Silvipasture and hortipasture Subabul/Agathi/Desmenthes+Congosignal/ Stylosanthes/Centrosema	• 22-29 t ha fodder yield can be obtained from coconut based silvipastures

The long term studies conducted on silvipastoral systems under IDRC project by IGFRI in red gravelly soils with five tree species (*Albizia amara, Albizia lebbeck, Acacia tortilis, Hardwickia binata* and *Leucaena leucocephala*) at Jhansi (1982 to 1994) and (*A. amara, A. lebbeck, A. tortilis, Dalbergia sissoo* and *L. leucocephala*) at Datia (1983 to 1993) with 3 spacing (4x3, 4x4 and 4x6m) revealed that there was no much difference in forage production from understorey up to 11-12 years from trees grown at these spacing (IDRC Report, 1982– 83 to 1993-94). The results showed that trees may be grown at the rate of 825 plants ha^{-1} up to 11-12 years and optimum understorey biomass production can be obtained. The average dry forage production obtained in 12 years from understorey vegetation at above sites under different silvipastoral systems when compared the production from control (pasture without tree) revealed that the forage yield decreased at the tune of 13.3, 9.8, 3.7 and 7.9% in *A. amara, A. tortilis, H binata* and *L. leucocephala*, respectively over control under Jhansi condition, while in case of *A. lebbeck* the production did not decrease. This might be due to high mortality of this tree species. Similarly, the average of 11 years data on dry forage yield at Datia, showed decrease in yield at the tune of 9.0, 13.4 and 8.3% in *A. tortilis, D. sissoo* and *A. amara*, respectively over control. The forage yield under *A. lebbeck* and *L. leucocephala* was not reported after 7th year onwards. This might be due to high mortality of these tree species. However, average dry forage production of initial 5 years was 3.13, 2.62, 2.99, 2.82, 2.36 t ha^{-1} in association with *A. amara, A. lebbeck, A. tortilis, D. sissoo* and *L. leucocephala,* respectively, where as in control (pasture without tree) the yield was 2.64 t ha^{-1}. These results showed that there was no adverse effect of tree components on forage production of understorey vegetation.

In another study carried out from 1984 to 1993 on ravines of Datia district of Madhya Pradesh having calcareous soils with pH value from 8 to 10 with plantation of 5 tree species (*A. tortilis, A. amara, A. lebbeck, D. sissoo* and *Dichrostachys cinerea*) under 3 spacing (4x3, 4x4 and 4x6 m) showed that *A. tortilis* was the best species for the growth and production in such habitats as the survival of other 4 tree species was very poor. Plantation of 4 grasses (*Brachiaria mutica, Bothriochloa intermedia, Dichanthium annulatum* and *Cynodon dactylon*) in the interspaces of trees at top, slope and base of the ravines revealed that on an average dry forage yield recorded was 3.86 t ha^{-1}. However, maximum forage yield of 6.76 t ha^{-1} was obtained on the base followed by slope (2.97 t ha^{-1}) and minimum was on the top of the ravine (1.83 t ha^{-1}) (IDRC Report, 1982-1994). A 13 year study on degraded lands with five tree species and three densities under silvipastoral systems revealed that the average total biomass production of 6.55 to 8.53 t ha^{-1} was obtained in association with *L. leucocephala* followed by *A. amara*. Average total biomass production of 6.91 t ha^{-1} was recorded when planting was done at 4x3 m spacing followed by 4x4 m and minimum with 4×6 m spacing. These results showed that for higher biomass (forage

+ leaf fodder + fuel wood) production, trees should be planted at the density of 825 plants ha^{-1} (4m x 3m) and maintained at least up to 12-13 years after that if understorey production starts decreasing felling of trees should be done (IDRC report, 1982-83, 1993-94). If harvesting of trees was done at this age, the biomass of about at least 2 to 3 t ha^{-1} $year^{-1}$ may be obtained. Thus, the total biomass production from silvipastoral system can be obtained by about two times higher than pasture alone. On the basis of more than 20 years studies in shallow red gravelly soils under semi-arid condition at Jhansi, Pathak et al (1996) reported that the degraded lands producing hardly up to 1.0 t ha^{-1} $year^{-1}$ have been improved to produce up to 10 t ha^{-1} $year^{-1}$ at a 10 year rotation through silvipastoral system. Besides yield improvement by 8 to 10 times, the quality of mixed forage has also improved by 6 to 7 times. Roy et al. (1998) reported the productivity of different MPTS based silvipastoral system at 8 years rotation on slightly degraded land in semi-arid region. The maximum productivity works out to be in *Albizia amara* based silvipastoral system (7.95 t ha^{-1} $year^{-1}$) closely followed by *Acacia tortilis* (7.60 t ha^{-1} $year^{-1}$). The productivity of *Hardwickia binata* based silvipastoral system was (3.32 t ha^{-1} $year^{-1}$) quite low due to slow growth of the tree. On similar site, the productivity of *Leucaena leucocephala* was found to be 6.24 t ha^{-1} $year^{-1}$ at eight year rotation.

The forage production from single (*Cenchrus* pasture alone), two (*Cenchrus* + *Ailanthus excelsa*) and three (*Cenchrus* + *A. excelsa* + *Dichrostachys nutan*) tier systems were compared with natural pasture at ICAR-CSWRI, Avikanagar, Rajasthan under semi-arid condition. Results showed that three tier silvipastoral system provided maximum average forage production (2.78 t ha^{-1} dry forage from pasture + 0.95 t ha^{-1} green tree leaves) followed by two tier silvipastoral system (2.09 t ha^{-1} from pasture + 0.62 t ha^{-1} green leaves from trees) and single tier (1.91 t ha^{-1}). The minimum forage yield of 1.66 t ha^{-1} was noted with natural pasture (Anonymous 1998). Mann et al (1989) recommended adoption of silvipasture system for more fodder in arid and semi-arid regions of Rajasthan. They reported that even during drought year in addition to 1.00-1.5 t ha^{-1} of good quality dry leaf fodder and 0.7 to 0.8 t ha^{-1} of grass can be obtained from silvipastoral systems. The system also provided fuel wood. At same location, Meena et al (2003) reported that intercropping of *C. setigerus* and *Vigna unguiculata* in the ratio of 1:2 in association with *A. excelsa* (100 trees ha^{-1}) at 8 years of tree growth was found most suitable combination for higher dry biomass production of 10.0 t ha^{-1} (3.93 t forage + 3.13 t leaf fodder + 2.91 t fuel wood). On the basis of 5 years production, Rathod et al (2000) reported highest dry forage yield of 10.2 t ha^{-1} with better quality in association with *L. leucocephala* as tree, *C. ciliaris* as grass and *C. ternatea* as legume in semi-arid region of Gujarat.

A fifteen years studies in Yamuna ravine near Agra, revealed that a silvipastoral system consisting of *A. nilotica* + *C. ciliaris* and *A. tortilis* + *C. ciliaris* planted at 3x3 m spacing produced on an average the total biomass yield of 5.0 t ha^{-1} $year^{-1}$ (2.5 t fuel + 2.5 t forage) and 6.9 t ha^{-1} $year^{-1}$ (4.2 t fuel + 2.7 t forage), respectively (Prajapati et al 1993). Similarly, studies at ICAR-CIRG, Makhdoom, near Mathura revealed that a 3 tier silvipastoral system consisting of *A. nilotica* + *L. leucocephala* + *C. ciliaris* produced total forage production of 8.0 t ha^{-1} (4.55 t top feed + 3.45 t pasture) gross energy (33, 866 M cal ha^{-1}) and crude protein yield (1270 kg ha^{-1}) in ravenous grazing lands followed by *L. leucocephala* + *D. cinerea* + *C. ciliaris* (3.48 t top feed + 4.02 t pasture = 7.50 t ha^{-1}), two tier system consisting of *L. leucocephala* + *C. ciliaris* (4.12 t top feed + 1.98 t pasture =

6.10 t ha^{-1} and *D. cinerea* + *C. ciliaris* (2.25 t top feed + 1.35 t pasture = 3.60 t ha^{-1}). The minimum dry forage (0.13 t top feed + 2.27 t pasture = 2.40 t ha^{-1}), gross energy (12,624 M cal ha^{-1}) and crude protein yield of 160 kg ha^{-1} were recorded with natural silvipasture (Bhattacharya and Sharma 1993).

The performance of nine different silvipastural system involving 3 NFTs (*Cassia siamea, Acacia holosericea* and *L. leucocephala*) and three pasture species (*S. hamata, S. humilis* and hybrid napier) were evaluated on degraded lands of North Eastern Ghat Zone of Orissa. Results showed that *C. siamea* and hybrid napier combination was found to be more suitable system for this region for higher biomass production. However, *C. siamea* and *S. hamata* were observed to conserve more soil moisture than other trees and pasture species (Mishra and Choudhury 2002). Under the northeast region, maximum dry forage yield of 13.54 t ha^{-1} was recorded from a combination of *Stylosanthes guianenssis* and *Setaria sphacelata* in association with *Alnus nepalensis* followed by the combination of *S. guianenssis* and *Panicum maximum* (11.30 t ha^{-1}) (Chauhan et al 1993). Singh (2005) reported that the total productivities under silvipastoral system increased from 8 to 30 t ha^{-1} after establishment.

PRIORITIES

The focus should be more appropriately placed on improving pastoral and agroforestry livelihoods and maintaining ecosystem health in these land use systems. The various agroforestry system identified and developed should reach the farmers. The maintenance of grassland integrity is also essential. Therefore, following points should be taken in to care while setting research and development priorities for grasslands and agroforestry:

- Rapidly changing systems with changing needs.
- Focus generally on human welfare and maintaining environmental goods and services.
- More emphasis on providing pastoral people and agroforesters with high quality information.
- Restoring and providing grasslands access to key resources, increasing mobility and flexibility, and ensuring security.
- Addressing gaps in our knowledge about how grassland and agroforestry systems work and how these systems can be improved.

REFERENCES

Ajit, Dhyani SK, Ramnewaj, Handa AK, Prasad R, Alam B, Rizvi RH, Gupta G, Pandey KK, Jain A and Uma 2013 Modelling analysis of potential carbon sequestration under existing agroforestry systems in three districts of Indo-gangetic plains in India. *Agroforesty Systems* DOI 10

Anonymous 1998 35 years of Research Central Sheep and Wool Research Institute, Avikanagar (Rajasthan) 1-128

Bhattacharya NK and Sharma K 1993 Silvipastoral system in relation to goat farming (eds) AS Gill R Deb Roy and AK Bisaria. *Extended Abstract (Late arrivals)*, NRCAF, Jhansi 1-12

Chauhan DS, Dhyani SK and Desai AR 1993 Productivity potential of grasses in association with *Alnus nepalensis* and pineapple under silvihortipastoral system of Agroforestry in Meghalaya. *Indian Journal of Dryland Agriculture Research & Development* **8**(1) 60-64

Dhyani SK, Palsaniya DR, Tewari R K and Chaturvedi O P 2010 Sustainable agroforestry production through contingency planning for aberrant weather in Bundelkhand region of Central India. *Indian Journal of Agroforestry* **12** (2) 1-5

Dwivedi GK, Sinha NC, Tomer PS and Dixit OP 1988 Nitrogen Economy Biomass Production and Seed Production Potential of *Chrysopogon fulvus* by Intercropping of Pasture Legumes. *Journal of Agronomy and Crop Science* **161** (2) 129–134

Gardener CJ 1980 Diet selection and liveweight performance of steers on *Stylosanthes hamata*-native grass pastures. *Australian Journal of Agricultural Research* **31** 379-392

Ghosh PK and Palsaniya DR 2014 Crop diversification for sustainable intensification and carbon management. National Symposium on Agricultural Diversification for sustainable Livelihood and Environmental Security, PAU, Ludhiana 3-5

Ghosh PK and Palsaniya DR 2014 Issues of forage resource development in changing scenario ***In:*** Integrated Nutrient Management for Quality Forage Production (eds) Kumar Sunil, Rai AK, Palsaniya DR, Dixit AK, Srinivasan R, Prabhu G and Kumari Anita, Indian Grassland and Fodder Research Institute, Jhansi- 284003, Uttar Pradesh, India1-6

Ghosh P K, Palsaniya DR, Rai AK and Sunil Kumar 2015 Strategies for higher nutrient use efficiency and productivity in forage crops. Nutrient Use Efficiency from Basics to Advances (eds) A Rakshit, HB Singh and A Sen *Springer India* 329-342

Ghosh PK, Sunil Kumar and Palsaniya DR 2013 Forage and livestock production strategies for Integrated Farming Systems. Integrated Farming Systems for Enhancing Resource Use Efficiency and Livelihood Security of Small and Marginal Farmers (eds) UK Behera, AK Chaudhary, Anchal Dass and AK Vyas New Delhi, India 143-156

Ghosh PK, Palsaniya DR and Srinivasan R 2016 Forage Research in India: Issues and strategies. *Agricultural Research Journal* **53** (1) 1-12

Herlocker D 1999 Rangeland ecology and resource development in Eastern Africa. GTZ, Nairobi, Kenya

Hodgkinson KC, Ludlow MM, Mott JJ and Baruch Z 1989 Comparative responses of the savanna grasses *Cenchrus ciliaris* and *Themeda triandra* to defoliation . *Oecologia* **79** 45-52

IDRC Report 1982-83 to 1993-94 Annual Report IDRC Silvipastoral Operational Project, IGFRI, Jhansi

IGFRI Vision 2050 Indian Grassland and Fodder Research Institute, Jhansi (UP)

IPCC 2007 Climate Change 2007 Impact, adaptation and vulnerability Contribution of working group II to the fourth assessment report of the Intergovernmental Panel on Climate Change (eds) Pary, et al. Cambridge University Press, Cambridge, 2007

Kang BT, Wilson GF, 1987 The development of alley cropping as a promising agroforestry technology (eds) Steppler HA and Nair PKR *Agroforestry: A decade of development*, ICRAF, Nairobi 227-243

Mann JS, Hassan B and Beniwal RK 1989 Adopt silvipasture system for more fodder in arid and semi arid regions of Rajasthan. *Indian Farming* **39**(2) 33-35

McArthur S 1998 Practical evidence supports cell grazing benefits. *Australian Farm Journal Beef,* September 8-9

McCosker T 1994 The dichotomy between research results and practical experience with time control grazing. *Australian Rural Science Annual*, Percival Publishing Sydney, Australia 26-31

McIvor JG, Noble AD and Orr DM 1998 Stability and productivity of native pastures oversown with tropical legumes. *North Australia Program Occasional Publication No. 1.* Meat Research Corporation, Sydney, Australia

Meena LR, Sharma SC and Mann JS 2003 Effect of Ardu densities and intercropping on performance of grass and legume in silvipasture system under semiarid condition of Rajasthan. *Range Management & Agroforestry* **24**(2) 123-126

Mishra PJ and Choudhury PR 2002 Evaluation of silvipasture systems in degraded lands of Orissa. *Indian Journal of Soil Conservation* **30**(3) 253-257

Mott JJ, Ludlow MM, Richards JH and Parsons AD 1992 Effects of moisture supply in the dry season and subsequent defoliation on persistence of the savanna grasses *Themeda triandra, Heteropogon contortus* and *Panicum maximum. Australian Journal of Agricultural Research* **43** 241-260

Nair PKR 2001 Agroforestry in our fragile world challenges and opportunities for sustainable development. *Forerunner to the Encyclopedia of Life Support Systems*, UNESCO, Paris, France, & EOLSS, UK 375–393

Norton BE 1998 The application of grazing management to increase sustainable livestock production. *Animal Production in Australia* **22** 15-26

Palsaniya DR, Asha Ram and Rai SK 2014 Role of legumes in Soil Fertility Management. ***In:*** Integrated Nutrient Management for Quality Forage Production (eds) Sunil Kumar, Rai AK, Palsaniya DR, Dixit AK, Srinivasan R, Prabhu G and Kumari Anita, Indian Grassland and Fodder Research Institute, Jhansi- 284003, Uttar Pradesh, India16-20

Palsaniya DR, Singh R, Tewari RK, Yadav RS and Dhyani SK 2012 Integrated watershed management for natural resource conservation and livelihood security in Semi-Arid Tropics of India. *Indian Journal of Agricultural Sciences* **82** (3) 241-247

Palsaniya DR, Singh R, Tewari RK, Yadav RS, Dwivedi RP, Kumar RV, Venkatesh A, Kareemulla K, Bajpai CK, Singh R, Yadav SPS, Chaturvedi OP and Dhyani SK 2008 Socio economic and livelihood analysis of people in Garhkundar-Dabar watershed of central India. *Indian Journal of Agroforestry* **10**(1) 65-72

Palsaniya DR, Singh R, Venkatesh A, Tewari RK and Dhyani SK 2010 Grass productivity and livestock dynamics as influenced by integrated watershed management interventions in drought prone semi-arid Bundelkhand, India. *Range Management and Agroforestry* 4-6

Palsaniya DR, Dhyani SK and Rai P 2011 Silvipasture in India: Present Perspectives and Challenges Ahead. Scientific Publishers, Jodhpur, India 207

Palsaniya DR, Singh Ramesh, Tewari RK and Dhyani SK 2012 Changing agro ecological scenario with climate change as perceived by farmers in Bundelkhand, India. *Indian Journal of Forestry* **35**(1) 21-28

Palsaniya DR, Singh Ramesh, Yadav RS, Tewari RK, Dwivedi RP, Kumar RV Venkatesh, Kareemulla A, Bajpai K, Singh CK, Yadav Rajendra, Chaturvedi OP and Dhyani SK 2009 Participatory Agro-ecosystem Analysis and Identification of Problems in Garhkundar-Dabar Watershed of Central India. *Indian Journal of Agroforestry* **11**(1) 91-98

Pathak PS, Gupta SK and Singh P 1996 IGFRI Approaches Rehabilitation of Degraded Lands, IGFRI, Jhansi 1-23

Prajapati MC, Nambiar KTN, Puri DN, Singh JP and Malhotra BM 1993 *Acacia nilotica* and *Cenchrus ciliaris* silvipasture for sustainable productivity of fuel and fodder in ravines ***In:*** Agroforestry for 2000 AD for the semiarid and arid tropics. Extended Abstract (Late arrivals) (eds) AS Gill R , Roy Deb and Bisaria AK, NRCAF, Jhansi 33-34

Rai P and Palsaniya DR 2015 Silvipasture Research and Development in India: Looking Back to Move Ahead. ***In:*** Agroforestry - Present Status and Way Forward eds. Dhyani SK ,Ram Newaj, Badre Alam and Inder Dev, *Biotech Books*, New Delhi, India 471-486

Rathod ND, Viradia RR, Valia MJ and Dudhat MS 2000 Evaluation of grass and legume in the Subabul under silvipastoral system. *Range Management & Agroforestry* **21**(1) 118-120

Roy MM, Nigam G and Kumar V 1998 Productivity of some multipurpose trees in silvipastoral system on highly degraded lands in semiarid region. *Indian Journal of Forestry* **21** 4-8

Singh G, Singh H, Bhojvaid PP 1998 Amelioration of sodic soils by trees for wheat and oat production. *Land Degradation and Development* **9** 536-545

Singh KA 2005 Impact of forage resource development on sustained land use and environment in north eastern hills region of India. *Range Management & Agroforestry* **26** 58-70

Sunil Kumar, Dixit AK, Palsaniya DR, Rai AK, Agrawal RK, Singh JB, Rai SK and Prabhu G 2013 Mitigation of Climate Change through Innovative Forage Production Systems. Indian Grassland and Fodder Research Institute, Jhansi- 284003, Uttar Pradesh, India 204

Waugh W 1997 Pastures under adverse conditions - handling what you have grazing systems in practice. ***In:*** Proceedings of the 12th Annual Conference of the Grassland Society of New South Wales 91-94 Dubbo, Australia, 1997 Orange, Australia Grassland Society of New South Wales

CHAPTER 5

Resource Conservation Based Climate Resilient Agroforestry System

Dhiraj Kumar, Inder Dev, Asha Ram, Ramesh Singh, AR Uthappa, KB Sridhar, Veeresh Kumar, Mahendra Singh, Naresh Kumar, Ram Newaj, Sudhir Kumar, Lal Chand and OP Chaturvedi*

ICAR-Central Agroforestry Research Institute, Jhansi (U.P), India

**Corresponding author's Email: dhiraj.iari@gmail.com*

ABSTRACT

The practice of harnessing benefits of agroforestry (AF) system is age old, but its adoption for other than the traditional benefits is quite evident now a days. The perceived benefits in terms of climate change mitigation by sequestering carbon in long lived pools are being explored by the researchers around the world. Integration of agroforestry (AF) with conservation agriculture (CA) has led to multifaceted benefits to the overall system sustainability and certainly adds the 4th dimension to the basic tenets of CA practice. The accrued improvement in terms of biodiversity improvement, ecosystem services, food security through regular source of income to farmers, risk aversion from climate change vagaries and environmental security is really a win-win situation. The role of CA based AF system for long term sustainability might be recognized through awareness creation, popularization, trainings, financial barriers removal, institutional strengthening etc. and it is the need of the hour.

Keywords: Agroforestry, Climate change, Conservation, Mitigation, Tillage

INTRODUCTION

Conservation agriculture (CA) relies on minimum disturbance of soil, permanent soil cover along with crop rotation for sustainable cultivation system. In the era of climate change where, we have limited land and other resources with adverse weather variability, conservation agriculture is one such answer which can produce sustainable food from less land with minimum adverse impact on environment. Most of the world countries are facing the problem of climatic variability, price volatility of inputs and poor soil in terms of salinization, erosion, nutrient deficiencies, soil organic matter decline etc. The conservation agriculture though advocated way back, but now spreading as a response to the concern of sustainable agriculture at global level, and it increased to cover about 11 per cent of global arable land (155 Mha) (Kassam et al 2014). Thus, it entails the very basic theme of conservation (soil, water, air and biodiversity) as compared to traditional agriculture (Derpsch et al 2010, Derpsch et al 2011). While on the other, agroforestry (AF) is the integration of woody perennials into the farmlands or rangelands and in doing so, diversifies and sustains production for increased benefits for farmers and the environment (Elevitch 2004). Thus, the judicious use of these environment friendly practices like CA and AF could serve as a building block for achieving the target of low resource based and climate resilient cropping system. Infact addition of AF to CA practices would add a 4th principle to 3 basic principles of CA i.e., tree-crop integration (Mowo and Kiwia 2009). In the CA practices, the major problem with small category farmers are the inadequate crop residues which are used as livestock feed, but when AF practice will be integrated along with CA then, the specific fodder trees will meet the fodder requirements of livestock and also adds lots of nutrients to the field. Thereby, the adoption of this practice is improving the soil quality and mitigating climatic abnormalities.

CONVENTIONAL AGRICULTURE VS CONSERVATION AGRICULTURE

CA is not the 'business as usual' but is an entirely different concept with respect to conventional agriculture. It is based on optimizing the yield and profit to have a balance in agricultural, environmental and economic benefits (Table 1).

Table 1: *Some important distinguishing features of conventional agriculture and CA*

Conventional agriculture	Conservation agriculture
Cultivating land, using science and technology to dominate nature	Least interference with natural processes
Excessive mechanical tillage and soil erosion	No-till or drastically reduced tillage (biological tillage)
High wind and soil erosion	Low wind and soil erosion
Residue burning or removal (bare surface)	Surface retention of residues (permanently covered)
Water infiltration is low	Infiltration rate of water is high
Use of *ex-situ* FYM/composts	Use of *in-situ* organics/composts
Green manuring (incorporated)	Brown manuring/cover crops (surface retention)
Kills established weeds but also stimulates more weed seeds to germinate	Weeds are a problem in the early stages of adoption but decrease with time
Free-wheeling of farm machinery, increased soil compaction	Controlled traffic, compaction in tramline, no compaction in crop area
Mono cropping/culture, less efficient rotations	Diversified and more efficient rotations
Heavy reliance on manual labour, uncertainty of operations	Mechanized operations, ensure timeliness of operations
Poor adaptation to stresses, yield losses greater under stress conditions	More resilience to stresses, yield losses are less under stress conditions
Productivity gains in long-run are in declining order	Productivity gains in long-run are in incremental order

Source: Sharma et al 2012

CHALLENGES OF CONSERVATION AGRICULTURE

Despite the many benefits which CA provides, its proper acceptability to the farmers especially the small holders is at stake. This is mainly due to the following reasons:

- The unavailability of the crop residues as these are either burnt in the fields or stored livestock feed, bedding, cooking, mulching materials etc.
- The problem of weed infestation is high under CA unless cared thereof, because the seeds lying dormant under the soil have better chance of re-emergence.
- The initial cost is high due to costly and sophisticated equipments and labour retirements.
- Improper availability of seeds of the cover crops.
- Lack of know-how and information on practice and implementation of CA.
- Above all, the mindset of the small category farmers for inclination towards tilling their field for better yield.

COMPLEMENTARITY AND SYNCHRONY BETWEEN CONSERVATION AGRICULTURE AND AGROFORESTRY

There are lots of facets through which the CA and AF share common functionalities. Infact to overcome the constraints accrued under CA, the benefits associated with AF can overcome it. The AF is a multifaceted subject, which along with CA will sustain the yield with minimal environmental damage. Some of the exciting areas where CA based AF system works in tandem are as follows:

MITIGATION AND ADAPTATION TO CLIMATE CHANGE

The sequestration of atmospheric CO_2 through agroforestry has been long considered as the climate change impact mitigation. As on the global level, agriculture occupies over one third of arable land (World Bank 2015), thus AF led a greater hope for increased sequestration of carbon on agricultural land. It is quite true that compared to monoculture or system where crop residues have been removed, agroforestry along with CA which adds litter and residues to the system can enhance soil organic carbon (SOC). Conservation tillage reduces the impact of rain drops and increases SOC through soil aggradation processes and reversal of soil degradation processes (Lal 1989, Carter1993).

Due to its multiple plant species and soil type in which agroforestry system prevails, it has huge potential as mitigation strategy to the climate change (Montagnini and Nair 2004, Albrecht and Kandji 2003, ICAR 2006). In the tropical condition for small agroforestry systems, it has been found that C-sequestration ranges from 1.5 to 3.5 Mg C ha^{-1} $yr^{-1}$and thus can be a viable strategy for carbon storage (Roshetco et al 2007, Montagnini and Nair 2004). Some of the researchers estimated CSP between 25–228 Mg C ha^{-1} (Dixon et al 1994; (Sathaye and Ravindranath1998). But these values vary depending on the biomass production in different regions (Pandey 2007). The carbon sequestration potential under agroforestry in different agro-climatic regions of India (Table 2) suggests that Indo-Gangetic plains having very huge potential of carbon sequestration and there is enough scope with us to maintain that level.

Table 2: *Estimates of carbon sequestration potential under agroforestry in different agro-climatic regions in India.*

Region	Average carbon sequestration rate
Degraded forest land	1.1 t ha^{-1} yr^{-1}
Central Himalayas	3.9 t ha^{-1} yr^{-1}
Indo-Gangetic plains	8.5-15.2 t ha^{-1} yr^{-1}
Strip plantation aged 5.3 yr in Haryana	15.5 t ha^{-1} yr^{-1}during the first rotation
Agricultural soils of Indo-Gangetic plains	12.4-22.6 t ha^{-1} yr^{-1}

Source: Maikhuri et al (2000)

The enhancement in SOC by 33.3 to 83.3 per cent for *Populus deltoides* and *Eucalyptus* hybrid with *Cymbopogon* sp., with greater increase in SOC under *Populus deltoides* system has been reported by Singh et al 1989. Gupta et al 2009 reported that in poplar based AF system, the trees even in first year of plantation, can sequester SOC to the tune of 6.07 t ha^{-1} yr^{-1} upto a depth of 30 cm than in subsequent years (1.95-2.63 t ha^{-1} yr^{-1}) with greater soil carbon storage in sandy clays than loamy sand. Both above and below ground biomass sequestered carbon is highly variable for an agroforestry system and is usually much higher than treeless land use system (Nair et al 2009a, Fialho and Zinn 2014). The carbon sequestration capacity differs across the geography of the area and plant species used in agroforestry system (Newaj and Dhyani 2008). The net annual carbon sequestration potential by agroforestry components (Table 3) gives estimates of potential species like populus, eucalyptus and bamboo for higher carbon sequestration rates in India. Especially the tropical home gardens have values of (16-36 Mg C ha^{-1} yr^{-1}) and have huge potential of carbon sequestration rates.

Table 3: *Estimates of net annual carbon sequestration by agroforestry components.*

Agroforestry component	Carbon sequestration (Mg ha^{-1} yr^{-1})
Populus spp.	8
Eucalyptus spp.	6
Bamboo spp.	6.51-8.95
Tropical home gardens	16-36

Source: Singh and Pandey (2011)

ECOSYSTEM SERVICES PROVIDED BY CONSERVATION AGRICULTURE WITH TREES (CAWT)

Ecosystem services are the benefits that humans and other species obtain from functioning ecosystems. Ecosystem services have an estimated value of $16-54 trillion yr^{-1}. Together these practices viz., reduced tillage (RT) or no-till (NT), permanent organic soil cover by retaining crop residues, and crop rotations, including cover crops aim to increase crop yields by enhancing several regulating and supporting ecosystem services (ESS). Though

CA was originally introduced to regulate wind and water erosion (Baveye et al 2011), it is now considered to deliver multiple ESS. Previous reviews indicate that CA can reduce water and wind erosion due to protection of the soil surface with residue retention and increased water infiltration and decreased runoff with NT (Verhulst et al 2010). Other ES benefits provided by CA includes nutrient cycling, carbon sequestration, and pest and disease control are quite variable, from positive to neutral or even negative depending on site-specific context, management, soil type, and climate. The integration of trees with CA has made agriculture more resilient. The CAWT provides wide spectrum of ecosystem services and some of the important services are discussed below. CAWT produces the soil conditions that result in reduced erosion and runoff and improved water quality compared to conventional practices. Likewise, water holding capacity and storage are enhanced with CA providing some buffer to crop production during drought conditions. SOM is almost invariably higher in the surface soil with CA practices compared to conventional practices and influences many other soil properties and processes involved in the delivery of ESS.

Different ecosystem services generated by CAWT

Provisional services	Regulating services	Cultural services
Food and fodder, water, oxygen, energy (hydropower, biomass fuels)	Carbon sequestration, climate and weather regulation, waste decomposition, water and air purification, pest and disease control, drought and flood risk, erosion control	Recreation, aesthetic
		Supporting services
		Nutrient dispersal and cycling, primary production

CLIMATE REGULATION

The net potential of CAWT to contribute to climate regulation and serve as a global warming mitigation strategy depends on the direction and magnitude of changes in soil C, nitrous oxide (N_2O) and methane (CH_4) emissions associated with its implementation compared to conventional practices. Collectively this is assessed in terms of the global warming potential of the farming practices which are soil, climate and management dependent (Robertson and Grace 2004). For example, if there is an increase in soil C that is greater than the combined increase in N_2O or CH_4 emissions (expressed as CO_2 equivalents), the net global warming potential decreases. The CAWT practices of retentions of trees, crop rotations and surface residue retention are intended to increase C inputs relative to conventional practices and NT (no till) to decrease decomposition through increased soil aggregation and a protection of soil C from decomposers. The balance of those two processes and the resulting soil C will vary with soil, climate and other management practices. Agroforestry plantations with CA are intensively managed, and produce far more biomass than the conventional forests. Ajit et al (2016) estimated average carbon sequestration potential of the agroforestry system, representing varying edapho-climatic conditions on farmers' field at country level was 0.21 Mg C $ha^{-1}yr^{-1}$.

REGULATING SERVICES

Soil

Incorporation of trees in the croplands can help in maintaining the physico- biochemical properties both under sequential and simultaneous agroforestry (Table 4). Tree litter and prunings improve soil fertility not only through the release of nutrients in the soil by mineralization but also by adding soil organic matter. However, it depends upon the quality and quantity of tree litter or prunings, soil type and climatic conditions of the area and increasing activity of microorganisms in the root zone. Trees generally have their roots well below the crop zone, use nutrients from the lower soil layers, resulting in increased nutrients and water use efficiency with least competition with crops. The CAWT practices of residue retention, NT, and certain crop rotations increase SOM in the top soil that, in turn, impact soil physical properties and process that reduces erosion and run-off and biological and chemical properties that could lead to improved N use efficiencies and fewer N losses to the environment. The amount of SOM that is needed to drive these changes in soil properties and processes and related ESS is not well known. Soil physical properties such as aggregation and macro porosity are important for determining rates of water infiltration, run-off, plant available soil water, erosion, and others. These factors are usually greater with CA and are related to reduced erosion and run-off. Though, there are fewer studies on water quality, in general, less sediment load and reduced N are observed with CA. Both CA and NT can reduce wind erosion due to the larger proportion of dry aggregates, less wind erodible fraction and greater crop residue cover of the soil surface (Singh et al 2012, Verhulst et al 2010). CA can also indirectly reduce erosion by water through the effects on soil properties and processes that increase water infiltration and reduce run-off. No-till management generally increases bulk density of the top soil, reduces total soil porosity, and even hydraulic conductivity compared to conventional tillage systems.

Water

Reducing run-off and water erosion with CAWT should result in lower transport of sediments, nutrients, pesticides/herbicides and higher water quality. With NT or conservation tillage, N in sediments and run-off water have been reduced by as much as 60 per cent (Kay et al 2009), though, there is a wide range with only 9 per cent reduction of N in run-off recently modelled by Liu et al (2013a). Nitrogen losses in sediments were consistently lower with NT in a study by Richardson and King (1995) but losses of soluble N differ with the type of crop, with less soluble N lost with wheat compared to maize or sorghum. There are consistent reductions in P losses to surface waters with NT and herbicides in run-off have been reduced by 40 to 70 per cent with NT (Richardson and King 1995).

Conservation agriculture practices result in more plant available water than conventional practices. This is a result of increased water infiltration and lower evaporation with reduced mixing of the surface soil, more residue cover and less exposure to drying compared to conventional tillage. Water holding capacity of the top soil is also generally higher due to increased SOM contents. Liu et al (2013b) found that soil moisture content was most affected by residues compared to tillage practice; soil moisture remained highest throughout the growing season with NT plus residue, intermediate levels of soil moisture with tillage plus residues, and lowest levels with NT without residues.The impact of higher soil moisture with NT on yields or water use efficiency is also reported with irrigation. Studies have shown a 20-50 per cent reduction in irrigation water with NT and residue retention with similar or higher crop yields (Gathala et al 2013).

Biodiversity

The combined effects of CAWT lead to a more diverse and stable soil community, including beneficial bacterial and fungal species that can suppress pathogens. The trees will act as a host for many beneficial organisms. The continuous addition of crop and tree litter creates a favourable environment for growth of various beneficial macro fauna and flora. Increased abundances of meso-fauna such as arthropods (Dubie et al 2011) and macro-fauna such as spiders and hymenopteran parasitoids have been found with NT (Rodriguez et al 2006). Abundance and diversity of populations of natural enemies of parasitic nematodes may be higher with NT. More diverse soil communities suppress the impacts of the pathogens (Verhulst et al 2010). Higher SOM levels with CA produce cooler and moister conditions favorable to pests and diseases and many fungal pathogens reside in crop residues of the host crop and thus may increase incidence and disease retention time in the soil (Cook 2006).

Table 4: *Soil processes and ecosystem services (ESS) between no-till (NT) and conventional tillage (CT) and surface residue retention vs no surface residue placement*

Soil properties, processes and ecosystem services	NT compared to CT	Residue retention
Aggregate stability	↑	↑
Bulk density	↑ but small number of studies showing opposite	↓
Total porosity	↓	↑
Macropores	↓↑ avg size larger	↑
Mesopores	↑	
Micropores	↑	
Hydraulic conductivity	↓ mixed results	↑
Infiltration	↑	↑
Runoff	↓	↓
Evaporation	↓	↓
Plant available water	↑	↑
Erosion	↓	↓
Total nitrogen	↑ follows pattern of soil organic matter	↑
Nitrogen availability (N mineralization)	generally ↓ at least in the short term and often long term	↑↓ depends on quality of residues
P, K, Ca, Mg	P ↑ in top soil layer. K ↑ in surface layers, in general. Ca, Mg few differences	K depends on type of crop residue
Cation exchange capacity	no effect	↑ but only in very top layer
pH	more often ↓	↓
Nutrient leaching	↓	↓

Source: Adapted from Verhulst et al 2010

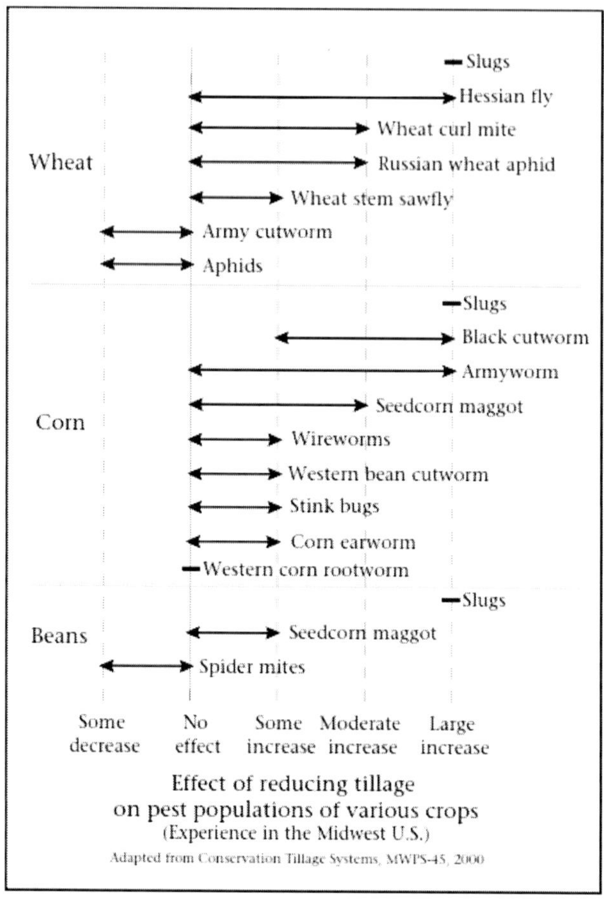

Effect of reducing tillage
on pest populations of various crops
(Experience in the Midwest U.S.)
Adapted from Conservation Tillage Systems, MWPS-45, 2000

Conservation agriculture (CA) increases biodiversity of both flora and fauna which helps to control insect pests. The integrated pest management (IPM) creates a balance between sustainable environmental practices and profitable farming. Both IPM and CA work on the same principles to increase biodiversity and conservation of natural resources. Mulching of ground with crop residue is a component of CA that promotes an increase in biological diversity not only below ground but also above ground. The number of beneficial insects was higher where, there was ground cover and mulch (Kendall et al 1995), and these help to keep the insect pests population in control.

However, there are some changes in reduced tillage systems that increase the potential problems of specific pests and it has been well documented in wheat, corn, and beans based CA. Those pests that overwinter in the soil or in crop residues and become active early in growth of the crop and benefitted most from reduced tillage. Although lower soil temperatures may cause these insects to develop more slowly than in tilled soil, they can be more numerous because they have not been exposed to tillage. Other insects may decrease after a number of years of no-till. This may be due to increased survival of beneficial insects, ants, ground beetles, rove beetles, and spiders, all of which can contribute to insect pest control (Andy McGuire 2000).

CA with trees encourages biocontrol agent's activity, because trees provide shelter during crop lean period and predators and parasitoids can survive on insect pests which harbour trees. Tree biomass which added to the soil also serve as food for the soil fauna and increase soil biota interaction and its diversity. Trees can serve as a barrier to insect movement and thereby reducing crop infestation.

The insect population, diversity fluctuation and insect damage can depend on the type of crop diversity, anatomy and morphology, age of the trees, microclimatic conditions created in CA and to what extent the companion crops are related. For example, space-biological diversity and relatedness may, through differences in phenophase and through provision of refuges, oviposition, hibernating and pupation sites, maintain the entomofauna in the system for a relatively longer time than in the time-biological diversity system with unrelated or related crop companions.

WAY FORWARD

From the above discussion on integration of CA with AF, it is quite obvious that both the systems will emulate the benefits of ecosystem services, carbon sequestration, increasing biodiversity, improving soil health etc. The pressure on cover crops in case of CA practices could be alleviated when fodder trees are planted together with crops. The reduction of soil erosion, improvement in infiltration rate, reduction of bulk density, addition of soil organic matter is quite evident. The improvement in crop diversification along with regular source of livelihood security to the small category farmers is possible only when we integrate trees along with crops by considering the system like agroforestry and CA.

REFERENCES

Ajit, Dhyani SK, Handa AK, Ram Newaj, Chavan SB, Alam B, Prasad R, Ram A, Rizvi RH, Jain AK, Uma, Tripathi D, Shakhela RR, Patel AG, Dalvi VV, Saxena AK, Parihar AKS, Backiyavathy MR, Sudhagar RJ, Bandeswaran C, Gunasekaran S 2016 Estimating carbon sequestration potential of existing agroforestry systems in India, *Agroforestry Systems*, doi:10.1007/s10457-016-9986-z

Albrecht A and Kandji ST 2003 Carbon sequestration in tropical agroforestry systems, *Agriculture, Ecosystems and Environment* **99** 15-27

Andy McGuire 2000 Conservation Tillage Systems and Management, Mid-West Plan Service, MWPS-45

Baveye PC, Rangel D, Jacobson AR, Laba M, Darnault C, Otten W, Radukovich R, and Camargo FAO 2011 From Dust Bowl to Dust Bowl: Soils are still very mucha frontier of science, *Soil Science Society of America Journal* **75** 2037–2048

Carter MR (eds) 1993 Conservation tillage in temperate agroecosystems, Lewis Publishers, Boca Raton, Fl, 390 pp

Cook RJ 2006 Toward cropping systems that enhance productivity and sustainability, *Proceedings of National Academy of Science* **103** 18389–18394

Derpsch R, Friedrich T, Kassam A, and Li HW 2010 Current status of adoption of notill farming in the world and some of itsmain benefits, *International Journal of Agricultural and Biological Engineering* **3** 1-25

Derpsch R, Friedrich, T, Landers JN, Rainbow R, Reicosky DC, Sa′ JCM, Sturny WG, Wall P, Ward R C, and Weiss K 2011 About the necessity of adequately defining no-tillage – a discussion paper, in: *Proceedings on 5th World Congress Conservation Agriculture*, 26-29 September 2011, Brisbane, Australia

Dixon RK, Brown S, Houghton RA, Solomon AM, Trexler MC, et al. 1994 Carbon pools and fluxes of global forest ecosystems, *Science* **263** 185-190

Dubie TR, Greenwood CM, Godsey C, and Payton ME 2011 Effects of tillage on soil microarthropods in winter wheat, *Southwestern Entomologist* **36** 11–20

Elevitch CR (eds) 2004 The overstory book: cultivating connections with trees, Second edition, Holualoa, Hawai'i. Permanent Agricultural Resources 526p

Fialho RC and Zinn YL 2014 Changes in soil organic carbon under Eucalyptus plantations in Brazil: a comparative analysis, *Land Degradation and Development* **25** 428–437 doi:10.1002/ldr.2158

Gathala MK, Kumara V, Sharma PC, Saharawat YS, Jat HS, Singh M, Kumar A, Jat ML, Humphreys E, Sharma, DK, Sharma S, and Ladha JK 2013 Optimizing intensive cereal-based cropping systems addressing current and futuredrivers of agricultural change in the northwestern Indo-Gangetic Plains of India, *Agriculture Ecosystems and Environment* 177 85–97

Gupta N, Kukal SS, Bawa SS, and Dhaliwal GS 2009 Soil organic carbon and aggregation under poplar based agroforestry system in relation to tree age and soil type, *Agroforestry Systems* **76** 27-35

ICAR 2006 Handbook of Agriculture, (5th edn) New Delhi: Indian Council of Agricultural Research

Kassam A, Friedrich T, Derpsch R, and Kienzle J 2014 Worldwide adoption of Conservation Agriculture, 6th World Congress on Conservation Agriculture, 22-25 June 2014, Winnipeg, Canada

Kay P, Edwards AC, and Foulger M 2009 A review of the efficacy of contemporaryagricultural stewardship measures for ameliorating water pollution problemsof key concern to the UK water industry, *Agricultural Systems* **99** 67–75

Kendall DA, Chin NE, Glen DM, Wiltshire CW, Winstone L, and Tidboald C 1995 Effects of soil management on cereal pests and their natural enemies, in: *Ecology and integrated farming systems* (eds) Glen DM, Greaves MP, Anderson HM, Proceedings on 13th Long Ashton International Symposium Chichester [England] Wiley Press; New York, NY pp. 83–102

Lal R 1989 Conservation tillage for sustainable agriculture, *Advances in Agronomy* **42** 85–197

Liu R, Zhang P, Wang X, Chen Y, and Zhenyao S 2013a Assessment of effects of bestmanagement practices on agricultural non-point source pollution in Xiangxi River watershed, *Agricultural Water Management* **117** 9–18

Liu Y, Gaoa M, Wua W, Tanveera SK, Wena X, and Liaoa Y 2013b The effects of conservation tillage practices on the soil water-holding capacity of anon-irrigated apple orchard in the Loess Plateau China, *Soil and Tillage Research* **130** 7–12

Maikhuri RK, Semwa RL, Rao KS, Singh K, and Saxena KG 2000 Growth and ecological impacts of traditional agroforestry tree species in Central Himalaya, India, *Agroforestry Systems* **48** 257-272

Montagnini F and Nair PKR 2004 Carbon sequestration: An underexploited environmental benefit of agroforestry systems, *Agroforestry Systems* **61** 281-295

Mowo J and Kiwia A 2009 Brief Report on The Conservation Agriculture With Trees Side Event: Proceedings Of The 2nd World Agroforestry Congress, 2009, Nairobi

Nair PKR, Kumar BM, and Nair VD 2009a Agroforestry as a strategy for carbon sequestration, *Journal of Plant Nutrition and Soil Science* **172** 10–23

Newaj R and Dhyani SK 2008 Agroforestry for carbon sequestration: scope and present status, *Indian Journal of Agroforestry* **10** 1–9

Pandey DN 2007 Multifunctional agroforestry systems in India, *Current Science* **92** 455-463

Richardson CW and King KW 1995 Erosion and nutrient losses from zero tillage on a clay soil, *Journal of Agricultural Engineering Research* **61** 81–86

Robertson GP and Grace PR 2004 Greenhouse gas fluxes in tropical and temperate agriculture: The need for a full-cost accounting of global warming potentials, *Environment, Development and Sustainability*, **6** 51-63

Rodriguez E, Fernandez-Anero FJ, Ruiz P, and Campos M 2006 Soil arthropod abundance under conventional and no tillage in a Mediterranean climate, *Soil and Tillage Research* **85** 229–233

Roshetco JM, Lasco RD, and Angeles MSD 2007 Small holder agroforestry systems for carbon storage, *Mitigation and Adaptation Strategies for Global Change* **12** 219-242

Sathaye JA and Ravindranath NH 1998 Climate change mitigation in the energy and forestry sectors of developing countries, *Annual Review of Energy and Environment* **23** 387-437

Sharma AR, Jat ML, Saharawat YS, Singh VP, and Singh R 2012 Conservation agriculture for improving productivityand resource-use efficiency: prospects and research needs in Indian context, *Indian Journal of Agronomy* **57**(IAC Special Issue): 131-140

Singh K, Chauhan HS, Rajput DK, and Singh DV 1989 Report of a 60 month study on fitter production, changes in soil chemical properties and productivity under Poplar (*P. deltoides*) and Eucalyptus (E. hybrid) interplanted with aromatic grasses, *Agroforestry Systems* **9** 37-45

Singh P, Sharratt B, and Schillinger WF 2012 Wind erosion and PM10 emissionaffected by tillage systems in the world's driest rainfed wheat region, *Soil and Tillage Research* **124** 219–225

Singh VS and Pandey DN 2011 Multifunctional agroforestry systems in India: Science based policy options. RSPCB Occasional Paper No 4, 2-35

Verhulst N, Govaerts B, Verachtert E, Castellanos-Navarrete A, Mezzalama M,Wall P, Deckers J, and Sayre KD 2010 Conservation Agriculture, Improving SoilQuality for Sustainable Production Systems? in: *Advances in Soil Science: Food Security and Soil Quality* (eds) Lal R, Stewart BA CRC Press, Boca Raton,FL, USA, pp. 137–208

World Bank 2015 Agricultural land (Percentage of land area) Available at http://data.worldbank.org/indicator/AG.LND.AGRI.ZS/countries? Display=graph (verified 16 September 2015)

CHAPTER 6

Fodder Tree based Agroforestry Systems in Hills of Uttarakhand

A Pattanayak*, JK Bisht, RP Yadav and BM Pandey

ICAR- Vivekananda Parvatiya Krishi Anusandhan Sansthan, Almora-263601, Uttarakhand, India

**Corresponding author's Email: director.vpkas@icar.gov.in*

ABSTRACT

Himalayan ecosystoms are complex, with frail environment, undulating terrain, and distinctive topography which are sensitive to regional and global changes. This region harbors a variety of agricultural practices out of which fodder tree based agroforestry systems are the paramount from sustainability and livelihood point of view. In hill region of Uttarakhand fodder tree based agroforestry systems evolved over a long period of time as farmers practicing these trees and crops combinations from generations. These systems fulfill diverse needs viz., food, fodder, fruit, timber needs of mountainous inhabitants. Agroforestry deals with the combination of tree species with crop plants, fisheries, animals, bee keeping etc. and it's based on the principle of optimum utilization of land. Agrihorti, Silvihorti, Hortipastoral and Silvipastoral systems are diversified land use options for agroforestry in the hill region. Improved system with five tree species i.e. *Grewia optiva, Quercus leucotrichophora, Bauhinia retusa, Melia azedarach, Morus alba* and four grasses i.e. *Setaria kazugulla, Setaria nandi*, Congo signal and Broad leaf paspalum were tested under silvipastoral system. *Quercus leucotrichophora* yielded (10675 kg/ha) significantly higher green biomass than others. Thus, for hilly terrain planting of interspatial woody perennials, with least negative influences on the agronomic crops, seems productive agroforestry system for settled farming. Therefore, agroforestry is a set of land use alternative, which if developed for resource poor farmers, can provide increased values and reduced risks and it should be made more popular in the rural areas.

Keywords: Agroforestry systems, Fodder trees, Himalaya, Livelihood, Sustainability

INTRODUCTION

The Himalayan region of India covers about 16.4 and 4% of the geographical area and population of the country, respectively. It is presently under heavy stress on account of large scale exploitation for fuel wood, timber and fodder mismanagement of forest resources. As compared to availability of land resources, the animal population in this region is quite high. As a result, the region is facing the acute shortage of fodder throughout the year, particularly during lean period (winter season). Due to continuous increase in human population, there is a great pressure for food and human settlement and other infrastructural development on arable land. This provides an opportunity to exploit such lands for production of forage and fuel to cater the ever increasing demand.

The Himalayan region has a long tradition of tree-based smallholder agroforestry practices, which are rich in the different species of trees. Several indigenous agroforestry systems based on people's needs and site-specific characteristics have been developed over the years by the smallholders for various uses (Yadav et al 2016a). Agro forestry is an age-old practice, which involves combining tree species with crop plants, fisheries, animals grazing etc. Agroforestry systems (AFS), in the different combinations of agrisilviculture, agrisilvipasture and silvipasture, are agro-ecological options which include in their concepts reference to the main components of sustainability, that is, the economic, the social, and the environmental components. The structural and functional aspects of tree species in traditional agroforestry systems greatly affect the overall productivity of the system. Generally, the overall productivity (crops + trees) in agroforestry systems is higher than that in sole cropping systems (Chaturvedi, 1981). Nair (1984) defined agroforestry as "a land use system that involves deliberate retention, introduction or mixture of trees or other woody perennials in crops and animal production to benefit from the resultant ecological and economical intersections. Each definition may have some limitation but the basic ideas coming from all the definition has common points,

- It is a land use system, which includes synergistic combinations of agriculture, forestry, horticulture, animal husbandry *etc.*
- Integration of trees with crops and/or animals, to reduce risk and to increase the total productivity.
- It should be stable and sustainable and
- Integration of trees with agriculture system will use sunlight, moisture and plant nutrients more efficiently than by mono-cropping of either agriculture or forestry crops.

The scientific management of land and water resources in hilly areas with the application of proper vegetative measurement has become important for getting long-term conservation and production needs. With the over exploitation and improper management of natural resources in hills, forest are gradually disappearing and grasslands are converted into wastelands causing acute fodder and fuel scarcity and danger of soil and water erosion. The over all scenario of the entire hill includes:

AREA AND CLIMATE

The Indian Himalayas cover an area of 53.7 m ha, which is 16.41% of total geographical area of the country. Out of 21 agro-ecological regions of the country, four regions are covered exclusively and one partly in the hill and mountain agro-ecosystem. These five agro-ecological zones have a wide variation in their climate, i.e. from cold arid to warm perhumid. Average annual rainfall in the region varies from 150 to 4000 mm. The mean annual temperature varies from 8 to 22°C. The growing period of different crop ranges from 90 to > 270 days in a year. Skeletal and calcareous to brown forest podzolic are the major soil groups of the region. These are alkaline to acidic in nature. The natural vegetation is tropical deciduous alpine, temperate and wet evergreen forest.

POPULATION AND LAND USE PATTERN

The population of the Himalayan region is around 39 million, which is about 4% of the total population of the country. The primary source of income here is agriculture, which contributes about 45% of the total regional income. This region is thinly populated (population density 627/1000 ha). However, the actual pressure on agricultural land in the hills is much higher since the net cultivated area is only 12% of the reporting area. The per capita availability of cultivated land in hill region is just 0.17 ha. Forest is the major land use in the Himalayas and it covers about 59% of total reporting area of hills. About 12% area of the region falls under permanent pastures and grazing lands.

AGRICULTURAL SCENARIO IN HIMALAYAN REGION

The land utilization pattern has been presented in Fig. 1. Forests cover the largest land area, being more than double to that of national percentage. A low level of developmental activities is pointed by a relatively less area put to non-agricultural use. Around 7-8 % of barren area, in North western Himalaya and north eastern Himalaya, indicates the need for developing alternative land use. NWH has a sizeable area under permanent pastures and other grazing lands, which is a great support to livestock. Culturable waste, almost equal in both regions, is probably due to migration from rural areas and calls for creating more opportunities for employment in rural areas to check migration and culturable waste. Total fallow in both the regions is less than the All India level. The net area sown (NWH-19 and NEH-14 %) in the region is far below the national average, which is due to high forest cover. The cropping intensity in NWH states is quite high compared to that of NEH as well as All India, which can be further enhanced by introducing crops and/or crop varieties suitable for the areas where keeping land fallow or mono-cropping is practiced.

The crop production systems in NWH region are based on agriculture (field crops), olericulture, horticulture or agri-horticulture, and agroforestry/ agri-horti-silvi-pastoral system whereas, in NE, sericulture is embedded with the agriculture and horticulture-cum-plantation crops based production system such as agri-seri-horticulture or agri-horti-sericulture system. Livestock is inseparable part of the system in the region. However, fishery and floriculture are also the parts of the production system in certain areas. Wheat, rice, maize, finger millet and barnyard millet are the major cereals and millets of the region and black gram, horse gram, ricebean, *rajmash* (a pulse version of French bean, which is a highly remunerative crop) and bhat (a variant of soybean, which is used as a pulse and has better quality fats and proteins to make it tastier and more digestible than the common soybean, which is primarily used as an oilseed) are the major pulse crops. Mustard and soybean constitute the major oilseed crops. Among vegetables, cole crops, cucurbits, capsicum, tomato, radish, pea, french bean, potato and onions are the major crops.

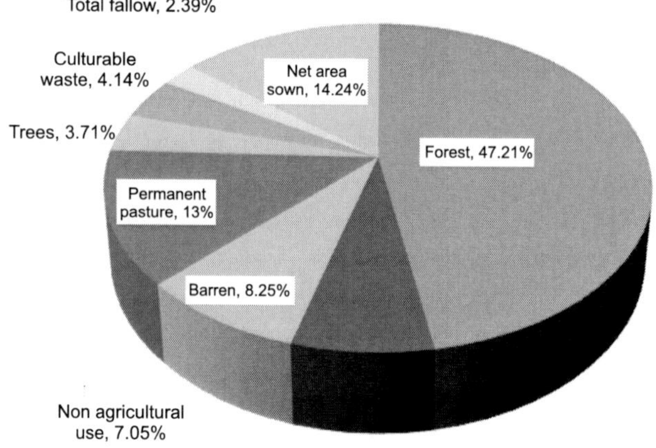

North Eastern Himalaya

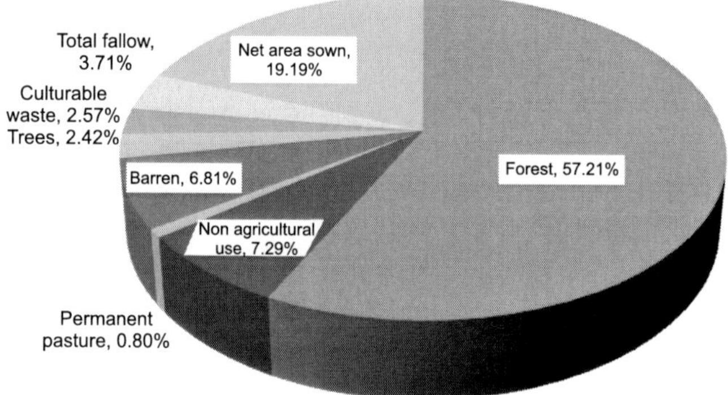

Figure 1: *Land utilization (% of total reported area) pattern in Indian Himalayas*

The region is riddled with a number of problems. From agriculture point of view, these can be categorized as regional and agriculture-related problems. Among regional problems, the major ones are - hill physiography makes places difficult to access and sparsely inhabitable, seismic sensitivity, water tower to plains yet a very little of water it generates is utilized locally, large magnitude of soil loss, sloping terrain and thin soil cover leads to slow recovery of the ecosystem from natural and human disturbances, high temporal variations in climate and uncertain weather, relatively low temperature throughout and high rate of migration of youths to greener pastures.

The agriculture related problems are - small and fragmented land holdings; low risk bearing capacity due to poor economic condition; largely rainfed agriculture, for example, the net irrigated area in hills of Uttarakhand is only 40,822 ha compared to 2,96,874 ha area in plains, *i.e.*, hills enjoy only 12.1% of the total net irrigated area in the state; modest soil fertility; age-old farming practices; low input use, *viz*., the fertilizer consumption in hill districts ranges from 65 to 666 thousand tonnes against 31,851 and 73,768 thousand tonnes in Haridwar and Udham Singh Nagar, respectively, which are plain districts (Mittal et al 2008); negligible farm mechanization; a certain degree of reluctance to agriculture as the preferred occupation due to low returns; unawareness of improved technologies; extremely insignificant extension infrastructure and inadequate extension efforts; huge gap in research farm and farmer's field yields; women dominant agriculture with scrimpy male contribution; inefficient and insufficient marketing base; wide weather and climatic variation presents a multitude of problems to agricultural researchers and planners to cope with; livestock are one of the main components of hill farming, but there is a great paucity of green and dry fodder; the majority of livestock are of local breed and there is an acute shortage of livestock improvement facilities and poultry hatcheries; and sizeable post-harvest loss to fruits and vegetables on account of lack of storage facility, poor transport facilities and processing units.

The climate change has put forth an entirely different scenario of biotic and abiotic stresses, *e.g.*, (i) heavy incidence of brown plant hopper (*Nilaparvata lugens*) was noticed for the first time in some rice-growing areas of Uttarakhand during 2010 (ii) high severity of yellow rust was observed during late-February to mid-March in wheat in 2011, and (iii) a very low winter rainfall had resulted in a dismal wheat crop in many rainfed areas during past four years (iv) adverse effect of increasing maximum temperature and sunshine hours of February months on wheat yield (v) reduction in crop productivity due to agricultural pest and plant pathogens in cold arid trans-Himalayan Ladakh region (Vaish et al 2011) (vi). The damages caused by wild animals, like monkeys and boars, has assumed menacing proportions in recent years.

The challenges to hill agriculture in the future are going to be enormous with fast depleting resources, unpredictable weather and changing climate, migration of youths due to unprofitable agriculture and feeding (Table 1) the ever increasing population.

Table 1: *Productivity requirement for 2050 vis-à-vis 2001-12 productivity of major hill crops*

Crops	Region	Requirement ('000 t)	Present area (000ha)	Required productivity (kg ha^{-1}) *	Present productivity (kg ha^{-1})	Productivity surplus/ deficit (%)
Cereal & millets	NEH	3426.9	1216.4	2817	1986.29	-29.49
	NWH	6178.9	2568.1	2406	1894.67	-21.25
Pulses	NEH	395.8	92.9	4260	1003.28	-76.45
	NWH	713.7	113.3	6299	784.33	-87.55
Oilseeds	NEH	791.6	158.8	13107	881.57	-93.27
	NWH	1427.4	109.5	13036	829	-93.64
Vegetables	NEH	2364.1	196.2	15798	9500	-39.87
	NWH	4262.6	238.1	17903	17400	-2.81
Projected population in 2050('000)- North Eastern Hill (NEH)-21689; North Western Hill (NWH)-39106.695						
# - Requirement (Cereal & millets @ 158 kg capita^{-1} annum^{-1}; Pulses @ 18.25 kg capita^{-1} annum^{-1}; Oilseeds @ 36.5 kg capita^{-1} annum^{-1}; Vegetables @ 109 kg capita^{-1} annum^{-1})						
* - On the basis of 2011-12 area			Source - (*Fertilizer Statistics 2012-13*)			

FORESTS

The non arable land constitutes about ~84%, and the forest occupies ~60% of the total geographical area (Alam 2016). The agrarian economy today depends on forest for energy supply, fodder, non-timber product, and livestock rearing. These all forests are under rapid pace of degradation. In hill region, farming has been intricately integrated with forestry. However, to increase the production of fuel wood, fodder, and other minor products and environmental protection, greater thrust is needed on agroforestry (Yadav and Bisht 2013).

SYSTEM OF AGROFORESTRY

Different components of agroforestry systems are identified as: a) Primary components: The components occupying the larger area of the total unit area and serving the major function i.e. production of primary output needed by the farmers was termed as primary component. b) Secondary components: The component occupying relatively lesser area of the total unit area compared to area under primary component and yielding secondary needs of the farmers was termed as secondary component (Yadav et al 2016b). Using above mentioned frame-work for identifying agroforestry systems, it can be classified (Fig. 2) as mentioned below:

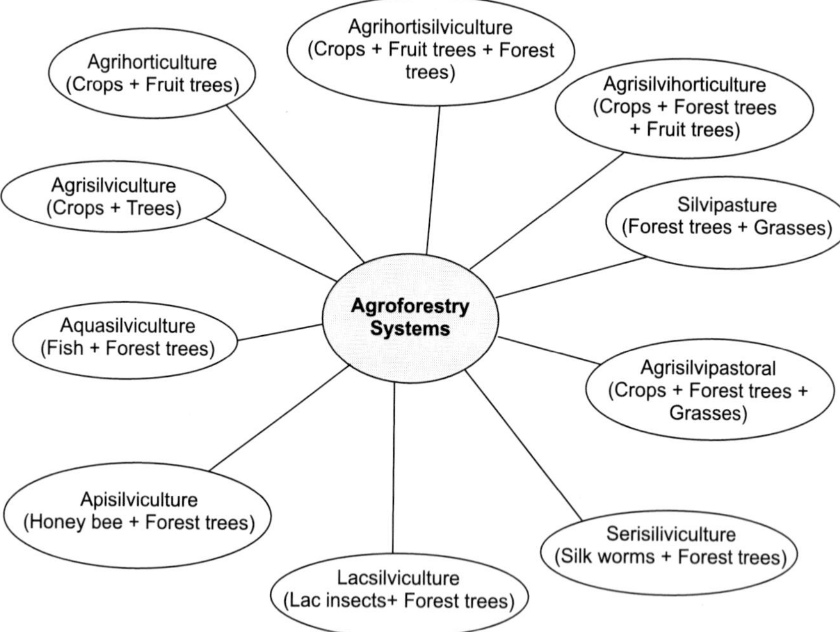

Figure 2: *Classification of agroforestry system based on structure and function (Source: Yadav et al 2016b)*

ADVANTAGES OF AGROFORESTRY

Appropriate agroforestry system have the potential to control erosion, maintain soil organic matter and physical properties, augment nitrogen fixation and promote efficient nutrient cycling (Young, 1989) besides livelihood security (Yadav et al 2016c). The main benefits of appropriate agroforestry system are as follows.

1. It fulfills the food, fuel and fodder needs of the man and animals
2. It conserves the soil
3. It can conserve the water
4. Soil improvement can be done
5. Checks the land slides
6. Maintains the soil nutrient status
7. Maintains the soil organic matter
8. Maintains the physical properties of the soil
9. Selecting appropriate model can reduce competition at the root zone between trees and crops in the agroforestry system
10. The proper management of tree can maintain relative humidity of a particular site
11. Trees can maintain the soil temperature even during fallow period
12. Agroforestry system can reduce the incidence of major pests and diseases due to diversity in crop plants
13. It generates the steady income for the farmers
14. It increases the socio-economic status of the farmers, as it is a labour-intensive programme. Thus, generate the new employment
15. Agroforestry is closely linked with environmental amelioration and improved quality of life through socio-economic up-liftment

AGROFORESTRY ECOSYSTEM SERVICES

AGROFORESTRY FOR THE CONSERVATION

In the absence of suitable vegetative cover or interventions, natural resources are depleting. This loss may be in term of soil loss through sedimentation and water loss. This way a huge amount of valuable top fertile soil is being lost. It has been estimated that on an average soil is displaced at the rate of about 16 t ha^{-1} yr^{-1} and washed in to the sea, which is much higher than the permissible limit of 4.5 tones /ha/year. This loss can be reduced by using Agroforestry for the land cover in barren lands of a watershed. Total annual accretion of litter fall in pecan nut based agri-horti system found 2.14 t $ha^{-1}yr^{-1}$ and relative abundance of nutrients in litter fall of pecan nut tree were in the order of C (901.9 kg ha^{-1} yr^{-1}) > N (57.44 kg ha^{-1} yr^{-1}) > K (43.29 kg ha^{-1} yr^{-1}) > P (3.21 kg ha^{-1} yr^{-1}) which are supposed to help in nutrient buildups of the soil (Yadav and Bisht 2014a).

AGROFORESTRY FOR WATERSHED MANAGEMENT

With the adoption of agro forestry we can utilize the entire land of a watershed in a proper way. As per need of the watershed we have to decide how to manage the land i.e. cropping management plan, pasture developments, forest plantation etc.

BIOMASS PRODUCTION AND CARBON STORAGE THROUGH AGROFORESTRY

The maximum production of biomass per unit area is a main objective of agroforestry. For the production of biomass we can use the bunds, canal roads, farm ponds, lakes, water logged area and ravines of a watershed in an economic way. Aboveground biomass production and carbon storage (Fig. 3) in some of the important agroforestry systems viz., fruit tree based agrihorticulture, oak high density plantation (1 m × 1 m) and pecan nut based agrihorticulture system was studied by Yadav et al (2015b, 2016d, 2016e). Due to population explosion and increase number of industries pressure on forest is increasing day by day. Because of this population pressure it is not possible to grow forest on agricultural lands. Through agroforestry we can grow trees on marginal lands with crops. This will also restore the ecological balances.

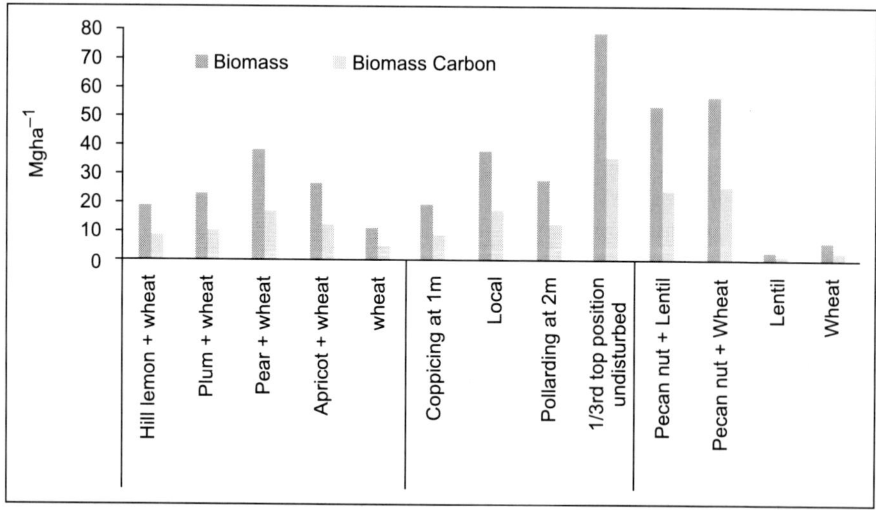

Figure 3: *Aboveground biomass and biomass carbon in different systems of Central Himalaya (Source: Yadav et al 2015b; 2016d; 2016e)*

AGROFORESTRY FOR LIVELIHOOD SECURITY

With the help of agroforestry in watershed we can promote the agro based cottage industry. This will include herbal drugs, paper pulp, fiber, poultry, piggery, aqua culture, dairy, beekeeping, sericulture and mushroom production. This will help in the income generation of the marginal farmers.

Table 2: *Pearson's correlation coefficients between various parameters of Central Himalayan watershed*

Correlation Parameter	1	2	3	4	5	6	7	8	9	10
Fodder consumption	1									
Annual Income	0.703**	1								
Livestock size	0.865**	0.488**	1							
Fuel consumption	0.174	0.273	0.179	1						
Fodder from forest	0.551**	0.313	0.785**	0.307	1					
Family size	-0.132	0.007	-0.006	-0.020	0.033	1				
Sex ratio	0.181	0.032	0.151	0.240	0.192	-0.201	1			
Literacy rate	0.176	0.340	0.056	0.361	0.024	0.016	0.354	1		
Land holding size	0.530**	0.378*	0.447*	0.235	0.403*	-0.467**	0.495**	0.432*	1	
Agroforestry awareness	0.168	0.325	0.058	0.375*	0.026	0.000	0.352	0.998**	0.434*	1

Note: **Correlation significant at the 0.01 level - *Correlation significant at the 0.05 level
Source: Yadav et al 2016c

In Indian Himalaya livelihood security is reliant on traditional farming practices that include agroforestry beside forest produce; forests provide 73-79 % of required energy from fuelwood and more than 81 % fodder (Yadav et al 2016c). The income of people was positively correlated with livestock rearing and farming. The adoption of this type of integrated farming which includes agroforestry with livestock (dairy and poultry/goat rearing) will improve the income of the people and sustainability of farming. Awareness of agroforestry was significantly related to literacy rate, land holding size and fuelwood consumption (Table 2). A significantly positive relationship was obtained among income, fodder consumption, fodder from forest, land holding size and livestock size (Yadav et al 2016c).

AGROFORESTRY FOR CLIMATE CHANGE MITIGATION

The mitigation measures address the causes of the problem which may increase the GHGs concentrations. For example, mitigation could be the measures to reduce energy consumption and promote clean technologies. Mitigation is a global issue, when a mitigation agroforestry project reduces emissions of GHGs, it benefits the whole world. Many agroforestry activities contribute to climate change mitigation (Fig. 4). C stocks can be increased by creating agroforestry through plantations (Yadav and Bisht, 2014b). The benefit of creating agroforestry plantations is the difference between the present growing stock and the baseline, as shown in the graph. Existing growing stocks can be conserved through reduced felling or no felling at all of trees from agroforestry farm lands. In this situation, the benefit of conserving is estimated with reference to the degradation or tree felling scenario. Emissions caused by agroforestry activities can be reduced, for example, by using less energy or fertilizers in various agroforestry operations. Biomaterials and bio-energy can be produced from agroforestry products to substitute materials or energy that generates GHGs. The first two activities refer to C sequestration in the agroforestry or any other ecosystem, while the last two refer to energy-related emissions (Yadav et al 2016b).

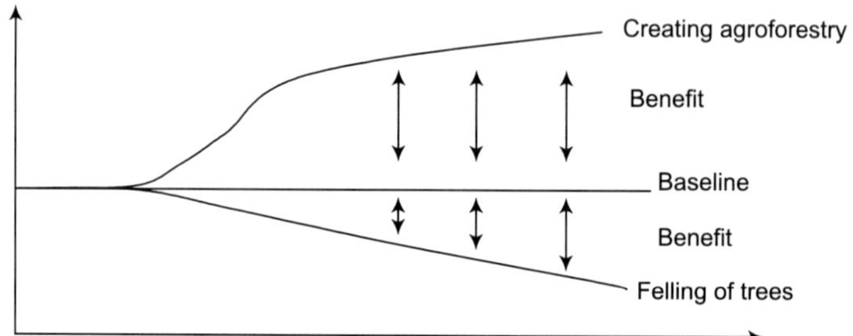

Figure 4: *Carbon sequestration through creating agroforestry plantation and reduced felling of trees from agroforestry farmlands (Yadav et al 2016b)*

PREVALENT AGROFORESTRY SYSTEM IN HIMALAYAN REGION

Several agroforestry systems are common in the hill region (Bisht and Srivastva 2006). Growing trees on the field bunds and around the fields is an age-old practice. Agrihorticulture, silvihorticulture, hortipastoral and silvipastoral systems are diversified land use options for agroforestry in the hill region. These agroforestry systems play vital role in the livelihood of the hill people (Yadav and Bisht 2013; Yadav et al 2016c). The study on some important agroforestry systems was conducted at experimental farm Hawalbagh (29° 36′ N and 79° 40′ E, 1250 m amsl) of Vivekananda Parvatiya Krishi Anusandhan Sansthan, Almora, India. The most common agroforestry systems in hills are as follows.

CROP BASED AGROFORESTRY SYSTEM

Some of the popular fodder trees like Bhimal (*Grewia optiva*), Kharik (*Celtis australis*), Quairal (*Bauhinia variegata*), Bakain (*Melia azedarach*), Timul (*Ficus roxburghii*), Beru (*Ficus palmata*) etc. are found on the sides of the agricultural fields and sloping land. The community forests are dominated by chir (*Pinus roxburghii*). At hill tops, oak (*Quercus leucotrichophora*) forests are pre-dominant. The common crop based agroforestry systems are, agri-silvi, agri-silvi-horti and silvi-horti-system. The major tree species grown in association with hilly crops are: *Albizia lebbeck*, *Grevillea robusta, Melia azedarach, Ailanthus excelsa, Moringa pterygosperma, Morus alba, Bauhinia purpurea, Grewia optiva, Populus xeriramericana, Eucalyptus globulus* etc. Agricultural crops like wheat, pea, potato, cauliflower, mustard and lentil etc. during winter and maize, tomato, ricebean, soybean etc. during rainy season are being grown.

In an agri-horti system study at ICAR-VPKAS, Almora four cropping sequences *viz.*, legume – legume (soybean – pea), legume – cereal (soybean – wheat), cereal – legume (maize – pea) and cereal – cereal (maize – wheat) were grown with and without pecannut trees. Study revealed that maize and soybean during *kharif* and wheat and pea during *rabi* can be grown successfully with pecannut tree (Bisht et al 1998) without significant reduction in the yield of the crop (Table 3). This also resulted in enhanced organic matter content of the soil.

Table 3: *Yield of field crops with and without pecannut tree (average of 5 years)*

Crop sequence	Grain yield (t ha^{-1})	
With Pecannut Tree	**Rabi**	**Kharif**
1. Maize - pea	1.56	3.19
2. Soybean - wheat	2.96	1.50
3. Maize - wheat	3.00	3.47
4. Soybean - pea	1.82	1.62
Without Pecannut Tree		
1. Maize - pea	1.75	3.75
2. Soybean - wheat	3.32	1.56
3. Maize - wheat	3.10	3.49
4. Soybean - pea	1.59	1.48

FORESTRY BASED AGROFORESTRY SYSTEM

The land use patterns of hills indicate that the maximum area (63%) is under the forest. In the forest, growing grasses under tree is a common practice. Studies conducted for about a decade at VPKAS, Almora have evolved several appropriate technologies for the development of grassland and other forage production systems for hilly region. These technologies include introduction of improved grasses and legumes through suitable methods. Pangola (*Digetaria decumbense*), Giant star (*Cynodon plectortachus*), *Panicum coloratum*, *Panicum repense* and *Setaria kazungula* were found promising for the region. Under specific situation (sloping sides, degraded land and field terraces), napier bajra hybrid (*Pennisetum glaucum*) gave the best results. Inclusion of legumes, Seratro and Kikuii (1:1) increased the forage yield by 64.6, 63.9 and 74.6%, respectively.

Closure of grassland for a particular period, fertilization, burning of old stubbles and adoption of suitable cutting management practice has also shown great potential of increasing fodder production (Bisht et al 1999). In hilly areas, the degraded pastureland and very poor quality grasslands are often seen under pine and deodar trees. Planting different grass species in different sectors of circle revealed that pangola grass produced the highest yield (Table4) both under pine and deodar trees. The other grass which performed equally well under trees at Hawalbagh was *Thysonalaena maxima.*

Table 4: *Performance of different improved and local grasses under pine and deodar*

Grass species	Green forage yield (t ha^{-1})	
	Pine	Deodar
1. Pangola	14.01	13.66
2. Rhodes	4.78	3.06
3. Para grass	0.48	2.04
4. Guine	0.41	2.55
5. Local	0.25	0.22

SILVI-PASTORAL SYSTEM

Silvi-pastoral system of forage production is of utmost importance for the degraded land sites. Growing of trees + pasture, increased the system productivity, enhanced fodder availability and checked the soil erosion. A silvi-pastoral system of 5.0 m x 2.0 m tree spacing gave green forage yield ranging from 1800-2450 g ml^{-1} yr^{-1} compared to 300g m^{-2} yr^{-1} under the existing system (*Hetropogon contortus* + *Chrysopogon fulvus*). The improved systems were consisting of Pangola + Quairal, Pangola + Kharik, Pangola + Bhimal and *Cenchrus ciliaris* + Kharik. Another improved system with five tree species i.e. bhimal, oak, katoch, bakain and shahtoot and four grasses i.e. Guinea (*Panicum maximum*), Rhodes (*Chloris gayana*), *Setaria (Setaria kazungula) and Panicum (Panicum coloratum)* is under study. In thiZs system, *Setaria Kazungulla* gave the highest green (21 t ha^{-1}) forage yield (Bisht et al 1999). Srivastva et al (2003) also reported the promising silvi- pastoral combination in Shivaliks this included *Grewia optiva*, *Bauhinia retusa, Morus alba, Quercus leucotrichophora* and *Melia azedarac*h as perennial trees and *Setaria kazungula, Panicum maximum, Panicum coloratum* and *Chloris guyana* as grasses.

ENERGY PLANTATION SYSTEM

Energy plantation has been found a viable agroforestry system for the hills. For such system, Quairal (*Bauhinia variegeta*), Bhimal (*Grewia optiva*), Bakain (*Melia azedarach*), oak (*Quercus leucotrichophora*), kharik (*Celtis australis*), *Beru (Ficus palmata)* and Paiya (*Prunus padam*) have been found suitable for energy plantation. The green matter yields, however varied with the varying tree species recorded on an average, 5 to 25 t ha^{-1} in 3rd year and 40 to 65 t ha^{-1}in the subsequent years. In case of oak energy plantation, combined application of chemical fertilizers (30 kg N+40 kg $P_2 0_5$ ha^{-1}) with manure (FYM @ t ha^{-1}) produced 14.0 % more leaf yield than chemical fertilizer (30 kgN+40 kg $P_2 0_5$) alone (12.4 t ha^{-1}) and 11 % higher than manure alone (12.7 t ha^{-1}).

BORDER ROW PLANTATION

It is a most common practice of planting fuel, fodder, timber trees and grasses on field boundaries. Bisht et al 1998 reported 6 to 8 kg green fodder and 2 to 3 kg fuel tree^{-1} in seven year old Bhimal (*Grewa optiva*) tree planted along the side of cultivated terraced at 2 m apart at north-west Himalayas.

HORTICULTURE BASED AGROFORESTRY SYSTEM

Fruit trees and field crops have been grown in various ways in the hills. As per pattern and configuration, their companion crops are known as intercrop, under planting, hedgerow planting or alley cropping. In intercropping, agricultural crops are generally grown between rows of fruit trees. The common agricultural crops planted under the orchards are given below (Singh et al 1990).

- **Species and condiments:** Turmeric, ginger, green chilies and coriander.
- **Tuber crops:** Coloacasia, sweet potato and potato
- **Cereals:** Oat, wheat and maize
- **Pulses:** Cowpea, green gram and soybean
- **Oil crops:** Ground nut
- **Grasses:** Various types of palatable grasses and legumes

Home garden/homestead gardens is an age old horticultural based agro forestry system. The commonly used horticulture based agro-forestry systems are agri-silvi-horticulture and silvi-horticulture. In this, farmers generally integrate cash crops like, ginger, turmeric, potato, tomato and chilies with fruit trees. In a study (Arora and Mohan 1986), long duration companion crops such as turmeric pineapple and an aromatic oil yielding lemon grass (*Cympopogon citratus*) were tried in a year old orchard of peach during 1982 to 86. On the basis of five years average yield, turmeric (2.55 t ha^{-1}) proved to be the most ideal companion crop under rainfed condition in peach orchard. Findings of Bisht *et al.* (2000 & 2009) on ginger and turmeric grown under ten year old fodder trees (bhimal, oak, quairal, kharik) have shown that yield of turmeric and ginger was affected significantly under different fodder trees. Turmeric and ginger produced significantly higher yield (Table 5) under oak.

Yield of turmeric and taro was affected significantly under different fodder trees. Turmeric gave significantly higher rhizome yield (Fig. 5) under *Quercus leucotrichophora*. Cultivation of turmeric in silvi-horti system was more profitable than taro and on an average gave about ₹ 41329 ha^{-1} more income than taro. Overall association of *Quercus leucotrichophora* with turmeric was found to be most appropriate as provided the maximum net return and B: C ratio (Bisht et al 2004).

Table 5: *Rhizomes yield of ginger and turmeric with different fodder trees*

Treatment	Yield (t ha^{-1})		
	1994-95	1995-96	Average
Turmeric			
1. Quairal	6.03	4.66	5.35
2. Kharik	12.77	8.57	10.67
3. Oak	15.00	9.08	12.04
4. Bhimal	7.70	5.70	6.70
5. Open	10.75	7.16	8.96

Ginger			
1. Quairal	5.78	3.15	4.47
2. Kharik	10.44	4.95	7.70
3. Oak	10.74	5.23	7.99
4. Bhimal	6.67	3.38	5.03
5. Open	9.12	3.91	6.52

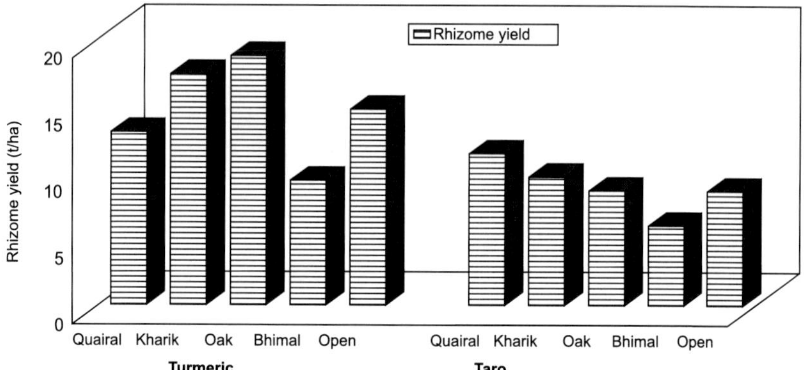

Figure 5: *Rhizome yield of turmeric and taro under different fodder trees and open fields*

HORTIPASTORAL SYSTEM

In hortipastoral production system four *rabi* viz., perennial rye, tall fescue, hima 14 and grassland manava grasses with and without peach tree were grown for utilization of marginal land (Fig. 6).

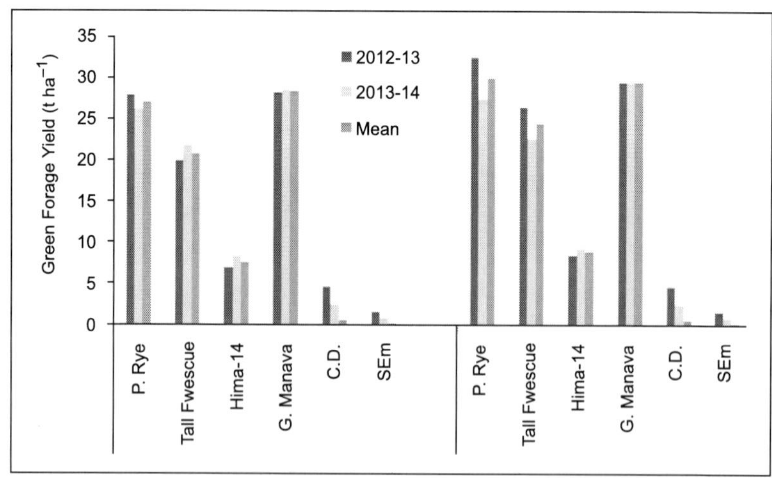

Fig. 6. *Green forage yield (q ha^{-1}) of winter grasses in peach based hortipastoral production system*

Source: Yadav et al 2015a

Grassland manava produced significantly higher mean green forage yield (28.41 t ha^{-1}) followed by perennial rye under peach tree, while in open perennial rye gave higher mean green forage yield 29.96 t ha^{-1} which was at par with the grassland manava 29.47 t ha^{-1}(Yadav et al 2015a).

AGROFORESTRY FOR ENTIRE FARMING SYSTEM

Borthakur et al (1981) have developed an agroforestry system as an alternate to jhuming or shifting cultivation in N-E hills region. In this model, lower 1/3rd area of hilly slop is terraced and utilized for raising field crops keeping farmers requirements in view. Rice, maize, ragi, mustard, sesame, black gram, ginger and turmeric may be grown on terraces. Fodder grasses and legumes suitable for terrace risers are *Stylosanthes hamata, Stylosanthes guianensis, Clitoria, ternatea* and thin Napier. Middle 1/3rd hill slop may be utilized to horti pastoral system leaving top 1/3 steep slopes for forests including fuel fodder plantation. *Citrus limon* could be inter cropped with cowpea, garden pea, french bean and sweet potato in agri horticulture system. This system may be coupled with runoff harvesting and recycling and other biological component. In this system scope of aquaculture, dairying, beekeeping and flower industries can be used. Incorporation of these systems will increase the income of the farmers.

NUTRITIONAL VALUE OF FODDER LEAVES

As per above mentioned discussion it is clear that fodder trees are paramount for livestock rearing particularly during winter period in hills. The nutritional analysis of leaves of fodder trees (Table 6) revealed that they are an important source of protein and other dietary supplements such as crude fiber and fat for livestock and can be used as substrates deficit in either of these nutrients for livestock feeding (Tilak et al 2016).

Table 6: *Nutritional value of fourteen locally preferred fodder tree species in mid Hills of Himalaya*

Fodder Trees	Moisture	Dry matter	Crude Protein	Crude Ash	Crude Fibre	Crude Fat	Nitrogen free Extract
1. *Bauhinia variegata* L.	22.12±0.86	77.88±0.86	17.60±0.34	8.31±0.08	22.14±0.20	1.38±0.03	50.6±0.4
2. *Celtis australis* L.	41.63±0.53	58.37±0.53	22.89±0.31	17.68±0.36	20.85±0.14	2.70±0.04	35.9±0.1
3. *Ehretia Laevis* Roxb.	50.82±1.03	49.18±1.03	14.82±0.44	15.51±0.21	25.25±0.25	1.87±0.02	42.5±0.3
4. *Morus alba* L.	55.40±0.69	44.60±0.69	24.69±0.60	18.75±0.38	16.77±0.22	1.25±0.03	38.6±0.8
5. *Robinia pseudoacasia* L.	47.72±0.66	52.28±0.66	26.71±0.27	5.61±0.18	31.17±0.57	1.06±0.01	35.4±0.7
6. *Quercus glauca* Thunb.	39.92±0.75	60.08±0.75	13.70±0.23	6.88±0.30	25.65±0.35	1.20±0.01	52.6±0.8
7. *Prunus cerasoides* D. Don	35.19±0.64	64.81±0.64	13.94±0.30	9.82±0.39	16.69±0.50	1.39±0.03	58.2±1.0
8. *Quercus leucotrichophora* A. Camus	27.55±0.35	72.45±0.35	10.56±0.40	5.73±0.32	31.67±0.59	3.08±0.05	49.0±0.6
9. *Alnus nepalensis* Don	45.20±0.87	54.80±0.87	15.81±0.58	7.88±0.15	25.96±0.25	1.90±0.01	48.5±0.6

10. *Debregeasia hypoleuca* (Hochst.) Wedd.	62.03±0.71	37.97±0.71	19.17±0.47	16.51±0.28	18.76±0.26	1.87±0.06	43.8±0.4
11. *Bauhinia retusa* L.	41.14±0.71	58.86±0.71	17.58±0.29	14.15±0.14	18.76±0.43	1.08±0.01	48.5±0.2
12. *Grewia optiva* J. R. Drumtnond ex Burret	46.86±0.76	53.14±0.76	10.07±0.65	10.91±0.23	33.04±0.75	0.90±005	41.6±0.1
13. *Melia azedarach* L.	60.20±1.05	39.80±1.05	20.30±0.39	19.06±0.28	12.72±0.40	1.94±0.02	49.3±0.5
14. Quercus serrata Murray	37.12±0.84	62.88±0.84	19.09±0.36	8.26±0.24	28.59±0.35	2.82±0.19	48.1±0.1

Source: Tilak et al 2016

The mixed farming and alley cropping system appears to be viable alternatives, for farmers practicing settled cultivation in hills base and plain areas, while for hilly terrain planting of interspatial woody perennials on terrace risers, with least negative influences on the agronomic crops, seems productive agroforestry system for settled farming.

CONCLUSION

Agroforestry is a set of land use alternative, which if developed for resource poor farmers, can provide increased values and reduced risks. This will help the poor farmer in overcoming the fuel and fodder shortage with increased productivity of the land. This will generate income to farmer in a drought year when crop is failed. This will have a positive effect on the environment as people will get their fuel and fodder from their fields itself thus, the pressure on forest will be reduced. Therefore, agroforestry should be made more popular in the rural areas through extension programmes and a top priority should be given in the watershed development.

REFERENCES

Alam A 2016 Long-term strategies and programmes for mechanization of agriculture in agro-climatic zone–I: Western Himalayan region,78–86 http://farmech.gov.in/06035-04-ACZ1-15052006.pdf

Bisht JK, Chandra S and Singh RD 2004 Performance of taro (*Colocasia esculanta*) and turmeric (*Curcuma longa*) under fodder trees in silvi-horti system of western Himalayas. *Indian Journal of Agricultural Sciences* **74**(6)291-294

Bisht JK, Srivastva AK, Gupta HS 2009 Sustainable fodder production management in NW Himalaya. *Technical Bulletin,* 32 (2/2009) VPKAS, Almora 1-62.

Bisht JK and Srivastva A K 2006 Alternate land uses and agro-forestry systems in hills. Sustainable production from agricultural watersheds in NWh Ed By H S Gupta, A K Srivastva and J C Bhatt VPKAS 441-456

Bisht JK , Subhash ChandrMani VP and Singh RD 1999 Fodder production and management strategies for hills, *Technical Bulletin,* VPKAS, Almora **13** (2/99) 36

Bisht JK, Subhash Chandra, Bhatnagar VK, Koranne KD, Evam Dube SD 1998 Pecannut Aur Khadya Phasal Sath-Sath, *Kheti* **51** (12) 17-19

Bisht JK, Subhash Chandra Chauhan, VS and Singh, RD 2000 Performance of ginger (Zingeber officinale) and turmeric (Curcuma longa) with fodder tree based silivi-horti system in hills, *Indian Journal of Agricultural Science* **70**(6) 578-82

Borthakur DN, Prasad SN , Gosh SP, Singh A, Awasthi RP, Rai RN, Verma A, Datta HH, Sachan JN and Singh MD 1981 Agro forestry based faming system as on alternation to Jhumming. Proceedings of the Agroforestry Seminar held at Imphal, India, 1979. New Delhi: ICAR.

Chaturvedi AN 1981 Poplar for planting. In: Uttar Pradesh Department Bull. India no-50 Lucknow 27

Nair PKR 1984 Soil productivity aspect of Agroforestry, ICRAF, Nairobi, Kenya

Srivastva AK, Misri B, Singh B 2003 Resource management in silvipastures in Shivaliks, *Indian Forester* **129** 446-455

Mondal T, Yadav RP, Bisht JK, Meena VS and Kumar S 2016 Nutritional analysis of locally preferred fodder trees leaf in mid hills of Himalaya. *In:* Natural Resource Management: Ecological Perspectives, Indian Ecological Society: International Conference-2016, SKUAST-Jammu 18-20 February 2016 (R. Peshin A K , Dhawan F, Bano K S Resam eds. **1** 436. http://iesconf.com/wp-content/uploads/2016/02/Compendium-Vol-1.pdf

Yadav RP, Bisht JK, Mondal T and Pattanayak A 2015 Forage Production in Peach Based Hortipastoral System in Indian Himalaya. International Grassland Congress-2015 held at New Delhi from 20th-24th November, 2015

Yadav RP and Bisht JK 2013 Agroforestry: a way to conserve MPTs in North Western Himalaya, *Research Journal of Agriculture and Forestry Sciences* **1** 8–13

Yadav RP and Bisht JK 2014 Agroforestry to combat climate change, *Kheti* **66** 11–14 (In Hindi)

Yadav RP and Bisht JK 2014 Litter fall and potential nutrient returns from pecan nut (Carya illinoinensis) in agroforestry system in Indian Himalaya, *International Journal of Herbal Medicine* **2** 51–52

Yadav RP, Bisht JK and Bhatt JC 2016 Biomass and allocation in different production system in the mid hills of central Himalaya, *Tropical Ecology* **57** (in press)

Yadav RP, Bisht JK and Pandey BM 2015 Aboveground biomass and C stock of fruit tree based land use systems in Indian Himalaya. Ecoscan **9** 779–783

Yadav RP, Bisht JK, Gupta B and Mondal T 2016 Toward the sequestration potential of agroforestry practices to combat climate change in Kumaon Himalaya, India, in *Conservation Agriculture* (eds.) JK Bisht, VS Meena, PK Mishra and A Pattanayak (eds.), DOI 10.1007/978-981-10-2558-7_11

Yadav RP, Bisht JK, Mondal T and Mishra PK 2016 Agroforestry practices for sequestration in the mid hills of central Himalaya, 477. *In* R. Peshin, A.K. Dhawan, F. Bano and K.S. Resam [eds.], *The Proceeding of the Natural resource management: ecological perspectives, Indian Ecological Society international conference*-2016, SKUAST-Jammu

Yadav RP, Bisht JK , Pandey BM , Kumar A and Pattanayak A 2016 Cutting management versus biomass and stock of oak under high density plantation in central Himalaya, India, *Applied Ecology and Environmental Research*, **14** 207–214

Yadav RP, Gupta B, Bhutia PL and Bisht JK 2016 Socioeconomics and sources of livelihood security in Central Himalaya, India a case study. *International Journal of Sustainable Development & World Ecology*. **23** 1-9 DOI: 10.1080/13504509.2016.1239233

Young A 1986 Effects of trees on soils. ICRAF Reprint No. 31, Nairobi, Kenya

CHAPTER 7

Agroforestry Systems in Hot Arid Region of India

JC Tewari, SPS Tanwar and OP Yadav*

ICAR - Central Arid Zone Research Institute, Jodhpur, India

**Corresponding author's E-mail: director.cazri@icar.gov.in*

ABSTRACT

In fragile ecosystems of hot arid regions, agriculture alone cannot be a dependable enterprise, hence desert dwellers with their traditional wisdom are integrating trees and livestock into farming systems since ages in order to confer stability and generate assured income. Growing of trees in hot arid regions is basically concerned with the management of trees for conservation and for limited production objectives like wood for fuel, poles and fencing material; leaves for fodder; and pod/seeds for many types use in human diet. Rainfall is primary governing factor for development of traditional agroforestry systems. These are primarily based on *Prosopis cineraria, Zizyphus sp., Acacia nilotica* and *Salvadora sp*. A number of improved agroforestry practices have been developed and standardized which mainly include agri-silviculture, agri-horticulture, silvi-pasture, horti-pasture, silvi-agri-pasture etc. Integrated farming system modules provide cushion against insecure production due to unpredictable weather situations of arid zone. Agroforestry plays multifarious role in arid landscape. It moderates microclimate and increases soil fertility, in which role of shelterbelt and *Prosopis cineraria* are worth mentioning. Trees improve water use efficiency, and protects soil erosion in otherwise loose and coarse-grained soils of arid regions. Agroforestry has immense capacity to mitigate the effect of greenhouse gases through carbon sequestration. Ecosystems of this part of the country are highly fragile with issues of sustaining livelihood in low rainfall and drought years. Agroforestry indeed provide a unique basis of life support in arid regions.

Keywords: Agroforestry systems, Arid region, Biodiversity, Integrated farming system, Livelihood

INTRODUCTION

Arid regions characteristically possess a combination of high temperature and low rainfall causing evaporation exceeding precipitation. According to conservative estimates, arid zones and their advances affect around 600 million people directly or indirectly. Arid regions are located in almost all the continents of the world. However, maximum extent is in Africa (46.1%) followed by Asia (35.5%) (UNEP 1992). Rest of the area (about 19.4%) is spread over in Australia, North America (Mexico and Southern part of USA) and South America minimal being in Europe (< 0.5%). The hot arid regions of India lie between 24° - 29° N' latitude, and 70° - 76° E' longitude, covering an area of 31.70 m ha constituting 11.8% area of the country. It spreads over in seven states viz. Rajasthan, Gujarat, Punjab, Haryana, Andhra Pradesh, Karnataka, and Maharashtra (Fig. 1). Amongst them, hot arid regions of Rajasthan, Gujarat, Punjab, and Haryana together constitute the Great Indian Desert, better known as the *Thar* desert of which 61% lies in western Rajasthan alone, therefore it is considered principal hot arid region of India (Table 1). The production and life support systems in arid western Rajasthan, are constrained by climatic limitations including low annual precipitation (100 – 300 mm); very high temperature during the summer season (mean maximum temperature 41°C) touching a maximum of 50 °C; short (December to mid-February) cool and dry winters (mean winter season temperature varies from 10° to 14 °C); high wind speed (30–40 km h^{-1}); high evapo-transpiration; and general low humidity (aridity index 0.045–0.190, Sharma and Tewari, 2005). Sand dunes are a dominant land formation of the region. More than 58% of the area is sandy and intensities of dunes vary from place to place. In general, soils contain 1.8–4.5% (clay), 0.4–1.3% (silt), 63.7–87.3% (fine sand) and 11.3–30.3% (coarse sand) with very high infiltration rate (7–15 cm h^{-1}). Because of the complete absence of any aggregation, the soils are highly erodible. They are poor in organic matter (0.04–0.12%), low in nitrogen content (0.07-0.20%) and low to medium in phosphorus content (0.05–0.10%) (Dhir 1997).

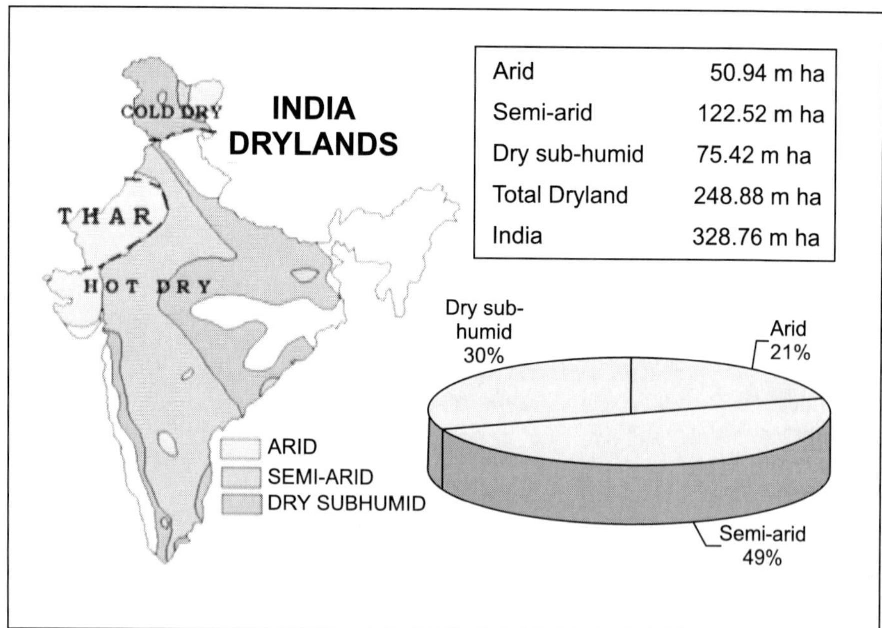

Figure 1: *Arid, semi-arid and dry sub-humid regions of India*

Table 1: *Distribution of hot arid regions in India*

S.No.	State	Area (m ha)	% of total
1	Rajasthan	19.61	61.00
2	Gujarat	6.22	19.60
3	Punjab & Haryana	2.73	9.00
4	Andhra Pradesh	2.15	7.00
5	Karnataka	0.86	3.00
6	Maharashtra	0.13	0.4
	Total	31.70	–

Vast arid expanse of Ghaggar is blank on Chalcolithic and Iron age maps of the Indian subcontinent. It indicated that human occupation in this region was thin, and depended upon hunting and limited pastoralism (Dhir 1982). From the analysis of available authentic records, the population of the region registered an increase of 4.9 times from 1901 to 1991. Population density in arid western Rajasthan increased from 48 person km^{-2} in 1971 to 127 person km^{-2} in 2011. This suggests that the population density is quite high in principal hot arid region of the country compared with the global average of 6–8 persons km^{-2} for arid zones. Similarly, livestock population is also very high (24.5 million according to 2012 livestock census) with a density of 161 animals km^{-2}. Human livestock ratio in this part of the country is 1.0: 1.3 in comparison to national average of 1.0: 0.5.

In fragile ecosystems of hot arid regions, agriculture alone cannot be a dependable enterprise, hence desert dwellers with their traditional wisdom are integrating forestry into farming systems since ages in order to confer stability and generate assured income (Narain and Tewari 2005). Rural folk in these regions have been practising arable cropping in association with the scattered trees on crop fields since time immemorial. Growing of trees in hot arid region is basically concerned with the management of trees for conservation and for limited production objectives like wood for fuel, poles and fencing material; leaves for livestock fodder; and pod/seeds for human diet. The role of trees to conserve the fragile ecosystems of hot arid regions has been well recognized (Mann and Muthana 1984). Trees also provide so continuum, the lifeline of hot arid region (Saxena 1997). Since climate change has become a reality, vulnerability of agricultural systems to it has received considerable attention from the scientific community in recent past (Fischer et al 2002; IISD 2003; Kurukulasuriya and Rosenthal 2003). Much of the available literature suggests that the overall impacts of climate change on agriculture especially in the tropics will be highly negative, although in a few areas there may be minor increases in crop yields in the short term (Maddison et al 2007). Recognizing the ability of agroforestry systems to address multiple problems and deliver multiple benefits, the Third Assessment Report of IPCC on Climate Change (IPCC 2001) states that agroforestry interventions provide the best "no regrets" adaptation measures in making communities resilient to the impacts of climate change. Boffa (1999) opined that multipurpose trees of traditional agroforestry systems are important for rural food security and income generation, and also for ensuring social and cultural stability to some extent. In addition, these systems actively maintain a rich pool of the forest genetic diversity on farmers' land. This paper describes structure and production functions of traditional and improved agroforestry systems, and scope of agroforestry in climate change adaptations and livelihood improvements.

TRADITIONAL AGROFORESTRY SYSTEMS

The image of arid western Rajasthan (or *Thar* desert) one has, of denuded, dense unbroken mantle of sand and inaccessible terrain condition is however, shattered upon entering the *Thar* desert in most of the parts barring extreme western fringes. One is welcomed by sparsely distributed trees and other woody taxa with underneath growth of arable crops (especially during *kharif* season, as agriculture is pre-dominantly rain-fed) and / or grasses in long stretches interspersed with distantly distributed village settlements and "*Dhanis*". It is a unique settlement pattern characteristic of *Thar* desert, meant for living of agriculturists families during active cropping period away from the villages but, nearby their crop field (Tewari et al 1999). The mixed crop-livestock- tree farming form the spectrum of economic activities in these areas ranging from settled agriculture (in true sense, settled agroforestry) to nomadism (particularly, pastoralism). Large tracts of lands have widely scattered trees/shrubs of various species in association with local food crops, providing grain and fodder as one of the best example of traditional agroforestry (Tewari, 1997).

On the basis of distribution and densities, traditional agroforestry systems of hot arid regions of Rajasthan have been described by Naraian and Tewari (2005). This was perhaps first report, where quantifications were done to describe the systems. This study indicated that, in general, rainfall is primary governing factor for evolution of traditional agroforestry systems in the region. Two transects, i.e. in upper and lower part of arid western Rajasthan belonging to different rainfall gradient were studied, the summarized results are mentioned here (Fig. 2).

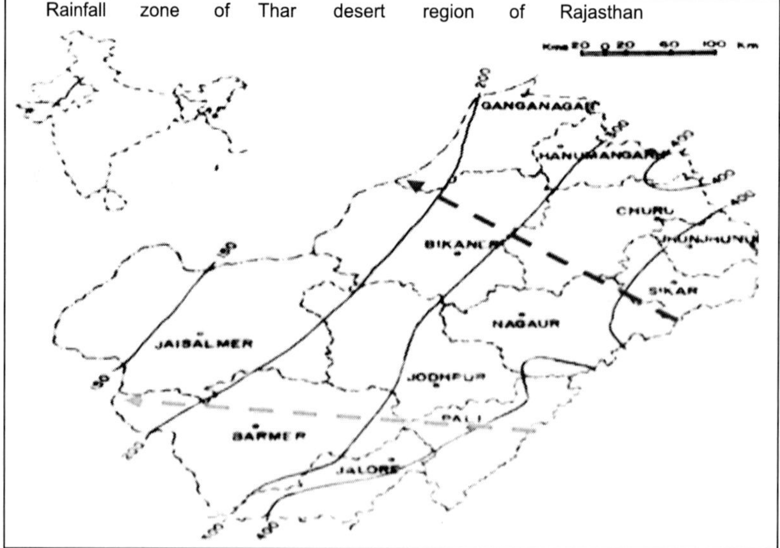

Figure 2: *Positions of Transects Studied to delineate Agroforestry Systems of Western Rajasthan*

TRADITIONAL AFS OF UPPER TRANSECT (SIKAR–BIKANER /SRI GANGANAGAR)

This transect extended from Danta-Ramgarh in Sikar district (transitional region between arid and semi-arid tracts) to western fringes of Bikaner/Ganganagar districts. On the basis of rainfall, four types of major traditional agroforestry systems have been identified (Table 2). Over 400 mm rainfall zone, the pre-dominant system was: *Prosopis cineraria – Acacia nilotica* based; between 300 and 400 mm rainfall zone *Prosopis cineraria* based; between 200 – 300 mm rainfall zone *Ziziphus* spp. – *Prosopis cineraria* based; and in less than 200 mm rainfall zone *Ziziphus* spp. – *Prosopis cineraria – Salvadora* spp. based were observed. With the decrease in rainfall, the density of woody component of the system declined substantially. *Prosopis cineraria* was found in all the rainfall zones. It was very interesting that density of *Prosopis cineraria* tends to decrease with decreasing rainfall from east to west ($r = 0.87$; $p < 0.0001$; Tewari et al 2014).

Table 2: *Woody species composition of traditional agroforestry systems of arid western Rajasthan (upper transect)*

Rainfall zone (mm)	Agroforestry system	Trees/shrubs (No. ha^{-1})	% density of prominent species
> 400	*Prosopis cineraria – Acacia nilotica* based	31.4	80.5
300 – 400	*P. cineraria* based	14.2	80.0
200 – 300	*Ziziphus* spp.* – *Prosopis. cineraria* based	91.7	100.0 (91.7% *Ziziphus* spp.)
< 200	*Ziziphus* spp* – *Prosopis cineraria – Salvadora* spp based	17.2	87.2 (65% *Ziziphus* spp.)

Source: Narain and Tewari (2005)

*in majority of cases the system was dominated by shrubs of *Ziziphus numularia.*

The main productive functions fulfilled by traditional agroforestry systems of western Rajasthan are production of cereals and other food grains, oil seeds, fuel wood, fodder, wood for agricultural implements, minor timber, minor tree products like fruits, gum, resins and vegetables etc. (Narain and Tewari 2005; CAZRI 2008). Pearl millet, dew gram and clusterbean were prominent crops of traditional agroforestry systems. *Prosopis cineraria – A. nilotica* based agroforestry system was found to be highly productive and *Ziziphus* spp. – *Prosopis cineraria – Salvadora* spp. system was least productive (Table 3). It was very interesting that yield of pearl millet, main cereal crop of the region, below the canopy of woody components in any system was not affected at all, rather in *Ziziphus* spp. – *Prosopis cineraria* and *Ziziphus* spp. – *Prosopis cineraria – Salvadora* spp. systems, it was substantially increased below the tree canopies in comparison to open spaces. However, legumes such as dew gram, clusterbean and green gram exhibited yield reduction below the tree canopies. The extent of reduction was 5.0 % (*Ziziphus* spp. – *Prosopis cineraria* system) to 19.5% (*Prosopis cineraria* system) in dew gram; 12.5% (*Ziziphus spp. – Prosopis cineraria – Salvadora* spp. system) to 18.2 % (*Ziziphus* spp. – *Prosopis cineraria* system) in cluster bean; 10.0 % (*Prosopis cineraria* system) to 33.7%

(*Prosopis cineraria* – *A. nilotica* system) in greengram; and 26.3% (*Prosopis cineraria* – *A. nilotica* system) to 46.1% (*Prosopis cineraria* system) in sesame.

Table 3: *Crop production (kg ha^{-1}) in traditional agroforestry systems in hot arid region of western Rajasthan (upper transect)*

AF systems Crops	P. cineraria - A. nilotica based	P. cineraria based	Ziziphus spp.* - P. cineraria based	Ziziphus spp.* – other spp. based
Pearl millet				
Below the canopy	1050.0	790.0	200.0	200.0
Away from canopy	1080.0	790.0	180.0	120.0
Moth bean				
Below the canopy	280.0	210.0	190.0	140.0
Away from canopy	380.0	260.0	200.0	160.0
Cluster bean				
Below the canopy	250.0	180.0	90.0	80.0
Away from canopy	290.0	210.0	110.0	100.0
Sesame				
Below the canopy	140.0	210.0	190.0	-
Away from canopy	190.0	390.0	260.0	-
Green gram				
Below the canopy	140.0	90.0	-	-
Away from canopy	210.0	100.0	-	-

Source: Narain and Tewari (2005)

* in majority of cases the system was dominated by shrubs of *Ziziphus numularia*

Foliage of trees constitutes nutritious components of animal feed in this region. The leaf fodder of some tree species is as nutritious as leguminous fodder and is comparable with grass production from the pastures. Fuel wood and leaf fodder are harvested by way of vertical lopping every year at the end of growing season (November onwards). Both fuel wood and leaf fodder production potential of *Prosopis cineraria* – *Acacia nilotica* based system was maximum, while it was minimum for *Ziziphus* spp.- *Prosopis cineraria* – *Salvadora* spp. system (Table 4). Thus, production from woody components of traditional agroforestry systems also follows the rainfall gradient like that of arable crops.

Table 4: *Fuel wood and top feed production from traditional agroforestry systems in hot arid region of western Rajasthan (upper transect)*

Agroforestry systems	Plant density (No. ha^{-1})	Fuel wood (t ha^{-1} yr^{-1})	Leaf fodder (t ha^{-1} yr^{-1})
P. cineraria-A. nilotica based	31.4	1.23	0.29
P. cineraria based	14.2	0.60	0.20
Ziziphus spp. - *P. cineraria* based	91.7	0.62	0.90
Ziziphus spp. - *P. cineraria* - *Salvadora* spp. based	17.2	0.18	0.14

Source: Narain and Tewari (2005)

TRADITIONAL AGROFORESTRY SYSTEMS OF LOWER TRANSECT (PALI –BARMER)

Four distinct traditional agroforestry systems have been identified in the lower transect that extended from extreme eastern part of Pali district to western parts of Barmer district. This transect encompasses rainfall gradient of <450 mm to 200 mm (Table 5). In south-eastern margin, where total annual rainfall is about 450 mm, the pre-dominant agroforestry system was *Azadirachta indica* - *Acacia nilotica* var. *cupressiformis* based. This system extended from extreme eastern part of Pali district to extreme western part of Barmer district. In the rainfall zone 450- 400 mm, *Prosopis cineraria – Azadirachta indica* based system was prevalent. The density of woody component in the system was found to have only 8.2 individuals per hectare with a very low total basal cover (0.6 m^2 ha^{-1}). In rainfall zone of 300- <400 mm, the major traditional agroforestry system was *Prosopis cineraria* - *Tecomella undulata* - *Salvadora spp.* based. The total basal cover contributed by prominent species was as high as 92% thereby indicating that maximum biomass was accumulated on prominent species as biomass corresponds to basal cover. In western part of Barmer district (rainfall zone <300-200 mm), the most prevalent system was *Prosopis cineraria – Salvadora oleoides- T. undulata* based.

Table 5: *Woody species composition in identified traditional agroforestry systems of arid western Rajasthan (lower transect)*

Rainfall zone	System	Total trees/ shrubs density (No. ha^{-1})	Total basal cover (m^2 ha^{-1})	Total canopy cover (% ha^{-1})
>450	*Azadirachta indica-Acacia nilotica* var. *cupressiformis* (12*)	15.9	1.1	2.3
<450-400	*P. cineraria-A. indica* (8)	8.2	0.6	1.5
<400-300	*P. cineraria-T. undulata-Salvadora* spp (8)	32.1	2.2	4.7
<300-200	*P. cineraria- Salvadora oleoides - T. undulata* (6)	33.8	4.3	4.8

*Values in parentheses are total number of tree/shrub species in the system

Source: Tewari et al (2014)

The major rainfed (*kharif*) crops grown in association with trees/shrubs were pearl millet, greengram, dew gram, clusterbean and sesame (Table 6). Pearl millet grain yield exhibited clear declining trend with decrease in total annual rainfall. The grain yield of pearl millet was found to be comparable both in open field and under the tree canopy cover in all the systems. However, under the tree canopy cover yield of greengram decreased 20.0-29.83%, dew gram 15.69-32.43%, clusterbean 20.0-36.61% and sesame 21.05-23.53%. The data clearly indicated that pearl millet crop production did not have any negative effect of tree canopy cover, however production of all short statured leguminous crops were adversely affected under the tree canopy cover.

Table 6: *Pattern of crop grain production (kg ha^{-1}) in identified traditional agroforestry systems (lower transect)*

System/crop	BC/AC*	Azadirachta indica- A. nilotica *var.* cupressiformis	P. cineraria- A. indica	P. cineraria –T. undulata – Salvadora spp	P. cineraria- S. oleoides – T. undulata
Pearl millet	BC	940	600	360	351
	AC	915	622	375	339
Green gram	BC	96	57	120	130
	AC	120	81	175	182
Moth bean	BC	75	86	100	105
	AC	111	102	162	135
Cluster bean	BC	107	104	90	62
	AC	75	130	125	92
Sesame	BC	95	104	120	–
	AC	121	136	152	–

*AC = Away from canopy; BC = Below canopy

Fuel wood and leaf fodder are main produce of woody component in the traditional agroforestry systems, which farmers collect by way of lopping and or/pollarding the trees and shrubs. The maximum production of fuel wood (4.9 t ha^{-1} yr.$^{-1}$) and leaf fodder (1.78 t ha^{-1} yr.$^{-1}$) was recorded in *Prosopis cineraria – Salvadora oleoides- Tecomella undulata* based agroforestry system, which was prevalent in western part of Barmer district (rainfall zone <300-200 mm; Table 7). It appeared that with the decrease in rainfall, the woody component of the system were better protected and appropriately managed by the farmers in this transect as it provides substantial amount of livelihood in relatively more environmentally stressed areas. This is further substantiated by the fact that prominent tree species contributed maximum amount of fuel wood and fodder in lowest rainfall situation and in general, contribution for fuel wood and fodder from prominent tree species in the system increased with decreasing rainfall. The people of the region have evolved agroforestry based drought protective mechanism through their ingenuity and centuries old experience, which has descended from one generation to other. Depending upon climatic, edaphic, socio-economic and cultural situation, drought hardy woody perennials, which are multipurpose in nature, are integrated in farming systems to develop productive systems in form of traditional agroforestry.

Table 7: *Fuel wood and leaf fodder production potential of different identified traditional agroforestry systems (lower transect)*

Systems	Fuel wood		Leaf fodder	
	Total ($t\ ha^{-1}\ yr^{-1}$)	Contribution of prominent species (%)	Total ($t\ ha^{-1}\ yr^{-1}$)	Contribution of prominent species (%)
A. indica – A. nilotica sub sp. *cupressiformis*	3.10	39.7	0.71	54.8
P. cineraria – A. indica	1.20	74.8	0.43	68.9
P. cineraria – T. undulata – Salvadora spp	4.73	52.3	1.70	68.9
P. cineraria – S. oleoides - T. undulata	4.90	70.0	1.78	73.5

IMPROVED AGROFORESTRY SYSTEMS

Annual crops are most affected by quantum and distribution of rainfall. To minimize the risk and uncertainty, farmers in arid region include perennial components in their farming system. In fact traditional farming systems of arid zone represent a crop – livestock - tree/shrubs – human continuum governed by rainfall distribution. The areas falling in <250 mm rainfall zone have predominance of grasses and shrubs; hence range/pasture development with livestock rearing is the major proposition (Bhati and Joshi 2007). In areas receiving rainfall between 250-350 mm besides grasses and shrubs, multipurpose tree species dominate and mixed farming encompassing agroforestry system, mixed cropping, livestock and pasture management are main livelihood options. In areas receiving rainfall >300 mm, crops and cropping system diversification, agroforestry and livestock rearing are major systems of sustenance of farmers These local survival systems some 60 years back were sustainable, but now become inadequate due to ever increasing human and livestock pressure and associated demands, and as well as technological advances of recent times (Tanwar 2016). In the light of increasing pressure on land resources systematic research on agroforestry systems was initiated in late 1970s at CAZRI, Jodhpur. Since then, in order to enhance overall productivity and economic returns of the farming communities, a number of improved agroforestry practices have been developed and standardized. Following improved practices have been found promising, remunerative, and easy to fit in existing farming systems of the region.

AGRI-SILVICULTURE

This involves co-growing of trees with annual crops. Unlike in forest these trees are managed for getting some economic product and/or reducing their harmful effect on the crops, if any or to enhance any symbiotic association, if any. In an earlier study, the effect of tree management practices on greengram and clusterbean production in 8 years old

Holoptelia integrifolia was studied by Muthana and Arora (1977) under different treatments viz (a) crop grown between unlopped trees, (b) crops under vertically lopped trees and (c) control i.e. crops without trees. Under unlopped trees, the grain yield was less which suggested that extended canopy had negative effect on grain production (Table 8). Thus, lopping not only provided tree leaf fodder but also resulted in higher crop production. As discussed earlier, a positive effect of *Prosopis cineraria* on the productivity of pearl millet is conspicuous in arid region.

Table 8: *Grain yield (Kg/ ha) under lopped and unlopped eight years old* ***Holoptelia integrifolia*** *trees (spacing 6m X 6m)*

Species	Unlopped trees	Lopped trees	Crop without trees
Mung bean	201.75	246.21	252.06
Clusterbean	93.90	244.14	253.14

Source: Muthana and Arora. (1977)

AGRI-HORTICULTURE

Arid region is endowed with a variety of fruit trees/shrubs that coexist with field crops of arid region both under rainfed as well as irrigated systems. Some of them like *Ziziphus mauritiana, Ziziphus numularia* var. *rotundifolia, Cordia myxa, Caparis decidua* and *Salvadora oleoides* are suitable for areas receiving rainfall <300 mm. Several other fruit trees such as *Embilica officinalis, Punica granatum, Aegle marmelos, Phoenix dactylifera* are suitable under irrigated situations. *Ziziphus mauritiana* can be successfully grown on a variety of silvi-agri-pastoral systems of arid land situations, not only to make various kinds of land forms productive and useful but also provide ecological stability in arid regions (Meghwal et al 2007). Leguminous crop (greengram) sown under five years old plantations of *ber (Ziziphus mauritiana* cv. Seb) produced 0.2 t ha^{-1} of grain and 0.8 t ha^{-1} quality ber fruits (400 plants ha^{-1}) from the same land unit even when seasonal rainfall was 200 mm (Table 9), thus rendering a drought proofing mechanism to the system (Gupta 1997).

Table 9: *Productivity of agri-horticultural system (Ber + green gram) in 200 mm seasonal rainfall (year 1989)*

Treatment	Fruit yield (kg ha^{-1})	Grain yield (kg ha^{-1})	Net profit (□ ha^{-1})
Sole crop	–	520	4800
Intercropped *with Ber*	800	200	8000

HORTI-PASTURE

Ber also fits well in horti-pasture systems of hot arid regions. The leaves make excellent fodder, crude protein 13-17% and fibre 15%. The branches and twigs are excellent fuel wood having calorific value 4878 K cal kg^{-1}. Root bark and leaves have 7% and 2% tannin, respectively and are sometimes mixed with other sources for tanning leather. Ber *spp. Ziziphus numularia* and *Ziziphus rotundifolia* both show compatibility with

Cenchrus ciliaris as well as *Lasiarus sindicus*. In an experiment, *Cenchrus ciliaris* grass was introduced between tree rows on a 16 years old Ber orchard (spacing 6m X 6m). For the next six years, on an average, across the years, dry grass production was 1.55 t ha^{-1} yr^{-1} (Table 10; Tewari et al 1999). Average fruit, leaf fodder and fuel wood production from *ber* plants was 2.77, 1.87 and 2.64 t ha^{-1} yr^{-1}, respectively.

Table 10: *Improved horti-pasture system (Ber + Cenchrus ciliaris) on farmers' field (base year 1992)*

Year	Dry grass (t ha^{-1})	Tree products (t ha^{-1})		
		Fuel wood	Leaf fodder	Fruit yield
1	1.50	3.10	2.13	3.10
2	1.67	2.74	1.80	2.74
3	1.42	2.46	1.71	2.49
4	1.66	2.49	1.63	2.85
5	1.48	2.43	2.01	2.80
6	1.57	2.63	1.94	2.67
Average	1.55	2.64	1.87	2.77

Source: Tewari et al (1999)

SILVI-PASTURE

Hot arid region of Rajasthan is reported to have 89 species of grasses and lot many species of shrubs and trees growing in the pasture lands. In an improved silvi-pasture, *Hardwickia binata,* a tree species native to central India, was taken as woody component at 3m x 3 m spacing with *Cenchrus ciliaris* grass. Results of nine years study revealed that average carrying capacity of the system was 4.1 sheep ha^{-1} yr^{-1} against 3.7 for sole *Cenchrus ciliaris* pasture and 1.6 for sole *Hardwickia binata* plantation (Table 11; Harsh et al 1992). In this type of improved silvi-pasture system, in addition to grass + top feed production which was in order of 3.06 t ha^{-1} yr^{-1}, fuel wood to the tune of 0.26 t ha^{-1} yr^{-1} was also harvested.

Table 11: *Production potential of Hardwickia binata + Cenchrus ciliaris based improved silvi-pasture.*

Year after planting	Grass yield (t ha^{-1})	Leaf fodder yield (t ha^{-1})	Total fodder yield (t ha^{-1})	Carrying capacity (sheep ha^{-1} yr^{-1})
1	0.82	–	0.82	1.9
2	1.23	–	1.23	2.9
3	1.64	–	1.64	3.9
4	2.05	–	2.05	4.9
5	1.64	–	1.64	3.9
6	1.64	–	1.64	3.9
7	1.24	2.31	3.55	8.5

8	0.84	0.66	1.50	3.6
9	0.84	0.66	1.50	3.6
Average	1.33	0.40	1.73	4.1

SILVI-AGRI-PASTURE

Diversified agroforestry systems found sustainable in hot arid regions. Such systems generally have three or more components. A recent study conducted on *Ailanthus excelsa* based silvi-agri-pastoral system revealed that the system has the potential to provide multitude of products under inhospitable environmental conditions of the region. Growth patterns of *Ailanthus excelsa* in association with pearl millet, dew gram and *Cenchrus ciliaris* indicated that wider spacing of trees (10m x 10m) was most appropriate for optimal growth of tree component (Fig. 3 to 5) and as well as production from herbaceous components. On an average the system productivity of herbaceous component for food grain (Table 12) and fodder (in the form of crop straw and grasses; Table 13) was found to be the best in wider spacing of *A. excelsa* (10m x 10 m). However, under closer (6m x 6 m) and moderate (8m x 8 m) spacing of the species, the productivity of herbaceous component was satisfactory. This indicates that with proper management, diversified agroforestry systems have the potential to produce multitude of products in fairly good quantum. As the system was quite young, the amount of leaf fodder production during fourth year is given in Table 14. Results indicate that system is capable of providing leaf fodder in appreciable quantity in years to come.

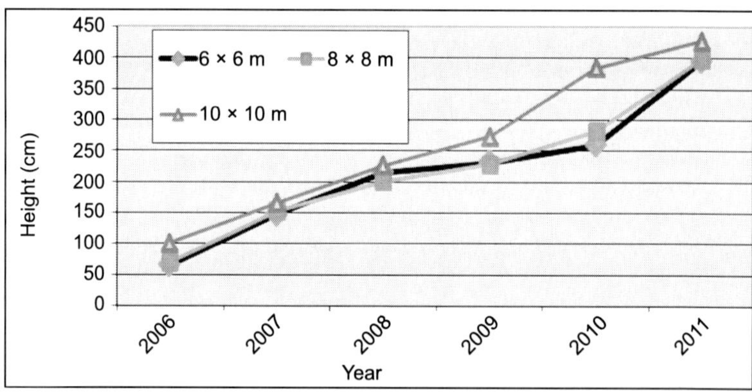

Figure 3: *Height (cm) of Ailanthus excelsa under different spacing in silvi-agri-pastoral systems*

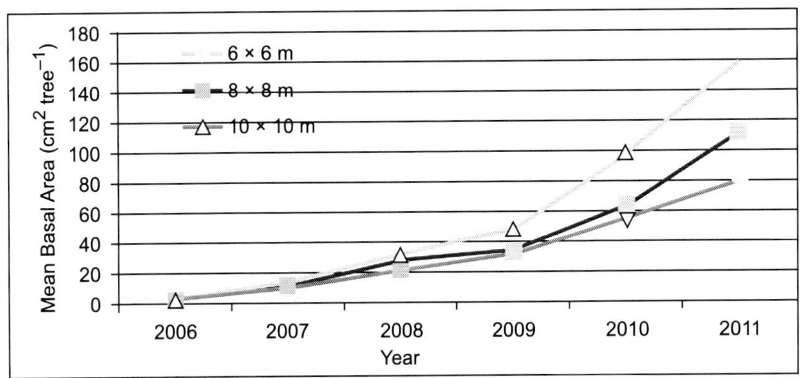

Figure 4: *Mean basal area (cm^2 $tree^{-1}$) under different spacing in silvi-agri-pastoral systems*

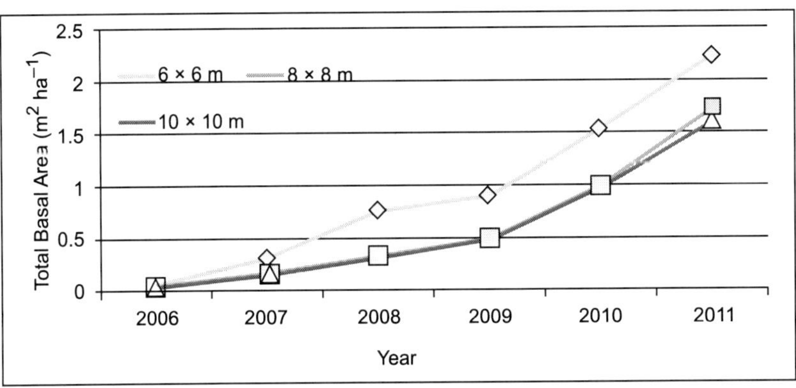

Figure 5: *Total basal area (m^2 ha^{-1}) of Ailanthus excelsa under different spacing in silvi-agri-pastoral systems*

Table 12: *Grain productivity (kg ha^{-1}) under different spacing of Ailanthus excelsa based silvi-agri-pastoral system*

Year	Pearl millet			Moth bean		
	6m × 6m	8m × 8m	10m × 10m	6m × 6m	8m × 8m	10m × 10m
2007-08	425	429	504	202	195	205
2008-09	528	565	650	24	22	29
2009-10*	–	–	–	–	–	–
2010-11	1312	1408	1552	233	250	256
Average	755	800.7	902	153	155.7	163.3

* Completely drought year

Table 13. *Crop straw/grass productivity (kg/ ha) under different spacing of Ailanthus excelsa based silvi-agri-pastoral system*

Year	Pearl millet			Moth bean			Cenchrus ciliaris		
	6m × 6m	8m × 8m	10m × 10m	6m × 6m	8m × 8m	10m × 10m	6m × 6m	8m × 8m	10m × 10m
2007-08	978	1008	1256	300	285	310	552	581	680
2008-09	1368	1292	1665	260	231	267	562	708	890
2009-10*	456	508	677	212	227	296	411	507	666
2010-11	3818	4154	4627	639	789	683	1750	1821	2174
Average	1655	1740.5	2056.3	352.8	383	389	818.8	904.3	1102.5

*Completely drought year

Table 14: *Canopy expansion and leaf fodder production by way of 70% vertical lopping of A. excelsa trees in silvi-agri-pastoral system*

Spacing (m)	Mean crown cover (m^2 tree^{-1})	Total canopy cover (% ha^{-1})	Green Leaf fodder production (□ ha^{-1})
6 × 6	5.21	14.43	69.8
8 × 8	9.71	15.14	155.4
10 × 10	11.55	11.55	184.0

INTEGRATED FARMING SYSTEM (IFS): A DIVERSIFIED AGROFORESTRY APPROACH FOR HOT ARID REGION

Integrated farming system approach has been widely advocated as a means to harmonize use of scarce resources of arid zone. Optimum combination of alternative land uses have been worked out for different rainfall zones of arid region (Table 15, Tanwar and Tewari 2016). This mainly includes the incorporation of perennial components like multipurpose trees and grasses along with agronomic crops. On the above considerations a 7.0 ha integrated farming system model has been developed at CAZRI based on twin strategy of system as well as crop diversification (rainfall zone 300-400 mm). This included arable cropping (15%), agri-silviculture (35%), agri-horticulture (20%), agri-pasture (10%) and silvi-pasture (20%). Crop diversification of pearl millet (40%) with pulses (30%), oilseed (10%) and clusterbean (20%) was found more profitable and sustainable. The perennial components were: *Prosopis cineraria*, *Hardwickia binata* and *Acacia tortilis* in agri-silviculture, *Ziziphus mauritiana* in agri-horticulture, and *Colosospermum mopane* and *Ziziphus rotundifolia* in silvi-pasture.

Table 15. *Diversified farming systems for various rainfall zones (% area) and their suitability parameters*

Land use (% area)	Rainfall situation (mm)			
	<250	250-350	350-450	450-600
Arable cropping	25	30	20	20
Agri-silviculture	40	40	40	30
Agri –Horticulture	–	15	30	40
Silvi- pasture	35	15	10	10
Livestock sustainability (ACU/year)	1.5 – 2.0	1.5 – 2.5	2.5 – 3.0	3.0 – 4.0
B:C ratio	1.5 – 2	1.5 – 2.5	2.5 – 3	2.5 – 3.5
Net returns (US$)	140 – 200	200 – 400	400 – 700	700 – 1000
Gestation period (Max.- Min)	8 – 6	6 – 4	4 – 3	4 – 3
Employment generation (man days yr^{-1})	200 – 300	300 – 500	400 – 600	500 – 700

Ten years of study of this model indicated that amongst farming system components B:C ratio was highest in the silvi-pasture (3.11) and was the lowest in arable farming (1.62) (Fig. 6). The system as per family and herd size was surplus in pulses, fruits, fuel-wood and milk, whereas, it was marginally deficient in cereals, fodder and concentrate, but substantially deficient in oilseeds. Over the decade (2001-10) this IFS model had showed reasonable IRR (21.6), net present worth (US$ 10560) and B:C ratio (2.05). This system showed resilience in low rainfall years (2004, 2009), long dry spells of mid-term drought (year 2008, 2011), delayed onset of monsoon (2006, 2012) besides recurring terminal droughts. In all such situations, produce from the perennial components like, fruit, fuel wood and fodder from trees as well as perennial grasses provided the much needed cash as well as fodder for sustenance of livestock.

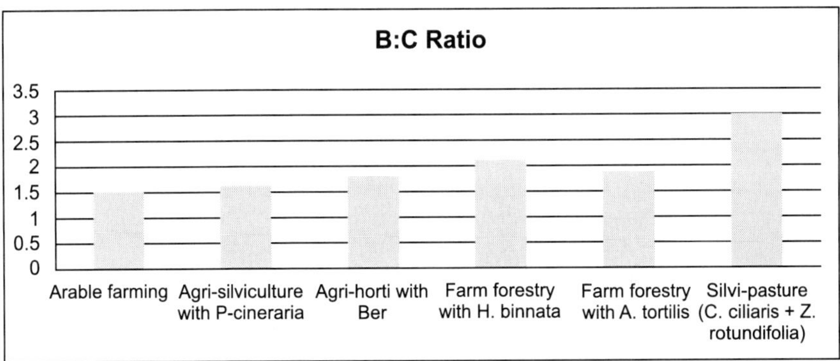

Figure 6: *Benefit cost ratio of various components of IFS model (Average of 10 years)*

MULTIFARIOUS ROLES OF AGROFORESTRY SYSTEMS IN ARID REGION

Moderation of Micro- Climate

It was observed that in arid environments, trees on agricultural fields bring favourable changes in the microclimate by influencing radiation flux, air temperature, wind speed and saturation deficit of understorey crops. All of these have a significant impact on modifying the rate and duration of photosynthesis and subsequent plant growth, transpiration, and soil water use (Monteith et al 1991). Following examples present microclimatic moderation effect in harsh climatic conditions of arid regions of western Rajasthan. Air temperature exhibited appreciable variation under the canopy of *Acacia tortilis* during monsoon period (Tewari et al 1989). A decline of 0.1 °C to 0.7 °C was evident beneath the canopy cover of *Acacia tortilis* than that of surrounding open areas at 07.00 hrs. Around 14.00 hrs, this decline was of the order of 0.6 °C to 2.0 °C. Similarly, it was found during monsoon season that the soil temperature just beneath the tree cover was lower by as much as 10 °C to 16 °C in top soil zone and 4 °C to 5 °C at 30 cm depth when compared to open field conditions, thereby indicating better soil-thermal regime. Seven year old *Acacia tortilis* trees provided more humidity on cluster bean grown with the trees. This finding proves the efficiency of agroforestry practices for maintaining favourable moisture status on arable lands (Ramakrishna and Shastri 1977).

IMPROVEMENT IN SOIL FERTILITY

It has been observed that arid zone soils are poor in nitrogen and phosphorus content. Organic carbon status in certain parts of arid regions has been depleted to an alarming state. While, fertilizers offer an easy way to replenish the soil fertility, at the current prices it is very unlikely that there will be any change in the investments made by farmers of arid region in fertilizer. In this context, agroforestry systems have attracted considerable attention as an attractive and sustainable pathway to improve soil fertility.

Soil fertility under the tree species (*Prosopis cineraria* and *Acacia nilotica*) and in open field conditions was studied intensively (Singh and Lal 1969). In general, organic carbon content decreased with depth but the same was slightly higher under the canopy of both the tree species than in open field condition. The total N, available P and K were maximum under the canopy of *Prosopis cineraria* followed by *Acacia nilotica*. The values for the same were minimum in open field condition. The soil fertility build up under the 14 year old stands of *Prosopis cineraria* and *P. juliflora* were found to be of higher order as compared to open field condition. Similarly, status of micro-nutrients was also higher under tree canopies than in the open field (Table 16).

Table 16: *Available macro and micro-nutrients under different tree based agroforestry systems*

Tree species	Macronutrients (kg ha^{-1})			Micronutrients (ppm)			
	N	P	K	Zn	Mn	Cu	Fe
Prosopis juliflora	231	7	333	0.89	9.3	0.58	3.3
Prosopis cineraria	221	11	479	1.44	10.8	0..89	2.8
Open field	199	6	3.2	0.19	7.0	0.38	3.5

HIGHER RAIN WATER USE EFFICIENCY

There are several mechanisms whereby agroforestry may use available water more effectively than the annual crops. Firstly, unlike in annual systems, where the land lies bare for extended periods, agroforestry systems with a perennial tree component can make use of the water remaining in the soil after harvest and the rainfall received beyond the crop season. Secondly, agroforestry increase the productivity of rain water by capturing a larger proportion of the annual rainfall by reducing the runoff and by using the water stored in deep layers. Thirdly, the changes in microclimate (lower air temperature, wind speed and saturation deficit of crops) reduce the evaporative demand and make more water available for transpiration. Tree canopies in agroforestry systems intercept the rain and reduce runoff (Khan et al 1995). It was found that in *Acacia tortilis* silvi-pasture stand, canopy interception was 21.4%, whereas in *Colophospermum mopane* silvi-pasture it was 31.1%. Rainfall interception was positively related with canopy cover and negatively related with throughfall. Average surface run off in *Acacia tortilis* based silvi-pasture stand was 53% higher than in *Colophospermum mopane* based silvi-pasture. This indicate that hydraulic response to rain is dominated by plant species character however, the percent annual runoff and soil erosion was very low in situations with trees on agricultural fields in comparison to bare soil condition.

EFFECTIVE SOIL EROSION CONTROL

Arid regions suffer from soil erosion both through water and wind. Higher temperatures and drier conditions lead to lower organic matter accumulation in the soil resulting in poor soil structure, reduction in infiltration of rain water and increase in runoff and erosion (Rao et al 1998), while the expected increase in the occurrence of extreme rainfall events will have adverse impact on the severity, frequency, and extent of erosion (WMO 2005). These changes will further exacerbate under climate change scenario as already serious problem is being faced by arid regions of the country. The woody component in agroforestry systems helps in reducing soil erosion, which is the most harmful abiotic stress in arid region. It was observed that if rows of trees are planted at right angle to wind direction, a tremendous amount of soil drop can be checked from agricultural fields. Shelterbelts on agricultural fields are a common agroforestry practice in arid region and are effective in controlling the soil erosion (Table 17; Gupta et al 1984). It is clear from the data that the soil loss was higher in agricultural field without trees as compared to field with agroforestry practices (shelterbelts). Further, higher soil loss during Year II (1980) as compared to Year I (1979) was due to high wind speed and relatively more windy days during Year II. There was heavy flood (unprecedented) during Year I resulting in severe soil erosion and loosening of soil in larger part, and perhaps this may also contributed to higher soil loss in Year II during the summer season when study was conducted.

Table 17: *Effect of different type of shelterbelts on soil erosion*

Type of shelterbelt	Total amount of soil loss (kg ha^{-1})		
	Year I	Year II	Mean
Prosopis juliflora	93.2	609.3	351.2
Cassia siamea	91.5	277.1	184.2
Acacia tortilis	106.0	494.1	300.0
Agricultural field without trees	262.7	831.0	546.8

Source: Gupta *et.al.* (1984)

ENHANCEMENT IN BIODIVERSITY

Although agroforestry may not entirely reduce deforestations, particularly in arid regions, it may act as an effective buffer to deforestation. Trees in agro-ecosystems in Rajasthan have been found to support threatened cavity-nesting birds, and offer forage and habitat to many other bird species. Biodiversity conservation may not be a primary goal of agroforestry systems. Nevertheless in some cases traditional agroforestry systems in arid western Rajasthan are found to support very good species diversity and also act as a buffer to parks and protected areas (Pandey 2007).

ROLE OF AGROFORESTRY IN CARBON SEQUESTRATION AND ADAPTATION TO CLIMATE CHANGE IN ARID REGIONS

Recognizing the ability of agroforestry systems to address multiple problems and deliver multiple benefits, the Third Assessment Report of IPCC on Climate Change (IPCC 2001) states that "*Agroforestry can both sequester carbon and produce a range of economic, environmental, and socio-economic benefits. For example, trees in agroforestry farms improve soil fertility through control of erosion, maintenance of soil organic matter and physical properties, increased N, extraction of nutrients from deep soil horizons, and promotion of more closed nutrient cycling.*" In order to assess the carbon sequestration potential of a 14 year old integrated farming system (IFS) model, which represented a diversified agroforestry system involving various land uses was studied at Jodhpur, India (Unpublished, preliminary estimates). Eight land use systems of the IFS model under study were arable component (crops alone grown with natural tree density), *Prosopis cineraria*+ crops (agri-silviculture), *Ziziphus mauritiana* + crops (agri-horticulture), *Hardwickia binata* + crops (agri-silviculture), *Colophospermum mopane* + *Cenchrus ciliaris* grass (silvi-pasture), *Ziziphus mauritiana* + *Cenchrus ciliaris* grass (horti-pasture), agri-pasture (rotation of *Cenchrus ciliaris* and crops for 3-5 years) and farm forestry. Irrespective of the land use system, the upper horizon (0-15 cm) had higher soil carbon and nitrogen stock. The maximum total carbon stock (vegetative + soil) was observed under *Acacia tortilis* plantation (41.04 Mg ha^{-1}) in farm forestry. Compared to 18.83 Mg/ha in arable cropping system, total carbon stock was 21.25, 23.65, 24.96 Mg ha^{-1} in tree-crop based systems i.e. *Prosopis cineraria*, *Ziziphus mauritiana*, *Hardwickia binata* trees, respectively + crops. In grass based systems i.e. agri-pasture, horti-pasture with *Ziziphus Mauritiana* and silvi-pasture with *C. mopane* carbon stock was in order of 22.38, 23.65, 26.73 Mg ha^{-1}, respectively. In *Hardwickia binata* + crops system, the additional stock was generated in vegetative part, while in *Ziziphus mauritiana*, the additional stock was through improvement in soil carbon.

The rate of carbon sequestration in different land use systems of IFS during initial 14 years varied between 1.19 -2.67 Mg ha^{-1} yr^{-1}. On overall basis, IFS added 6.07 Mg ha^{-1} carbon, which was 32% over the prevalent practice of growing crops in natural tree density (traditional agroforestry). It clearly indicated that agroforestry based integrated farming system approach not only provided adaptation measures to climate change, but also can be an important tool for mitigation through enhancing the carbon sequestration. Research on agroforestry as an adaptation to climate change and as a buffer against climate variability is in the process of evolving. There are conclusive evidences that these systems improved total system productivity and provide an insurance against complete crop failure especially in adverse climatic situations of drought and low rain fall. Agroforestry has immense capacity to mitigate the effect of greenhouse gases through carbon sequestration. This is because of greater efficiency of integrated systems in resource (nutrients, light and water) capture and utilization than sole culture (Yadav and Soni 2016)

IMPROVING LIVELIHOOD OPPORTUNITIES IN HOT ARID REGION OF INDIA

Hot arid regions are particularly notable for ethano-forestry practices and indigenous knowledge system on growing trees on farm lands. In fact, ecosystems of this part of country are highly fragile with issues of sustaining livelihood in low rain fall and drought years Multipurpose trees indeed provide a rational formula for drylands (Oldeman, 1983). For example, economics of previously discussed traditional agroforestry systems in upper transect of arid western Rajasthan indicated that net B:C ratio of such systems is generally on positive sides when density of true tree component is strong enough (Table 18). Cultivation of ber in hot arid regions offers solution to many problems. Tewari et al (2001) demonstrated that agroforestry, particularly improved ber based horti-pasture system can be major source of livelihood in many parts of hot arid region (Table 19). This system provided production in terms of human food, livestock feed and cooking energy needs to communities. Such multifunctional agroforestry systems provide social well-being to the people in addition to goods and services.

Table 18: *Economics of traditional agroforestry systems in upper transect of arid western Rajasthan (Base Year 2002)*

Agroforestry system	Expenditure (□ ha^{-1})	Returns (□ ha^{-1})			Gross returns (□ ha^{-1})	Net Returns (□ ha^{-1})	Net B:C ratio
		Crops	Fuel wood	Leaf fodder			
P. cineraria-A. nilotica based	1850	4103	1230	870	6203	4353	2.3
P. cineraria based	1550	3670	600	420	4690	3140	2.0
Z. spp. - *P. cineraria* based	1550	1506	620	600	2726	1176	0.7
Z. spp. - *P. cineraria* - *Salvadora* spp. based	1500	1400	500	500	2400	900	0.6

Table 19: *Economic returns (in terms of gross income) from a* ***Ber*** *based horti-pastoral system*

Particulars	Production (t/ha)	Returns (□)
Dry grass	1.55	3,100
Fuel wood	2.64	2,640
Leaf fodder	1.87	7,480
Fruit	2.77	33,240
Total	–	46,640

Source: Tewari *et.al.* (2001)

The most spectacular evidence in context of livelihood improvement potential of agroforestry in hot arid region came through *Acacia senegal* based localized traditional agroforestry system in parts of Barmer, Nagaur and Jodhpur districts (45 villages). *Acacia senegal* is a source of gum Arabic, which has very high commercial value. In addition to harvesting crop grain for food and crop straw for fodder, the farmers harvested substantial quantity of gum Arabic through small intervention by ICAR-CAZRI. From 2008-09 to 2015-16. Farmers in more than 45 target villages of arid western Rajasthan earned an additional income of more than ₹ 48.7 million by the sale of gum Arabic produced using CAZRI technology. Moreover, during 2008-09 to 2015-16, CAZRI generated revenue of more than ₹ 1.875 million through sale of CAZRI gum inducer (Table 20). The data given are only of CAZRI, Jodhpur. At the moment there are many NGOs and private players are in this business. Twenty three to twenty four thousands of *A. senegal* trees were treated per year by the gum inducer sale from CAZRI only, therefore it is hoped that more than double of this figure of trees would have been treated every year and consequently gum Arabic production must be double of figures provided here. This self-sustaining technology is now expanding in other parts involving other gum/resin yielding tree species in addition to *Acacia senegal*.

Table 20: *Gum Arabic production and economic returns in 45 target villages of Barmer, Jodhpur and Nagaur district of arid western Rajasthan*

Particulars	Year								Total
	2008-09	2009-10	2010-11	2011-12	2012-13	2013-14	2014-15	2015-16	
Number of *Acacia senegal* trees treated (in 000')	12.1	20.95	22.61	27.50	30.00	34.17	20.70	19.50	187.53
Production of gum Arabic by farmers (t)	5.45	10.48	7.67	11.00	12.00	13.67	8.28	7.80	76.35
Total income earned by farmers (₹ Lakhs)	27.23	52.38	38.33	77.00	84.00	95.69	57.96	62.40	487.19
Revenue generated by CAZRI (₹ Lakhs)	1.21	2.10	2.25	2.75	3.00	3.42	2.07	1.95	18.75

These few examples demonstrated the potential of agroforestry in hot arid regions for breaking the poverty and food insecurity circle. In addition to providing food and feed, there are numerous other non-timber tree products (excluding gums and resins) collected from trees on farm lands. *Prosopis cineraria* trees in addition to providing fuel and fodder give nutritious vegetable in form of pod, locally known as '*Sangri*'. Likewise seeds of *Acacia senegal* are also used in preparing vegetable. During drought years their value is increased tremendously. The majority of tree species scattered on agricultural fields in arid western Rajasthan are multipurpose by nature and they have ability to satisfy the expectations and aspiration of rural folk regarding production of their basic need- fuel, fodder, timber, food (fruits/seeds/pod) and other products such as exuded edible gums, and products of medicinal value to sustain their livelihoods. Thus, agroforestry systems provide unique basis of life support in arid western Rajasthan across deferent agro-eco-climatic regions.

CONCLUSION

Agroforestry systems are biologically more complex than other ways of using land alone through arable farming or forestry. The complexity of process that determine the response of crop/grass and tree to environment and their mutual interactions require large scale scientific assessment so that prediction could be made about system's performance in future. However, on the basis of available evidences from the researches carried out so far, it could be concluded that if compatible tree species were properly incorporated on farm lands, overall biomass production can be enhanced even under unfavourable climatic, edaphic and economic conditions of hot arid regimes.

Despite of inhospitable climatic and edaphic conditions, Indian hot arid regions are highly vegetated and woody species diversity is perhaps much higher in comparison of hot arid zones of other parts of world. Moreover, the tree species which have been introduced successfully in hot arid regions of the country from iso-climatic region of the world or from other drier parts were thoroughly evaluated on the basis of information available on climatic homologues, potential inherent plasticity and genetic variability before recommending them for agroforestry programmes, which is a very positive development for prospective users of innovative intensive Agroforestry models for arid regions.

REFERENCES

Bhati TK and Joshi NL 2007 Farming systems for the sustainable agriculture in Indian Arid zone. In: K.P.R. Vittal, R.L. Srivastava, N.L. Joshi, A. Kar, V.P. Tewari and S. Kathju (eds.) *Dryland Ecosystem: Indian Perspective*, Central Arid Zone Research Institute and Arid Forest Research Institute, Jodhpur, 35-52.

Boffa JM 1999 *Agroforestry parklands in sub-Saharan Africa*. Food and Agriculture Organization of the United Nations, Rome, 230.

CAZRI 2008 *Annual Progress Report* ICAR-Central Arid Zone Research Institute (CAZRI), Jodhpur, India, 166.

Dhir RP 1982 The human factors in ecological history. In: *Desertification and Development: Dryland Ecology in Social Perspective*. Academic Press, New York, 311–331.

Dhir RP 1997 Characteristics and behavioural aspects of arid and semi-arid zone soils. In: Yadav MS., Singh Manjit, Sharma SK, Tewari JC, and Burman U. (eds.) *Silvipastoral Systems in Arid and Semi-arid Ecosystems*. CAZRI, Jodhpur, India 39-46.

Fischer G, Shah M and Van Velthuizen H 2002 *Climate Change and Agricultural Vulnerability*. International Institute for Applied Systems Analysis. Report prepared under UN Institutional Contract Agreement 1113 for World Summit on Sustainable Development. Laxenburg, Austria.

Gupta JP 1997 Some alternative production systems and their management for sustainability. In: Gupta JP and Sharma BM (eds.) *Agroforestry for Sustained Productivity in Arid Regions*, Scientific publishers, Jodhpur, India, 31-39.

Gupta JP, Rao GGSN, Ramakrishna YS and Rao Ramana 1984 Role of shelterbelts in arid zone. *Indian Farming* **34**(7) 29-30.

Harsh LN, Tewari JC, Burman U and Sharma SK 1992 Agroforestry in arid region. *Indian Farming* (Special issue on environmentally sound biotechnology) **45** 32-37.

IPCC 2001 *Climate Change The Scientific Basis*. Contribution of the Working Group 1 to the Third Assessment Report of the IPCC. Cambridge, Cambridge University Press.

IISD 2003 Livelihoods and Climate Change Combining Disaster Risk Reduction, Natural Resource Management and Climate Change. Adaptation in a New Approach to the Reduction of Vulnerability and Poverty. *A Conceptual Framework Paper*, prepared by the Task Force on Climate Change, Vulnerable Communities and Adaptation, New York.

Khan MA, Tewari JC and Issac VC 1995 Hydrology of small forested catchments in the arid region of Rajasthan. *Annals of Arid Zone* **34**(4) 259-262.

Kurukulasuriya P and Rosenthal S 2003 *Climate Change and Agriculture: A Review of Impacts and Adaptations*. Climate Change Series Paper 91, World Bank, Washington, District of Columbia, 106.

Maddison D, Manley M and Kurukulasuriya P 2007 *The Impact of Climate Change on African Agriculture: A Ricardian Approach*. World Bank Policy Research Working Paper 4306.

Mann HS and Muthana KD 1984 *Arid Zone Forestry*, CAZRI Monograph No. 23, CAZRI, Jodhpur, 48.

Meghawal PR, Khan MA and Tewari JC 2007 *Growing Ber (Zizyphus mauritiana lam) for Sustainable Income and Employment in Arid and Semi-arid Regions*. CAZRI, Jodhpur, 26.

Monteith JL, Ong CK and Corlett JE 1991. Microclimatic interactions in agroforestry systems. *Forest Ecology and Management* **45**, 31– 44.

Muthana KD and Arora GD 1977. Studies on establishment and growth in an economic tree species viz. *Holoptelia integrifolia* in integrated land use patterns in arid environment. *CAZRI Annual Progress Report*, CAZRI, Jodhpur, 68.

Narain Pratap and Tewari JC 2005 Trees on agricultural fields: a unique basis of life support in Thar Desert. In: *Multipurpose Trees in the Tropics: Management and Improvement Strategies*. (eds.) Tewari VP and Srivastava RL. Arid Forest Research Institute, Jodhpur, 516-523.

Oldeman RAA 1983 The design of ecologically sound agroforests. In: Huxley P. A. (ed.) *Plant Research and Agroforestry*. International Council for Research in Agroforestry (ICRAF), Nairobi, Kenya. 173-207.

Pandey DN 2007 Multifunctional agroforestry systems in India. *Current Science* **92**(4) 455-463.

Ramakrishna YS and Shastri ASRAS 1977 Microclimate under *Acacia tortilis* plantation. *Annual Progress Report*, CAZRI, Jodhpur, 69-70.

Rao KPC, Steenhuis TS, Cogle AL, Srinivasan ST, Yule DF and Smith GD 1998. Rainfall infiltration and runoff from an Alfisol in semi-arid tropical India. II. Tilled systems. *Soil and Tillage Research* **48** 61-69.

Saxena SK 1997 Traditional agroforestry systems of western Rajasthan. In: *Agroforestry for Sustained Production in Arid Regions* (eds.) Gupta JP and Sharma BM, Scientific Publishers, India, 21-30.

Sharma AK and Tewari JC 2005 Arid zone forestry with special reference to Indian hot arid zone. In: *Encyclopaedia of Life Support System* (ELOSS), UNESCO, Paris

Singh KS and Lal P 1969 Effect of *Prosopis cineraria* and *Acacia nilotica* on soil fertility and profile characteristics. *Annals of Arid Zone* **8**:33-36.

Tanwar SPS and Tewari JC 2016 Agroforestry and integrated farming system. In: Kumar Suresh, Tanwar SPS and Singh Akhat (eds.) *Drought Mitigation and Management*, Scientific Publisher, India, 181-193.

Tewari JC 1997 Importance fodder trees of arid zone and their management. In: *Agroforestry for Sustained productivity in Arid Regions* (eds) Gupta JP and Sharma BM, Scientific Publishers, Jodhpur, India, pp. 147-153.

Tewari JC, Bohra MD and Harsh LN 1999 Structure and production function of traditional extensive agroforestry systems and scope of agroforestry in Thar desert. *Indian Journal of Agroforestry* **1** (1) 81-94.

Tewari JC, Harsh LN and Venkateshwarlu J 1989 Agroforestry rescarch in arid regions - A review. In: Singh RP, Ahlawat IPS and Saran Ganga (eds.), *Agroforestry Systems in India: Research and Development,* Indian Society of Agroforestry, IARI, New Delhi, 3-17.

Tewari JC, Ram Moola and Roy MM 2014 Livelihood improvements and climate change adaptations through Agroforestry in hot arid environment. In: *Agroforestry Systems in India: Livelihood Security and Ecosystem Services* (eds.) Dagar JC, Singh AK and Arunachalam A, Springer, India, 155-183.

Tewari JC, Tripathi D and Narain Pratap 2001 Jujube: A multipurpose tree crop for arid land farming systems. *The Botanica* **51** 121-126.

UNEP 1992 *World Atlas of Desertification*. UNEP, Edward Arnold (A Division of Hodder and Stoughton), London - New York - Melbourne - Auckland, 69.

WMO 2005 *Climate and Land Degradation.* WMO-No. 989 2005, World Meteorological Organization.

Yadav ND and Soni ML 2016 Tree- crop interactions and productivity of Agroforestry system in arid and semi-arid ecosystems. In: Bhatt RK, Burman U, Painuli DK, Singh DB, Sharma R and Tanwar SPS (eds.) *Climate Change and Agricultur Adaptation and Mitigation*, Satish Serial Publishing House, Delhi, 145-171.

CHAPTER 8

Prospects of Organic Farming in Hill Agroforestry Systems with Specific Reference to Sikkim Himalayas

RK Avasthe*, Matber Singh and Ashish Yadav

ICAR-National Organic Farming Research Institute, Tadong-737102, Gangtok, Sikkim, India

**Corresponding author's Email: ravisikkim@yahoo.co.in*

ABSTRACT

Organic agriculture and agroforestry are the holistic production management systems, which promotes and enhances agro-ecosystem health, including biodiversity, biological cycles, and soil biological activity. Agroforestry has the potential to improve livelihood as it offers multiple alternatives and opportunities to farmers to improve farm production and incomes and also provide productive and protective forest functions to the ecosystems, while protecting the natural environment across the globe. Multi-purpose tree species in agroforestry systems play a major role in organic agriculture production by supporting food production through increasing the soil's ability to support organic agriculture. A large number of forest and fruit producing trees are integral parts of traditional homestead and other agroforestry farming systems. In hills for sustainable development of agroforestry systems, the maintenance of traditional agroforestry systems and creation of new systems is required by designing context-specific silvicultural and farming systems to optimize food production, carbon sequestration and biodiversity conservation. Agroforestry systems are believed to have a higher potential for carbon sequestration than pastures or field crops. In addition to sequestering C in biomass and soil, these systems can contribute to both carbon conservation and carbon substitution.

Keywords: Biomass, Carbon Sequestration, Himalayas, Organic farming

INTRODUCTION

The entire rural economy of hills depends on agriculture, horticulture and livestock. Further, livelihood dependence on Non Timber Forest Products (NTFP), which is mostly organic in nature, is very high amongst rural people in hills. The process of diversification of staple-food production is triggered by rapid technological changes in agricultural production, improved rural infrastructure, and diversification in food-demand patterns. The diversification of organic agriculture with tree and livestock components will improve land use, generate employment and provide nutritional security to people.

Agroforestry has the potential to improve livelihood as it offers multiple alternatives and opportunities to farmers to improve farm production and income and also provide productive and protective (biological diversity, healthy ecosystems, protection, of soil and water resources, terrestrial carbon storage) forest functions to the ecosystems, while protecting the natural environment across the globe. The multi-functionality of the agroforestry systems has been realized over the last four decades. The technology interventions added through organic agriculture during the last five decades have changed the face of agriculture in many developed and developing countries tremendously, however has shown less impact on the hill farming. The productivity in hills is lower than both the regional and national averages for all the major field crops. Increasing per unit productivity will be the key in bridging this gap, which can be achieved with area expansion under high yielding varieties along with proper package of practices that will exploit the inherent production potential of varieties designed and/or selected for organic farming from the local gene pool; soil-test value based nutrient management and appropriate plant protection measures.

Land degradation in hills is among the major challenges confronting sustainable development. As a result, millions of people depending on forests and tree resources for their subsistence have become more vulnerable. Soil erosion caused by heavy rains in hills generally suffers from the continuous depletion of nutrients as farmers harvest without fertilizing adequately or fallowing the land. The presence of deep-rooted agroforestry systems can contribute to enhanced soil nutritional and physical conditions as well as higher soil microbiological activities. By providing nutrients to crops, agroforestry system can potentially help farmers improve their soils and income, thereby improving food security and contribute to reduce the hazards of erosion and desertification as well as conserve soil and water. It has been found that agroforestry systems increase nitrogen input by N2-fixing trees (NFTs), enhance availability of nutrients resulting from production and decomposition of tree biomass, and improve uptake and utilization of nutrients from deeper layers of soils by deep-rooted trees (Nair et al 2008).

World Agroforestry Centre defines agroforestry as a dynamic, ecologically based, natural resources management system that, through the integration of trees on farms and in the agricultural landscape, diversifies and sustains production for increased social, economic and environmental benefits for land users at all levels. Production potential is high under agroforestry system due to greater efficiency of tree species for photosynthesis; improved soil structure and fertility with increasing effects on crop yield; reduce losses from soil erosion and more closed cycling of organic matter and nutrients; creating better micro climatic conditions for the growth of agricultural crops.

ORGANIC FARMING

Organic agriculture is holistic production management system, which promotes and enhances agro-ecosystem health, including biodiversity, biological cycles, and soil biological activity. Organic production systems are based on specific and precise standards of production, which aim at achieving optimal agro-ecosystems, which are socially, ecologically and economically sustainable. The International Federation of Organic Agriculture Movements (IFOAM) defines "organic agriculture" as: "*a production system that sustains the health of soils, ecosystems and people. It relies on ecological processes, biodiversity and cycles adapted to local conditions, rather than the use of inputs with adverse effects. Organic agriculture combines tradition, innovation and science to benefit the shared environment and promote fair relationships and a good quality of life for all involved.*"

IFOAM has formulated four broad principles of organic farming, which are the basic roots for organic agriculture growth and development in a global context. They are:

PRINCIPLE OF HEALTH

Organic agriculture should sustain and enhance the health of soil, plant, animal, human and planet as one and indivisible. Health is the wholeness and integrity of living systems. It is not simply the absence of illness, but the maintenance of physical, mental, social and ecological well-being.

PRINCIPLE OF ECOLOGY

Organic agriculture should attain ecological balance through the design of farming systems, establishment of habitats and maintenance of genetic and agricultural diversity. Those who produce, process, trade, or consume organic products should protect and benefit the common environment including landscapes, climate, habitats, biodiversity, air and water.

PRINCIPLE OF FAIRNESS

Organic agriculture should build on relationships that ensure fairness with regard to the common environment and life opportunities. Fairness is characterized by equity, respect, justice and stewardship of the shared world, both among people and in their relations to other living beings.

PRINCIPLES OF CARE

Organic agriculture should be managed in a precautionary and responsible manner to protect the health and well-being of current and future generations and the environment. It should prevent significant risks by adopting appropriate technologies and rejecting unpredictable ones, such as genetic engineering. The multi-purpose trees in agroforestry systems play a major role in organic agriculture production in two ways: directly by providing edible products such as fruits, nuts, rhizomes, leaves, and tubers; and indirectly by supporting food production through increasing the soil's ability to support agriculture. A large number of fruit-producing trees are integral parts of traditional homestead and other agroforestry farming systems. Some under-exploited species in agroforestry systems are of immense cultural and economic values viz. medicinal plants. Some well-known agroforestry tree species grown on farms for uses such as fodder, food, fibre or fuel wood, are also used for medicinal purposes. The total area under agroforestry in the world is estimated as 1023 million ha (Nair et al 2009). Currently, Forest Survey of India reported 1,11,554 km^2 (3.39 per cent of country geographical area) area under agroforestry (FSI, 2014).

AGROFORESTRY SYSTEMS FOR HILLS

A large number of traditional organic agroforestry systems have been recognized for different hilly regions. The major practices that constitute the multitude of location-specific systems in the different zones are as follows:

SUB-TROPICAL AND MID HILL TEMPERATE ZONE

The major agroforestry systems generally practiced in the sub-tropical and mid hill temperate zones are as follows:

Agri-Horticultural System

In Sikkim, cultivation of sole maize, as an intercrop with ginger, turmeric, buckwheat, beans, pulse-type beans, tapioca and mandarin-ginger/turmeric system is most widespread in this system. In fact, maize occupies the highest acreage of 40,000 ha in the state. Local cultivars dominated the scenario of the various crops cultivated along with Sikkim mandarin and avocado. Sikkim mandarin + maize + ginger/turmeric intercropping witnesses very heavy application of organic sources that ranges from 40 to 100 t ha^{-1}. Prominent ginger cultivars are Bhaise and Gorubathane. Various kinds of vegetables are also cultivated mostly for self-consumption. Cultivation of flowers such as cymbidium orchids, gladioli, carnation, gerbera, tuberose and marigold is also gaining momentum.

Agri-Horti-Pastoral System

In this system intercropping of cereal, oilseeds, pulses, beans, millets, and spice like ginger and turmeric are cultivated in the interspaces of Sikkim mandarin, avocado, guava and lemon trees. Broom grass is used as a vegetative barrier for soil conservation and performed the role of winter fodder for the cattle, pigs and goats. Collection of ferns, nettles, and other wild edibles is also common practice.

Agri-Silvi-Pastoral System

This system is mainly comprised of agriculture crops along with forest trees. The locations of the trees are mostly on the bunds, but occasionally the trees were also found in-between and middle of the agricultural fields. The major tree based agri-silviculture systems included the tree species like *Grewia optiva*, *Celtis australis*, *Melia azedarach* and *Ficus roxburghii etc*., along with agricultural crops. Crops such as maize, wheat, pulses, buckwheat, oilseeds, beans and finger millet are grown in association with trees that are used a fuel wood or crude timber and winter fodder.

Horti-Silviculture System

This is perhaps the best kind of synergism recorded where large cardamom *(Amomum subulatum* Roxb.) thrives extremely well under the shade of more than 10 species of trees. It performs best under the shade of alder (*Alnus nepalensis*). Large cardamom occupies the largest cultivated area in Sikkim and revealed typical growth requirements that necessitated special mention amongst the spices. Other spices of Sikkim include ginger, turmeric, Red cherry pepper (*Capsicum frutescens* var. *cerasiforme*) and Bay leaf (collected from the forests, *Cinnamomum impressinervium*). Turmeric is also grown both as sole and intercrop (with maize) like ginger as stated earlier during pre-*kharif* season. Red cherry pepper a prized spice is cultivated during *kharif* and sold at premium prices from ₹ 80 to 200/- per kg.

Agri-Horti-Silvipastoral System

In this system maize, tapioca, ginger, beans, pulses, oilseeds and other crops are grown in association with tree species that fulfilled the fruit, fuel wood and fodder requirement. Ginger + maize on raised beds are very common practice. It is the old but reformed system of production, which almost does not use any form of chemical fertilizer or other chemicals and totally depends on organic sources for production. It is a cash crop and cultivated on more than 5000 ha area. Cultivation is done on raised beds. Maize seeds are sown either around the periphery of beds or in the space made between two beds and whole beds are covered with leaves and twigs of various forest trees, weeds, grasses available around the field, animal bedding and surplus rich straw as a mulch amounting 5-20 t ha^{-1} on dry basis. In the opinion of farmers, ferns, *Chromoleana odoratum*, and animal bedding are best mulch for higher productivity of ginger and *Cylindrica imperata* (*seeru*) and leaves of *Schima wallichii* (*chilaune*) are of inferior quality. The raised bed method of ginger cultivation is sustainable because of recycling of nutrients through organic manure with no use of agricultural chemicals. Even farmers get more yield of inter-crop maize compared to sole crop as a result of heavy quantity of manure application to ginger main crop. Maize also provides shade during the early period of ginger growth and after harvesting open space favours the development of bolder rhizomes. The mulch keeps the soil shades and warm, prevents weed infestation, minimizes soil erosion and protects the young plants from heavy rain. After decomposition manure and mulch not only meets the needs of growing plants, but also enriches the soil nutrient status (Patiram et al 1995). The organic manure and mulch loosen the soil for the proper development of rhizomes, so farmers prefer organic manuring for ginger cultivation.

Livestock-Based Mixed Farming

Animal rearing is common to perhaps all the systems irrespective of land holding size. Even landless residents maintain various animals. This placed tremendous pressure on the natural resources. The annual consumption of green fodder and fuel wood ranges from 300-400 tonnes and 70-80 tonnes, respectively, in the subtropical zone, where agriculture is more intense than the higher agro-ecological regions. Green fodder fed to the animals comprises foliage of various shrub and tree species. Most of the animals are traditional, low yielding breeds.

Sericulture-Based Farming

Silkworms are reared on the foliage of mulberry (*Morus laevigata*) and *Terminalia* spp. In the interspaces available crops such as pulses, oilseeds, oats, millets and broom grass are cultivated.

Bamboo-Based Farming

The general observation regarding poor rate of survival of vegetation under the shade of bamboo is an issue of larger discussion. Nevertheless, ginger, turmeric, large cardamom and rice bean grow well up to a distance of 11 to 15 m from the bamboo rows. 10 genera and 24 species of bamboo are reported in Sikkim. The tender shoots of *Chimonobambusa hookeriana* (Munro) Nakai, *Dendrocalamus hamiltonii* Nees & Arn. ex Munro., *Drepanostachyum intermedium* (Munro) Keng f, *Himalayacalamus falconeri* (Hook. f. ex. Munro) and *Phyllostachys bambusoides* Sieb. & Zucc. are consumed in different ways often considered a culinary delicacy (P. Singh, Pers. Commun. BSHC 2000, Gangtok).

Homesteads

The dwellings of the farming community of hills are either on own land holding or on the fringes, where vegetables, fruits, flowers, traditional medicinal plants, fodder requirements are grown in the surroundings. Different kinds of animals are also maintained. Gerbera, carnation, gladioli, tuberose in floriculture, mushroom production (apart from collection of wild species), apiary, and fishery in small tanks (grass carp and magur) are common as homesteads.

Tea Plantation

Tea plantation is another agroforestry system of hill region. *Albizia* species is mainly grown for shades. In Sikkim, the State Government maintains one tea (*Camellia sinensis*) plantation of >500 ha at the altitude range 1550 to 1800 m amsl. Wild cherry (*Prunus cerasoides*) is planted on the boundaries and as avenue trees.

TEMPERATE ZONE

Only three agroforestry systems are observed in the temperate zone *viz*., agri-horticultural, horti-silvi-pastoral and livestock-based mixed farming. The human population density is lower than the earlier two altitude zones. Adaptability of crops to the climate decreases at the higher altitudes and that reduces the farmers' options towards the subalpine zone.

Agri-Horticultural System

This is a very common traditional agroforestry system, where farmers grow horticultural and forest trees in various combinations on the agricultural farm along with agricultural crops. This is the general psychology of the farmers to prefer the horticultural tree species over the forest tree species. Cultivation of agricultural crop along with fruit trees (agri-horticulture) is an exclusive and unique practice in the Garhwal Himalayan region. Fruit trees are planted and retained by the farmers as associate crops in the different places on agricultural fields of this region. The presence of trees also create hindrance during agricultural operation and on the agricultural productivity, but their other benefits bound farmers to grow them in the suitable places. The status of natural regeneration of tree crop is very poor in the existing systems due to the deliberate removal of seedlings from the field during cultural and other operations. The agri-horticulture is a common practice by the farmers mostly having high land holding in order to produce fruits and crops from the same piece of land and to achieve more income. The horticultural trees are found scattered throughout the agricultural fields, while forest trees/MPTs are mostly seen on the bunds of the farmers' fields. The major agri-horti-silviculture systems includes *Grewia optiva* + *Malus domestica* (apple) + wheat; *Quercus leucotrichophora* + *Malus domestica* + wheat/potato; *Grewia optiva* + *Prunus domestica* (wild apricot) + barnyard millet *etc.* buckwheat, maize, barley and some vegetables are cultivated to utilize the interspace. Rice is grown up to an altitude of 2000 m amsl. Seed and table potato and maize are preferred.

Horti-Silvi-Pastoral

Maize, millets, and vegetables like potato (seed and table), cole crops, peas, beans, radish and *Brassica juncea* var. *rugosa* (*Raya sāg*) are grown along with mandarins (up to 2000 m), plum, peach, pomegranate and apple. Large cardamom grows well as an under growth of various tree species up to an altitude of 2700 m amsl.

Livestock-Based Mixed Farming System

Animals are an integral component of all the agroforestry systems. At the high altitudes maintenance of sheep and nomadic herds of *dzo's* (semi-domesticated, cross-bred offspring of yak reared for milk and a range of by-products) is recorded. Grazing pressure increases at the higher altitudes as most of the fodder needs are collected from the wild despite a blanket ban by the many State Governments.

SUBALPINE ZONE

In the higher altitudes the climate reduces the agroforestry systems to one in each zone. Horti-pastoral-transhumance and livestock-based mixed farming (beyond timberline)-transhumance are recorded in the sub-alpine and alpine zones, respectively. The semi-domesticated *dzo* herds graze at the higher altitudes and the survivors of the harsh winters of the higher and trans-Himalayas were re-gathered by the owners the next summer. The extremities of the climate reduce the options to animal-based ones, one in each zone.

Horti-Pastoral-Transhumance

Vegetables perform very well though the alternatives are few. Potato (seed and table), cole crops, peas, beans, radish and *Brassica juncea* var. *rugosa* (*Raya sāg*) produced yields comparable with any other place. Maize could grow well up to 2000 m amsl. Yaks, mules maintained for transportation and animals like sheep and *dzo's* are all nomadic grazers. At places yak herds of more than 200 can be recorded. Sikkim potato is famous for seed, which is harvested in August-September and reaches the plains for seed by November. In this belt farmers prepare the compost by using collected forest litter and animal bedding along with animal excreta in pits. In the month of December and January, after the ploughing of field, all the residue of crops and weeds are collected in heaps and covered with soil. The heaps are burned and residue left after burning spread over the field. 15 days before sowing of seed potato (January to March), the un-decomposed portion of the prepared composed are separated and good quality of manure is heaped near the field. The enriched compost is applied in furrows at the time of potato seed sowing. Pea is inter-cropped giving the space of 4-5 furrows of potato between each row to avoid the shading effect. In Lachen and Lachung valleys of north Sikkim, farmers burn the residues openly entirely on the field and then mixed with the soil by ploughing.

ALPINE ZONE

Livestock-Based Mixed Farming Trans-Humance

The system here is fully animal based with the economy largely driven by the yak both as a source of livelihood and mode of transport for humans and materials. The human population is small and is indulged in production of small quantities of potato, *Brassica juncea* var. *rugosa* (*Raya sāg*), and radish and also cabbage and peas up to 4200 m elevation. The timberline was recorded at 4200 m amsl. The vegetation is mostly herbaceous with woody species of *Rhododendron lepidotum*, *R. setosum*, *R. anthopogon*, *Juniperus prostrata*, *J. recruva*, *Betula utilis* and stumps of *Tsuga* sp. up to 4200 m. Large grazing herds of sheep and dzo's are maintained for meat, milk and other by-products. The pressure on natural resources and animal mortality rate is very high (Avasthe 1996).

A case study on large cardamom (*Amomum subulatum* Roxb.) based agroforestry systems for production, resource conservation and livelihood security in the Sikkim Himalayas (Avasthe et al 2011)

The largest agroforestry system in Sikkim is large cardamom (*Amomum subulatum* Roxb.)-based that supports as many as 23 tree species. India is the major producer of large cardamom, however the plantations are limited only to the sub-Himalayan Sikkim, Darjeeling district of West Bengal and some NE states. Normally large cardamom has been observed to thrive well in association with the Himalayan alder (*Alnus nepalensis* D. Don) and hence, this alliance is also called as the large cardamom-Alnus agroforestry system. Mainly five popular cultivars *viz*., Ramsey, Golsey, Seremna, Varlangey and Sawney are common in Sikkim. Several others include Ramla, Chivey, Ramsey, Ramnag, Madhusay, Seto Golsey, Slant Golsey, Red Sawney, Green Sawney and Mingney, and Varlangey Golsey. Generally August-September for low and mid altitudes and November-December for high altitude are the periods of harvesting. Large cardamom (*Amomum subulatum*) is the most important native cash crop of Sikkim grown under the forest, between 600 to 2000 m altitudes with an annual rainfall of 3000 to 5000 mm. It is a low volume, high value and non-perishable perennial cash crop, which form an integral component of sustainable ecosystem of Sikkim. The average yields range from 100 to 400 kg ha^{-1} of dry large cardamom (Pradhan and Avasthe, 2005). Presently, Nepal is the largest producer of large cardamom with 68% share, followed by India (22%) and Bhutan (9%). Sikkim contributes 88% of the annual production of India (4385 million t). The gross income from large cardamom cultivation in Sikkim increased from ₹ 30 million in 1989–90 to ₹ 240 million in 2001–02 and decreased to ₹ 169.5 million in 2006–07. The estimated large cardamom production in Sikkim during 2016-17 is 4433 t with total revenue of approximately ₹ 6520 million (Spices Board, Gangtok, 2016). The gross income of households and workers was found to be almost double in large cardamom dominated system as compared to traditional mixed forest system. Large cardamom based agroforestry practice retain the tree cover thus, conserving the soil, water and nutrients in the fragile mountain ecosystems. It was a low nutrient exhaustive agroforestry practice conforming to the mountain niche. Performance of large cardamom under the N_2-fixing *Alnus nepalensis*, a shade tree is a common practice, which proved to be highly beneficial in terms of production and nutrient cycling that found the system to be sustainable up to 20 years, and adoption of replantation thereafter for both *Alnus* and large cardamom for sustainability of this system.

BIOMASS PRODUCTION, CARBON SEQUESTRATION AND ECONOMIC SECURITY

Production of biomass (above ground, below ground and total biomass) by different components of the system is mainly affected by age of the perennial woody component, nature and distribution of different components of the system and other abiotic factors. Nayak (1996) also reported that above-ground biomass production was influenced due to species composition and distribution. Different management practices for perennial woody species such as training, pruning and other cultural practices for annual crops also affect the biomass production in agroforestry systems (Deshmukh 1998). Gupta and Singh (1981) reported differences in biomass production of agroforestry systems due to differences in soil conditions, phenology of dominant species, better root networking. Whereas, efficient use of limited resources for maintaining higher photosynthetic activities, leaf area index, better light interaction and water use efficiency were the biomass governing factors reported by Sehgal (1999) and Huxley (1983). Agroforestry systems are believed to have a higher potential to sequester carbon than pastures or field crops. In addition to sequestering C in biomass and soil, these systems can contribute to both carbon conservation and carbon substitution. In temperate regions, agroforestry practices have been estimated to have the potential to store C in the range of 15-198 Mg C ha^{-1}. In the tropics, agroforestry systems are estimated to have helped to regain 35% of the original C stock of the cleared forest, compared to only 12% by croplands and pastures. It has also been estimated that an increase of 1.0 Mg of soil carbon pool of degraded cropland soils may increase crop yields by 20-40 kg ha^{-1}(Nair et al 2008).

CONCLUSION

Agroforestry systems are of multifunctional value. In hills for sustainable development of agroforestry systems, the maintenance of traditional agroforestry systems and creation of new systems is required by designing context-specific silvicultural and farming systems to optimize food production, carbon sequestration and biodiversity conservation. Farmers' participatory approach should be followed for domestication of useful fruit tree species currently growing in the wilderness to provide more options for livelihoods support.

REFERENCES

Avasthe RK, Singh K K, Tomar JMS 2011 Large cardamom (*Amomum subulatum* Roxb.) based agroforestry systems for production, resource conservation and livelihood security in the Sikkim Himalayas *Indian Journal of Soil Conservation* **39** 2 155-160

Avasthe RK 1996 Socio-economic and ecological survey of a high altitude rural ecosystem: A case study from Muguthang, North Sikkim. In: *Proceedings of the National Seminar on Tribal Development: Options*. May 22-24 1996 at G. B. Pant Institute of Himalayan Environment and Development, Almora (U. P.) India

Deshmukh HK1998 Studies on the performance of wheat and soybean as intercrop with *Grewia optiva* pollarded at different heights. M.Sc. Thesis Dr YS Parmar University of Horticulture and Forestry, Nauni, Solan (H.P.) India **90**

FSI 2014 India State of Forest. Ministry of Environment and Forests, Government of India. Dehradun **252**

Gupta SR, Singh JS 1981 Influence of floristic composition on the net primary production and dry matter turnover in tropical grassland. *Australian Journal of Economics* **7** 363-374

Huxley PA 1983 Phenology of tropical woody perennials and seasonal crop plants with reference to their management in agroforestry systems, in *Plant Research and Agroforestry* (eds) PA Huxley 503-526

Nair PKR, Kumar BM, Nair VD 2009 Agroforestry as a strategy for carbon sequestration. *Journal of Plant Nutrition and Soil Scie*nce **172** 10-23

Nair PKR, Gordon AM and Mosquera-Losada MR 2008 Agroforestry, in Ecological Engineering. *Encyclopedia of Ecology* (eds) SE Jorgensen, BD Fath, Vol. 1, Elsevier, Oxford, UK 101-110

Nayak BK 1996 Studies on biomass productivity and nutrient content in *Eucalyptus tereticornis, Leucaena leucocephala* and *Melia azedarach* under high density plantation. MSc Thesis Dr YS Parmar University of Horticulture and Forestry, Nauni, Solan (H. P.)India. **112**

Patiram, Upadhyaya RC and Singh LN 1995 An appraisal of ginger (*Zingiber officinale* Rosc.) production in Sikkim, India. *Journal of Spices and Aromatic Crops* **4** 111-118

Pradhan Y and Avasthe RK 2005 Area and Food Production Statistics of Sikkim. Agricultural Census, Department of Food Security and Agricultural Development, Government of Sikkim **50**

Sehgal S 1999 Influence of coppicing and pollarding on water use efficiency, light transmission and root characteristics in agroforestry tree species. M.Sc. Thesis Dr YS Parmar University of Horticulture and Forestry, Nauni, Solan (H. P.) India.

CHAPTER 9

Role of Wood based Industries in Promotion of Agroforestry and Production of Quality Planting Material

RC Dhiman

ITC-PSPD (Unit Wimco Seedlings), Kashipur Road, Bagwala, Rudrapur, Uttarakhand, India

Corresponding author's Email: dhimanramesh@yahoo.com

ABSTRACT

Private sector is playing a pivotal role in promotion of agroforestry and planting stock production and supply chain for securing raw material production for their factories. The experiment started around four decades back by Wimco Ltd- a safety matches making company in North India is now matured and spread to different parts of the country. The sector has developed ultramodern production facilities for producing quality planting stock and providing technical know-how to grow good plantations along with agricultural crops and take back wood on harvesting. This is backed by a strong research and development, experienced extension staff to interact with farming communities on agroforestry related issues. The skill acquired by numerous employees and workers in these excellent nursery production facilities have started their own nursery production, on separation, as a mean for self-employment. These are now producing and supplying bulk of the planting stock to various agencies.

Keywords: Agroforestry, Planting stock, Private sector

INTRODUCTION

Forestry sector in India is passing through a critical phase when it is trying to balance the role of forests for goods and services in the changing scenario of climate change and carbon mitigation options. According to the last two Indian Forest Policies *viz.*, Indian Forest policy 1952 and Indian Forest Policy 1988, (including the Draft Report of the new Forest Policy under consideration with the Ministry of Environment, Forest and Climate Change, GoI), emphasis has been given to increase its forest and tree cover from the present level of 24 to 33 per cent of its geographical area. The latest State of Forests Report (based on satellite data between Oct 13 to Feb 14 carried out by sampling between 2008-2014, the Forest cover is 70.17 million ha (21.34% of the geographic area), whereas, the tree cover is 9.26 million ha (2.82%) (FSI 2015). Intense biotic pressure and forest fires are causing a huge damage to the exiting forest resources. Growing stock of the Indian forests is 73 cum ha^{-1} compared to a very high of 308 cum ha^{-1} in Cameroon. There is a complete paradigm shift in approach of forest department towards wood based industry. Whereas, 1952 National Forest Policy laid emphasis on production forestry to meet industrial and domestic wood based needs, the 1988 National Forest Policy called for greater substitution of wood wherever possible. The state forest departments are now taking undertaking from the wood based industries that they will not demand/claim wood from the state governments. The new Draft policy pending with the Ministry of Environment, Forests and Climate Change also endorsed the role of forest towards the non-wood production activities.

WIDENING GAP BETWEEN DEMAD AND SUPPLY OF TIMBER AND OTHER FOREST PRODUCTS

There are some overlapping figures on the production, demand and supply of wood and wood based products in the country. According to an estimate, there is a demand for 123 million cum of timber in the country. Against this demand, the production is estimated to be 1.873 million cum (ICFRE 2003), and 0.6 to 1.0 million cum Behera (2005). According to Sood (2014), only 3.2 million cum of wood were produced from Indian forests during 2011. According to Forestry Sector Report (2010), there was a production of 2.3491, 2.4054, 2.61502, 2.32402 and 2.18509 million cum timber during 2006, 2007, 2008, 2009 and 2010, respectively, averaging to around 2.375 million cum yr^{-1}.

The Court's actions coupled with shift in policies and programs and stronger economic growth over the last couple of decades has led to a paradigm shift of wood production from forest to agroforests and also from imports. The trees outside forests now constitute an important wood resource and its production is now significantly increasing for meeting the wood and timber based requirement of the industry and local people. The country is now spending huge foreign exchange in import of wood and wood related articles to meet the gap between their demand and supply. During 2015-16, the import of different articles in India was valued at ₹ 273708657.84 Lakhs, out of which wood and wood related articles constitute around 1.7 per cent of total imports. This does not include import of ₹ 133,446.04 lakhs under HS Code 13 (Lac, gums, resins and other vegetative saps and extracts) which are also largely a forest produce. There are five different HS codes under which wood and wood related articles are imported in the country. During the year 2015-16, import of wood and articles of wood, wood charcoal (HS Code 44) was ₹ 1550233.97 Lakhs; Cork and articles of cork (HS Code 45) was ₹ 3473.07 Lakhs; Manufactures of straw, of esprarto of other planting materials, basket ware and wickerwork (Code 46)- ₹ 1437.25 Lakhs; Pulp of wood or other fibrous cellulose material etc. (H S Code 46) - ₹ 1069885.10 Lakhs; and Paper and paper board, articles of paper or of paperboard (HS Code 48) ₹ 1620457.31 Lakhs (MoC 2016). The total fuel-wood consumption estimated in household sector is 248 million cum and about 13 million cum additional fuel-wood is consumed in hotels and restaurants, cottage industries and cremation of dead human bodies. This makes the total annual consumption of fuel-wood to be 261 million cum which comes from different sources. The production of fuel-wood from forests has been estimated to be 52 million cum (FSI, 2009) and remaining 209 million cum from farmland, community land, homestead, roadside, canal side and other wastelands (Forest Sector Report India, 2010).

According to FSI (2015), there was a drain of foreign exchange equivalent to ₹ 62472.94 million during 2010-11, ₹ 100202.50 million during 2011-12, ₹ 122194.75 million during 2012-13, ₹ 134229.01 million during 2013-14 and ₹ 131266.31 million during 2014-15 due to import of timber and main timber products. These figures are after making corrections for some exports for different countries under these HS codes. The average annual import value of wood and wood related products have increased from around 20 to 45 per cent over the last decade period as given in the following Table-1 (Inferred from MoC 2016). The values are percentage increased per annum based on average of two years imports under different HS codes. In absolute terms the increase in imports for different codes is given in the Fig. 1.

Table 1: *The annual growth (% in value terms) in import of wood and wood related articles during the last one decade*

Year	HS Code					
	44	45	46	47	48	Total
2014-16	18.60	24.32	-4.33	59.62	37.80	34.23
2012-14	32.66	24.32	15.88	12.25	10.08	19.12
2010-12	43.60	24.58	109.34	48.81	42.32	44.21
2008-10	32.47	20.88	49.96	30.86	37.18	33.97
2006-08	26.77	37.88	77.92	27.20	49.19	34.89
2004-06	56.15	24.67	64.73	33.73	42.38	45.42

Whereas, the basic principal of sustainable forest management remained the same, yet in practice we have left most of our forest area vulnerable to encroachments, fires, diversion to non-forest purposes etc. As a result 70 million ha forest area is hardly producing 3 million cum timber almost half of it from salvage marking. The rest comes from agroforests and imports as stated above. Dhiman (2013) has quantified that the commercial agroforestry being practiced in the country to 5 million ha (Eucalyptus 2.00 million ha, poplars 0.3 million ha, *Acacia* spp, 0.70 million ha, Casuarina 0.50 million ha and others as 1.50 million ha). This form of agroforestry is providing a large number of goods and services to the society and nation and is the source of bulk of wood production for wood based industry and domestic consumption. FSI (2015) estimated the growing stock inside forests as 4195.05 million cum and in trees outside forests as 1573.34 million cum. It is interesting to note that around 35 per cent of the total growing stock is on the trees existing on non-forest areas and many of them are grown under short rotation as cash crops in agroforestry by the farmers. For such a venture private sector has developed synergies with farming communities to grow wood raw material for their factories from farm fields.

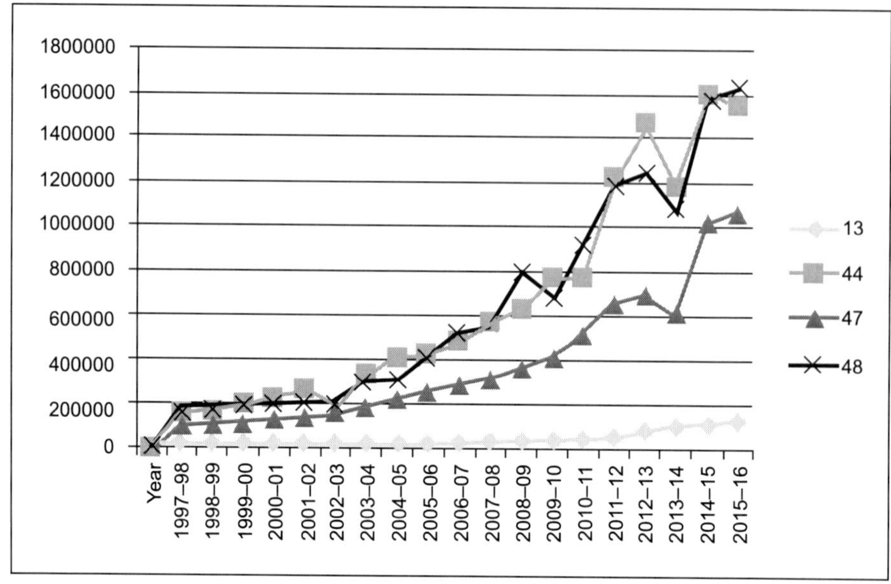

Figure 1: *Increasing imports of Wood and Wood Products (₹ Lakhs) for four HS Code items*

ROLE OF PRIVATE SECTOR

Private sector has now emerged as a key player in promotion of agroforestry and in planting stock production of agroforestry tree species. Private sector could not raise its captive plantations since they cannot hold land over very low ceiling limits fixed in different states. The engagement of private sector in forestry started around four decades back when WIMCO- a safety matches manufacturing company started developing poplar plantations on farmers' fields for match wood production (Dhiman 2012a and 2012b, 2015). Till that date forestry was holistically government controlled activity and was largely restricted to the main role of managing, protecting and conserving forest resources. Forestry sector being in concurrent list both state and central government were enacting laws and regulations for meeting the objectives of Indian Forest Policies enacted from time to time. There were a few public centric programs like Van Mahotsav, Social Forestry, Farm Forestry, Joint Forest Management (JFM) etc. schemes launched invariably in different states to encourage participation of the local people in forestry sector. However, the impact of these programs was not significant since farmers were not having an access to wood markets which were controlled by the government sector. With increased availability of wood from farm grown poplar, the wood marketing started outside government sector, no regulations for felling and transit of some tree species and relaxation of the same by some other states created an enabling environment for development of wood based industry and further expansion of agroforestry in many parts of the country. Subsequently ITC-PSPD, also started massive forestry program backed with its strong R&D and became a brand name for its Eucalyptus clones as Wimco Seedlings is known for poplar clones. Other wood based industry especially paper industries subsequently also started their forestry programs.

RESEARCH AND DEVELOPMENT ON AGROFORESTRY AND TREE SPECIES

Private sector is now a key player in research on developing new clones those are used for growing planting stock of agroforestry tree species. The major planting stock grown by various organizations including government sector belong to those clones developed by the private sector (Dhiman and Gandhi 2012c, 2014a, Kulkarni 2013). Wimco Seedlings (now unit of ITC-PSPD) is known for its poplar clones which are a major source of poplar culture in the country (Dhiman and Gandhi 2015, Dhiman et al 2010). Wimco has developed numerous clones of poplar (*P. deltoides)*, out of which 15 clones viz., WSL 22, WSL A/26, WSL 27, WSL 32, WSL 39, WSL A/49, Wimco 62, Wimco 81, Wimco 83, Wimco 96, Wimco 108, Wimco 109, Wimco 110, Wimco 111, and Wimco 112 are registered and three others *viz*., Udai, Kranti, Bahar are recognized by the International Poplar Commission (FAO). Wimco Seedlings has also developed widely accepted clones of *Eucalypts, Anthocephalus kadamba* and *Melia dubia* etc. There are some other clones of agroforestry tree species developed by some other government institutes but poplar and eucalyptus culture in agroforestry is largely dependent on Wimco Seedlings and ITC-PSPD clones. This is confirmed from a survey conducted in the state of Haryana by Dhiman and Chander (2012) that the major planting stock produced in all the government and private nurseries belong to Wimco introduced clones. The major details of different poplar clones and their share in Wimco's poplar program during 2012 are given in Table 2. Some of the clones *viz*., S7C8, S7C15 (Table 2) have now been replaced with new Wimco's clones. Wimco 110 is now the leading clone with 33% share of our poplar nursery production followed by WSL 22, Wimco 108 and others.

Private sector has also conducted a lot of research and published it in national and international periodicals (Kumar and Dhiman 2015, Kulkarni 2004) on tree-crop interactions (Dhiman 2012c), diseases and insect pests and their controls. The new emerging diseases and pests have been identified, documented and appropriate control measures have been worked out (Dhiman et al 2010). Based on research findings, the sector have been able to provide solutions to these problems faced by the farming community on regular basis. No research finding is useful, if it does not reach the end user. A lot of papers written by private sector managers include those meant for farming community, wood based industry and wood trades. These have been written in newspapers and also in Hindi vernaculars which are widely circulated and are widely consulted by the stakeholders. Unlike government institutes and state universities which now work at regional and local level, private sector operates on wider scale across states and hence the testing, demonstration and acceptance of its clones is better and this sector is, therefore, better recognized in promotion of agroforestry.

Table 2: *Clonal diversity in Wimco's promoted poplar programme*

Clone	Clone share (%) with in state				Total
	Uttar Pradesh	Uttarakhand	Punjab	Haryana	
G-48	19.88	47.18	52.58	27.89	36.22
Udai	14.71	6.44	10.02	19.3	11.9
S7C8	3.97	0.91	0.13	1.18	1.85
S7C15	13.93	0.6	2.63	0.66	5.76
WSL 22	21.87	15.77	16.94	20.07	18.76
WSL 32	2.85	4.36	0.34	4.36	2.92
WSL 39	10.19	13.83	5.08	4.73	9.05
WSL A/49	2.61	0.44	0	0	1.04
WSL A/26	0	1.16	2.47	1.31	1.07
Kranti	2.31	0.24	0.3	4.79	1.59
Bahar	1.42	0.04	0	0	0.51
Wimco 81	3.8	7.71	4	6.56	5.34
Wimco 108	0.56	0.46	1.85	4.57	1.36
Wimco 110	1.15	0.4	0.88	1.3	0.9
G-3	0.48	0	0	0	0.17
Wimco 83	0.09	0.82	1.16	1.23	0.7
Wimco 109	0	0.23	1.62	2.06	0.71
Others	0.09	0.42	0	0	0.15
Total	100	100	100	100	100

Table 3: *Poplar clonal culture by government and private sector*

State	Nursery (No. in 100 thousand)			Field planting (No. in 100 thousand)		
	Gov. sector	Private sector	Total	Gov. sector	Private sector	Total
J&K	4.00	9.00	13.00	0.00	12.00	12.00
Punjab	1.00	71.00	72.00	0.00	56.25	56.25
Haryana	20.00	114.00	134.00	0.00	51.00	51.00
H.P.	1.00	8.00	9.00	1.00	6.75	7.75
U.K.	2.00	90.00	92.00	2.00	51.00	53.00
U.P.	1.00	150.00	151.00	0.00	120.00	121.00
Bihar	4.00	0.00	4.00	0.00	4.50	4.50
Others	1.00	3.50	4.50	0.50	5.00	5.50
Total	34.00	445.50	479.50	3.50	306.50	311.00

PLANTING STOCK PRODUCTION AND SUPPLY CHAIN DEVELOPMENT

Private sector has now emerged as a key player in production and supply of planting stock which is planted by farmers on their farm fields (Fig. 3). The data on nursery and field planting of poplar during 2012 (Table 3) is an indicator of the dominant role of private sector in planting stock production system in the country. In Eucalyptus and poplar planting stock production, ITC- PSPD and Wimco Seedlings (now Unit of ITC-PSPD), respectively are the single major planting stock producers, yet bulk of the planting stock is produced by the individual nurserymen. In case of poplar nursery production, Ex-Wimcoites and in case of Eucalyptus clonal planting stock production Ex-employees of ITC-PSPD together are producing more planting stock than others including Wimco Seedlings and ITC-PSPD themselves.

Table 4: *Poplar plantations acquiring early crown density to qualify as forests*

Age year	Crown Dia (m)	Crown cover tree^{-1} (m^2)	Cum. crown cover (%) of 400 trees ha^{-1}	No. of trees req. for 10% crown cover
1	2.6	5.31	21.24	190
2	3.2	8.05	32.22	124
3	4.1	13.21	52.84	75
4	5.0	19.64	78.56	50
5	5.5	23.77	95.08	42
6	5.8	26.43	105.72	37
7	6.0	28.29	113.16	35
8	6.4	32.18	128.16	31

INCREASING TREE COVER

Unlike plantations made on forest land, farm grown plantations are made on final harvest density. In the former case there is repeated replacement of mortality and death of planted plants reducing the original planting density significantly when they are ready for harvest. On farms, short rotations tend to avoid this tendency of replacement of mortality as a result the farm grown plantations attain the status of forest as per its accepted definition much earlier than those on forest land. For example, data given in Table 4 gives an account of poplar plantations on farms which attain the status of forest in first year of its plantation itself.

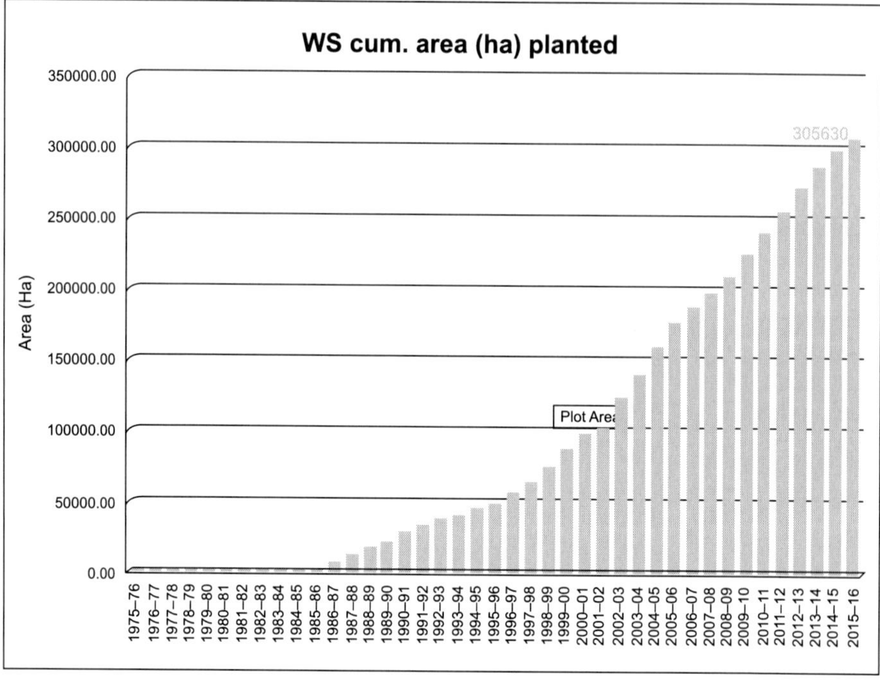

Figure 2: *Cumulative area planted through Wimco's supplied planting stock to farmers*

Wimco Seedlings/Wimco is the single organization outside government sector that has planted a maximum of over 0.3 million ha with poplar (146742 ha), eucalypts (155236 ha), kadamba and some other fast growing forest including some horticultural fruit trees (3652 ha). We consider agroforestry venture needs a single window solution and hence have been dealing with forestry, horticulture and some other plant species to meet the holistic needs of the farming community. Similarly ITC-PSPD has planted over 0.2 million ha till last year and along with its recently aligned Wimco Seedlings have covered an area over 0.5 million ha. It is the only company that is carbon plus for the last many years indicating that it plants more trees and use less raw material. For example, our match business has been able to use only less than 0.5% of total poplar wood production though Wimco Seedlings is the back bone of poplar culture in India. The major use of poplar wood is in panel industry (50%), Firewood (28.5%), paper pulp (14.86%), sports goods (1.37%), ice-cream spoons and sticks (0.71%), packing cases (0.65%) and others (3.5%) (Dhiman 2014a).

BALANCING WOOD PRODUCTION AND CONSUMPTION

Both farm grown tree plantations and wood based clusters are not uniformly distributed throughout the country. The licensing to wood based industry have been closely monitored by the Central Empowered Committee based on the wood production in each state. The licensing has been granted only to that extent to which that state has the wood availability within the state. Whereas, wood consumption was fixed by the CEC on one time data, wood production on farm fields is highly a variable factor decided by millions of farmers based on number of factors, open market conditions has been the major ones. Like agricultural crops more production and glut of a particular crop invariably results in crash in its prices and those crops which are less in production get good prices in those years. This cycle of price fluctuation in agricultural crops is short and a bit loner in farm grown trees. Poplar wood prices given in figure 3 indicate decline in the last couple of years. Poplar wood prices have also witnessed fall during 1995, 2003-2005 and now in 2014. During this phase of surplus supply of poplar wood number of wood based industrial units procures their raw material sourcing to species for which prices are lower in the market. For example, the traditional raw material for paper and pulp making have shifted from bamboo and babbar grass to Eucalyptus, Leucaena and Casuarina. Even these industrial units now procure poplar wood in large quantities during the crash in poplar prices.

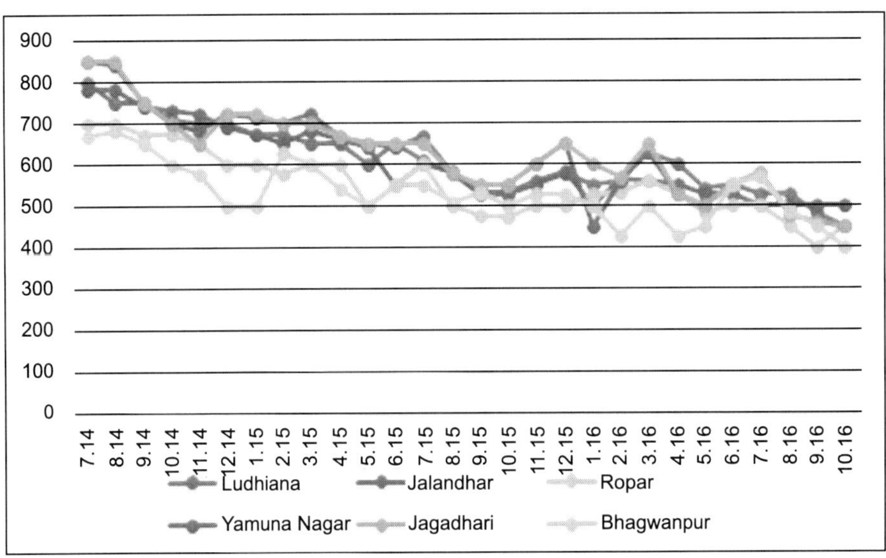

Figure 3: *Price trend for over-sized poplar wood in six mandies of North India*

There are therefore local, regional and national level imbalances in wood production and consumption which sometimes favor wood based industry as is happening these days and sometimes farmers as has happened 5-6 years back when poplar prices have touched over ₹ 10000 tonne^{-1} ex-markets. During a stage of surplus wood production in some locations, it is being exported to wood defect locations and thus wood based industry tries to check persistence of price crash for longer period in any location. A survey conducted for three major paper factories located in the poplar growing region regarding raw material procurement and consumption during March/April 2016 is given in Table 5. These factories have been major user of eucalyptus pulpwood in the past but have changed their procurement to poplar and its wastage from various units in the vicinity. This is simply because they are getting price benefit by using this material during this period of time. When these or any other units will find a price benefit in any other raw material they get shifted to that one. One change in raw material is taking place in wood based industry near ports where they prefer low cost imported wood than costly indigenous raw material. This clearly indicates that wood based industry balances the abundance and low supply by sourcing wood from surplus and low priced locations compared to costly ones. In the recent past when prices of poplar wood were very high in North India, all the above paper units were procuring wood from other low cost locations in India.

Table 5: *Raw material usage in three major paper making factories of poplar growing region*

Species		Bilt Yamunanagar			Century Lalkuan			Star Saharanpur			Total of 3 factories		
		Qty (t)	% of total	% of Wood	Qty (t)	% of total	% of Wood	Qty (t)	% of total	% of Wood	Qty (t)	% of total	% of Wood
Poplar													
	logs	140	24	24	1000	25	44	165	21	21	1305	24	36
	chips							165	21	21	165	3	5
	core	280	47	47	204	5	9	55	7	7	539	10	15
	Veneer waste	140	24	24	1050	26	47	165	21	21	1355	25	37
Total		560	94	94	2254	56	100	550	69	69	3364	62	92
Eucalyptus													
	logs	35	6	6				75	9	9	110	2	3
	chips							75	9	9	75	1	2
	core							25	3	3	25	0	1
	Veneer waste							75	9	9	75	1	2
Total		35	6	6				250	31	31	285	5	8
Others	Bamboo										0	0	
	Baggase				1800	44					1800	33	
	Waste paper										0	0	
Total		0			1800	44		0			1800	33	
G. total		**595**	**300**	**100**	**4054**	**100**	**100**	**800**	**100**	**100**	**5449**	**100**	**100**

SHARING OF KNOWLEDGE, MATERIAL AND EXPERIENCE WITH OTHER STACK-HOLDERS

Over the years private sector has attained a vast experience and knowledge in different aspects of agroforestry and planting stock production. They have a rich germ plasm of the patronized tree species. Some of them like Wimco Seedlings publish its annual research reports and share with main stakeholders, share their authentic germ plasm with those who needs it to conduct experiments, provide technical know-how on different aspects of AF to the tree growers and other. Their research establishmcnts are frequently visited by professional foresters, scientists, students, farmers and many other stake holders. They freely share their germ plasm, knowledge and experience with all other stack holders. This is possibly one of the main reason of their success in this venture.

SKILL DEVELOPMENT

Wimco Seedlings is brand name and known for its poplar clones and ITC-PSPD for its Euca clones. Many ex-employees of these companies on separations and even workers who had worked for some times with them have developed their own nursery production system after acquiring adequate skills. These individual nursery growers have developed planting stock production systems as a mean of self-employment and are producing plants at lower costs than the organized sector since they need not to bear heavy over-head costs of corporate sector. During 2015-16, there was a production of estimated 300 million clonal eucalyptus and around 10 million poplar saplings by the small nursery growers. These nursery growers have now become the dominant stake holder for production and supply of the planting stock to farmers and even sometimes to Government sector. During 2015-16, there were tenders from government sector for procurement of around 50 million eucalyptus clonal plants and around 20 million poplar saplings, most of the planting stock was supplied by these small nursery growers. However the quality of the planting stock production system in many of these nurseries remain under question mark.

AS CORPORATE SOCIAL RESPONSIBILITY ACTIVITY

CSR is an old concept but has now evolved as legal obligation for profit making companies (Dhiman 2014b). With the enactment of the Company Act 2013 and subsequently issuing of the Companies (Corporate Social Responsibility) Rules 2014 by the Ministry of Corporate Affairs, India has become the first country in the world to implement CSR provisions *w.e.f.* April 2014. According to the Act, companies having a net worth of ₹ 500 Crores or having a maximum turnover of ₹ 1000 crore or those having a net profits of at-least ₹ 5 crores have to mandatory spend 2% of the average of last three years profits in CSR activities. Schedule VII of section 135 enlists number of activities where, CSR money can be spend. Agroforestry has been included under CSR activity during 2014, though many companies were already spending their CST money in other activities like sustaining environment, poverty reduction, social development projects etc.

Industry is proactive and many wood based and other industries have started making agroforestry plantations from their profits through CSR activities. According to the Ministry of Corporate Affairs, GoI CSR expenditure of 460 listed companies, which have placed annual returns on CSR on their websites is estimated at ₹ 6337 crores during the year 2014-15 http://www.mca.gov.in /Ministry/pdf/Rajyasabha_unstarred_que999.pdf). The ITC (including that of Wimco Seedlings) CSR initiatives are well distributed in number of sectors those include agroforestry, farm forestry and watershed development. Wimco Seedlings is also supplying planting stock to some of the NGOs promoted by some companies for CSR activity. Reliance industries Ltd. in its annual Report for 2013-14 has spent ₹ 712 crores under CSR activities out of which 59% was spent on health, 23% on rural development (including promotion of environment conservation and making plantations for green cover surrounding operational areas), 11% on education and 7% on other activities. Some reports indicate that at least 1600 companies would be spending around ₹ 20000 to 22000 crore under CSR activities. It is estimated that implementation of CSR regulations may generate one lakh jobs to handle CSR activities. This amount will increase with higher economic growth which is happening presently. The regulations for spending are mandatory around the operation areas in India and if not complied companies need to give reasons thereof.

CONCLUSION

Planting stock production and raising plantations are not usual business activities for any businessman in the country. The initiative is hardly four decades old. During this phase there have been many lesson learnings. The experience of 1980's and 1990's when around 1000 companies mobilized between ₹ 100 billion to ₹ 250 billion from 2.7 to 15 million investors (Saigal et al 2002) has been very bad. The present engagement of private sector is restricted to a couple of good corporate houses to establish back-end and front-end synergy with farmers to sustain the raw material availability, for numerous small nursery growers as a mean of self-employment and for million farmers as mean of diversification of their land production systems to reduce adverse exposures from mono-cropping. The country is facing a policy paralysis for effective operational aspects of agroforestry. The three sector policies i.e., Indian Agroforestry Policy (2014), Indian Forest policy 1988 (including the proposed new Forest Policy) and Indian Agriculture Policy (2002) lay greater emphasis on this land use option for its potential to generate numerous goods and services. However the present success in agroforestry especially commercial agroforestry is purely a private-private relationship between industry and farmer. It is self-sustaining and a win-win situation for all especially the government sector which needs to create a conducive environment for regulating wood markets and removing restrictions on any activity that suppresses its activities, declare farm grown tree produce processing as agro-based industry and extend its advantage to this sector as well. There is hardly any database management system that forecasts trends in the planting, harvest and market prices, as a result famers many a times resist from making plantations as is presently happening in North India.

REFERENCES

Behera P 2005 Timber calculation. *The Indian Forester* 1513-1514.

Dhiman RC 2012a Transforming Rural Uttar Pradesh through Integrating Tree Culture on Farm Land: A Case Study of WIMCO's Poplar Programme, *LMA Convention Journal* **8**(1) 85-98.

Dhiman RC 2012b Status of poplar culture in India. *ENVIS Forestry Bulletin* 12(1) 15-32.

Dhiman RC 2012c Diagnosis of intercrops in poplar based agroforestry. *The Indian Forester* 138(7) **600-609**.

Dhiman RC 2013 Status and impact of commercial agroforestry in India. *Indian Journal of Agroforestry* **15**(2) 55-67.

Dhiman RC 2014a Poplars in India: Past, Present and Future in *Sustainable Forest Management for Multiple Values: A Paradigm Shift* (eds) PP Bhojveid and N Khandekar, Forest Research Institute, Dehradun, 183-206.

Dhiman RC 2014b Corporate Social Responsibility (CSR): An opportunity for wood based industry to strengthen the back-end synergy with farmers to produce industrial raw material. *Ply Gazette* 86-96.

Dhiman RC 2015 Contractual arrangements for sustaining agroforestry with special reference to WIMCO's poplar program in *Agroforestry: Present Status and Way Forward* (eds) SK Dhyani, Ram Newaj, Badre Alam and Inder Dev, New Delhi: Biotech Books 135-168.

Dhiman RC and Chander J 2012 Expanding poplar culture and plywood industry. *Ply World* (April-May) 20-29.

Dhiman RC and Gandhi JN 2012 Clonal development and diversity in WIMCO's poplar programme. *ENVIS Forestry Bulletin* **12(1)** 40-48.

Dhiman RC and Gandhi JN 2014 Status of eucalypt clonal culture in India in *Eucalypts in India* (eds) PP Bhojveid, S Kaushik, YP Singh, D Kumar, M Thapliyal and S Barthwal. ENVIS Centre on Forestry, National Forest Library and Information Centre, FRI, Dehradun 60-116.

Dhiman RC and Gandhi JN 2015a Cultivation of high value crop-Sweet corn (*Zea mays* Linn.) in poplar based agroforestry. *Indian Journal of Agroforestry* **17**(2) 70-75.

Dhiman RC and Gandhi JN 2015b Development and release of clones: A case study of 1999 poplar breeding population at Wimco Seedlings. *Journal of Tree* **34**(2) 1-11.

Dhiman RC, Kulkarni HD and Gandhi JN 2010 Spreading infestation of gall wasp (*Leptocybe invasa* gen. & sp. N.) on eucalypts in North India. *The Indian Forester* **136**(6) 791-803.

FSI 2009 The State of Forest Report. Forest Survey of India, Dehradun.

FSI 2015 The State of Forest Report. Forest Survey of India, Dehradun.

ICFRE 2003 Forestry Statistics India –2003. Indian Council of Forestry Research and Education, Dehradun, Uttranchal.

Kulkarni HD 2013 Pulp and paper industry raw material scenario - ITC plantation a case study. *IPPTA* **25**(1) 79-89.

Kulkarni HD 2004 Clonal forestry for industrial wood production. An ITC experience in Compendium on clonal forestry (eds) KT Parthiban, M Paramathma and KS Neelakantan, Coimbatore: NATP-FCRI Publication 92-113.

Kumar D and Dhiman RC 2016 Country report on poplars and willows period: 2012-2015. IPC (FAO) (http://www.fao.org/forestry/44756-09ec50609435431af805e892765 a686e3. pdf).

MoC 2016 Ministry of Commerce: Import and Export of commodities.

Saigal S, Arora H and Rizvi SS 2002 The New foresters: The role of private enterprise in the Indian forestry sector. Ecotech Services (India) Pvt. ltd (EPS), New Delhi.

Sood D 2014 Wood and Wood Products in India 2014. GAIN Report No. IN4049. Global Agricultural Information Network, USDA Foreign Agricultural Service, New Delhi.

CHAPTER 10

Potential of Short Rotation Forestry and Agroforestry for Climate Change Mitigation and Sustainability

Abhishek M Tripathi*, Radek Pokorny, Vikas Kumar[1], SK Chauhan[2], Michal V Marek

[1]Global Change Research Institute AS CR, v.v.i., Bělidla 986/4a, 603 00 Brno, Czech Republic

[1]Department of Silviculture and Agroforestry, College of Forestry, Kerala Agricultural University-680656 Trissur, Kerala, India

[2]Department of Forestry & Natural Resources, Punjab Agricultural University, Ludhiana-141 004, Punjab, India

*Corresponding author's Email: manicfre@gmail.com

ABSTRACT

The potential role of Short Rotation Forestry (SRF) and Agroforestry (AF) systems indicates conscious production of biological resources to mitigate climate change and also minimize the demand of sustainable production of biological resources, enhance food and energy security without deterioration of soil fertility and productivity in developed and under developing countries. Indeed, the primary goal of SRFi biomass production mainly for energy purpose, with the basic principle of growing fast-growing woody species (*Populus* and *Salix*) in a high density under Short Rotation Coppice (SRC) management. The harvest period of SRC generally varies from 2 to 6 years but also depends on genetic plant material, site quality, management practices, climatic and edaphic factors and purpose. Agriculture is one of the important sectors where the impacts of climate change surpass tolerance limits with implications for the livelihoods of millions of smallholder farmers dependent on this sector. SRF and AF, because of their capacity to support climatic and economic benefits, they are appraised as best "no regrets" measures in making societies adapt and become resilient to the impacts of climate change. SRF and AF systems can play a major role in the adaptation to climate change including changes in the reducing of the greenhouse gas emissions (mainly sequestration of carbon-C emissions), mediate microclimate (maintain the temperature, soil moisture under canopy), provide protection through provision of permanent cover, opportunities for diversification of the agricultural systems, water and climatic resources, enhance soil fertility, and provide protection from drought.

In addition, SRF and AF clearly offer economic and ecological advantages. This article suggests the potential of SRF and AF to mitigate the climate change and an option for sustainable land use system.

Keywords: Bioenergy, Biomass, Ecological services, Poplar, Short rotation coppice

INTRODUCTION

Short Rotation Forestry (SRF) and Agroforestry (AF) have the potential to check the soil depletion and enhance site quality through interactions among trees, agricultural crops, soil, and may be livestock. It can contribute to the improvement of rural welfare via a variety of direct (production of food and energy security) and indirect (soil nutrient improvement and water management, climate enhancement, carbon sequestration and climate change mitigation) mediums. These combined (SRF and AF) have been recognized as having greatest potential to several challenges including worsening poverty, food insecurity, low income base and disease pandemic arising due to a mix of agro-ecological factors (incessant drought, low soil fertility, low productivity, forest degradation and deforestation, environmental degradation) and other man-made problems (illiteracy, population explosion, unequal development, unfavorable development policies) in the developing and developed countries (Barolo and Chuda 2011). Factually, AF development is a traditional crop management in both developed and developing countries (Nair 1993) revolving around two issues-growing food crops in the forests and establishing tree-crop production systems on arable lands with major focus to food production (Kumar and Nair 2006). SRF development may be defined as the silvicultural practice underneath which high-density (5000 to 20000 trees ha^{-1}); sustainable plantations of fast-growing tree species (poplars, willows and eucalyptus) produce woody biomass on arable land or over fertile but degraded forest land (Tubby and Armstrong 2002, Mitchell et al 1999, Sage 1998). Trees are cultivated either as single stems (origin from seeds or hardwood cuttings) or as multi-stems (coppice systems), with a rotation period of maximum twenty to thirty years. The SRF cultivation should enhance the use of natural resources, cost effectiveness, and ecologically, physical, theoretical and practical knowledge (Ceulemans and Deraedt 1999). In developed and fast growing developing countries wood production is not sufficient for forest industries. SRF on arable land and on over fertilized land could offer a way to provide sufficient wood to forest industries and people in developed and fast growing developing countries in world with sufficient fuel while enhancing the forest cover (Christersson and Verma 2006). Under the SRF produced biomass can be used for pulp and paper, building materials, fodder and energy, it may replace wood from tropical forests and from protected forest covers and consequently help to conserve valuable natural forest for future generations. For SRF willows (*Salix* spp.) and poplars (*Populus* spp.) are more common, which are cultivated on agricultural lands and other land use systems. In AF poplars, eucalyptus, loboly pine and paulownia are cultivated in many countries such as China, India and Indonesia etc. (Nair 1993). In the current study, we are reviewing existing literature on tree plantations on farm lands of SRF such as poplar and short rotation coppice (SRC) management and AF such as poplar with annual crops (inter cropping). In addition, Ceulemans et al (1990), Verwijst et al (2004) and Dillen et al (2013) have reported that SRC denotes to a perennial, fast-growing, high-production woody crop that is harvested every two to six years and managed under a coppice management. According to Verwijst et al (2013) and Ceulemans et al (1990) two genera (*Populus* and *Salix*) from family Salicaceae can be the best trees to grown under SRC for heat and power generation. The main objective of this paper is to review briefly the potential risks and probable opportunities that are associated with the expected changes in global climates and current ideas about how SRF and AF systems could be used to mitigate the climate change or adapt the predicted impacts of climate change on forest and agriculture.

POTENTIAL IMPACT OF CLIMATE CHANGE ON AGRICULTURE AND FOREST

The role of forest and agriculture in the social and economic progress of developing and developed countries, forest fire, drought, CO_2 concentration changes and the vulnerability of forest and agricultural systems to the impacts of climate change have shown significant attention from the scientific community (Fischer et al 2002, Rosenthal and Kurukulasuriya 2003, Kirilenko and Sedjo 2007). The existing literature suggested that the overall impacts of climate change on forest and agriculture especially in the tropical regions will be highly negative, although in a few areas there may be minor increases in crop yields in the short period of time and vegetation can shift toward favorable climatic conditions (Kurukulasuriya and Mendelsohn 2008, Kirilenko and Sedjo 2007). The major processes of forest and agriculture that are directly influenced by climate change include effect on soil moisture, carbon and nutrient cycle, crop growth and production, forest migration, and incidence of weeds, pests, and diseases. These effects are expresses in terms of heat stress, increased evapotranspiration rate, increasing concentration of CO_2 shortened seasons, irregularities in the patterns of rainfall and increased photosynthesis and reduced water use due to higher CO_2 in the atmosphere. The consequences of numerous assessments of impacts of climate change on forest and agriculture (Karl et al 2009, Kirilenko and Sedjo 2007, Lavalle et al 2009) are usually in agreement and have recognized the following major challenges that can faces due to climate change in future forestry and agricultural systems: (i) shifting of vegetation and annual crops due to the increase in global average annual temperature, (ii) water requirements could be increased due to high evapotranspiration rate, (iii) drought can be a severe problem for the plants in most places, (iv) temporal and spatial changes in the land availability for forest and agriculture with tropical to temperate countries (v) enhancement in degradation of land resources from erosion due to the predicted disasters, and (vi) increased cost inputs due to high demand of resources (energy, food and timber), due to increasing population and deforestation of forests.

ADAPTATION CHALLENGES

Communities have shown interests to rapidly adapt to climate change but there are many challenges to come in practice. Societies have shown resilience and capacity to adapt to climate change through experimental approach, observations and other sectors are beyond that of a natural self-correcting process and also need supportive policies for rules and regulations to mitigate the impacts expected from climate changes while meeting household level daily needs such as food, energy and financial stability (Killeen and Solórzano 2008). Furthermore, for fast and large scale adoption, interventions should be refined, that are easy to adopt without requiring a complete shift from what societies are currently doing and that build on existing knowledge and practices through which societies have already adapted to the several risks related to variability in the climate (Conway and Schipper 2011, Bryant et al 2000). As societies are unprotected to unforeseen or have not expected changes in weather regularities and increased chance of risk, more strong adaptation plans are required to manage the extra risk. The couple of adaptation challenges (Rao et al 2007) are mentioned below:

(i) An effective control of climate related risks through innovative and sustainable techniques diversified the farm activities. The management of the heat stress on crops (forest and agriculture), animals and or both which requires use of crop diversities and management systems those do well under a broad range of soil and climatic conditions.

(ii) Move backwards land degradation through adoption of applications that reduce erosion and loss of organic carbon and maintaining soil fertility and productivity by keeping nutrient mining and sustaining soil fertility. Limiting greenhouse gas emissions and encouraging carbon sequestration by promoting management options that reduce tillage and use of fuels.

(iii) Protecting against pest and disease pressure and enhancing the resilience of societies by better targeting investments and improving their use efficiency and ensuring food and energy security.

ROLE OF SRF AND AF TO MITIGATE CLIMATE CHANGE

SRF crops may a substitute of fossil fuels (major source of greenhouse gas emissions) in order to mitigate greenhouse gas emissions and reduce the risk of climate change mainly global warming and in addition to fulfill the energy requirement (McKay 2011, Dhillon and von Wuehlisch 2013, Trenberth and Fasullo 2013). Emitted CO_2 from combustion can be balanced by amount taken up from plant growth and biomass production (Fig. 1). AF concerns with soil and water conservation besides improving soil fertility (McCarthy et al 2011), and provides both subsistence and income for increasing population under the condition of land and water stress (Godfray et al 2010, Lasco et al 2014, Kumar 2016a). AF may provide an opportunity to prevent the adverse impacts of climate change through the joint action of adaptation and mitigation (Lin 2007a). In an AF study (Torquebiau and Kwesiga 1996) have observed fallows with *Sesbania sesban*, decreased soil bulk density and improved water infiltration explaining better early growth of the subsequent crop. Tree roots can reach 7 m deep in two years and represent 1.7 to 2.9 Mg ha^{-1} after two years, i.e. about 0.6 to 1 Mg C ha^{-1}. SRF production on unfertilized lands showed the highest energy use efficiency as compared with other annual and perennial bioenergy crops (Nassi O Di Nasso et al 2010, Verwijst and Telenius 1999, Kumar 2016b). In European Union (EU), the reduction of greenhouse gas emissions via the displacement of fossil fuels by 'carbon lean' alternatives is one of the biggest drivers behind the establishment of SRF mainly SRC. Biomass through SRF can be a good sink for CO_2 sequestration. The complex problem of soil carbon emissions and dynamics are excluded completely. Cultivating soils with a high carbon content for the production of energy crops, using intensive crop management regimes including inorganic fertilizers, irrigation and pesticides, inefficient fuel processing and fuel transport logistics, sub optimum conversion technologies, all of these can reduce the potential for biomass systems to reduce carbon emissions from a 'complete life cycle' perspective (Fig. 1 and 2). We accept that SRF and AF interventions provide the best "no regrets" adaptation measures in making societies resilient to the impacts of climate change and do discuss the same in relation to the challenges posed by the changing and variations in climate.

MODERATING THE MICRO-CLIMATE

SRF and AF systems play a critical role in moderating the micro-climate where the full potential of many crops and varities can only be realized when climatic conditions are close to optimum. A climatic condition has direct impact on production, growth and development of crops (Stigter 2015, and Lin 2010). SRF and AF systems with appropriate trees shade offers a significant impact to alter the effects of heat stress and maintain the temperature locally.

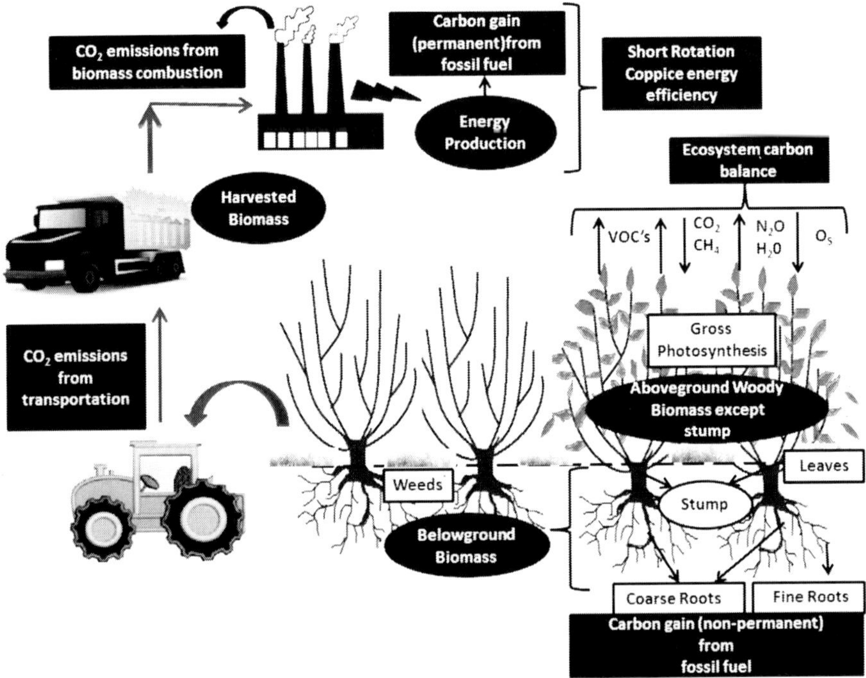

Figure 1: *Pictorial description of life cycle assessment (LCA) of short rotation coppice (SRC), carbon balance (C), and VOC's (Volatile Organic Compounds)*

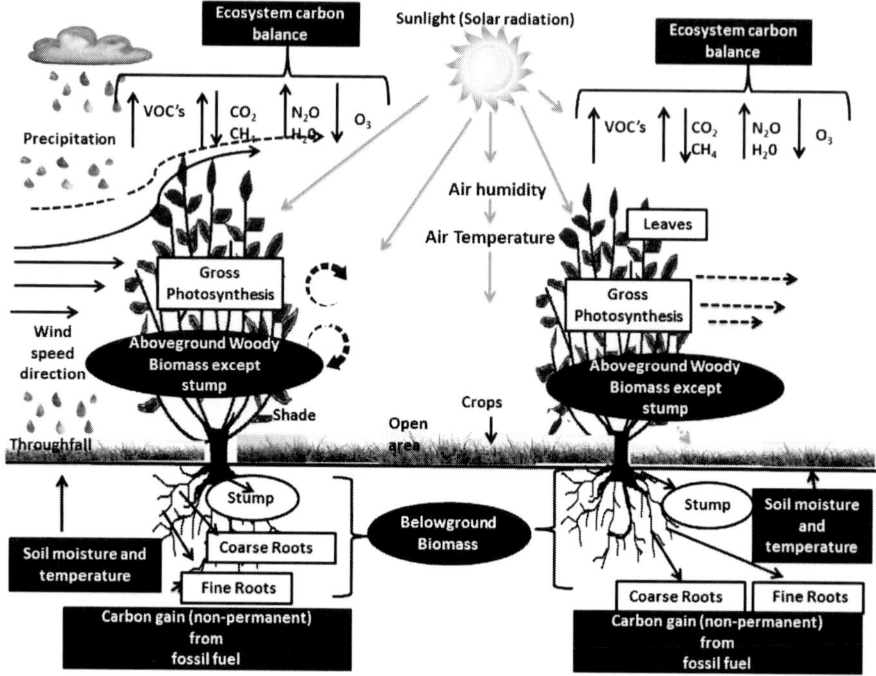

Figure 2: *Pictorial description of microclimate modification and life cycle assessment (LCA) of agroforestry (AF) carbon balance (C), and VOC's (Volatile Organic Compounds)*

Trees on farm under SRF and AF management bring favorable changes in the microclimatic conditions by influencing air temperature, radiation flux, wind velocity, relative air humidity, saturation deficit of beneath the canopy crops all of which will have a significant impact on modifying the rate and duration of photosynthesis and subsequent plant growth, transpiration, and soil water use (Monteith et al 1991, Novák and Havrila 2006, Fischer et al 2013, Sharma et al 2016). Normally, shade will create micro-climates with lower seasonal means in high temperature and solar radiation as well as smaller fluctuations. For instance, the beneficial aspects of micro-climatic alterations are extensively used shade trees to protect heat sensitive cash crops like coffee, cacao, ginger and cardamom from high temperatures, wind breaks and shelter belts to slow down the wind speed to reduce evaporation and physical damage to crops, mulches to reduce soil temperature and various crop tree mixes to reduce erosion and maximize resource use efficiency (Steffan-Dewenter et al 2007). In a study, Lin (2007b) have observed that shade of trees buffer high and low temperature extremes by as much as 5 °C which can favor coffee and cocao plantations for optimum growth and production. Crop cover of the production systems may not reach the high canopy closure to affect relative humidity (De Frenne et al 2015). SRF and AF systems probably yielded higher transpiration rate which enhanced evaporative cooling effect resulting in lower soil temperatures. Mourtzinis et al (2016) has observed that the plant transpiration may be demonstrated to modify climatic conditions at the land surface and in the soil during the vegetative season. There are several ways by which climate change manifests soil degradation. Higher temperatures and drier conditions lead to lower organic matter accumulation in the soil resulting in poor soil structure, reduction in infiltration of rain water and increase in runoff and erosion (Rao et al 2007) while the expected increase in the occurrence of extreme rainfall events will adversely impact on the severity, frequency, and extent of erosion (WMO 2008). These changes will further exacerbate an already serious problem the continent is facing.

SUSTAINABLE DIVERSIFICATION OF ENERGY, AGRICULTURAL SYSTEMS AND SOCIO ECONOMICS

SRF and AF have been identified for two to three decades as a sustainable agricultural practice (Jug et al 1999, Verwijst et al 2013, Nair et al 2010, Wilson and Lovell 2016, Rao et al 2007). Traditional AF is one of the oldest practices adopted by the farmers to reduce/spread the risks and capitalize on the opportunities associated with variable climate through better exploitation of potential synergies and complementarities among different farm enterprises (Below et al 2010). Integrated AF systems can be a right pathway for sustainable diversification of crops yield. The fast growing trees (willows and poplar) have become a major tree component on many farms or arable land in Europe for SRF under short rotation coppice management (McKay 2011) and poplar in South Asia for AF under intercropping management (Chauhan et al 2011, Chauhan et al 2013). In Africa and Latin America fruit yielding trees are very common for home gardens with a diverse range of vegetable and have a promising contribution to food security by providing products throughout the year round (Maroyi 2009, Das and Das 2013, Das and Das 2005, Wezel and Bender 2003). The SRF crops on former agricultural lands potentially lead to significant soil nutrient, biomass production for bioenergy, sustainable land use, enhance the biodiversity such as providing habitat for wildlife, reducing pollutant in water, phytoremediation and as potential carbon sink (Smeets and Lewandowski 2005, Bromilow et al 2007, Pandey et al 2016, McKay 2011). SRF and AF could be used to manage flood risk, reduced cultivation of soil associated with a change of land management from annual crops to SRF, beneficial in reducing soil erosion and increasing the organic matter content of the soil, they also could be grown on flood land using modest amounts of irrigation (compared to fruit and vegetable crops) and also have potential to reduce fertilizer input on agricultural land as compared to conventional crops (Amichev et al 2010, Pandey 2007, Sileshi et al 2007). SRF and AF address to maintain sustainability through conserving forest ecosystems, energy sources, limitations of fossil fuels and farmland biodiversity simultaneously by diversifying farm production to enhance the coping capacity of small farmers to climate risks, rapidly increasing demand from growing economies such as China and India may reflect a new era in energy provision (Styles and Jones 2008, Lasco et al 2014, Pandey 2007). According to (Torquebiau 1992) AF under home gardens can contribute to food and nutrition of households where he observed that up to 44 per cent of calorie and 32 per cent of protein uptake are met by the products from home gardens. In addition, home gardens and intercropping in AF generate good source of cash income for households (Méndez et al 2001, Chauhan et al 2015) and SRC provides biomass bioenergy, if suitable market structures are put in place (Mitchell et al 1999, Aylott et al 2008, Faasch and Patenaude 2012), SRF is more common in developed countries (European Union,

America and Australia etc.) while AF is more popular in developing countries (China, India, Indonesia, Latin America etc.) for energy, timber, and food production (Rockwood et al 2004, Albrecht and Kandji 2003, Torquebiau 2000). Plantation view of SRF and AF are shown in Fig. 3.

Figure 3: *(A & B) shows agroforestry (AF) plantation with poplar in India (C & D) show short rotation forestry (SRF) with poplar in Czech Republic*

It is now being recognized that expanding market opportunities for smallholders particularly in niche markets and high value products are critical to the success of agroforestry innovations (Russell and Franzel 2004). The major constraints to the growth of the small holder tree product sector in Africa are forest policies, physical and social barriers to smallholder participation in markets, the overall lack of information at all levels on markets for agroforestry products, the challenges to outgrowing schemes and contract farming. Notwithstanding these constraints, there are promising developments including contract fuel wood schemes, small-scale nursery enterprises, charcoal policy reform, novel market information systems, facilitating and capacity building of farmer and farm forest associations, and collaboration between the private sector, research and extension (Russell and Franzel 2004, Torquebiau 1992, Trinh et al 2003, Galhena et al 2013).

ADOPTION OF SRF AND AF SYSTEMS

Since last few decades, research has demonstrated that SRF and AF systems can generate considerable environmental and socio economic benefits while meeting the instant requirements of food, energy and income, when decision with carefully selected suitable species that meet end user requirements and when the competition for growth resources between the tree and the crop component is appropriately managed. In the past, environmental services such as carbon sequestration, maintain the water requirement and ecological conservation have been recognized as global environmental services and few extent are being financed. In Indonesia, Cacho et al (2003) and Tomich et al (2002) predicted that how carbon sequestration payments would change the relative returns to alternative land use systems using the data from the Alternatives to Slash and Burn (ASB) programme. Usually, the adoption decision by farmers for SRF and AF is governed by many issues which are mentioned below (Sileshi et al 2007, Ajayi and Kwesiga 2003): (i) positive influence on farmers adoption action, (ii) negative effects (something inconclusive), (iii) no clear direct or indirect effect, and (iv) systemic influence on all types of households in a particular areas (developed and developing countries).

For sustained high biomass productivity SRF is adopted along with cultivation and replanting, after which effects on the soil of repeated disturbance would have major long term effects on both physical and chemical soil properties. The available literature suggests that there are a lot of opportunities to make the SRF and AF systems more beneficial. There is a need to quantify existing SRF and AF systems to determine the interaction between component species, to characterize the trees used, and then to purify the systems in view of soil, climate, and socio-economic boundaries. There is a need for future soils research, which should include chemical, physical and biological elements and the long term balance of these elements can only be assessed if the harvesting practice is defined at the start. The farmers' perceptions 'play an important part in adoption decisions and must be taken into consideration' in policy development for SRF or SRC and AF (Charles et al 2013, Verchot et al 2007, Dallemand et al 2008, Briggs 2012). For intercropping AF, detailed studies on the competition and complementarity between trees and underneath canopy agricultural crops for solar radiation, space, and soil factors are needed. Future systems should integrate trees that are tolerant to strong light and high evaporative demand and crops that are adapted to shade and relatively high humidity.

CONCLUSION

The SRF and AF can be a robust tool for climate change mitigation in terms of economic gain, and reducing the carbon emissions from different sources. It has a great potential to sequestrate the carbon, maintain the microclimate under canopy, enhance the soil fertility and SRF especially SRC can be one of the best alternative of energy sources to replace the fossil fuels. There are many possible limiting factors to achieve food and energy security, maximum growth rate of trees, and demonstrating the potential of the practice. SRC for SRF and inter cropping for AF can be a good opportunities of sustainable sources in world as per the country needs and mitigate the climate change. SRF can be a good system for land change and land use to produce bioenergy and AF for microclimate modification to sensitive crops which are very sensitive to optimal temperature such as coffee. The larger role of SRF under SRC management in offsetting the terrestrial CO_2 emissions and mitigating global climate change has been established unequivocally, which calls for more intensified research on terrestrial C sequestration through AF. More studies are needed to describe benefits of AF in developed countries while use of SRF plantations in tropics and developing countries.

ACKNOWLEDGMENTS

We are thankful to the Global Change Research Institute (AS CR, v.v.i., Czech Republic) for valuable input and Indian Society of Agroforestry and ICAR-Central Agroforestry Research Institute, Jhansi, India for the invitation to contribute a paper in a National Symposium on "Agroforestry for environmental challenges, sustainable land use, biodiversity conservation and rural livelihood options.

REFERENCES

Ajayi OC and Kwesiga F 2003 Implications of local policies and institutions on the adoption of improved fallows in eastern Zambia, *Agroforestry Systems* 327–336

Albrecht A and Kandji ST 2003 Carbon sequestration in tropical agroforestry systems, *Agriculture, Ecosystems and Environment* **99**(1–3) 15–27

Amichev BY, Johnston M and Van Rees KCJ 2010 Hybrid poplar growth in bioenergy production systems: Biomass prediction with a simple process-based model (3PG), *Biomass and Bioenergy* **34**(5) 687–702

Aylott MJ, Casella E, Tubby I, Street NR, Smith P and Taylor G 2008 Yield and spatial supply of bioenergy poplar and willow short□rotation coppice in the UK, *New Phytologist 178*(2) 358-370

Barolo M and Chuda M 2011 Final public international conference of the Benwood project Short Rotation Forestry and Agroforestry : an Exchange of Experience between CDM Countries and Europe, Published by the Benwood project consortium Organizing Committee (Benwood consortium)

Below T, Artner A, Siebert R and Sieber S 2010 Micro-level practices to adapt to climate change for African small-scale farmers, *A review of Selected Literature. Environmental Production Technology Division 953*

Bromilow R, Karp A, Riche A, Richter G, Watts C, Whitmore A, Finch J and Robinson M 2007 The benefits and negative impacts of biomass crops on resource protection 1–18

Bryant CR, Smit B, Brklacich M, Johnston TR, Smithers J, Chjotti Q and Singh B 2000 Adaptation in Canadian agriculture to climatic variability and change, *Climatic Change* **45**(1)181-201

Cacho OJ, Marshall GR and Milne M 2003 Smallholder agroforestry projects: potential for carbon sequestration and poverty alleviation, Rome, Italy: The Food and Agriculture Organization of the United Nations (FAO). 81

Ceulemans R, Stettler RF, Hinckley TM, Isebrands JG and Heilman PE 1990 Crown architecture of Populus clones as determined by branch orientation and branch characteristics, *Tree Physiology 7*(1-2-3-4) 157-167

Ceulemans R and Deraedt W 1999 Production physiology and growth potential of poplars under short-rotation forestry culture, *Forest Ecology and Management* **121**(1–2) 9–23

Charles R, Munishi P and Nzunda E 2013 Agroforestry as Adaptation Strategy under Climate Change in Mwanga District, Kilimanjaro, Tanzania, *International Journal of Environmental Protection* **3**(11) 29–38

Chauhan SK, Gupta N, Walia R, Yadav S, Chauhan R and Mangat PS 2011 Biomass and carbon sequestration potential of poplar-wheat inter-cropping system in irrigated agro-ecosystem in India, *Journal of Agriculture Science Technology* **1** 575-586

Chauhan SK, Sharma R, Singh B and Sharma SC 2015 Biomass production, carbon sequestration and economics of on-farm poplar plantations in Punjab, India, *Journal of Applied and Natural Science* 7(1) 452-458

Chauhan SK, Dhillon WS, Singh N and Sharma R 2013 Physiological behaviour and yield evaluation of agronomic crops under agri-horti-silviculture system, *International Journal of Plant Research* *3*(1) 1-8

Christersson L and Verma K 2006 Short-rotation forestry - A complement to "conventional" forestry, *Unasylva* **57**(1) 34–39

Conway D and Schipper ELF 2011 Adaptation to climate change in Africa: Challenges and opportunities identified from Ethiopia, *Global Environmental Change* **21**(1) 227–237

Dallemand EJF, Petersen JE and Karp A 2008 "Short Rotation Forestry, Short Rotation Coppice and perennial grasses in the European Union : Agro-environmental aspects, present use and perspectives

Das T and Das AK 2005 Inventorying plant biodiversity in homegardens: A case study in Barak Valley, Assam, North East India, *Current Science* **89**(1) 155–163

Das T and Das AK 2013 Vegetative and reproductive phenology of some multipurpose tree species in the homegardens of Barak Valley, northeast India, *International Journal of Biometeorology* **57**(1)185–196

Dhillon RS and von Wuehlisch G 2013 Mitigation of global warming through renewable biomass, *Biomass and Bioenergy* **48** 75–89

Dillen SY, Djomo SN, Al Afas N, Vanbeveren S and Ceulemans R 2013 Biomass yield and energy balance of a short-rotation poplar coppice with multiple clones on degraded land during 16 years, *Biomass and Bioenergy* ***56*** 157-165

Faasch RJ and Patenaude G 2012 The economics of short rotation coppice in Germany, *Biomass and Bioenergy* **45** 27–40

Fischer G, Shah M and Velthuizen H Van 2002 Climate Change and Agricultural Vulnerability, *World* 152

Fischer M, Trnka M, Kučera J, Deckmyn G, Orság M, Sedlák P, Žalud Z and Ceulemans R 2013 Evapotranspiration of a high-density poplar stand in comparison with a reference grass cover in the Czech–Moravian Highlands, *Agricultural and Forest Meteorology* ***181*** 43-60

De Frenne P, Rodríguez-Sánchez F, De Schrijver A, Coomes DA, Hermy M, Vangansbeke P and Verheyen K 2015 Light accelerates plant responses to warming, *Nature plants* ***1*** 15110

Galhena D, Freed R and Maredia KM 2013 Home gardens: a promising approach to enhance household food security and wellbeing, *Agriculture and Food Security* **2**(1) 8

Godfray HCJ, Beddington JR, Crute IR, Haddad L, Lawrence D, Muir JF, Pretty J, Robinson S, Thomas SM and Toulmin C 2010 Food security: the challenge of feeding 9 billion people, *Science* ***327***(5967) 812-818

Jug A, Makeschin F, Rehfuess KE and Hofmann-Schielle C 1999 Short-rotation plantations of balsam poplars, aspen and willows on former arable land in the Federal Republic of Germany. III. Soil ecological effects, *Forest Ecology and Management* **121**(1) 85-99

Karl TR, Melillo JM and Peterson TC 2009 *Global Climate Change Impacts in the United States*, Cambridge University press

Killeen TJ and Solórzano L 2008 Conservation strategies to mitigate impacts from climate change in Amazonia. *Philosophical transactions of the Royal Society of London. Series B, Biological sciences* **363**(1498) 1881–1888

Kirilenko AP and Sedjo RA 2007 Climate change impacts on forestry, *Proceedings of the National Academy of Sciences of the United States of America* **104**(50) 19697–702

Kumar BM and Nair PKR 2006 Tropical Homegardens, *Advances* 129(October) 383

Kumar V 2016 Multifunctional Agroforestry Systems in Tropics Region, *Nature Environment and Pollution Technology* **15**(2) 365-376

Kumar V 2016 Bio-fuel production using agroforestry systems: an overview (In press)

Kurukulasuriya P and Mendelsohn R 2008 A Ricardian Analysis of the Impact of Climate Change on African Cropland, *AfJARE* **2**(1)23

Lasco RD, Delfino RJP, Catacutan DC, Simelton ES and Wilson DM 2014 Climate risk adaptation by smallholder farmers: the roles of trees and agroforestry, *Current Opinion in Environmental Sustainability 6* 83-88

Lavalle C, Micale F, Houston TD, Camia A, Hiederer R, Lazar C, Conte C, Amatulli G and Genovese G 2009 Climate change in Europe. Impact on agriculture and forestry. A review: *Agronomy for Sustainable Development 29*(3) 433-446

Lin BB 2007 Agroforestry management as an adaptive strategy against potential microclimate extremes in coffee agriculture, *Agricultural and Forest Meteorology* **144**(1–2) 85–94

Lin BB 2007 Agroforestry management as an adaptive strategy against potential microclimate extremes in coffee agriculture, *Agricultural and Forest Meteorology* **144**(1–2) 85–94

Lin BB 2010 The role of agroforestry in reducing water loss through soil evaporation and crop transpiration in coffee agroecosystems, *Agricultural and Forest Meteorology* **150**(4) 510–518

Maroyi A 2009 Traditional homegardens and rural livelihoods in Nhema, Zimbabwe: a sustainable agroforestry system, *International Journal of Sustainable Development & World EcologyOnline) Journal Alfred Maroyi International Journal of Sustainable Development & World Ecology* **16**(1) 1350–4509

McCarthy N, Lipper L and Branca G 2011 Climate-smart agriculture: smallholder adoption and implications for climate change adaptation and mitigation, *Mitigation of Climate Change in Agriculture Working Paper 3* 1-37

McKay H 2011 Short rotation forestry: review of growth and environmental impacts, *Forest Research Monograph 2* 1-211

Méndez VE, Lok R and Somarriba E 2001 Interdisciplinary analysis of homegardens in Nicaragua: Micro-zonation, plant use and socioeconomic importance, *Agroforestry Systems* **51**(2) 85–96

Mitchell CP, Stevens E and Watters MP 1999 Short-rotation forestry – operations, productivity and costs based on experience gained in the UK, *Forest Ecology and Management* **121**(1–2) 123–136

Monteith JL, Ong CK and Corlett JE 1991 Microclimatic interactions in agroforestry systems, *Forest Ecology and Management* **45**(1–4) 31–44

Mourtzinis S, Ortiz BV and Damianidis D 2016 Climate Change and ENSO Effects on Southeastern US Climate Patterns and Maize Yield, *Scientific Reports 6*

Nair PKR 1993 *An introduction to agroforestry*, Springer Science & Business Media

Nassi o di Nasso N, Guidi W, Ragaglini G, Tozzini C and Bonari E 2010 Biomass production and energy balance of a 12□year□old short□rotation coppice poplar stand under different cutting cycles, *GCB Bioenergy* *2*(2) 89-97

Novák V and Havrila J 2006 Method to estimate the critical soil water content of limited availability for plants, *Biologia* **61**(19) 289–S293

Pandey DN 2007 Multifunctional agroforestry systems in India, *Current science* ***92***(4) 455-463

Pandey VC, Bajpai O and Singh N 2016 Energy crops in sustainable phytoremediation, *Renewable and Sustainable Energy Reviews* **54** 58–73

Nair PKR, Nair VD, Kumar BM and Showalter JM 2010 Chapter five-carbon sequestration in agroforestry systems, *Advances in Agronomy* ***108*** 237-307

Rao KPC, Verchot LV and Laarman J 2007 Adaptation to Climate Change through Sustainable Management and Development of Agroforestry Systems, *Journal of SAT Agricultural Research* **4**(1) 1–30

Rockwood DL, Naidu CV, Carter DR, Rahmani M, Spriggs TA, Lin C, Alker GR, Isebrands JG and Segrest SA 2004 Short-rotation woody crops and phytoremediation: Opportunities for agroforestry? In *New Vistas in Agroforestry Springer* Netherlands 51-63

Rosenthal S and Kurukulasuriya P 2003 Climate Change and Agriculture A Review of Impacts and Adaptations, *Climate Change Series* (**91**) 1–77

Russell D and Franzel S 2004 Trees of prosperity: Agroforestry, markets and the African smallholder, *Agroforestry Systems* 345–355

S Briggs 2012 Agroforestry: a new approach to increasing farm production

Sage RB 1998 Short rotation coppice for energy: Towards ecological guidelines, *Biomass and Bioenergy* 39–47

Sharma R, Chauhan SK and Tripathi AM 2016 Carbon sequestration potential in agroforestry system in India: an analysis for carbon project, *Agroforestry Systems*

Sileshi G, Akinnifesi FK, Ajayi OC, Chakeredza S, Kaonga M and Matakala PW 2007 Contributions of agroforestry to ecosystem services in the Miombo eco-region of eastern and southern Africa, *African Journal of Environmental Science and Technology* **1**(4) 68-80

Smeets E and Lewandowski I 2005 *The impact of sustainability criteria on the costs and potentials of bioenergy production*,

Steffan-Dewenter I, Kessler M, Barkmann J, Bos MM, Buchori D, Erasmi S, Faust H, Gerold G, Glenk K, Gradstein SR and Guhardja E 2007 Tradeoffs between income, biodiversity, and ecosystem functioning during tropical rainforest conversion and agroforestry intensification, *Proceedings of the National Academy of Sciences* **104**(12) 4973-4978

Stigter KCJ 2015 Agroforestry and (Micro) Climate Change 119–145

Styles D and Jones MB 2008 Life-cycle environmental and economic impacts of energy-crop fuel-chains: an integrated assessment of potential GHG avoidance in Ireland, *Environmental Science and Policy* **11**(4) 294–306

Tomich TP, de Foresta, H Dennis, R Ketterings, Q Murdiyarso D, Palm C, Stolle F and van Noordwijk M 2002 Carbon offsets for conservation and development in Indonesia? *American Journal of Alternative Agriculture* **17**(03) 125-137

Torquebiau E 1992 Are tropical agroforestry home gardens sustainable? *Agriculture, Ecosystems and Environment*, **41**(2)189–207

Torquebiau EE and Kwesiga E 1996 Root development in a Sesbania sesban fallow-maize system in Eastern Zambia, *Agroforestry Systems* **34** 193–211

Torquebiau EF 2000 A renewed perspective on agroforestry concepts and classification, *Comptes Rendus de l'Academie des Sciences - Serie III* **323**(11) 1009–1017

Trenberth KE and Fasullo JT 2013 An apparent hiatus in global warming ? *Earth's Future* **1**(1)19–32

Trinh LN, Watson JW, Hue NN, De NN, Minh NV, Chu P, Sthapit BR and Eyzaguirre PB 2003 Agrobiodiversity conservation and development in Vietnamese home gardens, *Agriculture, Ecosystems & Environment* ***97***(1) 317-344

Tubby I and Armstrong 2002 Establishment and management of short rotation coppice, *Practice Note-Forestry Commission* **7**1–12

Verchot LV, Van Noordwijk M, Kandji S, Tomich T, Ong C, Albrecht A, Mackensen J, Bantilan C, Anupama KV and Palm C 2007 Climate change: linking adaptation and mitigation through agroforestry. *Mitigation and Adaptation Strategies for Global Change* **12(**5) 901-918

Verwijst T, Lundkvist A, Edelfeldt S and Albertsson J 2013 Development of sustainable willow short rotation forestry in northern Europe, *Biomass now-sustainable growth and use* **10**(5772) 55072

Verwijst T and Telenius B 1999 Biomass estimation procedures in short rotation forestry, *Forest Ecology and Management* **121**(1–2)137–146

Wezel A and Bender S 2003 Plant species diversity of homegardens of Cuba and its significance for household food supply, *Agroforestry Systems* **57**(1) 39–49

Wilson MH and Lovell ST 2016 Agroforestry-The next step in sustainable and resilient agriculture, *Sustainability (Switzerland)* **8**(6) 1–15

WMO 2008 Guide to Meteorological Instruments and Methods of observation. *Guide to Meteorological Instruments and Methods of Observation* 8-1 to I.9-1.

CHAPTER 11

Biomass Production and Carbon Sequestration Potential under different Landuse Systems of Western Himalaya

Sanjeev Kumar, BP Mishra, KS Verma[1], Naresh Kumar[2] and AK Handa[2]*

College of Horticulture and Forestry, CAU, Pasighat-791102 A.P., India

[1]Dr. Y.S. Parmar University of Horticulture and Forestry, Solan-173230 H.P., India

[2]ICAR-Central Agroforestry Research Institute, Jhansi-284003 U.P., India

**Corresponding author's Email: drsanjeevagf@gmail.com*

ABSTRACT

The present study was undertaken at Nauni (30° 51' N and 76° 11' E), Solan, Himachal Pradesh which represents sub-humid sub-tropical to sub temperate climate with an altitude of 1250 m amsl. The biomass, carbon stocks and relative carbon sequestration potential of different landuse systems viz. horti-pastoral (HP), silvi-pastoral (SP), agri-horticulture (AH), agri-silviculture (AS), agri-horti-silviculture (AHS) and natural grassland (NG) system of Western Himalaya were studied. The experiment was laid in split plot design and each landuse system was replicated thrice. Different landuse systems had significant variations in their total biomass production levels over natural grassland. The SP system produced the highest biomass, whereas minimum was in NG. The total carbon stocks (plant + soil) revealed the superiority of SP system followed by AHS system with their respective values of 101.69 and 54.31 t ha^{-1}. Thus, silvi-pastoral landuse system can be a better option in non-arable lands, whereas in arable lands agri-horti-silviculture system shall be a better landuse system for carbon sequestration.

Keywords: Carbon stock, Natural grassland, Relative carbon sequestration, Silivi-pastoral system

INTRODUCTION

Over the past few decades, acceleration in the human-induced changes in the climate of the earth has become the focus of scientific and social scrutiny. Scientific evidences suggest that increased atmospheric CO_2 could have some positive effect such as improved plant productivity (Idso and Kimball 2001; Keutgen and Chan 2001). However, negative changes in the global climate are often the most consequential processes associated with an increased concentration of CO_2 in the atmosphere (Smith 1993 and Dixon 1995). Carbon sequestration is the process involved in carbon capture and the long-term storage of atmospheric carbon dioxide (Sedjo and Sohngen 2012). Sequestration methods include: plant sequestration, soil sequestration, geo-sequestration, ocean sequestration and mineral carbonation. The carbon sequestration potential of three biomes viz., agricultural lands, biomass croplands and deserts have been estimated as 0.85-0.90, 0.50-0.80 and 0.80-1.30 Pg yr^{-1}, respectively, that might be sustained for a period of 25-50 years. The carbon sequestration potential of tropical agroforestry systems is estimated between 12 and 228 Mg ha^{-1} with a mean value of 95 Mg ha^{-1}. Therefore, based on earth's area that is suitable for practices ($585\text{-}1215\times10^6$ ha), 1.1-2.2 Pg C could be stored in the terrestrial ecosystems over the next 50 years (Albrecht and Kandji 2003). Removing atmospheric carbon and storing it in terrestrial biosphere is one of the options, which have been proposed to compensate greenhouse gas emission. Agricultural lands are believed to be a major potential sink and could absorb large quantities of carbon if trees are reintroduced to these systems and judiciously managed together with crops and/or animals. Thus, the importance of agroforestry as a landuse system is receiving wide recognition not only in terms of agricultural sustainability but also in issues related to climate change or carbon sequestration. Agroforestry landuse systems can be better climate change mitigation option because in addition to its high carbon mitigation potential, it is also linked with other benefits such as food security, increasing farm income, restoring and maintaining above and below ground biodiversity, maintaining watershed hydrology and soil conservation. In India, agroforestry is being promoted as an alternate landuse system to deal with the problems related to landuse sustainability and environmental amelioration. Numerous agroforestry systems both natural as well as developed in different agro-climatic regions of India have been found highly productive and environmentally suitable.

METHODOLOGY

The investigation was conducted at the Dr Y S Parmar University of Horticulture and Forestry, Nauni, Solan, Himachal Pradesh. The study site is located at 30° 51′ N latitude and 76° 11′ E longitudes, at an elevation of 1250 m asl representing transitional zone between subtropical and sub-temperate region. The soils of the site were gravely sandy loam and classified as inceptisols and typic entrochrepts. Parent material was a mix of sand stone, conglomerate and dolomite. Six landuse systems viz. natural grassland (NG), horti-pastoral (HP), silvi-pastoral (SP), agri-silviculture (AS), agri-horticulture (AH) and agri-horti-silviculture (AHS) were selected for study and each landuse system was taken as one treatment. The experiment was laid in split plot design using these six landuse systems and each landuse system was replicated thrice. The detail of the selected landuse systems is given in Table 1. The data for biomass of tree, crop, grasses and surface litter was recorded following the standard procedures. The tree biomass was portioned into stem branch and leaf and recorded separately. For the crop and grass biomass estimation, 1m x 1m quadrates were laid out and the crop plants and grasses occurring within the borders of the quadrates were cut at ground level and collected samples were weighed, sub sampled and oven dried at 65 ^{0}C temperature to a constant weight. The biomass was converted into carbon by multiplying with a factor of 0.45 (Woomer 1999). The data obtained were analyzed using split plot design by Gomez and Gomez (1984).

Table 1: *Experimental details indicating specific tree-crop combination and their distribution*

Landuse systems	Code	Tree-crop combination	Plot size (m)	Net cropped area (ha)	Area under trees (ha)	Area under grasses	No. of trees (ha^{-1})	Year of planting
Natural grass land	NG	Grass	50 × 10	–	–	1.0	–	–
Horti-pastoral	HP	Plum + grass	50 × 10	–	–	–	286	1985
Silvi-pastoral	SP	Leuceana + Grewia + Grass	50 × 10	–	0.2	0.8	2500	1990
Agri-silviculture	AS	Grewia + Bauhinia + Wheat	50 × 10	0.9	0.1	–	400	1990
Agri-horticulture	AH	Peach + Wheat	50 × 10	0.9	0.1	–	200	1990
Agri-horti-silviculture	AHS	Plum + Morus + wheat	50 × 10	0.9	0.1	–	332	1990

RESULT AND DISCUSSION

Aboveground and Belowground Biomass

Different landuse systems have shown significant variations in their total biomass production levels. The stem biomass was highest in SP system followed by AHS (Table 2). Branch biomass was also significantly higher in SP system, but reflected a trend different to that of stem biomass among the different systems. Contrary to stem biomass the second highest value for branch biomass was shown by HP system. Similar to stem and branch biomass, leaf biomass was also highest in SP system. Whole tree biomass was significantly higher in SP system with a value of 53.48 t ha^{-1}. It was followed by AHS, AS, HP and AH. Landuse systems viz., AS, AH, and HP showed statistically alike biomass values. The stem, branch and leaf biomass of a tree depends upon the number of factors viz., growth habit of the tree species, soil on which it grows, age of the tree, its management and also its compatibility with the associated crop plant species. The highest stem, branch and leaf biomass obtained under SP system can be explained to the fact that in this system the tree species namely *Leucaena* and *Grewia* were planted at high density in mixture and were managed using pollarding techniques. Therefore, more number of stems per hectare in SP system over the remaining systems resulted into higher stem, branch and leaf biomass. The second highest biomass value was obtained for AHS system integrating plum and mulberry as woody perennial which reflected the influence of system structure. The variation in crop and grass biomass can be explained due to the tree-crop interaction effects. The reduced biomass may be due to more competition for resources like nutrients, moisture and light compared to the tree species. Variations in crop (wheat) biomass were statistically significant, yet AH and AHS systems produced exactly equal amount (2.87

t ha^{-1}) of biomass. The variations in the crop biomass can be explained due to tree-crop interaction effects. Grass biomass was significantly higher in NG system. Litter biomass was significantly higher in SP system followed by NG and HP with their respective value of 3.34, 2.74 and 2.64 t ha^{-1}. AH system showed the minimum litter biomass (1.42 t ha^{-1}), which was statistically similar to the value shown by AHS. The total aboveground biomass followed the trend almost similar to the stem biomass among the different systems. It was significantly higher in SP system followed by AHS, AS, HP and AH, with their respective values of 58.74, 21.93, 19.59, 17.22 and 16.70 t ha^{-1}. NG system produced the lowest value of 4.83 t ha^{-1}. The belowground biomass observed as fine roots was highest in AHS followed by HP with their respective values of 1.04 and 1.03 t ha^{-1}, although statistically alike. The remaining systems also showed almost similar amounts of fine roots biomass, which were statistically non-significant with one another. Total biomass (above and belowground) was highest in SP system (59.72 t ha^{-1}) followed by AHS (22.97 t ha^{-1}) and AS (20.53 t ha^{-1}). The minimum biomass (5.79 t ha^{-1}) was observed under natural grasslands (Table 2).

Table 2: *Biomass production levels of different landuse systems (t ha^{-1})*

System Component	Natural grassland (NG)	Horti-pastoral (HP)	Silvi-pastoral (SP)	Agri-silviculture (AS)	Agri-horticulture (AH)	Agri-horti-silviculture (AHS)	$CD_{0.05}$
Above ground							
Stem	0	4.20	27.02	7.83	6.41	11.19	2.27
Branch	0	7.34	20.73	6.49	4.98	4.41	2.58
Leaf	0	1.51	5.36	1.51	1.03	1.79	1.21
Whole Tree	0	13.07	53.48	15.49	12.41	17.31	4.28
Crop	0	0	0	2.44	2.87	2.87	0.28
Grass	2.08	1.52	1.93	0	0	0	0.38
Litter	2.74	2.64	3.34	1.66	1.42	1.44	0.21
Total (a)	4.83	17.22	58.74	19.59	16.70	21.93	0.27
Below groundc							
Fine roots (b)	0.97	1.03	0.98	0.94	0.93	1.04	NS
Grand Total (a+b)	5.79	18.25	59.72	20.53	17.63	22.97	4.22

CARBON STOCKS IN DIFFERENT LANDUSE SYSTEMS

All the systems showed significantly higher stem carbon stock values over HP (1.89 t ha^{-1}). SP system had the highest stem carbon (12.16 t ha^{-1}), which was followed by AHS, AS, AH, and HP systems. Similarly, branch carbon (9.33 t ha^{-1}) and leaf carbon were, also, highest in SP system (2.41 t ha^{-1}). Whole tree carbon stocks were significantly higher in SP system with a value of 24.07 t ha^{-1} whereas, AH showed the lowest value of 5.59 t ha^{-1}. Among different landuse systems studied, crop (wheat) component was present in three arable landuse systems viz. AS, AH, and AHS. The carbon stocks varied significantly between these three landuse systems. AS system had significantly lowest value of 1.10 t ha^{-1}, whereas AH and AHS systems depicted the same value of 1.29 t ha^{-1}. The grass component was also present in the three landuse systems (NG, HP and SP) in non-arable lands only. All these three systems had significantly varied carbon stock values. The maximum was in NG (0.94 t ha^{-1}). Litter carbon stock was significantly higher in SP system with its maximum value of 1.50 t ha^{-1}. A perusal of the data on total carbon stocks for different land use systems (Table 3) further revealed highest carbon storage (26.43 t ha^{-1}) in SP system. This system stored about 12 times more carbon than NG and about 3.4 times of HP system. Similarly, carbon stored in SP system was about 3.00, 3.51 and 2.55 times higher over AS, AH and AHS systems, respectively. The minimum amount (2.17 t ha^{-1}) of carbon stored was in NG system. Carbon stored through fine roots has been referred as below ground carbon stock of a system. AHS system stored highest carbon (0.47 t ha^{-1}) through roots. Further examination of the data for whole system carbon stocks depicted maximum value (26.87 t ha^{-1}) in SP system, which was significantly higher than the values obtained for remaining systems. NG system showed the lowest value of 2.61 t ha^{-1} (Table 3). These findings showed that variability in the carbon stocks of different landuse systems depends primarily on its complexity. Albrecht and Kandji (2003) have also reported that carbon variability in plant biomass can be high within complex systems and productivity depends on several factors including the age, the structure and the management of the system. Similar findings of carbon storage were observed by Beer et al (1990).

Table 3: *Carbon stocks in different landuse systems (t ha^{-1})*

System Component	Natural grassland (NG)	Horti-pastoral (HP)	Silvi-pastoral (SP)	Agri-silviculture (AS)	Agri-horticulture (AH)	Agri-horti-silviculture (AHS)	CD $_{0.05}$
Above ground							
Stem	0	1.89	12.16	3.52	2.89	5.04	1.02
Branch	0	3.30	9.33	2.92	2.24	1.99	1.16
Leaf	0	0.68	2.41	0.68	0.46	0.81	0.55
Whole Tree	0	5.88	24.07	6.97	5.59	7.63	1.93
Crop	0	0	0	1.10	1.29	1.29	0.13
Grass	0.94	0.68	0.87	0	0	0	0.17
Litter	1.23	1.19	1.50	0.75	0.64	0.65	0.09
Total (a)	2.17	7.75	26.43	8.82	7.52	9.87	1.92
Below ground							
Fine roots (b)	0.44	0.46	0.44	0.42	0.42	0.47	NS
Grand Total (a+b)	2.61	8.21	26.87	9.24	7.93	10.34	1.90

CO_2 MITIGATION BY DIFFERENT LANDUSE SYSTEMS

SP system showed the maximum value (98.36 t ha^{-1}) of CO_2 mitigated, which was significantly higher than the values obtained for remaining systems. The system types viz. HP, AS, AH and AHS having their respective values of 30.06, 33.81, 29.04 and 37.83 t ha^{-1}

showed statistically non-significant differences with one another and hence were found statistically similar. NG system showed the minimum (9.54 t ha^{-1}) CO_2 mitigation (Table 4). In India a number of studies have indicated that the tree component in agroforestry has a capacity for biomass production at least as great as that of natural vegetation. The results obtained are similar to as reported by Kaur et al 2002; Puri and Swami 2001; Kaur et al 2000, Maikhuri et al 2000.

Table 4: *CO_2 mitigation levels of different landuse systems (t ha^{-1})*

System Component	Natural grassland (NG)	Horti-pastoral (HP)	Silvi-pastoral (SP)	Agri-silviculture (AS)	Agri-horticulture (AH)	Agri-horti-silviculture (AHS)	$CD_{0.05}$
				Above ground			
Stem	0	6.92	44.50	12.90	10.56	18.43	3.74
Branch	0	12.09	34.14	10.69	8.20	7.26	4.25
Leaf	0	2.49	8.83	2.49	1.70	2.95	1.99
Whole Tree	0	21.53	88.08	25.51	20.44	29.00	7.05
Crop	0	0	0	4.03	4.72	4.72	0.45
Grass	3.44	2.49	3.17	0	0	0	0.62
Litter	4.50	4.37	5.49	2.74	2.34	2.38	0.36
Total (a)	7.94	28.36	96.75	32.26	27.50	36.12	7.03
				Below ground			
Fine roots (b)	1.60	1.68	1.62	1.55	1.54	1.71	NS
Grand Total (a+b)	9.54	30.06	98.36	33.81	29.04	37.83	6.95

RELATIVE CARBON LOSS OR GAIN IN SYSTEMS

The net loss or gain in total carbon stocks of different landuse systems has been calculated over the actual value of NG system considering as 100 and (Table 5). It was observed that SP AHS AS HP and AH system had net gains in the total above ground carbon stocks with respect to NG. It indicates that landuse system involving woody perennial species either fruit or fodder or any other tree have higher potential to sequester the carbon. The finding also revealed that carbon stocks can be enhanced by integrating woody perennial species along with the annual or herbaceous component.

Table 5: *Relative loss or gain in the carbon stocks (t ha^{-1}) of different landuse systems*

Landuse systems	***Relative loss or gain***
Natural grassland (NG)	100.00 (±0.00)
Horti-pastoral (HP)	315.19 (+215.19)
Silvi-pastoral (SP)	1031.43 (+931.43)
Agri-silvi-culture (AS)	354.58 (+254.58)
Agri-horti-culture (AH)	304.49 (+204.49)
Agri-horti-silviculture (AHS)	396.71 (+296.71)

RELATIVE CARBON SEQUESTRATION POTENTIAL

The relative net gain and loss in total carbon stocks of different landuse systems were further used to find out the carbon sequestration potential (CSP) of each system taking the value of NG system as 1. The data thus obtained (Fig. 1) have inferred that CSP of SP system was 10 times higher than the NG followed by AHS, AS and HP systems. The lowest was observed in case of AS system, however it was also three times higher than the NG system. On the basis of CSP different landuse systems showed the following descending order: silvi-pastoral>agri-horti-silviculture>agri-silviculture>hortipastoral>agri-horticulture.

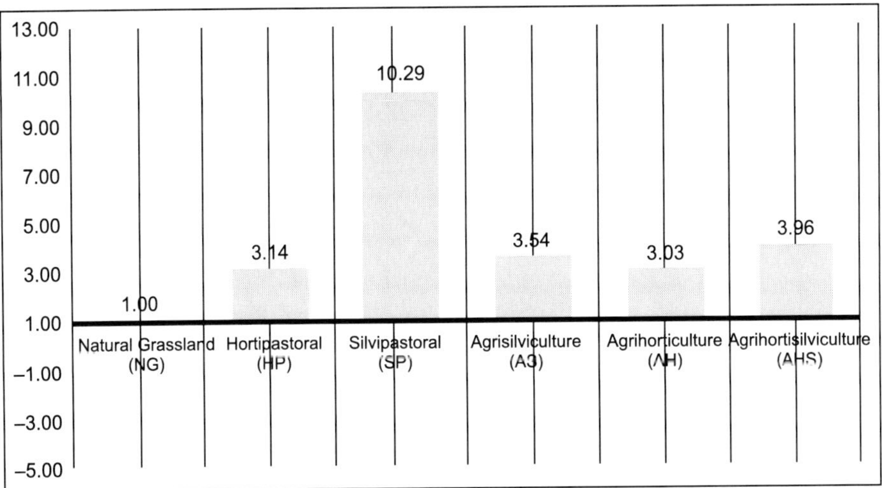

Figure 1: *Relative carbon sequestration (t ha^{-1}) potential of different landuse systems*

CONCLUSION

Silvi-pastoral system gave maximum biomass, carbon stock and CO_2 mitigation potential followed by agri-horti-silviculture system. These findings showed that silvi-pastoral landuse system can be a better option in non-arable land, whereas in arable lands agri-horti-silviculture system shall be a better landuse system for carbon sequestration.

REFERENCES

Albrecht A and Kandji ST 2003 Carbon sequestration in tropical agroforestry system. *Agriculture, Ecosystems and Environment* **99**: 15-27.

Beer J, Bonnemann A, Chavez W, Fassbender HW, Imbach AC and Martel I 1990 Modelling agroforestry systems of Cocoa (*Theobroma cocoa*) with laurel (*Cordia alliadora*) and poro (*Erythrina poeppigiana*) in Costa Rica V- productivity indices, organic material models and sustainability over a period of ten years. *Agroforestry Systems* **12:** 229-249.

Dixon RK 1995 Agroforestry systems: sources or sinks of greenhouse gases? *Agroforestry Systems.* **31**:99-116.

Gomez KA and Gomez AA 1984 Statistical procedure for agriculture research. John Willy & Sons, New York 680 p.

Idso SB and Kimball BA 2001 CO_2 enrichment of sour orange tree. *Environmental* and Experimental Botany **48:** 147-153.

Kaur B, Gupta SR and Singh G 2000 Carbon storage and nitrogen cycling in silvopastoral systems on moderately alkaline soils in northern India . *Applied Soil Ecology* **15**: 283-294.

Kaur B, Gupta SR and Singh G 2002 Soil carbon, microbial activity and nitrogen availability in agroforestry systems on sodic soils in northwestern India. *Agroforestry Systems* **54:** 21-29.

Keutgen N and Chan K 2001 Response of citrus leaf photosynthesis, chlorophyll fluorescence, macronutrient and carbohydrate content to elevated CO_2 *Journal of Plant Physiology* **158:** 1307-1316

Maikhuri RK, Rao KS, Saxena KG and Semwal R L 1999 Traditional crop diversity based nutrition and prospects for sustainable development in central Himalaya. *Himalayan Paryavaran* **6**: 36-44.

Montagnini F and Nair PKR 2004 Carbon sequestration: an under exploited environmental benefit of agroforestry systems. *Agroforestry Systems* **61:** 281-295.

Puri S and Swami SL 2001 Growth and biomass production in *Azadirchta indica* seedling in response to nutrient and moisture stress *Agroforestry Systems* **51:** 57-68.

Sedjo R and Sohngen B 2012 Carbon sequestration in forests and soils *Annual Review of Resource Economics* **4**: 127–144.

Smith I M 1993 CO_2 and climate change: an overview of the Science *Energy Conversion* and *Management* **34**: 729-735.

Woomer PL 1999 Impact of cultivation on carbon fluxes in woody savannas of Southern Africa. *Water air and Soil Pollution* **70**(14): 403-412.

CHAPTER 12

Carbon Sequestration Potential of Fast Growing Short Rotation Tree Species based Agroforestry Systems in Terai Region of Central Himalaya

S Sarvade*, HS Mishra[1], Rajesh Kaushal[2], Sumit Chaturvedi[1], Rahul Singh[3], Chaman Lal[3] and Varun Attri[3]

College of Agriculture, Balaghat, Jawaharlal Nehru Krishi Vishwa Vidyalaya, Jabalpur, Madhya Pradesh, India

[1]Department of Agronomy, G. B. Pant University of Agriculture & Technology, Pantnagar, Uttarakhand, India

[2]Central Soil and Water Conservation Research & Training Institute, Dehradun, Uttarakhand, India

[3]College of Forestry, Dr. Y. S. Parmar UHF, Nauni-173 230 (HP), India

**Corresponding author's Email: somanath553@gmail.com*

ABSTRACT

Tree species viz., *Populus deltoides* (poplar), *Eucalyptus camaldulensis* (eucalyptus), *Melia azedarach* (melia) and *Leucaena leucocephala* (leucaena) are widely used on farm lands for multiple uses in Terai region of Central Himalaya, India. Poplar trees were recorded highest height (9.57 m) at very dense/close tree spacing (3.0×1.0m), whereas it showed inverse relation with diameter at breast height (0.112 m at 3.0×2.5m tree spacing). Poplar showed significant fast growth (13.11 m^2 $tree^{-1}$ basal area and 2.357 m^3 $tree^{-1}$ volume) at relative wider tree spacing (3.0×2.5m) as compared to other tree species. The higher total tree biomass (3.521 t ha^{-1}) and carbon density (1.655 t ha^{-1}) were also recorded by poplar trees at 3.0×1.0m spacing. Total crop biomass and carbon density were increased with increasing tree spacing under all Short Rotation Fast Growing Tree species (SRFGTs) except eucalyptus at 3.0×1.5m tree spacing, where higher total crop biomass and carbon density were recorded. Soil bulk density, organic carbon and organic matter were improved at very dense planting of SRFGTs. The higher soil organic carbon (67.494 t ha^{-1}) was recoded under poplar at very dense tree spacing. Poplar based agroforestry system with very dense tree spacing recorded higher total system biomass (3.916 t ha^{-1}) and carbon density (69.326 t ha^{-1}).

Keywords: Aboveground and belowground biomass, Climate change mitigation, Soil bulk density, Soil organic carbon

INTRODUCTION

The greenhouse gases (GHGs) *viz*., carbon dioxide, methane, water vapour, nitrous oxide and chlorofluorocarbons absorbs heat from re-radiated in outer space results in to warming of the earth surface (IPCC 2006). Their concentration in an atmosphere has considerably increased over the last century and is set to rise further (Albrecht and Kandji 2003). Warming of the earth is one of the important issues in the developing world which germinates the seeds of climate change. It affects differently i.e. positively and negatively on natural systems of the world. The increased atmospheric CO_2 could have some positive effects such as improved plant productivity (Keutgen and Chen 2001, Bonan 2008, Bala et al 2010), however, changes in the global climate (rising temperatures, higher frequency of droughts and floods) are mostly associated with an increased concentration of CO_2 in the atmosphere (USDA NRSC 2000). The sustainable development along with farm productivity and food security is going to slow down day by day. CO_2 contributes highest among all the green house gases, in global warming. Deforestation is the largest source of global greenhouse gases emission. Demand and supply gap in fuel wood, timber and fodder, shifting cultivation, skewed development, poverty are the main causes of deforestation. It creates extra pressure on the present natural resources (Gibbs et al 2007).

Agroforestry being complex system has potential to combat these problems (Albrecht and Kandji 2003, Nair 2012). Short rotation fast growing tree species (SRFGTs) are the tree species grown with a rotation period of 6 to 12 years and having high productivity at least 10 to 30 m^3 ha^{-1} yr^{-1} (Sarvade et al 2014b). Poplar, eucalyptus, melia and leucaena are the most widely practiced SRFGTs with wheat for industrial timber in terai region of Northern India (Sarvade et al 2014a). It helps to bridge the gap between demand and supply of industrial timber (Mishra et al 2010). These commercial tree species have great potential to fix atmospheric carbon (Joshi et al 2013). The five years poplar stored 34.075 t ha^{-1} carbon in poplar based agroforestry system of central zone of Punjab, India (Chauhan et al 2012). Poplar, eucalyptus and melia plantations from Northern plains of India, stored 10.74, 9.36 and 3.96 t ha^{-1} carbon, respectively and showed their potential for plantation on farmers' fields except melia (Chauhan et al 2009). The different fast growing species with their planting geometry at specific locality factors affects the growth and carbon storage potential (Nair et al 2009, 2010, Saha et al 2010, Nair 2012, Arora et al 2014). Information on the effects of planting density on C sequestration by fast growing MPTs is scarce (Jandl et al 2007). As the tree species and their geometry or/and different spacing designs affect significantly on their growth and CO_2 sequestration potential, present investigation was carried out to examine the CO_2 sequestration potential of SRFGTs based agroforestry systems in terai region.

METHODOLOGY

Study Area

Study was carried out in the years of 2009-2010 at agroforestry research station, Haldi, G. B. Pant University of Agriculture and Technology, Pantnagar, in Udham Singh Nagar district of Uttarakhand. The study site is situated at 29^0 00' N latitude and 79^0 30' E longitude with an altitude of 243.84m asl. Area lies in foothills of the Shivalik ranges in narrow strip called 'terai' region. The silty-clay-loam soils are developed from calcareous, medium to moderately coarse textured materials under pre-dominant influence of tall vegetation and moderate to well drain conditions. The soils are weakly developed with mollic epipedons and classified under order Mollisol, sub-order Udoll, great group Hapludoll, sub-group typic Hapludoll (Deshpande et al 1971a, b). Climate of the area is humid and sub-tropical, experiences hot and dry summers, and cold winters. The dry season starts from early October and runs up to mid-June, and the monsoon season starts in the third or fourth week of June and lasts until September or the first week of October. The mean annual rainfall was 1266 mm and minimum-maximum temperature ranged from 6.64 to 37.33^{0}C during the period of investigation. Relative humidity of the area ranged from 43.77 to 80.29 per cent (Fig. 1). SRFGTs (T_1: *Populus deltoides* Bartr. ex. Marsh; T_2: *Eucalyptus camaldulensis* Dehn; T_3: *Melia azedarach* Linn. and T_4: *Leucaena leucocephala* de Wit) were planted in the year of 2007-08 with four spacing treatments *viz.*, S_1: 3×1m, S_2: 3×1.5m, S_3: 3×2m and S_4: 3×2.5m. Wheat (*Triticum aestivum* Linn.), variety PBW-502 was cultivated under four SRFGTs with recommended cultural practices during the first week of November, 2009 and harvested in April, 2010.

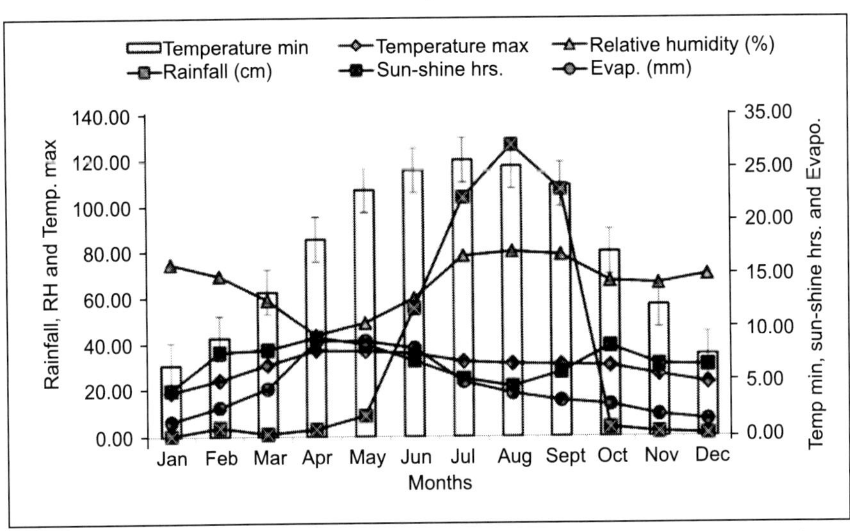

Figure 1: *Monthly averageweather data of the study area (2007–2010)*

ESTIMATION OF VEGETATION BIOMASS AND CARBON STOCK

The volume of an individual tree species was calculated by using Pressler formula (Pressler 1895) used by Koul and Panwar (2008). The aboveground biomass was calculated by multiplying volume by wood specific gravity (Devagiri et al 2013, Pande et al 2010, Sheikh et al 2012) and biomass expansion factor of 1.3 (IPCC 2006). The belowground biomass of trees was determined by multiplying aboveground biomass with factor 0.25 following IPCC (1996). The aboveground and belowground biomass was converted in to carbon stock by using conversion factor of 0.5 given by IPCC (2006) and Nair (2012). The production of wheat crop was converted in to aboveground biomass by using conversion factor and equation given by Wani et al (2010). The belowground biomass was determined by using shoot-root ratio (0.13). The carbon content in aboveground and belowground biomass was calculated by multiplying biomass by 0.45 (Wani et al 2010). The collected and processed soil samples were analysed for organic carbon (OC) by using Walkley and Black (1934) method. Soil bulk density (g cc^{-1}) was determined by pycnometer and/or weighing bottle method (Singh 1980). The amount of carbon stored in soil (t ha^{-1}) was determined by following equation given by Joa Carlos et al (2001). Total carbon stock (t ha^{-1}) of the system was determined by summing up carbon stored in tree-crop components (above and belowground) and soil organic carbon. The data on tree and crop growth, biomass production, carbon storage, bulk density and soil organic carbon analyzed to examine the effect of tree species and their spacing on carbon sequestration potential of overall system by analysis of variance (ANOVA) using JMP/ SAS 10.0 statistical software. Significant differences were tested at $P \leq 0.05$ using Tukey HSD test.

RESULTS AND DISCUSSION

Growth and Structure of FGSRTs

Height and diameter at breast height showed inverse relation for all tree species at all spacing treatments (Fig. 2). The closer spacing (S_1) showed highest values of height for all FGSRTs. Poplar tree grows fast as they showed significant differences in case of growth parameters. The lowest values of height and dbh were recorded for melia tree species. Poplar planted at closer spacing grows taller, whereas dbh growth was greater at wider spacing (S_4). Poplar trees produce significantly higher volume of timber (Table 1). Wider spacing of poplar trees benefited to produce higher volume, however it statistically similar with S_2 and S_3 spacing treatments. The basal area of poplar and eucalyptus was lower at spacing S_3, whereas higher basal area of poplar (13.11 m^2 $tree^{-1}$) was recorded at wider spacing and eucalyptus at closer spacing. Melia at closer spacing showed 77.35% lower basal area than poplar trees at wider spacing. Basal area of leucaena decreased with increasing spacing. The study showed that the poplar trees grown fast than the other three FGSRTs. Fast growth rate of poplar was also reported by Chauhan et al (2009) in agroforestry systems of northern plains of India. Singh et al (1989), Kohli et al (1997), Chauhan et al (2012) and Arora et al (2014) also reported the excellent growth rate of exotic tree species especially poplar tree species. The high productivity of poplar up to 50 m^3 ha^{-1} yr^{-1} has been reported by Gill et al (2009) in irrigated poplar based agroforestry systems of north-western India. The closely planted tree species grow faster due to withstand the competition for light, moisture and nutrients (Bhardwaj et al 2001).

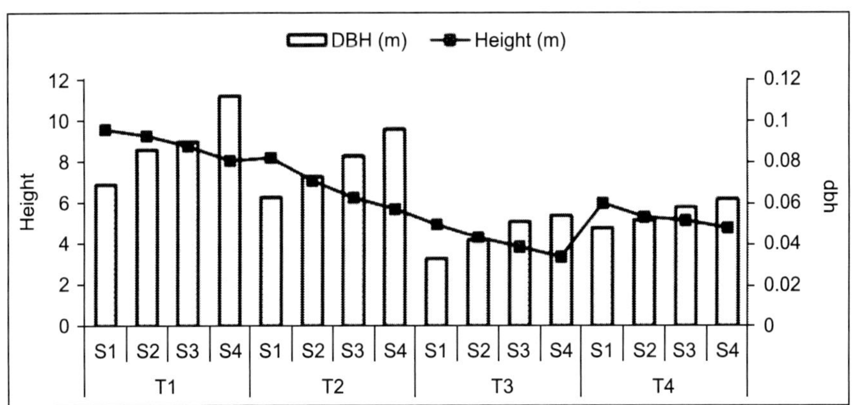

Figure 2: *Tree height and diameter at breast height (dbh) influenced by their spacing*

Table 1: *Growth, basal area, biomass production and carbon density of tree components*

FGSRTs and spacing	Basal area (m^2 Tree^{-1})	Volume (m^3 tree^{-1})	AGB (t ha^{-1})	BGB (t ha^{-1})	AGC (t ha^{-1})	BGC (t ha^{-1})
T1						
S1	12.37a	1.625b	2.817a	0.704a	1.324a	0.331a
S2	12.97a	2.173ab	2.511ab	0.628ab	1.180ab	0.295ab
S3	10.54a	2.022ab	1.753bc	0.438bc	0.824bc	0.206bc
S4	13.11a	2.357a	1.634cd	0.409cd	0.768cd	0.192cd
T2						
S1	10.60a	0.861c	2.415ab	0.604ab	1.135ab	0.284ab
S2	9.40ab	0.786cd	1.470cd	0.367cd	0.691cd	0.173cd
S3	9.03ab	0.610cde	0.855de	0.214de	0.402de	0.100de
S4	9.61ab	0.592cde	0.664ef	0.166ef	0.312ef	0.078ef
T3						
S1	2.97c	0.054e	0.161ef	0.040ef	0.076ef	0.019ef
S2	3.08c	0.051e	0.102ef	0.025ef	0.048ef	0.012ef
S3	3.36c	0.051e	0.077ef	0.019ef	0.036ef	0.009ef
S4	3.06c	0.036e	0.044f	0.011f	0.021f	0.005f
T4						
S1	5.98bc	0.184de	0.600ef	0.149ef	0.280ef	0.070ef
S2	4.74c	0.166de	0.357ef	0.089ef	0.168ef	0.042ef
S3	4.48c	0.174de	0.282ef	0.071ef	0.133ef	0.033ef
S4	4.12c	0.153e	0.197ef	0.049ef	0.093ef	0.023ef
P value	0.5682	0.0328	0.0001	0.0001	0.0001	0.0001

Where, AGB is aboveground biomass, BGB is belowground biomass, AGC is aboveground carbon, BGC is belowground carbon. Values in the same column followed by the same letter are not significantly different at $P \leq 0.05$ level by using Tukey-Kramer HSD test.

BIOMASS PRODUCTION AND CARBON STORAGE BY FGSRTs AND ANNUAL CROP

Poplar planted at closer spacing produce higher aboveground and belowground biomass, whereas it was statistically at par with poplar planted at S_2 and eucalyptus planted at closer spacing treatment. Melia planted at wider spacing showed 98.44 % reduction in biomass (aboveground and belowground) production as compared to poplar at closer spacing. The biomass production by wheat crop was non-significantly affected by tree species and their spacing treatments (Table 2). Crop sown under poplar at wider spacing produces higher aboveground and belowground biomass. Increasing tree spacing showed positive effect on crop aboveground and belowground biomass production for all tree species except eucalyptus, it produces highest biomass at S_3 spacing treatment. Carbon storage followed the trend of biomass production. The carbon (aboveground and belowground) storage potential was decreased with increasing tree spacing for all four FGSRTs (Table 1). The significantly highest carbon was stored by poplar trees planted at closer spacing; however it was statistically similar with poplar planted at S_2 and evcalyptus planted at closer spacing. Melia planted at wider spacing stored 98.41 % lowest carbon followed by leucaena (92.98 %) and eucalyptus (76.43 %) than poplar planted at closer spacing. Carbon stored by crop was higher under poplar at wider spacing. Whereas carbon stored by crop was 36.73 % lower under melia trees planted at S_2 spacing treatment compared to poplar at wider spacing. Biomass production and carbon storage capacity of vegetation also depends on the growth rate and growth properties of tree species. Poplar trees grow faster and produce more biomass and store high carbon in standing biomass (Chauhan et al 2012, Arora et al 2014). The biomass production and carbon storage were influenced directly or indirectly due to growing nature of tree species and their spacing. Closer spacing increases tree-crop competition for light, moisture and nutrients and reduces biomass production and carbon storage capacity (Sharma et al 2001, Kaushik and Singh 2001, Kumar et al 2006, Chauhan and Dhiman 2007).

Table 2: *Biomass production and carbon stored by wheat crop*

FGSRTs and spacing	AGB (t ha^{-1})	BGB (t ha^{-1})	AGC (t ha^{-1})	BGC (t ha^{-1})
T1				
S1	0.349a	0.045a	0.157a	0.020a
S2	0.396a	0.052a	0.178a	0.023a
S3	0.497a	0.065a	0.224a	0.029a
S4	0.611a	0.079a	0.275a	0.036a
T2				
S1	0.403a	0.052a	0.181a	0.024a
S2	0.421a	0.055a	0.189a	0.025a

S3	0.464a	0.060a	0.209a	0.027a
S4	0.396a	0.052a	0.178a	0.023a
		T3		
S1	0.410a	0.053a	0.184a	0.024a
S2	0.386a	0.050a	0.174a	0.023a
S3	0.553a	0.072a	0.249a	0.032a
S4	0.564a	0.073a	0.254a	0.033a
		T4		
S1	0.467a	0.061a	0.210a	0.027a
S2	0.527a	0.069a	0.237a	0.031a
S3	0.494a	0.064a	0.222a	0.029a
S4	0.561a	0.073a	0.253a	0.033a
P value	0.2794	0.2794	0.2794	0.2794

Where, AGB is aboveground biomass, BGB is belowground biomass, AGC is aboveground carbon, BGC is belowground carbon. Values in the same column followed by the same letter are not significantly different at $P \leq 0.05$ level by using Tukey-Kramer HSD test.

SOIL PROPERTIES AND ORGANIC CARBON

The effect of FGSRTs and spacing treatment was non-significant on soil bulk density ($P > 0.05$), however eucalyptus planted at wider spacing recorded higher (1.42 g cc^{-1}) soil bulk density (Table 3). The closer spacing showed reduction in soil bulk density. The lower value of soil bulk density (1.31 g cc^{-1}) was recorded from poplar, melia and leucaena plots planted at closer spacing, it showed 7.75 % reduction. Soil organic carbon and soil organic matter were recorded higher (1.727 and 2.977 %, respectively) from poplar planted at closer spacing (Table 3), however it was non-significantly ($P > 0.05$) affected by FGSRTs and spacing treatments. Soil organic carbon was significantly ($P < 0.05$) influenced and recorded significantly highest (67.494 t ha^{-1}) from poplar planted at closer spacing (Table 3). Melia planted at wider spacing recorded lower (52.49 % lower than poplar planted at closer spacing) soil organic carbon (32.068 t ha^{-1}). Improvement of soil properties under tree species also helps to foster the tree growth and, biomass production and carbon storage potential (Dutta and Agrawal 2002). They also reported that the soil nutrients status also the factors limiting plant growth. Poplar shed their leaves during the wheat sowing period; it helps to add organic matter and nutrients to the soil (Gill et al 2009, Sarvade et al 2014b). Fang et al (2005) reported potential of poplar in ecological and site productivity benefits. Mishra et al (2002, 2004) also reported increase of SOC under plantation of fast growing tree species. The closer spacing of tree species adds more organic matter and improves soil properties than wider spacing. The soil organic carbon (t ha^{-1}) increased with improving soil properties such as organic carbon (%) and bulk density (Fig. 3).

Table 3. *Soil properties and organic carbon in FGSRTs based agroforestry systems*

FGSRTs and spacing	Bulk density (g cc^{-1})	Organic Carbon (%)	Soil Organic Matter (%)	Soil Organic Carbon in t ha^{-1} (E)
T1				
S1	1.31d	1.727a	2.977a	67.494a
S2	1.33cd	1.050b	1.810b	41.670b
S3	1.35bcd	1.217ab	2.098ab	49.103ab
S4	1.37abc	1.040b	1.793b	42.580b
T2				
S1	1.34bcd	1.040b	1.793b	41.729b
S2	1.39ab	1.043b	1.799b	43.308ab
S3	1.37abc	0.967b	1.667b	39.504b
S4	1.42a	0.853b	1.471b	36.198b
T3				
S1	1.31d	1.193ab	2.057ab	46.699ab
S2	1.34bcd	1.320ab	2.276ab	52.997ab
S3	1.32cd	1.013b	1.747b	39.797b
S4	1.37abc	0.783b	1.350b	32.068b
T4				
S1	1.31d	1.230ab	2.121ab	48.142ab
S2	1.35bcd	0.970b	1.672b	39.106b
S3	1.35bcd	1.113b	1.919b	44.824ab
S4	1.39ab	1.067b	1.839b	44.296ab
P value	0.5717	0.0511	0.0511	0.0424

Values in the same column followed by the same letter are not significantly different at $P \leq 0.05$ level by using Tukey-Kramer HSD test.

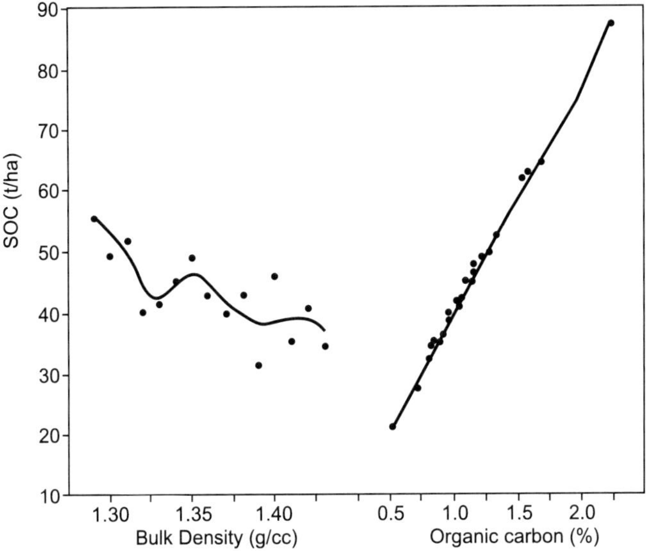

Figure 3: *Relation of Soil Organic Carbon (t/ha) with Organic Carbon (%) and Soil Bulk Density*

TOTAL BIOMASS PRODUCTION AND CARBON STORAGE POTENTIAL

Total tree biomass (aboveground + belowground) was significantly ($P < 0.05$) affected by tree species and reduced with widening the planting distance (Table 4). Poplar plated at closer spacing produced highest total biomass (3.52 t ha^{-1}), however it was statistically similar with poplar with S_2 spacing (3.139 t ha^{-1}) and eucalyptus planted at closer spacing (3.018 t ha^{-1}). The highest (98.44 %) reduction in biomass production was showed by melia planted at wider spacing as compared to poplar planted at closer spacing. Total biomass produced by annual wheat crop was non-significant ($P > 0.05$), whereas wider spacing recorded higher crop biomass from all tree species except eucalyptus. Eucalyptus planted at S_3 spacing recorded higher crop biomass (0.524 t ha^{-1}), however poplar planted at wider spacing recorded higher (0.691 t ha^{-1}) crop biomass as compared to all other tree species. Total system biomass (total tree + total crop) was affected significantly ($P < 0.05$) and recorded higher (3.916 t ha^{-1}) biomass from poplar planted at closer spacing. Total carbon density of vegetation follows the pattern of their biomass production. Carbon density of poplar tree species planted at closer spacing was significantly higher ($P < 0.05$); however it was statistically similar with poplar planted at S_2 spacing and eucalyptus planted at S_1 spacing. Total carbon stored by annual crop was affected non-significantly ($P > 0.05$), whereas wider spacing of tree species improves carbon storage in wheat crop (Table 4). Total system carbon density (tree + crop + soil) was statistically higher (69.326 t ha^{-1}) under poplar planted at closer spacing ($P = 0.0433$). However poplar at S_3, melia at S_1 and S_2, and leucaena at S_1 and S_3 spacing was statistically at par with poplar planted at closer spacing for total system carbon density. Melia planted at wider spacing recorded lower total system carbon density, whereas it showed 53.29 % lower carbon density compared to poplar at closer spacing. The total system carbon storage also depends on the factors mentioned above, which affects directly or indirectly. It is mainly affected by carbon storage in soil and tree component of the system.

Table 4: *Total biomass (t ha^{-1}) production and carbon (t ha^{-1}) sequestration potential of FGSRTs based agroforestry systems*

FGSRTs and spacing	Total tree biomass (A)	Total crop biomass (B)	Total system biomass (A+B)	Total tree carbon density (C)	Total crop carbon density (D)	Total system carbon (C+D+E)
			T1			
S1	3.521a	0.395a	3.916a	1.655a	0.178a	69.326a
S2	3.139ab	0.448a	3.587ab	1.475ab	0.202a	43.347b
S3	2.191bc	0.562a	2.753bc	1.030bc	0.253a	50.385ab
S4	2.043cd	0.691a	2.734bc	0.960cd	0.311a	43.852b
			T2			
S1	3.018ab	0.456a	3.474ab	1.419ab	0.205a	43.353b
S2	1.837cd	0.475a	2.312cd	0.863cd	0.214a	44.385b
S3	1.069de	0.524a	1.593de	0.502de	0.236a	40.242b
S4	0.830ef	0.448a	1.278de	0.390ef	0.202a	36.789b
			T3			
S1	0.201ef	0.463a	0.664e	0.095ef	0.208a	47.002ab
S2	0.127ef	0.437a	0.564e	0.060ef	0.197a	53.253ab
S3	0.096ef	0.625a	0.721e	0.045ef	0.281a	40.124b
S4	0.055f	0.638a	0.693e	0.026f	0.287a	32.380b
			T4			
S1	0.745ef	0.528a	1.273de	0.350ef	0.238a	48.729ab
S2	0.447ef	0.596a	1.043e	0.210ef	0.268a	39.584b
S3	0.353ef	0.558a	0.911e	0.166ef	0.251a	45.240ab
S4	0.249ef	0.634a	0.883e	0.116ef	0.285a	44.67b
P value	0.0001	0.2794	0.0005	0.0001	0.2794	0.0433

Values in the same column followed by the same letter are not significantly different at $P \leq 0.05$ level by using Tukey-Kramer HSD test.

CONCLUSION

Fast growing and short rotation tree species planted at close spacing produce high biomass and show carbon storage potential in agroforestry systems of northern plains of India. Poplar-wheat agroforestry system is very widely used system in Tarai region of India. Poplar planted at closer spacing grows fast and produce more biomass. Leaf shading by poplar at closer spacing during wheat sowing season made some hindrance on crop growth but showed better results on soil health. Overall, poplar based agroforestry system with closer spacing produce high biomass and showed carbon storage potential.

ACKNOWLEDGEMENT

Authors are grateful to the Agroforestry Research Centre, Pantnagar for allowing and providing facilities. Thanks to Department of Science and Technology, Government of India for providing financial assistance. Acknowledgements are also due to various field assistants for their help data collection activities.

REFERENCES

Albrecht A and Kandji ST 2003 Carbon sequestration in tropical agroforestry systems, *Agriculture, Ecosystems and Environment* **99** 15-27

Arora G, Chaturvedi S, Kaushal R, Nain A, Tewari S, Alam NM and Chaturvedi OP 2014 Growth, biomass, carbon stocks, and sequestration in an age series of *Populus deltoides* plantations in Tarai region of central Himalaya, *Turkish Journal of Agriculture and Forestry* **38** 1-11

Bala G, Gopalakrishnan R, Jayaraman M, Nemani R and Ravindranath NH 2010 CO_2-fertilization and potential future terrestrial carbon uptake in India, *Mitigation and Adaptation Strategies for Global Change* **16** 143-160

Bhardwaj SD, Panwar P and Gautam S 2001 Biomass production potential and nutrient dynamics of *Populus deltoides* under high density plantation, *Indian Forester* **127** (2) 144-153

Bonan GB 2008 *Ecological climatology: Principles and Applications.* 2nd ed Cambridge University Press, Cambridge, New York 563

Chauhan SK, Gupta N, Yadav RS and Chauhan R 2009 Biomass and carbon allocation in different parts of agroforestry tree species, *Indian Forester* **135**(7)981-993

Chauhan SK, Brar MS and Sharma R 2012 Performance of poplar (*Populus deltoides* bartr.) and its effect on wheat yield under agroforestry system in irrigated agro-ecosystem India, *Caspian Journal of Environmental Sciences* **10** (1) 53-60

Chauhan VK and Dhiman RC 2007 Atmospheric humidity and air temperature studies in wheat-poplar based agroforestry system, *Indian Forester* **133** 73-78

Deshpande SB, Fahrenbacher JB and Beavers AH 1971 Mollisols of Tarai region of Uttar Pradesh, Northern India.Morphology and mineralogy, *Geoderma* **6** 179-193

Deshpande SB, Fehrenbacher JB and Ray BW 1971b Mollisols of Tarai region of Uttar Pradesh, Northern India. Genesis and classification, *Geoderma* **6** 195-201

Devagiri GM, Money S, Singh S, Dadhawal VK, Patil P, Khaple A, Devakumar AS and Hubballi S 2013 Assessment of above ground biomass and carbon pool in different vegetation types of south western part of Karnataka, India using spectral modelling, *Tropical Ecology* **54**(2) 149-165

Dutta RK and Agrawal M 2002 Effect of tree plantations on the soil characteristics and microbial activity of coal mine spoil land, *Tropical Ecology* **43**(2) 315-324

Fang S, Xu X, Yu X and Li Z 2005 Poplar in wetland agroforestry: a case study of ecological benefits, site productivity, and economics, *Wetlands Ecology and Management* **13** 93-104

Gibbs HK, Brown S, Niles JO and Foley JA 2007 Monitoring and estimating tropical forest carbon stocks: making REDD a reality, *Environmental Research Letters* **2** 1-13

Gill RIS, Singh B and Kaur N 2009 Productivity and nutrient uptake of newly released wheat varieties at different sowing times under poplar plantation in north-western India, *Agroforestry Systems* **76** 579-590

[IPCC] Intergovernmental Panel on Climate Change 1996 Land use change and forestry, *in*: Revised 1996 IPCC guidelines for national greenhouse gas inventories, Cambridge University Press, Cambridge UK **5** 5.1-5.20

[IPCC] Intergovernmental Panel on Climate Change 2006 Agriculture, Forestry and Other Land Use, in: Guidelines for national greenhouse gas inventories IGES, Japan **4** 4.7-4.79

Jandl R, Lindner M, Vesterdal L, Bauwens B, Baritze R, Hagedorn F, Johnsong DW, Minkkinen K and Byrne KA 2007 How strongly can forest management influence soil carbon sequestration? *Geoderma* **137**(3-4) 253-268

Joa Carlos de MS, Carlos CC, Warren AD, Lal R, Filho SPV, Piccolo MC and Feigl BE 2001 Organic matter dynamics and carbon sequestration rates for a tillage chronosequence in a Brazilian Oxisol, *Soil Science Society of America Journal* **65** 1486-1499

Joshi NR, Tewari A and Singh V 2013 Biomass and carbon accumulation potential towards climate change mitigation by young plantations of *Dalbergia sissoo* Roxb. and *Eucalyptus hybrid* in Terai Central Himalaya, India, *American Journal of Research Communication* **1** (4) 261-274

Kaushik N and Singh J 2001 Performance of pearl millet- wheat in poplar based agri-silvicultural system in sandy soils of southern Haryana, *Indian Journal of Agroforestry* **3**(1) 51-54

Keutgen N and Chen K 2001 Responses of citrus leaf photosynthesis, chlorophyll fluorescence, macronutrient and carbohydrate contents to elevated CO_2, *Journal of Plant Physiology* **158** 1307-1316

Kohli RK, Singh HP and Batish DR 1997 Phytotoxic potential of *Populus deltoides* Bartr Ex Marsh I Comparative contribution of different parts, *Indian Journal of Forestry* **20** 300-304

Koul DN and Panwar P 2008 Prioritizing land: management options for carbon sequestration potential, *Current Science* **95** 685-663

Kumar K, Nandal DPS and Dahiya D 2006 Effect of fertilizer levels on yield and nutrient uptake of wheat intersown with poplar, *Indian Journal of Forestry* **29**(1) 55-61

Mishra A, Sharma SD, Pandey R and Mishra L 2004 Amelioration of a highly alkaline soil by trees in northern India, *Soil Use and Management* **20**(3) 325-332

Mishra A, Sharma SD and Khan GH 2002 Rehabilitation of degraded sodic lands during a decade of *Dalbergia sissoo* plantation in Sultanpur district of Uttar Pradesh, India, *Land Degradation and Development* **13**(5) 375-386

Mishra A, Swamy SL, Bargali SS and Singh AK 2010 Tree growth, biomass and productivity of wheat under five promising clones of *Populus deltoides* in agri-silviculture system, *International Journal of Ecology and Environmental Sciences* **36** (2-3) 167-174

Nair PKR 2012 Carbon sequestration studies in agroforestry systems reality-check, *Agroforestry Systems* **86** 243-253

Nair PKR, Kumar BM and Nair VD 2009 Agroforestry as a strategy for carbon sequestration, *Journal of Plant Nutrition and Soil Science* **172** 10-23

Nair PKR, Nair VD, Kumar BM and Showalter JM 2010 Carbon sequestration in agroforestry systems, *Advances in Agronomy* **108** 237-307

Pandey U, Kushwaha SPS, Kachhwaha TS, Kunwar P and Dadhwal VK 2010 Potential of Envisat ASAR data for woody biomass assessment, *Tropical Ecology* **51**(1) 117-124

Pressler M 1895 Das Gesetz der Stambildung Leipzig P 153

Saha SK, Nair PKR, Nair VD and Kumar BM 2010 Carbon storage in relation to soil size-fractions under tropical tree-based land use systems, *Plant and Soil* **328** 433-446

Sarvade S, Mishra HS, Kaushal R, Chaturvedi S, Tewari S and Jadhav TA 2014 Performance of wheat (*Triticum aestivum* L.) crop under different spacings of trees and fertility levels, *African Journal of Agricultural Research* **9**(9) 866-873

Sarvade S, Mishra HS, Kaushal R, Chaturvedi S and Tewari S 2014 Wheat (*Triticum aestivum* L.) yield and soil properties as influenced by different agri-silviculture systems of Terai Region, Northern India, *International Journal of Bio-resource and Stress Management* **5**(3) 350-355

Sharma NK, Samra JS and Singh HP 2001 Poplar (*Populus deltoides*) based agroforestry systems for an alluvial soil under irrigated condition in western Uttar Pradesh, *Indian Forester* **127** 61-69

Sheikh MA, Kumar S and Kumar M 2012 Above and below ground organic carbon stocks in a sub-tropical *Pinus roxburghii* Sargent forest of the Garhwal Himalayas, *Forestry Studies in China* **14**(3) 205-209

Singh K, Chauhan HS, Rajput DK and Singh DV 1989 Report of a 60 month study on fitter production, changes in soil chemical properties and productivity under Poplar (*P. deltoides*) and Eucalyptus (*E hybrid*) inter-planted with aromatic grasses, *Agroforestry Systems* **9** 37-45

Singh RA 1980 *Soil Physical Analysis*, Kalyani Publishers, New Delhi 118-124

[USDA NRCS] United States Department of Agriculture Natural Resources Conservation Service 2000 Growing carbon: new crop that helps agricultural producers and the climate too, *Soil and Water Conservation Society*

Walkley AJ and Black A 1934 Estimation of soil organic carbon by chromic acid titration method, Soil Science **37** 29-38

Wani N, Velmurugan A and Dadhwal VK 2010 Assessment of agricultural crop and soil carbon pools in Madhya Pradesh, India, *Tropical Ecology* **51**(1) 11-19

CHAPTER 13

Carbon Sequestration Potential and Economics of *Dalbergia sissoo* + Paddy Based Agroforestry System

Subrata Sharma*, Poornima Malviya, ML Sahu and LD Koshta

Department of Forestry, JNKVV, Jabalpur -482004 M. P., India

**Corresponding author's Email: sharmasubrata@yahoo.com*

ABSTRACT

The experiment was taken up with 17 years old agrisilviculture system (*D. sissoo* + paddy-wheat), where 4 pruning treatments (viz; no pruning, 25 per cent, 50 per cent and 75 per cent pruning) and one open were carried out in main plot and three levels of fertilizer and seed rate of paddy variety IR 36viz; T1 recommended dose of fertilizer and seed rate, T2 (T1+ 25 per cent more nitrogen than recommended dose, and T3 (T1 +25 per cent more seed rate than recommended dose) in sub plot; replicated five times. Paddy + shisham 25 per cent pruning (agroforestry system) recorded significantly higher carbon sequestration (100.89 t ha^{-1}) as compared to crop alone (5.81 t ha^{-1}) and tree alone (74.3 t ha^{-1}). Different levels of fertilizer dose and seed rate showed no significant effect on carbon sequestration. Significantly higher grain yield of paddy (21.54 q $ha^{-1}$was recorded under open conditions i.e. crop grown without tree, whereas in the crop grown with tree under no pruning recorded significantly lowest yield grain yield of paddy (10.92 q $ha^{-1)}$. Among different pruning treatments 75 per cent pruning recorded higher grain yield in paddy (18.75 q ha^{-1}) and was significantly superior to other pruning treatments. The per cent reduction in grain yield under no pruning, 25 per cent, 50per cent and 75 per cent as compared to open was 49.30 per cent, 39.46 per cent, 26.46 per cent, and 12.95 per cent in paddy, respectively. The net monetary return was significantly higher in the crop under 25 per cent pruned trees (₹ 59670 ha^{-1}) and minimum in open. Managed agroforestry system (i.e. pruning treatment) recorded higher monetary return (₹ 52592 ha^{-1}) as compared to crop alone (₹ 19467 ha^{-1}) and tree alone (₹ 39490 ha^{-1}).The per cent reduction in grain yield under no pruning, 25 per cent, 50 per cent and 75 per cent pruning as compared to open condition was 49.3 per cent, 39.46 per cent, 26.46 per cent and 12.95 per cent, respectively

Keywords: Agri-silviculture, Carbon sequestration, Grain yield, Monetary return

INTRODUCTION

Among the greenhouse gases increasing carbon-dioxide concentration is a blistering challenge for the scientific world. Trees are the most convenient alternative for sink of CO_2 concentration in atmosphere. Recently agro-forestry has gained increased attention as a strategy to sequester carbon and mitigate global climate change. However, little has been reported regarding carbon sequestration potential of agro-forestry systems in Madhya Pradesh. It is important to divulge the productivity, monetary gain and carbon sequestration potential of traditional agro-forestry system at regional level. Agro-forestry seems to be a potential and low cost method to sequester atmospheric carbon and is being recognized as a strategy for climate change mitigation (Alavapati and Nair 2001). Community assisted active forest management can substantiate a significant aid to increase carbon sequestration. Hence, the present investigation has been conducted to evaluate the carbon sequestration and economics of *D. sissoo* + paddy based agro-forestry system.

METHODOLOGY

The treatment combination involved five pruning treatments of *D. sissoo* viz. no pruning, 25 per cent, 50 per cent, 75 per cent pruning and one open (tree alone) in main plot and paddy variety IR 36 with three treatments viz; F1 recommended dose of fertilizer and seed rate, F2 (T1+ 25 per cent more nitrogen than recommended dose, and F3 (T1 +25 per cent more seed rate than recommended dose) in sub plot.

The diameter at breast height (1.37 m from ground) was measured at seventeen years of tree growth by the following formula:

$$\text{Basal area} = 0.00007854 \times (\text{dbh})^2$$

$$\text{Volume of tree} = \text{Basal area} \times \text{Tree height}$$

A 5 cm thick disc of wood sample was collected from main bole. Paddy plants were uprooted and fresh wt. of all tree and crop samples was taken and oven dried at 70°C till constant weight. The powdered samples of wood and paddy crop were burnt in a muffle furnace at 550°C for 4-6 hours and residue was weighted a ash content. Carbon content in all the samples of tree and crop has been estimated as per the given formula

$$\text{Carbon sequestration} = \text{Biomass} \times \text{carbon per cent}$$

From the total dry biomass, total carbon content and amount of CO_2 sequestrated in trees was calculated. For calculating CO_2 equivalent, total carbon was multiplied with a factor of 3.6663 on the basis of atomic weight ratio of CO_2 to C. For calculating CO_2 sequestration rate of individual tree species, total CO_2 was divided by age of trees (17 years). Economics of the system was calculated by multiplying the produce with estimated market price for the duration.

RESULTS AND DISCUSSION

The average carbon sequestration potential of *D. sissoo* tree alone in the system was 74.3 t ha^{-1}. Owing to high energy conversion, *D. sissoo* resulted in higher biomass and carbon accumulation (Bohre et al 2012). Maximum carbon was sequestered by 25 per cent pruned *D. sissoo* which was at par with unpruned and 50 per cent pruned trees. Chourasia (2012) reported highest biomass (580 kg $tree^{-1}$) and highest carbon sequestration potential (261 kg

C tree^{-1}) in *D. sissoo* Pires et al (2002) also reported dbh, volume to be inversely associated with tree pruning intensities The average carbon sequestered by agri-silviculture system (*D. sissoo* + paddy) was 67.6 t ha^{-1} being significantly maximum in the crop sown under 25 per cent pruned trees and it varied non-significantly from the crop sown under 50 per cent and unpruned trees. The carbon sequestration of paddy was significantly lower in the crop sown under open condition (5.81 t ha^{-1}). Warren and Patwardhan (2008) have reported the total carbon shared in forest including soil to be 9578 Mt. Fertilizer dose and seed rate had non-significant effect on carbon sequestration. As regards to net monetary return from *D. sisoo* trees, the monetary return was significantly higher from 25 per cent and 50 per cent pruned trees which varied non-significantly from each other. The monetary return was significantly lower from 75 per cent and unpruned trees (Table 1). The net gain in paddy varied non-significantly as effect of seed rate and fertilizer dose. Managed agroforestry system (i.e. pruning treatment + paddy) recorded higher monetary return (₹ 52592 ha^{-1}) as compared to crop alone (₹ 19467 ha^{-1}) and tree alone (₹ 39490 ha^{-1}). The grain yield (q ha^{-1}) expressed significant variation as effect of different pruning treatments. Significant maximum grain yield of paddy was recorded under open condition (21.54 q ha^{-1}) and was significantly superior to all the pruning treatments and no pruning. Among pruning intensities, significantly maximum grain yield was recorded under 75 per cent pruning (18.75 q ha^{-1}) followed by 50 per cent pruning (15.84 q ha^{-1}) and 25 per cent pruning (13.04 q ha^{-1}) and no pruning. No pruning recorded significantly lowest grain yield (10.92 q ha^{-1}). The per cent reduction in grain yield under no pruning, 25per cent, 50 per cent and 75 per cent pruning as compared to open condition was 49.3 per cent, 39.46 per cent, 26.46 per cent and 12.95 per cent, respectively (Fig. 1). Different level of seed rate and fertilizer dose showed significant effect on grain yield. F2 i.e. 25 per cent more nitrogen dose than recomended dose of fertilizer and seed rate gave significantly higher grain yield (17.37 q ha^{-1}) at par with F3 i.e. 25 per cent more seed rate than recommended dose of seed rate andfertilizer (16.21 q ha^{-1}). F1 i.e. recommended dose of seed rate and fertilizer dose recorded significantly lowest grain yield (14.47 q ha^{-1}). The per cent increase in yield by increasing 25 per cent more nitrogen than recommended dose of fertilizer and seed rate (F2) and 25 per cent more seed rate than recomended seed rate and fertilizer dose (F3) as compared to recommended dose of seed rate and fertilizer dose (F1), was 20.0 and 12.0 per cent, respectively.

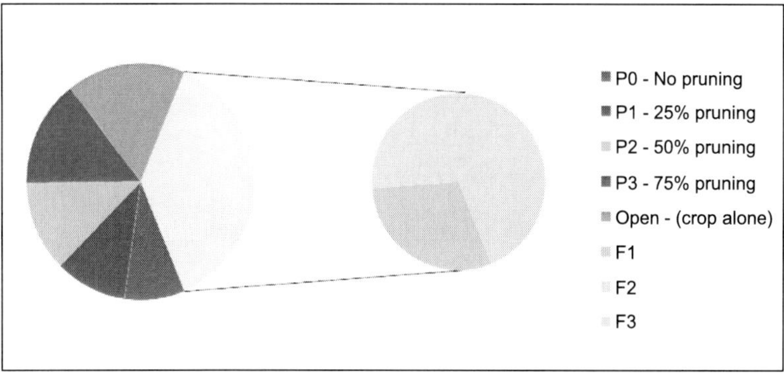

Figure 1: *Grain yield of paddy as influenced by different pruning intensities, fertilizer dose and seed rate in agri-silviculture system*

Table 1: *Carbon sequestration potential and net monitory return under Agrisilviculture system (D. sissoo + Paddy)*

Fertilizer doses and seed rate / Pruning intensity	Carbon sequestration potential (t ha^{-1})					Net monetary return (₹ ha^{-1})				
	F1	F2	F3	Mean	Tree alone	F1	F2	F3	Mean	Tree alone
P_0 - No pruning	85.66	100.69	87.11	91.15	72.65	48300	61174	48500	52658	39259
P_1- 25per cent pruning	94.22	106.95	101.51	100.89	98.56	56191	66182	56636	59670	50233
P_2- 50per cent pruning	94.70	84.47	83.06	87.41	79.44	58221	56552	55130	56634	45273
P_3- 75per cent pruning	67.18	50.96	40.13	52.76	46.52	47551	42683	34184	41473	23192
OPEN	6.13	5.45	5.84	5.81		14025	23095	21281	19467	-
MEAN	69.6	69.7	63.5	67.6	74.3	44857	49937	43146	-	39490

Carbon sequestration	SEm±	CD (P = 0.05)	Net monetary return	SEm±	CD (P = 0.05)
Pruning Intensities	9.1	27.3		5469	16393
Fertilizer doses and seed rate	6.7	NS		4742	NS

CONCLUSION

Carbon sequestration potential (74.3 t ha^{-1}) and net monetary gain was significantly higher under *D. sissoo* + paddy based managed agro-forestry system as compared to crop or tree alone due to high and efficient energy conversion by the system. The per cent reduction in grain yield under no pruning, 25 per cent, 50 per cent and 75 per cent pruning as compared to open condition was 49.3 per cent, 39.46 per cent, 26.46 per cent and 12.95 per cent, respectively

REFERENCES

Alavapati JRR and Nair PK 2001 Socioeconomic and institutional perspectives of agroforestry, in *World forests, Society and Environment – Markets and Policies* (eds) Palo, M and JU Usivuori, 52-62 Kluwer Academic Publishers, Dordrecht.

Bohre P, Choubey OP, and Singhal PK 2012 Biomass accumulation and carbon sequestration in *D. sisoo* Roxb, *International Journal of Bioscience and Bio Technology* **4**3-43

Chourasia S 2012 Carbon sequestration study on *Dalbergia sissoo* planted in different geometry, M. Sc. Thesis, JNKVV, Jabalpur 15-33

Pires BM, Reis M, Das GF, and Reis G 2002 Pruning effect of growth of *Eucalyptus grandis* on Southeastern Brazil, *Brazil Florestal* **21** 12-21

Warren A and Patwardhan A 2008 "Carbon sequestration Potential of trees in and around Pune City", Retrieved from www.ranwa.org on 17.12.2008

CHAPTER 14

Biomass and Carbon Potential in Agroforestry Systems along Altitudes in Tehri District of Uttarakhand

KK Vikrant*, DS Chauhan and RH Rizvi[1]

Department of Forestry & NR, HNB Garhwal University, Srinagar, Garhwal, Uttarakhand – 246174, India

ICAR-Central Agroforestry Research Institute, Jhansi, Uttar Pradesh– 284003, India

**Corresponding author's Email: kundankumarvikrant@gmail.com*

ABSTRACT

Agroforestry land use systems contribute substantially to mitigate the global climate change. The main objective of this study is to assess the potential of agroforestry systems in mitigation of greenhouse gases through carbon sequestration. This study involves three agroforestry systems viz. Agri-silviculture (AS), Agri-horti-silviculture (AHS) and Agri-horticulture (AH) in Kritinagar block of Tehri district in Uttarakhnad. The region represents three main agro-climatic zones viz. sub-tropical zone/ foot hill (286-1200m), sub-temperate zone/ middle altitude (1200-2000m) and temperate zone/ high altitude (2000- 2800m). Along 1200-2000m altitude, agri-silviculture system contains maximum biomass (3.68 t ha^{-1}), carbon (0.163 tha^{-1}) and CO_2 mitigation potential (0.593 t ha^{-1}). *Grewia oppositifolia* based agroforestry system contain maximum biomass (0.128 t ha^{-1}), carbon (0.0575 t ha^{-1}) and CO_2 mitigation potential (0.209 t ha^{-1}) followed by *Celtis australis* (0.124 t ha^{-1}) biomass (0.056 t ha^{-1}) carbon and (0.205 t ha^{-1}) CO_2 mitigation potential. The finding evinced that agri-silviculture system was mostly adopted by farmers in the study area and *Grewia oppositifolia* was dominant species which has maximum biomass, carbon and CO_2 mitigation potential. Thus, *Grewia oppositiafolia* based agri-silviculture systems are not only remunerative to the farmers of this region but also contributing towards CO_2 mitigation vis-à-vis reducing climate change effect.

Keywords: Agroforestry systems, Carbon sequestration, Climate change, CO_2 mitigation

INTRODUCTION

Agroforestry is a viable option for sequestering carbon on agriculture land because it can sequester significant amounts of carbon while leaving the bulk of the land for agricultural production. These practices represent a category of conservation activities that can sequester large amount of carbon while providing a multitude of additional benefit (Montagnini and Nair 2004). Agroforestry systems are believed to have a higher potential to sequester C than pastures or field crops (Sanchez 2000). This conjecture is based on the notion that tree incorporation in croplands and pastures would result in greater net aboveground as well as belowground C sequestration (Palm et al 2004). Since agroforestry is mostly practiced by subsistence farmers in developing countries, there is an attractive opportunity for those farmers to benefit economically from agroforestry if the C sequestered through agroforestry activities is sold to developed countries.

Agroforestry is a collective name for land-use systems and technologies where woody perennials (trees, shrubs, palms, bamboos, etc.) are deliberately used on the same land-management units as agricultural crops and/or animals, in some form of spatial arrangement or temporal sequence. In agroforestry systems there are both ecological and economical interactions between the different components (Lundgren and Raintree 1982). In Garhwal Himalaya, agroforestry is permanent feature of agriculture landscape (Semwal and Maikhuri 1996, Bhatt and Todaria 1999). It forms integral relationships with the farmer community to supplement fuel, fodder, fruits and fibers, on the one hand and balance the environment on the other. Farmers have common practice to integrate crops, trees, livestock to solve the problem of acute shortage of fuel, fodder and other goods (Bhatt 2002). The farmers have little opportunities to select the tree species, and therefore, they accept whatever is available on their land (Bhatt et al 2010). The assortment of trees on the edges of the agricultural field is farmer friendly and compatible. Selection of intercrops depends mainly on edapho-climatic conditions of the area, farmer's need and resource availability (Saroj and Dadhwal 1997).

In India, carbon sequestration in different agroforestry systems occurs both belowground, in the form of enhancement of soil carbon plus root biomass and aboveground as carbon stored in standing tree biomass. Some of the earliest studies of potential carbon storage in agroforestry systems and alternative land use systems estimated a sequestration potential of 68-228 Mg C ha^{-1} over 96 M ha of land in India (Dixon et al 1994). But this value varies in different regions depending on the biomass production. Studies done by Jha et al (2001) showed that agroforestry could store nearly 83.6 t C ha^{-1} up to 30 cm soil depth, 26% more carbon compared to cultivation in Haryana plains. However, the magnitude of carbon sequestration from forestry activities would depend on the scale of operation and the final use of wood. There is no reliable data available for carbon sequestration potential of prevalent agroforestry systems in Garhwal Hiamalaya. So this case study was carried out in Kritinagar block in Tehri district of Garhwal Himalaya under the main study entitled "*Carbon stock along altitudes in traditional agroforestry system in Garhwal Himalaya*".

MATERIALS AND METHOD

The present research works were conducted in Kritinagar block in Tehri district of Garhwal Himalaya. Kritinagar block located between 30°.15 to 30°.30N latitude and 78^{0}.58 to

$78^0.59$E longitude (Figure 1). It is a town in Tehri Garhwal district of Uttrakhand in India, situated at the bank of Alaknanada river. January is the coldest month whereas temperature becomes hottest usually during June. Rainfall, occurs almost throughout the year and maximum during the monsoon period i.e. from July to September. Agriculture is the main occupation of the people. However, intensive cultivation is not possible as major part of the area is mountainous. Agricultural activities are common on gentle hill slopes and in relatively plain areas. Rice, wheat, mandua, barley, maize and sawan are the principal crops grown in this area.

Agro-climatically the area has been divided into three main zones viz. sub-tropical zone/ foot hill (286-1200m), sub temperate zone/ middle altitude (1200-2000m) and temperate zone/ high altitude (2000- 2800m) depending on the topography, altitude and temperature condition (Singh and Singh, 1992).

SAMPLING PROCEDURE

In Kritinagar block, 10% sample villages were selected and in each village 10 random sample plot (10x10 m^2) were taken for field data collection. Total five villages were selected for present study under this block, Based on the altitudes two villages situated in lower altitude (286 – 1200 m), three villages in middle altitude (1200 – 2000 m) and there was no settlement area on high altitude (2000 – 2800 m). A survey was conducted in this block during July, 2014. Primary data such as name of the village, altitude, GPS location, exiting agroforestry systems, tree species, and agricultural crops was recorded from agroforestry field. From each village, randomly 10 sample plot of 10×10 m^2 for trees and 1x1m^2 for crops were marked for detailed study in the agroforestry field.

All the tress falling in the plot were enumerated and stem biomass was estimated. The diameter at breast height was measured with caliper and height with Ravi's altimeter. Form factor was calculated with Spiegel Relaskope to find out the tree volume using the formula given by Presseler 1865; Bitlerlich 1984. The specific gravity was used to determine biomass of the stem using maximum moisture method (Smith 1954). Total number of branches irrespective of size were counted on each of the sample tree, then these branches were categorized on the basis of diameter into three group viz<6cm, 6-10cm, and >10cm. Fresh weight of two sampled branches from each group was recorded separately. The following formula was used to determine the dry weight of branches: $B_{dwi} = B_{fwi}/1+M_{cbdi}$. Leaves from the branches were removed, weighed and oven dried separately at constant temperature of 65 °C. The total tree biomass was the sum of stem, branch and leaf biomass. The tree biomass was converted into carbon by multiplying with factor of 0.45 (Woomer 1999).

Crop biomass was estimated in 1 × 1m quadrates. All the crop plants occurring within the borders of the quadrant were cut at ground level and collected samples were weighted, sub sampled and oven dried at 65 °C. Crop biomass was converted to carbon by multiplying with a factor of 0.45 (Woomer 1999). Underground biomass of roots was calculated by multiplying the aboveground biomass by a factor of 0.26.

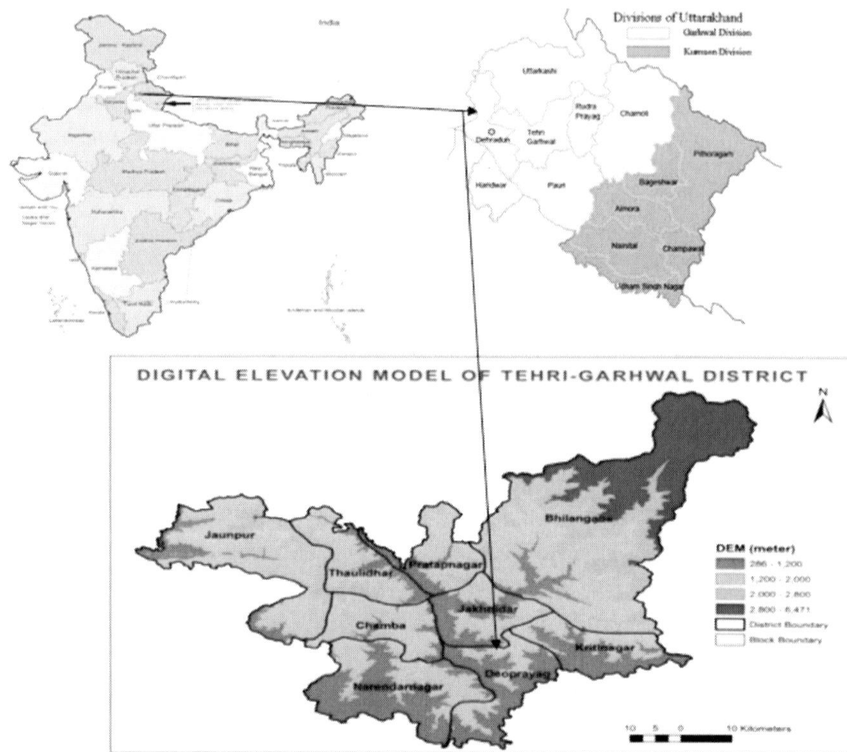

Figure 1: *Geographical location of study area*

RESULTS AND DISCUSSION

Based on the field survey, the common existing agroforestry systems in Garhwal Himalaya were agri-siviculture (trees + agricultural crops), agri-horticulture system (edible fruit trees + agriculture crops), agri-horti-silviculture system (trees including edible fruit trees and forest trees + agricultural crops). Major tree species (*Grewia oppositifolia, Quercus leucotrichophora, Celtis australis, Melia azedarach*) were found along altitudes (Table 1). The *Grewia oppositifolia* is one of the species thrives on the diverse altitudinal range.

Table 1: *Tree species under agroforestry systems in different altitudes of Kritinagar block*

Village	Altitude (m)	Agroforestry systems	Major tree species	Major crops
Timal gaon	1200–2000	Agrisilviculture, Agrihortisilvi	*Grewia oppositifolia, Toona ciliata, Syzygium cumini, Mangifera indica*	*Amaranthus spinosus* L., *Glycine max, Zea mays, Echinochloa frumentacea*
Dagar	1200–2000	Agrihortisilvi	*Ficus bengalensis* L., *Melia azedarach, Citrus sinensis* (L.) Osbeck, *Prunus cerasoides, Juglans regia, Toona ciliata, Musa paradisiaca*	*Echinichloa frumentacea, Vigna mungo, Curcuma longa*
Maikhandi	286–1200	Agrisilvi, Agrihortisilvi	*Grewia oppositifolia, Toona ciliata, Celtis australis, Melia azedarach, Musa paradisiaca, Bauhinia variegata, Ficus palmata*	*Curcuma longa, Vigna mungo, Echinochloa frumentacea*
Riskoti	1200–2000	Agrisilvi, Agrihortisilvi	*Grewia oppositifolia, Ficus roxburghii, Prunus cerasoides, Bauhinia variegata, Celtis australis, Mangifera indica*	*Vigna mungo, Glycine max, Zea mays, Amaranthus spinosus*
Jakhand	286–1200	Agrisilvi, Agrihortisilvi	*Celtis australis, Toona ciliata, Grewia oppositifolia, Citrus limon, Psidium guajava, Celtis australis*	*Vigna mungo, Glycine max, Zea mays, Amaranthus spinosus*
Dhaulangi	286–1200	Agrisilvi, Agrihortisilvi	*Ficus roxburghii, Pinus roxburghii, Syzygium cumini, Celtis australis, Toona ciliata, Melia azedarach*	*Amaranthus spinosus*

Along 1200-2000 m altitude in agrisilviculture (AS) system contains maximum 3.68 t ha^{-1} biomass, 0.163 t ha^{-1} Carbon and 0.593 t ha^{-1} CO_2 mitigation potential, followed by agri-horti-silviculture (AHS) system contains 0.13 t ha^{-1} biomass, 0.012 t ha^{-1} carbon and 0.045 t ha^{-1} CO_2 mitigation potential was recorded (Table 2). Along 286-1200m altitude agri-horti-silviculture system contain maximum 1.37 t ha^{-1} biomass, 0.062 t ha^{-1} carbon and 0.227 t ha^{-1} CO_2 mitigation potential, followed by agrisilviculture system contain 0.52 t ha^{-1} biomass, 0.082 t ha^{-1} carbon and 0.299 t ha^{-1} CO_2 mitigation potential was recorded (Table 3). As far as dominant species based system is concerned, *Grewia oppositifolia* contain maximum 0.128 t ha^{-1} biomass, 0.057 t ha^{-1} carbon and 0.209 t ha^{-1} CO_2 mitigation potential, followed by *Celtis australis* contain 0.124 t ha^{-1} biomass 0.056 t ha^{-1} carbon and 0.205 t ha^{-1} CO_2 mitigation potential and *Quercus leucotrichophora* contain 0.101 t ha^{-1} biomass, 0.045 t ha^{-1} carbon and 0.165 t ha^{-1} CO_2 mitigation potential was recorded. Considering the tree density, maximum 330 trees ha^{-1} were recorded in Agrisiviculture system as compare to other systems along altitudes. In different agroforestry systems, predominant tree species were *Grewia oppositifolia, Celtis australis, Quercus leucotrichophora,* and tree density was recorded in order of AS> AHS > AH respectively, which were considered suitable for this agro-climatic region. It is generally known that variability in the carbon stocks of different land use types depends primarily on its complexity. This result show that farmers integrate (raised/protected) forest and fruit tree species with agricultural crops for multiple benefits. Biomass in agroforestry systems depends on a number of factors including the growth habit of the species, site quality, age, management practices, interaction between trees & understory crops (Jana et al 2009, Kanime et al 2013). In hills, agroforestry, generally fulfills local farmers' requirements of fodder, fuelwood, fiber, fertilizer and small timber by looping/cutting of the trees and leaves collection. Agroforestry systems also help in locking higher amount of carbon and their CO_2 mitigation capacity can be enhanced by substituting suitable tree and annual crop species in existing agroforestry systems. C mitigation potential of agroforestry system is dependent on the tree density, age structure and C concentration in different components (Swami and Puri 2005). The present results are in line with findings of Albercht and Kandji (2003) where a high variability in plant biomass with complex agroforestry system was recorded in tropical agroforestry systems.

Table 2: *Biomass, Carbon and CO_2 mitigation potential in agroforestry systems.*

Altitudes (m)	Land use systems	Density (tress ha^{-1})	Total biomass (t ha^{-1})	Total carbon (t ha^{-1})	CO_2 mitigation potential (t ha^{-1})
1200-2000	Agrisilvi	330	3.68	0.163	0.593
	Agrihortisilvi	97	0.13	0.012	0.045
286-1200	Agrisilvi	200	0.524	0.082	0.299
	Agri-hortisilvi	135	1.37	0.062	0.227
	Agrihorti	15	0.326	0.016	0.06

Table 3: *Carbon stock of different species in agroforestry system*

Species	Stem biomass (Kg/100 m^2)	Branch biomass (Kg/100 m^2)	Leaf biomass (Kg/100 m^2)	Total biomass (t/ha)	Total Carbon (t/ha)
Celtis australis	46.27	39.37	13.15	0.1245	0.056
Syzgium cumini	20.27	4.33	2.08	0.0336	0.015
Embilica officinalis	8.27	5.24	0.81	0.0180	0.008
Melia azadirach	12.27	8.66	3.375	0.0306	0.014
Toona ciliata	6.61	3.63	2.21	0.0157	0.007
Pinus roxburghii	3.83	1.13	0.28	0.0066	0.003
Ficus roxburghii	1.70	1.7	0.75	0.0052	0.002
Ficus palmata	17.79	3.32	2.99	0.0304	0.014
Ficus auriculata	12.15	7.47	0.78	0.0257	0.011

Leauceana leucocephala	1.28	0.73	0.45	0.0031	0.001
Grewia oppositifolia	49.08	37.89	14.51	0.1279	0.057
Quercus leucotrichophora	18.47	39.09	22.67	0.1011	0.045
Mangifera indica	0.16	27.46	5.72	0.0420	0.019

CONCLUSION

From this study, it is concluded that agri-silviculture system is most dominant systems, which contribute maximum biomass, carbon and CO_2 assimilation. This system was mostly adopted by farmers of Garhwal Himalaya. Agroforestry land use systems help to meet the diverse needs of food, fodder, fuelwood and timber besides producing significant biomass. These agroforestry systems also make an important contribution towards C stock by atmospheric CO_2 sequestration in tree biomass and soils. Due to a ban on felling green/ live trees in the entire Himalayan region of India, agroforestry systems can be a good source of significant C credit to the farmers.

ACKNOWLEDGEMENTS

This research was supported by Rajiv Gandhi National Fellowship funded by University Grant Commission. We thank our colleagues from Central Agroforestry Research Institute, Jhansi who provided insight and expertise that greatly assisted this research. The authors are also thankful to Head, Prof. N.P. Todaria, Department of Forestry and Natural Resourses for their encouragement and guidance during the course of present work.

REFERENCES

Albrecht A and Kandji ST 2003 Carbon sequestration in tropical agroforestry systems. *Agriculture, Ecosystems and Environment,* **99** 15-27

Lundgren and Raintree 1982 Sustained Agroforestry, in *Agricultural Research For Development: Potentials and challenges in Asia* (eds) Nestel B Isnar, The Hague, The Netherlands 37–49

Semwal RL, Maikhuri K 1996 Structure and functioning of traditional hill agro-ecosystems of Garhwal Himalaya, *Biological Agriculture and Horticulture* **13** 267–289

Saroj PL, Dadhwal K S1997 Present Status and Future Scope of Mango Based Agroforestry Systems in India, *Indian Journal of Soil Conservation* **25**(2) 118–127

Singh JS, Singh SP 1992 Forests of Himalaya: Structure, Functioning and Impact of Man, Gyanodaya Prakashan, Nainital, Uttarakhand, India

Dadhwal KS, Narain P, Dhyani SK 1989 Agroforestry Systems, in The Garhwal Himalayas of India, **7** 213–225

Bhatt VP, Purohit V, Negi V 2010 Multipurpose Tree Species Of Western Himalaya With An Agroforestry Perspective For Rural Needs. *Journal of American Science* **6**(1) 73–80

Dixon RK, Brown S, Houghton RA, Solomon AM , Trexler MC 1994 Carbon pools and fluxes of global forest ecosystems, *Science* **263** 185-190

Jana BK, Biswas S, Majumder M, Roy PK, Mazumdar A 2009 Carbon sequestration rate and aboveground biomass carbon potential of four young species, *Journal of Ecology and the Natural Environment* **1** 15–24

Jha MN, Gupta MK, Raina AK 2001 Carbon Sequestration: Forest soil and land use management, *Annals of Forestry* **9** 249-256

Kanime N, Kaushal R, Tewari SK , Raverkar KP, Chaturvedi S, Chaturvedi O P 2013 Biomass production and carbon sequestration in different tree-based systems of Central Himalayan Tarai region, *For Trees Livelihoods* **22** 38–50

Palm CA, Van NM, Woomer PL, Alegre J , Arevalo L , Castilla C, Cordeiro DG, Hairiah K, Parton WJ, Ricse A, Rodrigues V, Sitompul SM 2004 Carbon losses and sequestration following land use change in the humid tropics. Alternatives to Slash and Burn: *The Search for Alternatives*, Columbia University Press

Smith DM 1954 Maximum moisture content methods for determining specific gravity of small wood samples. Forest Product Laboratory, Forest Service, U.S. Department of Agriculture, Report No. 2014 p. 1-8.

CHAPTER 15

Carbon Sequestration Potential of few Selected Tree Species

AR Manju Hashini, M Sangareswari Nagajothi, A Balasubramanian, S Radhakrishnan and B Palanikumaran*

Forest College and Research Institute, Mettupalayam, Coimbatore, TamilNadu-641301, India

**Corresponding author's Email: kumaranbass@gmail.com*

ABSTRACT

A study was conducted to know the carbon sequestration potential of selected tree species at Forest College and Research Institute, Mettupalayam. Twenty six tree species were selected and categorized into different height classes as 0-3 m, 3-6 m and 6-9 m from various institutions of Coimbatore. The growth parameters such as height, DBH were recorded and biomass carbon was estimated. The calculated biomass was then converted into mass of carbon sequestered. Maximum biomass accumulation was recorded in height class 6m to 9m found in *Delonix regia* (157.64 kg tree^{-1}) and the minimum was recorded in *Gmelina arborea* (0.19 kg tree^{-1}) in the height class 0-3m.

Keywords: Biomass carbon, Carbon sequestration, Climate change.

INTRODUCTION

Climate change or global warming is largely dictated by increased emission carbon dioxide (CO_2) in atmosphere (Singh 2013). Awareness of the effects of carbon dioxide emissions from fossil fuel use on the global climate system has sparked research into strategies to mitigate the effects of these emissions. The mitigation of climate change demands, determined commitment of scientists to develop strategies to effectively manage the issues of the changing climate through carbon sequestration. Carbon sequestration is thought to be a promising solution for reducing atmospheric carbon dioxide, an important greenhouse gas. Among different modes of carbon sequestration developed by different scientists, tree carbon sequestration found to be more promising and less cost consuming effort. About two third of terrestrial carbon is sequestered in the standing forest, forest under storey plants, leaf and forest debris. A considerable interest has been generated about carbon sequestration through afforestation and reforestation activities (Vashum and Jay Kumar 2012). Natural forest can sequester carbon permanently but it loses carbon on account of harvest or natural death. Hence, plantation recognized importance in mitigating climate change has led countries to study their carbon budgets and initiate the assessment of enhancing and maintaining carbon sequestration of their resource.Concern about rising atmospheric concentrations of greenhouse gases (IPCC 2011) has prompted the search for methods of sequestering carbon in plant biomass. Recently biomass is being increasingly used to help quantify pools and fluxes of green house gases (GHG) from terrestrial biosphere associated with land use and land cover changes (Cairns et al 2007). In India, planted forests of short- rotation tree species have huge carbon sequestration potential and elaborate some possible opportunities for sustainable carbon forestry (Updegraff et al 2004). In urban settings, 14 per cent of all carbon is stored in trees, making urban trees an important source of carbon sequestration (Freilicher 2012). It is very important to know the carbon sequestration potential of particular species for urban plantation to neutral the carbon. With the backdrop of all these studies the research was designed with the objective to evaluate growth and biomass carbon accumulation in selected trees for carbon sequestration potential.

METHODOLOGY

The trees were selected from various educational institutions in and around the Coimbatore city. The selected trees species were *Acacia auriculiformis, Bauhinia racemosa, Cassia fistula, Cassia siamea, Delonix regia, Gmelina arborea, Grevelia robusta, Leuceana leucocephala, Melia dubia, Pongamia pinnata, Peltophorum pterocarpum, Spathodia campanulata, Aleurites fordii, Polyalthia longifolia, Albizzia lebbeck, Tecoma stans, Terminalia arjuna* and *Kaya senagalensis.* These tree species were grouped based on height class as 0-3 m, 3 -6 m and 6-9 m. The following procedures were adopted to assess the carbon sequestration potential of selected tree species. The parameters namely height (m), DBH (cm) were recorded to analyse the biomass carbon.

VOLUME ESTIMATION

The volume of trees was estimated using the following formula (Chaturvedi and Khanna 1982) and expressed in cubic centimetre (cm^3).

$$V = \pi r^2 h$$

where, (V = Volume, r = Radius & h = Total height)

ESTIMATION OF BIOMASS IN TREES

The biomass of the trees was estimated using non-destructive sampling method. Above ground biomass, below ground biomass & total biomass estimated.

ABOVE GROUND BIOMASS (AGB)

AGB (kg $tree^{-1}$) = Volume of tree (m^3) x Wood density (kg m^{-3})

Note: The wood density of tree species was unavailable, the standard average value0.6 gm cm^{-3} were taken.

THE BELOW GROUND BIOMASS (BGB)

Below ground biomass includes all biomass of live roots excluding fine roots having < 2 mm diameter. The Below ground biomass was calculated by multiplying AGB by 0.26 factors as the root: shoot ratio given by Ravindranath and Ostwald 2008 method.

Below ground biomass (kg tree^{-1}) or (t tree^{-1}) = AGB (kg tree^{-1}) or (t tree^{-1}) × 0.26

TOTAL BIOMASS

Total biomass is the sum of the above and below ground biomass. Total biomass (TB) = above ground biomass + below ground biomass. Total biomass was estimated using Ravindranath and Ostwald (2008) method.

Total biomass (kg tree^{-1}) or (t tree^{-1}) = AGB + BGB

CARBON ESTIMATION

Generally, for any plant species 50% of its biomass is considered as carbon (Pearson et al 2005) i.e., carbon storage = biomass × 50% or biomass/2

RESULTS AND DISCUSSION

The carbon stocking potential of selected trees in study area was assessed. For this study total volume, above ground biomass, below ground biomass and total biomass obtained and the results are given here under.

CARBON CONTENT

The total biomass carbon and volume of selected tree species of different height classes were estimated and presented in Table 1 and Fig. 1, respectively. In height class 0–3 m, the maximum carbon content was registered in *Cassia siamea* (3.05 kg tree^{-1}) followed by *Acacia auriculiformis* (2.95 kg tree^{-1}) and *Cassia fistula* (2.59 kg tree^{-1}) (Table 2). The minimum biomass carbon was observed in *Gmelina arborea* (0.19 kg tree^{-1}) for the same height class. Which shows that the maximum total biomass ultimately resulted in highest biomass carbon. *Delonix regia* recorded the maximum biomass carbon with the value of 68.67 kg tree^{-1} followed by *Cassia fistula* (44.62 kg tree^{-1}) and *Acacia auriculiformis* (31.87 kg tree^{-1}) for the height class 3 m to 6 m. The minimum biomass carbon was observed in *Gmelina arborea* (4.54 kg tree^{-1}) for the same height class. For the height class 6 to 9 m, the maximum biomass carbon accumulation was found in *Delonix regia* (157.64 kg tree^{-1}) and the minimum was recorded in *Gmelina arborea* (15.82 kg tree^{-1}). Among different height classes studied the maximum biomass carbon accumulation was registered in 6 to 9 m height class. The biometric values are directly related to total biomass of the tree. Biometric estimation is useful for non-destructive estimation of biomass which is mostly used for yield prediction and carbon sequestration (Murali et al 2005). Similar approach was followed in present study to estimate the biomass of tree species. The DBH/height indicated a linear relationship in contribution to biomass accumulation. However, wood density was important and directly influence on above ground biomass (Brown and Lugo 2002).

Table 1: *Tree Biomass (kg tree^{-1}) of species in different height class interval*

S.No	Species	Height class (0-3 m)			Height class (3-6 m)			Height class (6- 9 m)		
		AGB	BGB	Total	AGB	BGB	Total	AGB	BGB	Total
1	*Acacia auriculiformis*	4.68	1.22	5.90	50.58	13.15	63.74	102.46	26.64	129.09
2	*Bauhinia racemosa*	0.34	0.09	0.43	10.73	2.79	13.52	35.96	9.35	45.31
3	*Cassia fistula*	4.10	1.07	5.17	70.83	18.42	89.24	170.44	44.31	214.75
4	*Cassia siamea*	4.84	1.26	6.10	13.61	3.54	17.15	54.19	14.09	68.28
5	*Delonix regia*	3.50	0.91	4.40	109.00	28.34	137.34	250.22	65.06	315.28
6	*Gmelina arborea*	0.30	0.08	0.37	7.20	1.87	9.07	25.11	6.53	31.64
7	*Grevelia robusta*	3.41	0.89	4.29	9.20	2.39	11.59	43.89	11.41	55.31
8	*Leucaena leucocephala*	2.18	0.57	2.75	8.26	2.15	10.40	47.89	12.45	60.34
9	*Melia dubia*	0.84	0.22	1.05	10.22	2.66	12.88	35.53	9.24	44.77
10	*Pongamia pinnata*	3.92	1.02	4.94	11.74	3.05	14.79	51.52	13.39	64.91
11	*Spathodia campanulata*	0.58	0.15	0.73	18.22	4.74	22.96	202.33	52.61	254.94
12	*Peltophorum pterocarpum*	0.87	0.23	1.10	16.63	4.32	20.96	163.54	42.52	206.06
13	*Aleurites fordii*	2.81	0.73	3.54	16.04	4.17	20.21	–	–	–
14	*Polyalthia longifolia*	2.91	0.76	3.67	31.35	8.15	39.51	–	–	–
15	*Albizzia lebbeck*	-	-	-	30.97	8.05	39.02	190.05	49.41	239.46
16	*Tecoma stans*	-	-	-	13.93	3.62	17.56	36.36	9.45	45.81
17	*Terminalia arjuna*	1.95	0.51	2.45	46.09	11.98	58.07	-	-	-
18	*Kaya senagalensis*	2.58	0.67	3.25	-	-	-	144.80	37.65	182.45

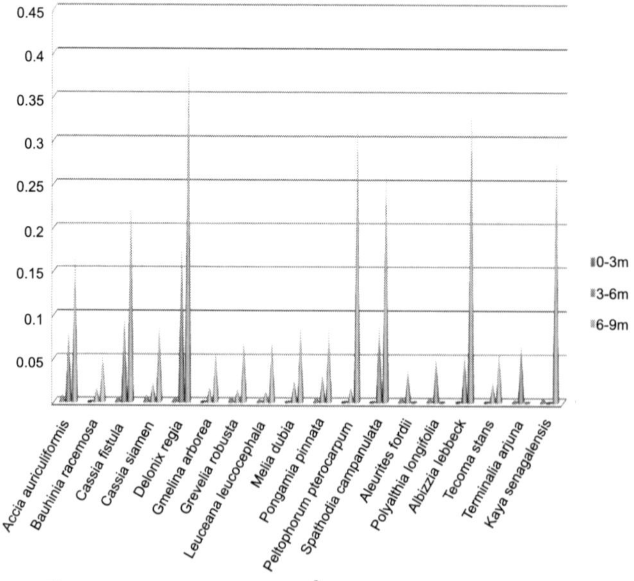

Figure 1: *Tree Volume (m^3) at different height class*

Table 2: *Carbon content (kg tree^{-1}) of trees in different height class interval*

S.No	Species	Height (0-3 m)	Height class (3-6 m)	Height class (6- 9 m)
1	*Acacia auriculiformis*	2.95	31.87	64.55
2	*Bauhinia racemosa*	0.22	6.76	22.66
3	*Cassia fistula*	2.59	44.62	107.38
4	*Cassia siamea*	3.05	8.58	34.14
5	*Delonix regia*	2.20	68.67	157.64
6	*Gmelina arborea*	0.19	4.54	15.82
7	*Grevelia robusta*	2.15	5.80	27.66
8	*Leucaena leucocephala*	1.38	5.20	30.17
9	*Melia dubia*	0.53	6.44	22.39
10	*Pongamia pinnata*	2.47	7.40	32.46
11	*Peltophorum pterocarpum*	0.37	11.48	127.47
12	*Spathodia campanulata*	0.55	10.48	103.03
13	*Aleurites fordii*	1.77	10.11	-
14	*Polyalthia longifolia*	1.84	19.76	-
15	*Albizzia lebbeck*	-	19.51	119.73
16	*Tecoma stans*	-	8.78	22.91
17	*Terminalia arjuna*	1.23	29.04	-
18	*Kaya senagalensis*	1.63	-	91.23

CONCLUSION

The maximum biomass and carbon content was observed in the height class of 6 to 9 m class interval. The lowest biomass was found in *Gmelina arborea* (31.64 kg tree^{-1}).

ACKNOWLEDGEMENT

Researchers deeply acknowledge Confederation of Indian Industries (CII) for provided fellowship and support. Research work was carried out under the scheme of vision carbon neutral.

REFERENCES

Brown S and Lugo AE 2002 The storage and production of organic matter in tropical forests and their role in the global carbon cycle, *Biotropica*, **14** 161-187

Cairns MA, Olmsted L, Granados Jand Argaez J 2007 Composition and above ground biomass tree biomass of a dry semi evergreen forest of Mexico's Yucatan peninsula,*Forest Ecology and Environment*, **186** (1-3) 125-132

Chaturvedi AN and Khanna LS 1982 Forest Mensuration, International Book Distributors, Dehra Dun, India

Freilicher M 2012 Carbon sequestration and urban trees. The Citizen Forester, Massachusetts Urban and Community Forestry Program, 1-2 pages

IPCC 2011 Climate change 2007 Impacts, adaptation and vulnerability. Fourth Assessment Report. Parry ML, Canziani OF, Palutikof JP, van der Linden P Jand Hanson CE (Eds.). Cambridge University Press, Cambridge, UK.

Murali KS, Bhat DM and Ravindranath NH 2005 Biomass estimation equations for tropical deciduous and evergreen forests. *International Journal of Agricultural Resources, Governance and Ecology*, **4**(1) 81-92

Pearson TRH, Brown S and Ravindranath NH, 2005 Integrating carbon benefits estimates into GEF Projects, 1-56

Ravindranath NH and Ostwald M 2008 Approaches to estimate carbon stock changes. *In*: Advances in Global Change Research 29, Carbon Inventory Methods, Springer Publications

Singh HS 2013 Tree wealth in the non-forest areas of Gujarat. Trees outside Forest (TOF) - Third Tree Counting-2013, Gujarat Forest Department, Gandhinagar. pp. 34.

Updegraff K, Baughman MJandTaffSJ2004 Environmental benefits of cropland conversion to hybrids poplar: economic and policy considerations, *Biomass and Bioenergy*, **27** 411-428

Vashum KT and Jayakumar S 2012 Methods to Estimate Above-Ground Biomass and Carbon Stock in Natural Forests - A Review. *Journal of Ecosystem and Ecography,* *2*(4) http://dx.doi.org/10.4172/2157-7625.100

CHAPTER 16

Biomass and Carbon Stock Under High Density Spacings of Different Tree Species

Kavita Satyawali*, Sumit Chaturvedi, Neema Bisht, Seema Maikhuri, Salil Tewari and Rajesh Kaushal

College of Agriculture, Govind Ballabh Pant University of Agriculture and Technology, Pantnagar, U. S. Nagar-26314, Uttarakhand, India

** Corresponding author's Email: kavita.satyawali19@gmail.com*

ABSTRACT

The biomass and carbon accumulation in different above and below ground components of 6 yr old *Eucalyptus camaldulensis* and *Melia azedarach* planted at four spacings i.e. 3.0 m × 1.0 m, 3.0 m × 1.5 m, 3.0 m × 2.0 m and 3.0 m × 2.5 m were evaluated in Tarai region of Central Himalayan. The total biomass (330.7 t ha^{-1}) and carbon stock (160.4 t ha^{-1}) was recorded to be higher in *E. camaldulensis* as compared to *M. azedarach*. As the spacing increased, individual stems increased in diameter and total biomass; however, total biomass production per hectare decreased. Maximum total tree biomass (286.0 t ha^{-1}) and total carbon storage (131.2 t ha^{-1}) was observed with narrow spacing of 3.0 m × 1.0 m followed by 3.0 m × 2.0 m > 3.0 m × 2.5 m and minimum in 3.0 m × 1.5 m spacing. Biomass allocation in above ground components followed the order viz., bole (49.08 and 47.09%) > branch (48.40 and 47.92%) > twig (44.06 and 43.85%)> leaves (40.40 and 39.69%) whereas, in below ground components as stump root (47.80 and 46.72%) > lateral root (46.90 and 45.75%) > fine root (45.90 and 44.88%), in *Eucalyptus* and *Melia*, respectively. Mean carbon concentration in all the aboveground components of both the species varied from 40-50%. However, soil carbon stock was found higher (61.54 t ha^{-1}) under *M. azedarach* which decreased with soil depth. Among the different spacings, maximum soil carbon stock was accumulated with 3.0m × 1.0m spacing and decreases with each successive increase in spacings. These results showed that there were significant differences in total biomass accumulation among the tree species or different tree spacings under high density. Thus, the study indicates superiority of *E. camaldulensis* over *M. azedarach* for biomass production and carbon accumulation at dense spacing, whereas, soil carbon stock was found to be greater under *M. azedarach*.

Keywords: Biomass allocation, carbon accumulation, carbon stock, soil carbon stock, tree density

INTRODUCTION

A tree represents an important carbon sink, thus playing a key role in the global cycle (Canadell et al 2008). Trees act as a sink for CO_2 by fixing carbon during photosynthesis and storing excess carbon as biomass (Gurveen et al 2013). As forest trees are able to store more carbon per unit area than any other terrestrial ecosystem, these may act as an effective mitigation option for climate warming. The mitigation potential of the trees, based on a biomass-demand based scenario, using short term commercial forestry option estimated to be 122 Tg carbon for the period 2000-2012 (Ravindranath et al 2002).Tree growth serves as an important means of capturing and storing atmospheric carbon in vegetation, soil, and biomass products (Kanime et al 2013). Most terrestrial carbon storage is in tree trunks, branches, foliage, and roots which is often called biomass. The quantity of tree biomass per unit area of land constitutes the primary data needed to understand the flow of material and water through forest ecosystem (Swank and Schreuder 1974). Biomass is an essential aspect of studies of carbon cycle (Ketterings et al 2001, Cairns et al 2003), especially tree biomass, including the trunk, branches, foliage and roots. Most of the total carbon is stored in the above ground components of the tree. The carbon sequestration capacity of plantations may be influenced by many factors. Fast growing multi-species plantations are likely to demonstrate a greater productivity and can enhance soil carbon sequestration (Wang et al 2013, Fonseca et al 2014, Huang et al 2014).

Soil organic carbon (SOC) also plays a very significant role in the global carbon cycle, as it is the largest terrestrial carbon pool. Soil carbon stocks are determined by the balance between inputs of carbon through litter fall and roots, and loss of carbon, mainly through decomposition of soil organic matter (Gonzalez et al 2013). In general, plant biomass increases gradually with stand age, but soil carbon might show different trends. Roots are most effective pathway for building up soil organic stocks than foliar litter. The total carbon stock under tree based system is estimated to be 861 ± 66 Pg C in 2011of which 44% was stored in soil (to 1 meter depth), 42% in live biomass (above and below ground), 8% in deadwood and 5% in litter (Pan et al 2011). To assess the potential of additional carbon sequestration by the tree management as part of the climate change mitigation strategies, it is necessary to understand the carbon storage in tree biomass, soil and the interaction between these compartments. Therefore, present investigation was carried out to gather valuable information for estimating potential of biomass and carbon storage and their allocations in *Eucalyptus camaldulensis* and *Melia azedarach* tree species under short rotation high density plantation.

METHODOLOGY

The field experiment was carried during 2006-07 to 2012-13 at Agroforestry Centre, G.B.P.U.A & T., Pantnagar, Uttarakhand. The experiment was laid out in split plot design with three replications. The treatment consists of six year old two tree species and four spacings. The soil of *Tarai* region have developed from calcarious, medium to moderately coarse textured materials under pre-dominant influence of tall vegetation and moderate to well drain conditions with neutral in reaction (soil pH), high in organic carbon (%). Twelve trees of each species (3 trees of each spacing) were selected and their biomass was estimated by destructive method. After felling, above ground parts viz. branches, twigs and leaves were separated from the main stem. The stem was then cross cut into logs of appropriate length depending upon the general form of the stem. Fresh weight of each log was taken in the field for each sample tree with the help of heavy weight spring balance. Three cross sectional discs representing top, middle and bottom portion of trees were taken from each tree for determining moisture content. Similarly, branches, leaves, twig and roots components were taken in amount of 200 g for calculating the fresh and dry weight. Moisture content of logs was determined by drying the representative disc sample (500g) and other components (200g) in the oven at constant temperature 60 °C. The moisture content of the oven dried sample was expressed as a percentage on oven dry weight and was calculated by the formula given by Husch et al 1972.

The total biomass of *Eucalyptus* and *Melia* trees was calculated by addition of all above and below ground components *viz.,* stem, branches, leaf, twigs, stump root, lateral root and fine root. Carbon accumulation is approximately 50 per cent of biomass accumulation. Total carbon stock potential of *Eucalyptus* and *Melia* trees was obtained by addition of carbon stock by above and below ground components of tree species and soil carbon stock upto 30cm of depth and was depicted as t ha^{-1}. Carbon concentration in above given components of *Eucalyptus* and *Melia* was determined by combustion method and is expressed in percentage (Gallardo and Merino 1993). Oven dried biomass samples was grinded in Willey mill and 5gm of powdered sample was taken in silica crucible for combustion in muffle furnace at 550±50 °C for 4-5 h till complete ashing. Ash free dry weight *i.e.* carbon content was calculated for each component by subtracting ash weight from initial weight of powdered sample *i.e.* 20g and expressed in per cent. Biomass in different above and below ground components and their carbon concentration was multiplied to estimate carbon stock component wise. Sum total of carbon stock by all above ground components taken as total carbon stock.

Soil samples were analyzed for soil organic carbon (SOC) by Walkley and Black method (1934). Bulk density of soil was determined using metal core samplers of known volume. Soil samples were collected from the three different depths 0-30 cm, 30-60 cm, and 60-90 cm. The samples were then oven dried separately at 105±10 °C for 48 h. The oven dried weight of sample divided by volume of core sampler gave the bulk density of soil which was depicted as g cm^{-3} (Blake and Hartge 1986).The amount of carbon stored per hectare was obtained by multiplying the values of soil depth (cm), bulk density (Mg m^{-3}), and the percentage of SOC content (Joa Carlos et al 2001). Total C stock (biomass C + soil C) of different diameter of *Eucalyptus* and *Melia* trees were obtained by addition of carbon stock by above and below ground components of tree species and soil carbon stock was only considered upto 30cm of depth as per the guidelines of the Intergovernmental Panel on Climate Change (IPCC). The total carbon stock by *Eucalyptus* and *Melia* based system was depicted as t ha^{-1}. The data obtained during the course of investigation was analyzed using standard statistical procedure for split-plot design except for the data obtained for soil organic carbon, bulk density, soil organic carbon stock in which the two factorial was used. Significance of treatment means was tested by, "F-test" (Panse and Sukhatme 1967).

RESULTS AND DISCUSSION

Tree Biomass

Biomass allocation pattern reflected that more biomass was allocated to above ground components *viz.*, stem, branches, foliage as compared to below ground part which resulted in higher shoot and root ratio (Table 1 and 2). Contribution of above ground biomass followed the pattern: bole (65.40 and 59.75%) > branch (13.81 and 6.56%) > leaves (1.87 and 1.20%) > twig (0.84 and 1.0 %), whereas, contribution of below ground biomass follows the pattern *i.e.* stump root (10.82 and 15.76%) > lateral root (5.80 and 5.40%) > fine root (1.27 and 0.14%) in *Eucalyptus* and *Melia*, respectively. The results are further consistent with the earlier findings of Puri et al 2002 and Swamy et al 2003 who reported that in nutrient rich soil, more of biomass is allocated to above ground parts.

Table 1: *Above ground biomass production (t ha^{-1}) of tree species and spacing under high density plantation.*

Treatments	Above ground biomass production				Total above ground biomass
	Bole	Branch	Leaf	Twig	
A. Tree species					
Eucalyptus	216.3 (65.40)	45.7 (13.81)	6.2 (1.87)	2.8 (0.84)	271.5
Melia	119.4 (59.75)	33.1 (6.56)	2.4 (1.20)	2.0 (1.00)	157.1
SEd±	6.24	2.04	0.15	0.17	4.24
C.D. (P=0.05)	26.12	8.57	0.63	0.73	17.75
B. Tree spacing					
3.0 m × 1.0 m	184.3 (64.44)	45.9 (16.04)	3.73 (1.30)	2.5 (0.87)	236.6
3.0 m × 1.5 m	149.6 (60.61)	41.9 (16.97)	4.55 (1.84)	2.2 (0.89)	198.4
3.0 m × 2.0 m	171.5 (63.87)	36.8 (13.70)	4.58 (1.70)	2.1 (0.78)	215.1
3.0 m × 2.5 m	166.5 (64.06)	33.2 (12.77)	4.50 (1.73)	2.8 (1.07)	207.0
SEd±	14.85	2.92	0.21	0.30	15.4
C.D. (P=0.05)	NS	6.37	0.47	NS	NS
Interactions (A×B)	NS	NS	S	NS	NS

S = Significant; NS = Non significant; *Values in parentheses are the % contribution of total tree biomass*

Table 2: Below ground biomass production (t ha^{-1}) of tree species and spacing under high density plantation.

Treatments	Below ground biomass production			Total below ground biomass
	Stump root	Lateral root	Fine root	
A. Tree species				
Eucalyptus	35.8 (10.82)	19.2 (5.80)	4.2 (1.27)	59.2
Melia	31.5 (15.76)	10.8 (5.40)	0.29 (0.14)	42.7
SEd±	2.24	0.46	0.31	2.40
C.D. (P = 0.05)	NS	1.93	1.31	10.04
B. Tree spacing				
3.0 m × 1.0 m	31.7 (11.08)	15.1 (5.27)	2.47 (0.86)	49.3
3.0 m × 1.5 m	30.6 (12.39)	15.6 (6.32)	2.15 (0.87)	48.3
3.0 m × 2.0 m	36.5 (13.59)	14.6 (5.43)	2.19 (0.81)	53.4
3.0 m × 2.5 m	35.9 (13.81)	14.8 (5.69)	2.17 (0.83)	52.8
SEd±	1.80	1.06	0.16	2.04
C.D. (P = 0.05)	3.92	NS	NS	NS
Interactions (A×B)	NS	NS	NS	NS

S = Significant; NS = Non significant; *Values in parentheses are the % contribution of total tree biomass*

Maximum biomass of different tree components was recorded in the narrow spacing of 3.0 m × 1.0 m which decreases with increase in spacing to 3.0 m × 1.5 m. Hegazy et al 2008 also found that the main stem biomass increased markedly with decreasing the distances between trees and recorded 40.77, 16.64 and 12.55 t ha^{-1} for distance 0.7, 1.4 and 2.1 m spacing, respectively. These results agree with the findings of Poggiani et al. 1983 who found that the stems of *Eucalyptus grandis* showed higher per hectare biomass yields at 3 yr of age when spacing was reduced from 3 m × 2 m to 2 m × 1 m. In present study the proportion of biomass stored in roots were found to be 18-21 % of the total tree biomass, which are comparable than the values 19-21 % reported for 5-8 yr old poplar plantations (Lodhiyal et al 1995, Bargali et al 1992).The variations in tree biomass as compared to other studies are attributed to a number of factors, such as growth conditions, site quality, age, density, structure, and management practices (Oelbermann et al 2004, Swamy and Puri 2005, Kanime et al 2012, Goswami et al 2013).

TREE CARBON STOCK

Tree Carbon Concentration (%)

Carbon concentration in different parts of various species depends upon the ash content which further varies in different components of the trees viz., stem, branch, leaf etc. (Fig. 1).

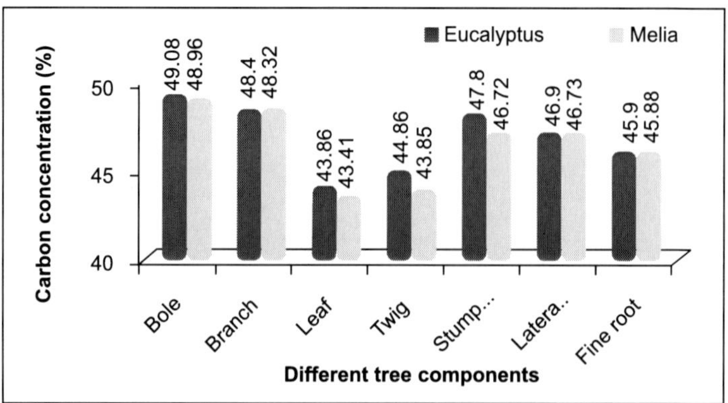

Figure 1: *Carbon concentration in different components of Eucalyptus and Melia tree*

Carbon concentration in different above ground components varied from 40-50 %. The maximum carbon concentration in above ground components was observed in bole (49.08 and 47.09 %) followed by branch (48.40 and 47.92%) > twig (44.06 and 43.85%) > leaves (40.40 and 39.69%) in *Eucalyptus* and *Melia*, respectively (Fig. 1 & 2). Similarly, in below ground components, the maximum carbon concentration was observed in stump root (47.80 and 46.72%) followed by lateral root (46.90 and 45.75%) > fine root (45.90 and 44.88%) in *Eucalyptus* and *Melia*, respectively. Carbon allocation studies in different components of some important tree species made by Negi et al 2003 reported that carbon content of *Dalbergia sissoo* in leaf was 38.93%, while in wood the value was 44.45%. In *Populus deltoides*, leaf carbon content was 32.26% and in the wood the value was 45.29%. Even studies suggested that stem wood biomass accounted for 49.7–51.7 % of the total biomass. The similar trend was followed by tropical forests of India by Swamy and Puri 2005 for *Gmelina arborea.* Carbon concentration in the biomass is vital indicator to determine the carbon sequestration in any plant. Mc Brayer and Cromack 1980 and also Gallardo and Merino 1993 predicted that generally carbon assumed to constitute about 50% of ash free dry mass. The literature also revealed that carbon content in different tree parts has been generally assumed to be 45 % –50% of the dry weight (Wang and Feng 1995).

Above and Belowground Carbon

Carbon stock is dependent on the tree density and carbon concentration in different components. Carbon storage in plant can be high in complex agroforestry system and productivity depend on several factors such as species, age, structure and way how the systems are managed (Swamy and Puri 2005, Oelbermann et al 2004). Components of above ground carbon were significantly affected by tree species. Contribution of above ground carbon was recorded in the order: bole (66.27 and 59.95%) > branch (13.77 and 16.82%) > leaf (1.31 and 1.02%) > twig (0.79 and 0.94%), whereas, below ground parts contributed as: stump root (10.66 and 15.65%) > lateral root (5.61 and 5.28%) > fine root (1.19 and 1.44%) in *Eucalyptus* and *Melia*, respectively (Table 3 & 4 and Fig. 3 & 4).

Table 3: *Above ground carbon storage (t ha^{-1}) of tree species under high density plantation*

Treatments	Above ground carbon storage				Total above ground carbon
	Bole	Branch	Leaf	Twig	
A. Tree species					
Eucalyptus	106.3(66.27)	22.1(13.77)	2.11(1.31)	1.28(0.79)	132.3
Melia	56.3(59.95)	15.8(16.82)	0.96(1.02)	0.89(0.94)	74.1
SEd±	3.38	1.00	0.05	0.07	2.34
C.D. (P = 0.05)	14.16	4.21	0.20	0.31	9.82
B. Tree spacing					
3.0 m × 1.0 m	85.4(65.09)	21.8(16.61)	1.20(0.91)	0.83 (0.63)	109.3
3.0 m × 1.5 m	73.1(61.84)	20.0(16.92)	1.60(1.35)	1.07 (0.90)	95.8
3.0 m × 2.0 m	84.2(64.22)	17.9(13.65)	2.14(1.63)	1.05 (0.80)	105.3
3.0 m × 2.5 m	82.4(64.27)	16.2(12.63)	2.21(1.72)	1.38 (1.07)	102.3
SEd±	7.48	1.38	0.07	0.14	7.80
C.D. (P = 0.05)	NS	3.02	0.17	0.31	NS
Interactions (A × B)	NS	NS	S	NS	S

S = Significant; NS = Non significant; *Values in parentheses are the % contribution of total tree biomass*

Table 4: *Below ground carbon storage (t ha^{-1}) of tree species and spacing under high density plantation.*

Treatments	Below ground carbon storage			Total below ground carbon
	Stump root	Lateral root	Fine root	
A. Tree species				
Eucalyptus	17.1(10.66)	9.01(5.61)	1.92(1.19)	28.0
Melia	14.7(15.65)	4.96(5.28)	1.36(1.44)	19.8
SEd±	1.06	0.21	0.15	0.11
C.D. (P = 0.05)	NS	0.88	NS	4.77
B. Tree spacing				
3.0 m × 1.0 m	13.3(10.13)	6.63(5.05)	1.07(0.81)	21.9
3.0 m × 1.5 m	13.7(11.59)	7.08(5.98)	0.96(0.81)	22.3
3.0 m × 2.0 m	16.8(12.81)	7.02(5.35)	1.04(0.79)	25.7
3.0 m × 2.5 m	17.2(13.41)	7.20(5.61)	1.03(0.80)	25.8
SEd±	0.78	0.51	0.07	0.95
C.D. (P = 0.05)	1.70	NS	NS	2.08
Interaction (A × B)	NS	NS	NS	NS

S = Significant; NS = Non significant; *Values in parentheses are the % contribution of total tree biomass*

Spacing also significantly affected the different above and belowground carbon except bole, lateral and fine roots. In the present study, the contribution of bole accounted for the largest amount of carbon (nearly 50%) from total tree biomass, which is comparable with other findings for different species (Redondo 2007, Fonseca et al 2012, Rizvi et al 2011). Above, below and total carbon stock under *Eucalyptus* and *Melia* at all the spacing ranges from 74.1-132.3 t ha^{-1}, 19.8-28.0 t ha^{-1} and 93.9-160.4 t ha^{-1}, respectively. The results are comparable with the findings of Albrecht and Kandji 2003 reported that agroforestry can store carbon in the range 12-228 Mg ha^{-1}. Albretch and Kandji 2004 also reported significant carbon stock under tropical agroforestry conditions. The results obtained are support the earlier findings of Dey 1996 and Fang et al 2007.

Total Carbon Stock

Carbon storage in different agroforestry systems occurs in the form of enhancement of soil carbon plus root biomass and above ground as carbon stored in standing biomass. Total dry matter accumulation and carbon storage were significantly affected by the tree species and was recorded maximum in *Eucalyptus* (330.7 and 160.4 t ha^{-1}, respectively) as compared to *Melia* which recorded 199.8 and 93.9 t ha^{-1}, respectively (Table 5). However, different tree spacing's were not found significant in obtaining dry matter accumulation and carbon storage. Soil carbon stock doesn't show any significant differences either among the tree species or tree spacings. However, interaction among tree species and spacing's was found significant.

Table 5: *Total carbon stock (t ha^{-1}) of tree species under high density plantation*

Treatments	Total biomass accumulation	Total carbon storage	Soil carbon stock	Total carbon stock (soil+vegetation)
A. Tree species				
Eucalyptus	330.7	160.4	61.14	221.5
Melia	199.8	93.9	61.54	155.5
SEd±	5.98	3.04	3.52	4.73
C.D. (P = 0.05)	25.02	12.73	NS	24.47
B. Tree spacing				
3.0 m × 1.0 m	286.0	131.2	72.08	203.3
3.0 m × 1.5 m	246.8	118.2	65.39	183.6
3.0 m × 2.0 m	268.5	131.1	57.12	188.2
3.0 m × 2.5 m	259.9	128.2	50.76	178.9
SEd±	14.21	7.21	1.33	7.66
C.D. (P = 0.05)	NS	NS	NS	16.69
Interactions (A × B)	NS	S	S	NS

S = Significant; NS = Non significant

Total carbon stock was significantly affected by both the tree species as well as different tree spacing's. Maximum carbon stock was recorded in *Eucalyptus* (221.5 t ha^{-1}) and was significantly superior to *Melia* which obtained 155.5 t ha^{-1} of carbon (Fig. 5). Among the different tree spacings, maximum total carbon stock (203.3 t ha^{-1}) was recorded with narrow spacing of 3.0 m × 1.0 m followed by 3.0 m × 2.0 m spacing, whereas, minimum total carbon stock was reported with wider spacing of 3.0 m × 2.5 m. However, interaction between tree species and different tree spacings was found to be non-significant.

Soil organic carbon stock decreased in surface to deeper soil depth in the order is 0-30 cm > 30-60 cm > 60-90 cm and the same pattern was followed in soil organic carbon. The soil organic carbon stock is dependent upon tree, soil type, site and history of the site. Moreover, the SOC values vary with tree species, soil type and density of the tree species.

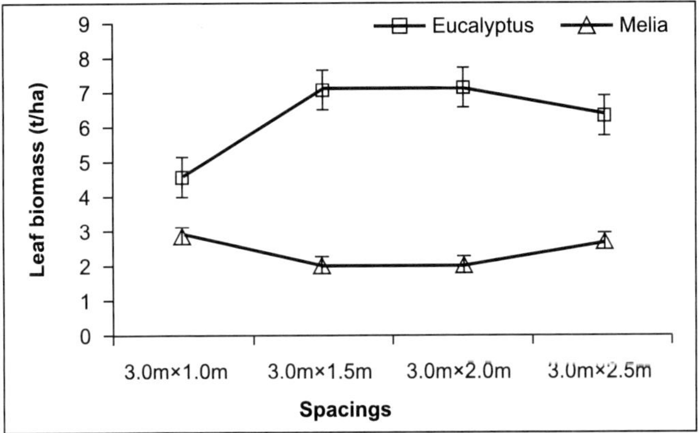

Figure 2: *Interaction between tree species and spacing in obtaining leaf biomass (t ha^{-1})*

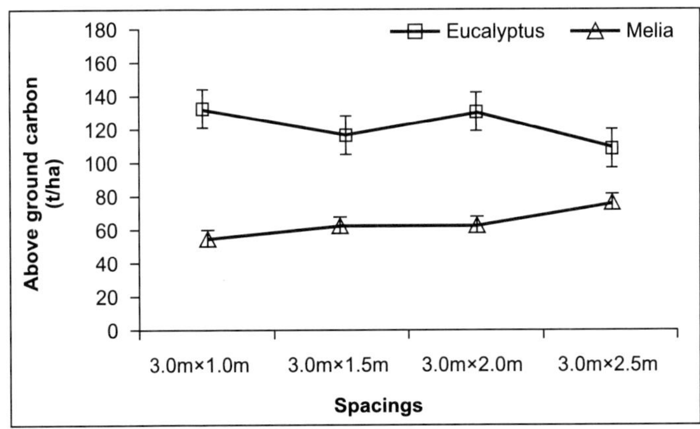

Figure 3: *Interaction between tree species and spacing in obtaining above ground carbon (t ha^{-1})*

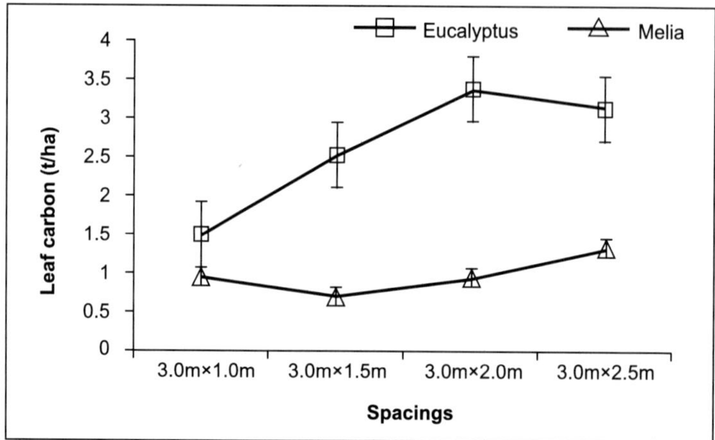

Figure 4: *Interaction between tree species and spacing in obtaining leaf carbon (t ha^{-1})*

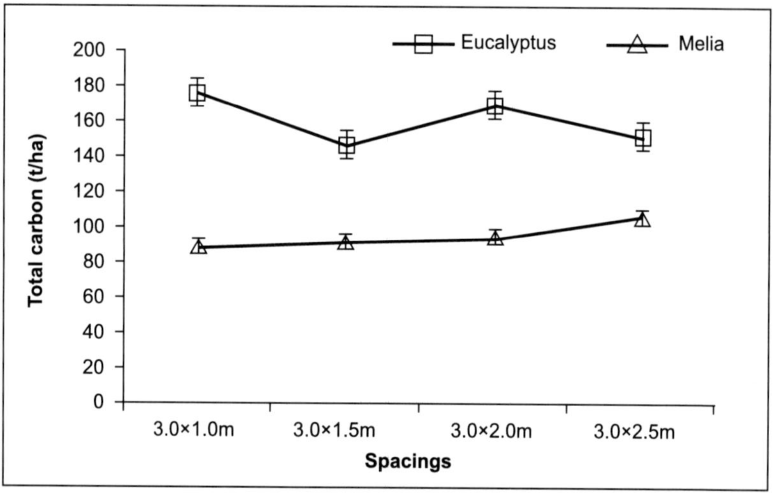

Figure 5: *Interaction between tree species and spacing in obtaining total carbon stock (t ha^{-1})*

CONCLUSION

The finding suggests that *Eucalyptus* has higher potential for biomass carbon sequestration than *Melia* under high density short rotation plantation. However, soil health and soil carbon stocks were more under the *Melia* tree species. So, planting of fast growing tree species under high density can serves dual function, i.e. promotion of biodiversity as well as sequester carbon.

REFERENCES

Albercht A, Kanji ST 2003 Carbon sequestration in tropical agroforestry Systems *Agriculture, Ecosystems and Environment* **99** 15-27

Albrecht A and Kandji SD 2004 Carbon sequestration in tropical agroforestry system, *Agriculture Ecosystem and Environment* **99** 1-3 15-27

Arora G, Chaturvedi S, Kaushal R, Tewari SK and Chaturvedi OP 2013 Growth, Biomass, Carbon Stocks and Sequestration in Age Series *Poplus deltoides* Plantations, *Turkish Journal of Agriculture and Forestry* 1307-94

Bargali SS, Singh SP and Singh RP 1992 Structure and function of an age series of *Eucalyptus* plantations in central Himalaya. I. Dry matter dynamics, *Annals of Botany* **69** 405-411

Blake GR and Hartge KH 1986 Methods of soil analysis, American society of agronomy, Madison, *Wiscosin, part- I, Agronomy* **9** 363-375

Canadell JG and Raupach MR 2008 Managing forests for climate change mitigation, *Science* 320 1456-1457

Day AN 1996 Biomass studies on *Populus deltoides* Marsh, under different land use systems. M.Sc. Thesis, UHF, Nauni, Solan (H.P.) 144pp

Fang S, Xue J and Tang L 2007 Biomass production and carbon sequestration potential in poplar plantations with different management patterns, *Journal of Environmental Management* **85** 672–679

Fonseca F, Figueiredo T and Martins A2014 Carbon storage as affected by different site preparation techniques two years after mixed forest stand installation, *Forest Systems* **23**184–92

Fonseca W, Alice FE and Rey-Benayas JM 2012 Carbon accumulation in aboveground and belowground biomass and soil of different age native forest plantations in the humid tropical lowlands of Costa Rica, *New Forests* **43** 197–211

Fonseca W, Benayas JMR and Alice FE 2011 Carbon accumulation in the biomass and soil of different aged secondary forests in the humid tropics of Costa Rica, *Forest Ecology and Management* **262** 1400-1408

Gallardo A and Merino J 1993 Leaf decomposition in two Mediterranean ecosystem of Southwest Spain: influence of substrate quality, *Ecology* **74** 152-161

Gonzalez M, Augusto L, Galtter-Budynek A, Xue J, Yauschew-Raguenes N, Guyon D, Trichet P, Delerue F, Niollet S, Amnderasson F 2013 Contribution of understorey species to total ecosystem aboveground and belowground biomass in temperate *Pinus pinaster* Ait. Forests, *Forest Ecology and Management* **289** 38-47

Goodale CL, Apps MJ, Birdsey RA, Richard A, Field CB, Heath LS, Houghton RA, Jenkins JC, Kohlmaier GH, Kurz W2002 Forest carbon sinks in the Northern Hemisphere, *Ecological Applications* **12** 891–899

Goswami S, Verma KS, Kaushal R 2013 Biomass and carbon sequestration in different agroforestry systems of a Western Himalayan watershed, *Biological Agriculture and Horticulture* DOI: 10.1080/01448765.2013.855990

Hegazy SS, Ibrahim MA, Hamad Al-Mefarrej and Lotfy El-Juhany I 2008 Effect of Spacing on the Biomass Production and Allocation in *Conocarpus erectus* L, Saudi *Journal of Biological Sciences* **15** 2 315-322

Husch B, Muller CI and Beers TW 1972 Forest mensuration, Ronald press co New York 41-54

Huang XM, Liu SR, Wang H, Hu ZD, Li ZG and You YM 2014 Changes of soil microbial biomass carbon and community composition through mixing nitrogen-fixing species with *Eucalyptus urophylla* in sub tropical China, *Soil Biology and Biochemistry* **73** 42–48

Joa Carlos de MS, Carlos CC, Warren AD, Rattan L, Solismar PVenske F, Marisa CP and Brigitte EF2001 Organic matter dynamics and carbon sequestration rates for a little chronosequence in Brasilian Oxisol, *Soil Science Society of America Journal* **65** 5 1486-1499

Kanime N, Kausal Rajesh, Tewari Salil, Raverkar Chaturvedi S and Chaturvedi OP 2013 Biomass Production and Carbon sequestration in Different Tree Based Systems of Central Himalayan Tarai Region, *Forests Trees and Livelihoods* **22**(1) 38-50

Ketterings QM, Coe R, Van Noordwijk M, Ambagau Y and Palm CA 2001 Reducing uncertainty in the use of allometric biomass equations for predicting above ground tree biomass in mixed secondary forests, *Forest Ecology and Management* **146** 199-209

Lodhiyal LS, Singh RP and Singh SP 1995a Structure and function of an age series of poplar plantations in Central Himalaya: I. Dry matter dynamics, *Annals of Botany* **76** 191-199

Mc. Brayer JE and Cromack K 1980 Effect of snow peak on oak litter breakdown and nutrient release in Mannesota forest, *Pedobilogia* **20** 47-54

Negi JDS, Manhas RK and Chauham PS 2003 Carbon allocation in different components of some tree species of India: a new approach for carbon estimation, *Current Science* **85** 11 1528-1531

Oelbermann M, Voroney RP, Gordon AM2004 Carbon sequestration in tropical and temperate agroforestry systems: a review with examples from Costa Rica and Southern Canada. *Agriculture Ecosystems and Environment* **104** 359-377

Pan Y, Birdsey RA, Fang J, Houghton R, Kauppi PE et al 2011 A large and persistent carbon sink in the world's forests, *Science* **333** 988-993

Panse VG and Sukhatme PV 1967 Statistical Methods for Agriculture Workers.4thed., New Delhi, ICAR, pp359

Puri S and Sharma A 2002 Productivity of wheat under *Populus deltoids* agroforestry system, *Pakistan Journal of Agricultural Sciences* **40** 1-2:77-81

Rasse DP, Rumpel C, Dignac MF 2005 Is soil carbon mostly root carbon? Mechanisms for a specific stabilisation, *Plant Soil* **269** 1–2 341–356

Ravindranath NH, Somasekhar BS, Gadgil M 2002 Forestry for sustainable Biomass Production and Carbon Sequestration in India, *Mitigation Adaptation Strategy of Global Change* **6** 233-256

Redondo A 2007 Growth, carbon sequestration, and management of native tree plantations in humid regions of Costa Rica, *New Forests* **34** 253–268

Swamy SL and Puri S 2005 Biomass production and carbon sequestration of *Gmelina arborea* in plantation and agroforestry system in India, *Agroforestry System* **64** 181-195

Swamy SL, Puri S, Singh AK 2003 Growth, biomass, carbon storage and nutrient distribution in Gmelina arborea Roxb. Stands on red lateritic soils in Central India, *Bioresource Technology* **90** 109-126

Swank WT and Schreuder HT 1974 Camparisons of three methods of estimating surface area and biomass for a forest of young eastern white pine forest, *Science* **20** 91-100

Walkley AJ and Black IA 1934 Estimation of soil organic carbon by chronic acid titration method, *Soil Science* **37** 29-38

Wang QK, Wang SI, Zhong MC 2013 Ecosystem carbon storage and soil organic carbon stability in pure and mixed stands of *Cunning hamialanceolata* and *Michelia macclurei*, *Plant and Soil* **370** 295-304

Watson RT, Noble IR, Bolin B, Ravindranath NH, Verardo DJ and Dokken DJ 2000 IPCC special report on land use, land-use change and forestry, Cambridge Univ. press, New York.

CHAPTER 17

Climate Change and Adaptation in Agriculture: A Study of Himachal Pradesh

MS Pathania* JS Guleria, Babli Joshi[1], and Inder Dev[2]

1. Deptt. of Agricultural Economics, CSK HPKV, Palampur (H.P), India

**Corresponding author's Email: msp1960@rediffmail.com*

[1]Deptt. of Sociology, SCVB Govt. College, Palampur (H.P) India

[2]ICAR-Central Agroforestry Research Institute, Jhansi (U.P) India.

ABSTRACT

The impact of climate changes is most seriously felt in mountain regions because of the most vulnerable ecosystems. In Himachal Pradesh, evidences of climate change could be clearly observed by changes in receding snowfall in the Himalayas, disrupted rainfall pattern and increasing temperature. Agricultural production system will be affected by the short and long term climate changes. The study observed that different factors responsible for climate change in the state indicated decreasing trend in the amount of precipitation, decrease of snowfall and decreasing trend of flow of water in different rivers. The contribution of secondary sector to SGDP has increased. The productivity of different crops is less than their production potential. Farmers' perceptions regarding climate change were increase in temperature during summers, unpredictable rain fall, uneven rainfall distribution, prolonged summers, delayed and less snowfall, delayed onset of winter, delayed onset of SW monsoon, short duration winters, etc. The age, farming experience, education level, innovativeness, consciousness to environment and exposures to mass media had a positive and significant relationship with farmers' perceptions to climate change. Most worrying scenario is that amount of total precipitation has decreased over the period and has a direct bearing on production of crops. Therefore, there is a need of developing suitable varieties for rainfed cultivation that will resistant to climate change and adjust the cropping pattern with changing trend of climatic factors. Both short term and long term strategies need to be formulated to cope up with the climate change.

Keywords: Climate changes, Cropping pattern, Crop productivity, Farmers' perceptions

INTRODUCTION

Climate change is one of the most important environmental challenges facing humanity with its implications for agricultural productions, natural resources, water supply and health etc. Climate change is due to increasing emission and accumulation of GHGs. Climate change is likely to have the greatest negative impact on the poorest sections of population Adger et al (1999). The poorest people will be affected first and most by climate change (IPCC, 2007) due to their livelihood insecurity. Climate change is a threat to alter the spatial distribution of infectious diseases, crop productivity and increase the frequency and severity of natural hazards FAO (2007), Cline, (2007), and Dasgupta et al (2007). Agriculture is most sensitive to climate change and has both negative and positive impact on its production. Agriculture is the backbone of Indian economy as it provides employment to 59% people, contributed 18% in GDP and 12% to exports of agriculture sector. Agricultural productivity in general could decline between 10-25% by 2080 and the decline in yield in rainfed agriculture could be as much as 50% due to change in climatic factors (Cline 2007). Marginal and small farmers with rainfed farming are most vulnerable to the impacts of climate change on crop productivity through failure of crops, increase in diseases and pest attack. This increased livelihood insecurity. This traps the farmers in indebtedness, migration and dependency on food which will further aggravate the vulnerability associated with subsistence production. This makes the farmers more sufferers and creates a vicious cycle of poverty Ibarraran et al (2009).

The impact of climate changes is most seriously felt in mountain regions because of the most vulnerable ecosystems. The Himachal Pradesh has low to high mountains and four different agro climatic zones with an altitudinal range 350 to 7000 m above mean sea level. The state is located between 30° 22′ and 33° 12′ N latitude and 75° 47′ and 79° 4′ East longitude. The livelihood of people in the region depends on the availability of natural resource as most of the population is dependent on agriculture, horticulture and animal husbandry. Any change in the natural resources of the region due to climate change will have far reaching repercussions. In Himachal Pradesh, evidences of climate change could be clearly observed by changes in receding snowfall in the Himalayas, disrupted rainfall pattern and increasing temperature. This poses shortage of water which have serious threat to productivity and sustainability of crop cultivation in the state. The issues of environment and natural resources have become the focused one in the state. These changes in climatic parameters have both negative and positive impact on agriculture. Agricultural production system will be affected by the short and long term climate changes and will have implications on rural livelihoods. Keeping in view the implication of climate changes on agriculture the present study was undertaken to analyse changes in different parameters affecting climate and determine the ability of farmers to detect climate change

METHODOLOGY

The study was conducted in Himachal Pradesh. The study is based on both secondary and primary data. The secondary data collected from the state Government publications *Economic Review & Statistical Outlines,* Himachal Pradesh Department of Economics and Statistics, Shimla-H P. and other different published sources. Information was compiled for estimating the changes in rainfall, temperature, population, area, production, productivity of crops etc. The sample study has been conducted in Nagrota Bagwan and Baijnath block of district Kangra in Himachal Pradesh. Multistage stratified random sampling technique was adopted for this study. Three villages were selected randomly from each block. In all 80 farmers from these villages were selected. The selected respondents were personally interviewed with the help of a well structured schedule. Data were collected about different perception of climate change. Data collected from both secondary and primary sources were tabulated and analysed through simple mathematical and statistics techniques. Trend analysis was done to analyse the variability and adaptation of the factors. Correlation analysis was also done for parameters.

RESULTS AND DISCUSSION

Climate Change Scenarios in Himachal Pradesh

Average Rainfall

The average rainfall the over period in the state shows a declining trend. The decrease of rainfall in some districts and uneven distribution of rainfall in other may affect the production and productivity of crops. Table 1 shows the mean distribution of rainfall across different districts over a period from 1995-96 to 2011-12. There was high variability in the rainfall over the period in all the districts. The variability around mean was as high as 67% in Bilaspur and lowest 32% in Shimla. The trends in rainfall showed that at the state level, there is a decrease of about 1.46 mm per annum in rainfall over the period. The maximum decrease was visualized in Kangra (13 mm per annum) and Mandi (7.50 mm per year) and Shimla (5.44 mm per year). The decrease in rainfall has a direct bearing on production of crops Mean rainfall of the state was 1122 mm. Western part of state situated in foothills of Dhauladhar ranges of the Himalayas has the highest annual precipitation ranging 1844 mm to 2601mm. In India, Dharamshala and Palampur of district Kangra receive highest precipitation next to Chirapunji. The areas like Lahaul and Spiti, Kinnaur receive normal precipitation of 300 mm, during winter season. Northern part of the state is having by and large temperate climate receiving precipitation between 332 mm to 1088 mm annually.

Table 1: *Distribution of rainfall in Himachal Pradesh, 1995-96 to 2011-12*

Districts	Mean (mm)	CV (%)	Trend in rainfall (mm/year)
Bilaspur	284.61	66.55	–3.44
Chamba	765.12	40.10	25.90
Hamirpur	268.32	39.22	2.86
Kangra	347. 30	44.75	–12.89
Kinnaur	457. 85	56.76	–4.28
Kullu	365.40	37.53	–1.19
Mandi	302.05	48.14	–7.50
Shimla	380.14	32.16	–5.48
Sirmaur	212.82	37.80	–2.96
Solan	276.84	38.85	–3.14
Una	213.25	46.10	–1.45
H.P.	349.70	33.88	–1.38

Effect of Rainfall on Crop Performance

Table 2 showed the relationship between rainfall and cereal production and productivity in different districts. Very high and significant positive correlation was observed between rainfall and production as well as rainfall and productivity in Bilaspur, Shimla, Kangra and the state as a whole. In all other districts too, there was positive, but non-significant correlation of rainfall with production as well as yield of cereals. The mean temperature of the state remains between 10.4 °C to 22.7 °C. The north east part which constituted Lahaul & Spiti, parts of Chamba, parts of Kullu, Shimla and Kinnaur district experience annual mean temperature below 14 °C, while in south-western part of state winter starts from November and temperature starts shooting up during March. Maximum and minimum temperature in this part varies between 19.5 °C to 31.9 °C and 0.4 °C to 17.0 °C, respectively with mean annual temperature up to 24 °C. Table 2 showed distribution of temperature in the state. It was observed from the table that highest temperature was noticed in Una centre followed by Sirmaur (Nahan) and Mandi. The variability in temperature was noticed between 42 (Kullu) to 32 % (Shimla).

Table 2: *Correlation between rainfall & cereals production & productivity and temperature distribution*

Districts	Production and rainfall	Productivity & rainfall	Temperature distribution	
			Mean (°C)	CV (%)
Bilaspur	0.68*	0.67*	30.3	41.84
Chamba	0.42	0.23	23.5	36.14
Hamirpur	0.42	0.46	30.3	41.84
Kangra	0.37*	0.45*	16.7	32.16
Kinnaur	0.46	-0.27	10.6	40 .27
Kullu	0.24	0.49*	24.4.	41.62
Mandi	0.62*	0.66*	24.5	37.23
Shimla	0.56	0.39	21.6	32.44
Sirmaur	0.42	0.51	26.8.	42. 10
Solan	0.33	0.30	17.2	35.56
Una	0.55*	0.35*	30.3	41.84
H.P.	0.67*	0.59*	16.8	30.06

* *Significant at 5% level*

Changes in Snowfall Trends

This study was conducted by RS Rana, Scientist, and his team, CSKHPKV Palampur and published by CRIDA* to analyse the trend of snowfall and discharge flow of major river basins of Himachal Pradesh. The study of trend analysis for snow fall data recorded for 21 recording sites over 22 years from 1984 to 2005. The study indicates a decrease of annual snowfall amount at the rate of 36.8 mm averaged over all sites. The average annual snowfall trends for all observation sites located at 2030m above sea level to 4000m above sea level indicated an average decrease of 8.71mm in last 22 years. The trends showed more rate of decrease of snowfall from 2030 to 4079m above sea level during recent decade (1995-2004) as compared to 1984-94 reflecting decreasing rate of snowfall in all the elevations. Monthly analysis of snow fall amounts indicates a sharp decrease in monthly average snowfall over past 22 years from 21 observation sites during September to December except slight increase during October. The snowfall amount shows increasing trends during January and February reflecting delay of snowfall in winter season. There was trends indicated significant decrease of snowfall amount during March and May. Such trends in snowfall occurrence increase the opportunity of growing more crops during March to October in higher temperate wet and dry agro climate region of Himachal Pradesh.

River Discharge Flow Trends in Himachal Pradesh

Changes in water in different region noted that annual water surplus showed decreasing trends over decades. The water surplus decrease was observed more in *kharif* season compared to *rabi* season. A sharp reduction was observed during March to May and November to December. The reduction of rainfall and snow in high reaches caused water surplus in kharif to water deficit by more than 45mm in *rabi.*

Major River Basins of Himachal Pradesh

Water is unique gift of nature and an important natural resource supporting life on Earth. The water resources of Himachal Pradesh feed not only state but lower plain region like Punjab, Haryana and Rajasthan etc. There are four main rivers basins namely Satluj, Beas, Chenab and Ravi. Satluj river basin covers more than 48% of total geographical area of the state. All the rivers are snow fed. Himachal Pradesh has vast glacial reserves some of which contain huge amount of perpetual snow and ice. The total number of glaciers in Himachal Pradesh is estimated to be 2554 with reserve of 387.35 km^3 covering an area of 4160.58 km^2. These rivers are of high economic importance to India as these provide water for drinking, agricultural as well as for the generation of hydro-power. These river basins are home to a variety of natural resources including vast flora and fauna. The inflow of these rivers over the period (1984 to 2003-04) revealed a decreasing trend in annual inflow indicating reduction in hydro- electric power generation and water for drinking and irrigation.

OTHER FACTORS AFFECTING THE CLIMATE

Trends of Population and Land Resource Use Pattern

The population in Himachal Pradesh was 34.60 lakh in 1971 census which increased to 68.65 lakh during 2011 showing thereby an increase of 98 %. The increasing population over time needs more food and water. But, change in climate may affect the requirement of food and water in the long run. The land put to different productive purpose has decreased due to burgeoning needs of infrastructure and over-head facilities for economic development in the state. The decline in forest and pasture area over period indicate negative signal to environment in the future.

Cropping Pattern and Productivity

The main foodgrain crops grown in the state are: rice, wheat, maize and pulses. Wheat is the most important *rabi* crop and ranks first in the state which alone accounts for 36-40 % of the total cropping area. Table 3 indicates that the % share of area to the total food grains under this crop increased from 36 in 1985-86 to 40% in 2013-14. Maize is grown in *kharif* season in the state and the area under this crop decreased from 315 in1985-86 to 292 thousand hectare in 2013-14 to area under total food grains. This crop is mostly grown under rain fed conditions in the state. Paddy is another important crop of *kharif* season and is generally grown on irrigated lands. The area under this crop decreased over the years. This crop is water loving and shows a declining trend which could be attributed to the scarcity of water and undistributed rainfall during critical periods. Area under pulses and oilseeds showed a declining trend. There was a slight decrease in area under food grains over the years. This could be due to shift in area from food grains to the cultivation of cash crops. The irrigated cropped area decreased over time. This indicate decline in water availability in different sources of irrigation over the years. This can also be due to affect of climatic factor like rainfall and temperature.

The productivity of cereal crops has increased over time and has been depicted in Table 3. The average productivity of wheat increased significantly, while, the productivity of maize showed an increasing trend over the years. This could be attributed to the adoption of improved varieties, increased use of inputs, mechanization etc by the farmers. The productivity of paddy is not showing any trend. As against this, the pulses had a negative growth in productivity because of their cultivation on unirrigated, marginal lands and low use of inputs compared to wheat, maize and paddy. The productivity of these crops is less than their production potential. Productivity of cereals and pulse crops can be increased by adopting the recommended technology. To meet the increasing demand of food the productivity of food crops needs to be increased by the introduction of HYVs, intensive use of land and other critical inputs. The intensive use of land and water affect their quality which in long run may affect the environment by emission and accumulation of GHGs. The decline in the productivity in some of the crop may have affect of climatic factors and needs modification of cropping system.

Table 3: *Cropping pattern & productivity of crops*

Crop	1985-86		1990-91		2000-01		2007-08		2013-14	
I. Cereals	**A**	**P**	**A**	**P**	**A**	**P**	**A**	**P**	**A**	**P**
Wheat	373.2	1.30	377.3	1.38	377.3	0.69	366.59	1.53	371.06	1.45
Maize	315.1	1.52	311.9	1.53	311.9	2.29	300.15	2.27	292.14	2.32
Paddy	94.8	1.38	86.2	0.95	86.2	1.53	78.5	1.56	78.34	1.74
Barley	31.5	1.09	27.7	1.21	27.7	0.83	23.57	1.30	18.62	1.25
Other cereals	29.1	0.39	16.5	0.22	16.5	.0.87	9.29	0.73	8.00	0.83
II. Pulses										
Gram	5.8	0.61	2.4	0.50	2.4	1.07	1.45	0.94	0.42	0.93
Other	38.0	0.23	33.5	0.14	33.5	0.61	32.42	0.37	33.,12	1.17
Total food grain	881.8	1.38	855.5	1.29	855.5	1.36	811.98	1.77	799.68	1.77
III. Oilseeds	19.3	0.28	16.1	0.40	16.1	0.39	13.92	0.35	15.54	0.41

Note: A = Area (000 ha.) and P = productivity (Tonnes/ha)

FARMERS' PERCEPTIONS TO CLIMATE CHANGE

The impacts of climate change are most seriously felt in Himalayan regions of Himachal Pradesh. The farmers of the state perceive the different quantum of climate change on hill agriculture. Farmers' perceptions regarding change in climate has been presented in Table 4. The climate change described by farmers as the increase in temperature during summers, unpredictable rain fall, uneven distribution rainfall, prolonged summers, delayed and less snowfall, delayed onset of winter, decrease in snowfall, delayed onset of SW monsoon short duration winters, etc. were the main opinions of the farmers regarding climate change in the state. Farmers' perceptions clearly portray picture of a shift of high water requiring crop like basmati to low water requiring crops like maize and local paddy. The farmers are replacing fruit crops with off-season vegetables. The results indicated that most of the farmers perceived that the rainfall and temperature distribution has undergone a significant change. The majority of farmers believed that the rainfall levels had decreased. Similarly, the overall perception on changes in precipitation and that there are pronounced changes in the timing of rains. A sizeable respondent witnessed uneven distribution and unpredictable behavior of the rains. Therefore, there is a need to adjust the cropping pattern with changing trend of climatic factors and development of crop varieties that will resistant to climate change. Similar results were reported by Bhagat et al (2007).

Table 4: *Farmers' perceptions (%) of climate change*

S.No	Particular	No change	Decreased	Increased
1	Temperature	28	—	72
2	Temperature during summers	22	—	78
3	Temp. above normal in winters	30	10	60
4	Precipitation	12	76	12
5	Unpredictable rain fall	13	5	82
6	Uneven distribution rainfall	11	5	84
7	Delayed onset of SW monsoon	18	18	64
8	Early monsoon withdrawal	18	24	58
9	Degradation of water quality	25	8	67
10	Degradation of land quality	22	8	70
11	Long dry spell	16	18	66
12	Occurrence of drought	40	10	50
13	Prolonged summers	22	-	78
14	Delayed onset of winter	20	12	68
15	Short winter periods	22	10	68
16	Snowfall during winters	12	82	6
17	Delayed and less snowfall	12	8	80
18	Temporal weather displacement cycle	20	24	56

FACTORS AFFECTING FARMERS' PERCEPTIONS TO CLIMATE CHANGE

The analyses examined the factors influencing the farmers' perceptions to climate change and are depicted in Table 5. The results revealed that the age, farming experience, education level, innovativeness, consciousness to environment and exposures to mass media had a positive and significant relationship with farmers' perceptions to climate change. The other factors given in the table has also positive relationship, but non-significant relationship with farmers' perceptions to climate change.

ADAPTATION TO CLIMATE CHANGE

The various adaptation strategies being used by farmers in response to changing climatic are presented in Table 6. Adaptations made by respondents revealed that use of water conservation techniques was considered to be one of the most important adaptations in response to climatic change. Change in time of farm operation, adjusting the cropping sequence, soil conservation measures, pre-monsoon dry seeding and use of short duration crop varieties were another options. To cope with climate variability, farmers have developed a wide range of management practices. The abrupt climate fluctuations such as less and untimely rainfall and drought had made the farmers to harvest rainwater for irrigation during water scarcity. Some of the sample farmers adopted zero tillage technology of showing in paddy-wheat cropping system to make best use of residual moisture of paddy field. Mortimore and Adams (2001) and Smit and Skinner (2002) also reported similar results.

Table 5: *Relationship of different variables & farmers' perceptions to climate change*

S.No	Variables	Correlation 'r'
1	Age of farmers	0.45*
2	Education level	0.54*
3	Land holding size	0.16
4	Farming experience	0.68*
5	Innovativeness	0.71*
6	Consciousness to environment	0.56*
7	Awareness of natural resources conservation	0.34
8	Exposure to mass media	0.49*
9	Contact with extension personnel	0.12

* Significant at 5 % probability level

Table 6: *Adaptations to climate change by farmers*

S.No	Adaptation	%
1	Use of short duration crop varieties	41.4
2	Use of water conservation techniques	74.3
3	Soil conservation measures	55.0

4	Pre-monsoon dry seeding	45.2
4	Change in time of farm operation	71.2
5	Intercropping	64.5
6	Rainwater harvesting	24.0
7	Zero tillage to conserve soil moisture	16.4
8	Agro forestry.	37.5

CONCLUSION

The high variability was noted in rainfall over the period in all the districts. Very high and significant positive correlation was observed between rainfall and production as well as rainfall and productivity of cereal crops in Bilaspur, Shimla, Kangra and the state as a whole. In all other districts too, there was positive but non-significant correlation of rainfall with production as well as yield of cereals. The study indicated decrease of snowfall. The flow of water in different rivers over time showed a decreasing trend. There was decrease in water availability for hydro- electric power generation and irrigation was also observed. The land put under different agricultural purpose declined over period indicate negative signal to environment in the future years. Area under pulses and oilseeds showed a declining trend. There was a slight decrease in area under food grains over the years. The productivity of wheat increased significantly, while, the productivity of maize showed an increasing trend over the years. The productivity of paddy is not showing any trend. The pulses had a negative growth in productivity. The productivity of these crops is less than their production potential. Farmers' perceptions regarding climate change were : increase in temperature during summers, unpredictable rain fall, uneven distribution rainfall, prolonged summers, delayed and less snowfall, delayed onset of winter, delayed onset of SW monsoon, short duration winters, etc. The age, farming experience, education level, innovativeness, consciousness to environment and exposures to mass media had a positive and significant relationship with farmer's perceptions to climate change.

WAY FORWARD

The major implications of climate change are largely for production and productivity of crops. These changes have implications for food production and food security. Cereals command maximum cropped area in the state. These crops may the most affected by the climate change as more than 80% of the cropped area is rainfed. Most worrying scenario is that amount of total precipitation has decreased over the period in all the districts. The decrease in the rainfall has a direct bearing on production of crops. Therefore, there is a need of developing suitable varieties for rainfed cultivation under late sown condition. There are signals of climate change in Himachal Pradesh which have clearly seen in receding precipitations and rise of temperature in different regions of the state. So, there is a need to minimize negative impacts of climatic factors through, adoption of technology and land and water management for crop production. The study revealed that significant numbers of farmers believe that temperatures have increased and the precipitation has declined along with late onset and early withdrawal of monsoon with long dry spells. Therefore, there is a need to adjust the cropping pattern with changing trend of climatic factors and need of development of crop varieties that will resistant to climate change. There is a dire need to improve farmers' ability to use water more efficiently and better management of their fragile soil to accommodate the shocks of climate change. Both short term and long term strategies need to be formulated to cope up with the rainfed cultivation and increase their productivity in the state.

REFERENCES

Adger N and Kelly M 1999 Social Vulnerability to Climate Change and the Architecture of Entitlements Mitigation and Adaptation *Strategies for Global Change* **4** 253-266

Anonymous 2010 Impact Adoption and Vulnerability of Indian Agriculture to Climate Change, ICAR Net work project, edited annual report by CRIDA (ICAR) Hyderabad 258-269

Bhagat RM, Singh RS, Prasad R, Harbans L, Kalia V and Sood C 2007 Project Report: Impact *Vulnerability and Adaptation to Climate Change,* submitted to network project on Climate Change, ICAR, New Delhi

Cline William R 2007 Global Warming and Agriculture: New Country Estimates Show Developing Countries Face Declines in Agricultural Productivity Center for Global Development

Dasgupta S, Laplante B, Meisner C, Wheeler D and Yan DJ 2007 *The Impact of Sea Level Rise on Developing Countries: A* Comparative Analysis, Policy Research Working Paper No. 4136. Washington, D C World Bank

Department of Economics and Statistics, *Economic Review*, Himachal Pradesh Department of Economics and Statistics, Shimla-HP (Various issues)

Department of Economics and Statistics. *Statistical Outlines,* Himachal Pradesh Department of Economics and Statistics, Shimla-H P (Various issues)

IPCC 2007 Climate Change 2007 Impacts, Adaptation and Vulnerability, in Parry M Canziani O, Palutikof J, Linden P, Hanson C (eds), *Contribution of Working Group II to the Fourth Assessment Report on Climate Change*, Cambridge University Press, Cambridge

FAO 2007 Adaptation to Climate Change in Agriculture, Forestry and Fisheries: Perspectives, framework & priorities, Rome *Interdepartmental Working Group on Climate Change*

Government of India 2007 Water Resources *India 2007* Publication Division. Ministry of Information and Broadcasting, 851

Ibarraran MI, Ruth M, Ahmad S and London M 2009 Climate Change and Natural Disasters: Macroeconomic Performance and Distributional Impacts, *Environment, Development and Sustainability*. **11** 549-569

Mendelsohn R and Dinar A 1999 Climate Change, Agriculture, and Developing Countries Does Adaptation Matter?, *World Bank Research Observe*, **14**(2) 277-93

Mortimore MJ and Adams WM 2001 Farmer Adaptation, Change and 'Crisis' in the Sahel. *Global Environmental Change* **11**(1) 49-57

Partap U and Partap T 2002 Warning Signals from the Apple Valleys of the Hindu Kush-Himalayas- Productivity Concerns and Pollination Problems (Shreshtha, A B M ed.) *Centre for Integrated Mountain Development (ICIMOD),* Kathmandu, Nepal

Smit B and Skinner M 2002 Adaptation Options in Agriculture to Climate Change: A Typology. *Mitigation & Adaptation Practices & Policies for Global Change* **7**(1) 85-114 .

CHAPTER 18

Agroforestry Systems for Climate Change Adaptation and Mitigation: A Review

Sourav Gupta*, Jyotimala Sahu, RN Sharma, US Thakur, Sushil Yadav and BS Kasana

College of Agriculture, Rajmata Vijayaraje Scindia Krishi Vishwa Vidyalaya, Gwalior – 474002 M.P., India

**Corresponding author's Email: souravgupta654@gmail.com*

ABSTRACT

Agriculture is the human enterprise that is most vulnerable to climate change. Tropical agriculture, particularly subsistence agriculture is particularly vulnerable, as smallholder farmers do not have adequate resources to adapt to climate change. While agroforestry may play a significant role in mitigating the atmospheric accumulation of greenhouse gases (GHG), it also has a role to play in helping smallholder farmers adapt to climate change. Agriculture faces the unprecedented task of feeding a world population of 9 billion people by 2050 while simultaneously avoiding harmful environmental and social effects. Agroforestry interventions, because of their ability to provide economic and environmental benefits, are considered to be the best "no regrets" measures in making communities adapt and become resilient to the impacts of climate change. The important elements of agroforestry systems that can play a significant role in the adaptation to climate change include changes in the microclimate, protection through provision of permanent cover, opportunities for diversification of the agricultural systems, improving efficiency to use of soil, water and climatic resources, contribution to soil fertility improvement, reducing carbon emissions and increasing sequestration, and promoting gender equity.

Keywords: Agroforestry, Climate change, Climate smart agriculture, Green house gases, Mitigation

INTRODUCTION

Agriculture shapes our planet in profound ways. Roughly 38 per cent of the land surface of the earth is used to grow food, making agriculture the largest anthropogenic land use (Foley 2011). Expansion in agricultural land is the leading cause of deforestation and native habitat loss (Norris 2008, Jacobson and Kar 2013), a situation that has led to declines in wildlife, including birds (Johnson et al 2011), insects (Thomas et al 2004), and mammals (Daleszczyk et al 2016), some of which are now considered endangered species (Norris 2008). Nutrient leaching from fertilizer results in the eutrophication of waterways, leading to oxygen deficient "dead zones" in water bodies around the world (Boesch and Brinsfield 2000, McIsaac et al 2001).

Developing countries are going to bear the brunt of climate change and suffer most from its negative impacts. Global conventions are not sufficiently effective to halt the increase of atmospheric greenhouse gases (GHG) concentrations, and we now accept that the primary drivers of climate change are not going to stop. Mitigation efforts will therefore only provide a partial softening of the effects of climate change. Local climates and terrestrial ecosystems will change, threatening biota and human livelihoods. Yet, even as climate changes, food and fiber production, environmental services and rural livelihoods must improve, and not just be maintained. Agriculture is the largest human-caused contributor to the greenhouse gas emissions implicated in climate change (Foley 2011, Robertson et al 2000). In addition to environmental and human impacts, there are disconcerting implications for the resilience of our agricultural systems (Foley 2016). Worldwide, just fifteen crops produce 90 per cent of food calories, with wheat, rice and maize alone supplying 60 per cent (FAO 2016). A majority of these crops are grown in vast tracts of annual monocultures which have a high risk for pest and disease outbreaks (Altieri 1999, Malézieux et al 2009). The Irish potato famine of 1845–1850 contributed to the deaths of over a million people and is a stark reminder of what can happen when disease destroys a single crop that is relied upon too heavily (O'Boyle 2006). These monocultures require yearly replanting, high inputs, and weed control (Davis et al 2012), and it has been suggested that this cycle of plant-fertilize-spray tends to serve the interests of the large agribusiness companies who supply the inputs for this system more than furthering the goal of feeding the world (Jacobsen et al 2013). In the near future, our agricultural systems will also have to adapt to a changing climate that is expected to bring more extreme weather events like droughts and floods, in addition to increases in outbreaks of diseases and pests (Frison et al 2011).

CLIMATE CHANGE AND CLIMATE SMART AGRICULTURE

Concept of climate smart agriculture by FAO: "Trees are the key component for climate change adaptation and mitigation". The paper is based on review of the collected articles revealed that agroforestry as a climate change adaption tool to protect against changing temperature and precipitation regimes. The potential green house gas mitigation benefits could be achieved with implementing agroforestry system into present agricultural system. Methods of agricultural adaptation to climate change were documented across scales of climate stress and scales of the operation (Table 1).

Table 1: *Examples of stress factors affecting smallholder farmers in the tropics with indications of the impact of climate change on the stress factors*

Stress factor	Climate signal
Land access	No
Markets (inputs, outputs; access, prices)	Yes
Knowledge (basic principles, innovative cap.)	No
Technologies (strategic & tactical interventions)	No
Water (drought, flooding, irrigation, drainage)	Yes
Soil fertility	Yes
Pest & disease	Yes
On-farm labor (household size, off-farm activities,illness)	No
Weeds	Yes
Potential production of germplasm used	Yes
Dissatisfied customers	No

Source: Verchot et al 2007

Over the past two decades climate change has evolved from a debate about whether the planet is really warming to an increased focus on how to mitigate and adapt to its impacts, due mainly from the growing acceptance among scientists, policy makers, and even the general public that climate change is real and happening. This acceptance is based on the overwhelming evidence presented by the scientific community through intensive monitoring of global climatic systems, extensive observations on changes in terrestrial and aquatic systems, and predictive modeling (IPCC 2007, Hansen et al 2007). From the instrumental records of global temperature since 1850, the Fourth Assessment Report of the Intergovernmental Panel on Climate Change (IPCC) estimated that the current global mean surface temperature was about 0.42°C to 0.54°C above the 1961-1990 annual average (IPCC 2007). It also ranks eleven of the last twelve years (1995-2006) among the 12 warmest years. After an extensive review of the available literature on evidence of climate change, the Working Group I to the Fourth Assessment Report of the IPCC concluded that "the warming of the climate system is unequivocal, as is now evident from observations of increases in global average air and ocean temperatures, widespread melting of snow and ice, and rising global average temperature".

By Fifth Assessment Report (IPCC 2014), warming of the atmosphere and ocean system is *unequivocal*. Many of the associated impacts such as sea level change (among other metrics) have occurred since 1950 at rates unprecedented in the historical record. It is *extremely likely* (95-100 per cent probability) that human influence has been the dominant cause of observed warming since 1950, with the level of confidence having increased since the fourth report. It is likely (66-100 per cent probability) that 1983-2013 was the warmest 30-year period for 1400 years. The global surface temperature increase by the end of the 21st century is likely (66-100 per cent probability) to exceed 1.5°C relative to the 1850 to 1900 period for most scenarios, and is likely (66-100 per cent probability) to exceed 2.0°C for many scenarios. Global mean sea level will continue to rise at a rate very likely (90-100 per cent probability) to exceed the rate of the past four decades. Changes in climate will cause an increase in the rate of CO_2 production. Increased uptake by the oceans will increase the acidification of the oceans. A concise overview of Working Group I's findings was summaries also detailed the range of forecasts for warming, and climate impacts with different emission scenarios. Compared to the previous report, the lower bounds for the sensitivity of the climate system to emissions were slightly lowered, though the projections for global mean temperature rise (compared to pre-industrial levels) by 2100 exceeded 1.5°C in all scenarios.

POTENTIAL IMPACTS OF CLIMATE CHANGE ON AGRICULTURE

The major processes of agriculture that are directly influenced by climate change are soil, water, carbon and nitrogen cycles, crop growth and development, and incidence of weeds, pests, and diseases. These effects are manifested in terms of increased heat stress, increased evapo-transpiration, shortened seasons, and increased photosynthesis and reduced water use due to higher CO_2 in the atmosphere. The results of various assessments of impacts of climate change on agriculture (Rosenzweig and Hillel 1998, Mendelsohn and Neumann 1999, Reilly et al 2001, Das 2003) are generally in agreement and have identified (Table 2) the following as the major challenges that the future agricultural systems will face from climate change:

i. A northward shift in the domain of crops due to the increase in average annual temperature in the northern hemisphere
ii. An increase in crop water requirements to meet the increased evapo-transpiration demands
iii. Spatial and temporal changes in the land available for agriculture with tropical countries being more disadvantaged
iv. Increased degradation of land resources from erosion due to the projected increase in extreme events
v. New management challenges such as dealing with changes in the new pest and disease complexes
vi. Increased cost of inputs due to steep increases in energy and other input costs including taxation
vii. Competition for land by emerging initiatives like bio-fuel production. This may further lead to an increase of food prices with adverse impacts on accessibility by the poor
viii. Decrease in biodiversity and extinction of some species
ix. Erosion of communities' capacity to invest in agriculture due to losses from the increased frequency and intensity of extreme weather (drought, floods, etc.) conditions

Table 2: *Projected changes in climate and their impact on agriculture*

Phenomena and direction of change	Likelihood of occurrence	Major projected impacts on agriculture
Warmer and fewer cold days and nights; warmer/more frequent hot days and nights over most land areas	Virtually certain (>99% chance)	Increased yields in colder environments decreased yields in warmer environments increased insect outbreaks
Warm spells/heat waves: frequency increases over most land areas	Very likely (90-99% chance)	Reduced yields in warmer regions due to heat stress; wild fire danger increase
Heavy precipitation events: frequency increases over most areas	Very likely	Damage to crops; soil erosion, inability to cultivate land due to water logging of soils
Area affected by drought: increases	Likely (66-90% chance)	Land degradation, lower yields/crop damage and failure; increased livestock deaths; increased risk of wildfire
Intense tropical cyclone activity increases	Likely	Damage to crops; windthrow (uprooting) of trees; damage to coral reefs
Increased incidence of extreme high sea level (excludes tsunamis)	Likely	Salinization of irrigation water, estuaries and freshwater systems

Source: Rao et al 2007

CLIMATE SMART AGRICULTURE AND AGROFORESTRY

In 2009, the Food and Agriculture Organization of the United Nations coined a new term, "climate smart agriculture" (CSA). CSA is not a specific tool or technology, but an approach to improving the productivity and sustainability of agriculture under climate change, for the purposes of advancing food security and strengthening the resilience of rural livelihoods (FAO 2013). Climate smart agriculture integrates the three dimensions of sustainable development (economic, social and environmental) by jointly addressing food security and climate challenges. It is composed of three main pillars:

1. Sustainably increasing agricultural productivity and incomes
2. Adapting and building resilience to climate change
3. Reducing and/or removing greenhouse gases emissions, where possible" (FAO 2013)

CSA is a donor term that covers a wide range of practices, from fisheries and aquaculture to urban farming, efficient livestock production, conservation agriculture, and most importantly agroforestry. Because of its inclusive nature and "smart" sounding name, other donor agencies and international actors have adopted CSA as an approach to agriculture under climate change. Some would argue that, because of its high potential for building resilience to climate change, sequestering carbon dioxide, and strengthening rural farmer incomes and livelihoods, agroforestry is the most promising component of CSA.

ROLE OF AGROFORESTRY IN ADAPTING TO CLIMATE CHANGE

Agroforestry, the integration of trees and shrubs with annual crops production, is an age old management system practiced by farmers to provide shade, a steady supply of food and/or income throughout the year, arrest degradation and maintain soil fertility, diversify income sources, increase and stabilize income, enhance use efficiency of soil nutrients, water and radiation, and provide regular employment. A wide range of agroforestry systems now exist with a potential to improve productivity, favorably influence microclimate, prevent soil degradation and restore soil fertility and diversify income generating opportunities (Table 3).

Table 3: *Some agroforestry systems and practices*

System	Practice	Combination	Components
Agro-silvicultural systems	1.Improved fallow	trees planted during non-forest phase, if land not expected to revert to forest	w: fast growing h: agricultural crop
	2.Taungya	crops during tree seedling stage	w: plantation species h: agricultural crops
	3. Alley cropping	trees in hedges, crops in alleys	w: coppice trees h: crops
	4. Tree gardens	multispecies, dense, mixed	w: vertical structure, fruit trees h: shade tolerant
	5. Multipurpose trees on cropland	trees scattered, boundaries	w: multipurpose trees h: crops
	6. Estate crop combinations		w: coffee, coconut, fruit trees h: shade tolerant
	7.Homegardens	multistorey combinations around homes	w: fruit trees h: crops
	8. Trees in soil conservation, reclamation		w: multipurpose fruit trees h:crops
	9. Shelterbelts, windbreaks, live hedges	around farmland plots	w: trees h:crops
	10. Fuelwood production	firewood species around cropland plots	w: firewood species h: crops
Agro-silvi-pastoral systems	14.Homegardens with animals	around homes	w: fruit trees a : present
	15.Multipurpose woody hedgerows	trees for browse, mulch, soil protection	w: coppicing fodder trees a, h: present
	16. Aquaforestry	trees lining ponds	w: leaves forage for fish
Silvi-pastoral systems	11.Trees on rangelands	scattered trees	w: multipurpose, fodder h: present a: present
	12. Protein banks	trees for protein-rich cut fodder	w: leguminous trees h: present a: present
	13. Estate crops with pasture	Example cattle under coconut palms	w: estate crops h: present a: present

Note: w: woody species, a: animals, h: herbaceous (crop) species

Source: Schoene et al 2007

Recognizing the ability of agroforestry systems to address multiple problems and deliver multiple benefits, the IPCC Third Assessment Report on Climate Change (IPCC 2001) states that "Agroforestry can both sequester carbon and produce a range of economic, environmental, and socioeconomic benefits. For example, trees in agroforestry farms improve soil fertility through control of erosion, maintenance of soil organic matter and physical properties, increased N, extraction of nutrients from deep soil horizons, and promotion of more closed nutrient cycling." We believe that agroforestry interventions provide the best "no regrets" adaptation measures in making communities resilient to the impacts of climate change and do discuss the same in relation to the challenges posed by the changing and variable climate.

Performance of mitigation options in agroforestry will depend on the relative influence of tree species selection and management, soil characteristics, topography, rainfall, agricultural practices, priorities for food security, economic development options, among others. In order to improve carbon sequestration, or to reduce carbon emissions, several options are available (Table 4), but all are related to development needs of local communities.

Table 4: *Feasibility and limits of some AF practices*

Objectives of AFS	Example of feasibility	Limits of the influence factors	References
Increased soil fertility Low input fertilizer	Using nitrogen fixing trees. Windbreaks and erosion control, long fallows	Integrated nutrient management (balance with other sources of nutrients)	(Kumar and Nair 2012, Garrity et al 2010)
Increase availability of water	Conservation agriculture (CSA) with trees, shade trees to reduce evaporation	Tree density, selection of species	(Spracklen et al 2012)
Wood (fuel wood) — production	Thinning, coppicing, woodlots	Long term viability. Land tenure	(Kursten 2000)
Fruit production	Fruit trees of high market value (e.g. shea butter)	Age of trees in parklands. Productivity	(Assogbadjo et al 2012)
Ecosystem health	Shade trees (e.g. in Cocoa farms), promotion of agrobiodiversity, Forest corridors	Dominance of cash crops, limited lands for food crops	(DeSouza et al 2012)
Animal husbandry	Fodder trees, grassland management	Pressure on selected fodder trees, available land for animals	(Lusiana et al 2012)

AGROFORESTRY AS A TRANSFORMATIVE SOLUTION

One multifunctional approach for our food system is agroforestry, the intentional combination of trees and shrubs with crops or livestock. Agroforestry has been recognized for nearly half a century as a sustainable agricultural practice (Nair and Garrity 2012), and the concept of integrating trees into the agricultural landscape is as old as the practice of cultivating land. The beneficial outcomes of agroforestry include reductions in nutrient and pesticide runoff, carbon sequestration, increased soil quality, erosion control, improved wildlife habitat, reduced fossil fuel use, and increasing resilience in the face of an uncertain agricultural future (Davis et al 2012, Winans et al 2015, Dixon et al 1994, Jordan and Davis 2015, Baah-Acheamfour et al 2014, Caudill et al 2015, Brandle and Schoeneberger 2014).

AGROFORESTRY PRACTICES AND PRODUCTS

In addition to the environmental benefits, agroforestry can supply products such as timber, crops, fruits, nuts, mushrooms, forages, livestock, biomass, Christmas trees, and herbal medicine (Gold et al 2015). A diverse portfolio of products would allow revenue streams to be spread out over the short-term (crops, forage, livestock, mushrooms, certain fruits like currants), medium-term (nuts, fruits suchas apples or persimmons, biomass, medicinal plants), and long-term (lumber, increased property value). This diversity (Fig. 1) of products can also reduce risk for farmers, though it may require creative marketing (Gold et al 2015).

Alley Cropping

Alley cropping involves growing field crops between rows of trees (Campbell et al 1991). The trees can be grown for timber or fruits and nuts, while the alley crops can include a variety of grains, vegetables, or forages cut for hay. The crops provide short-term income while the trees provide longer-term revenue. The tree and crop species also may interact in ways that allow increased production due to the different niches that the trees and crops occupy (Cardinael et al 2015). Under arid and semi- arid conditions bamboo based agroforestry system is one of the most viable system for livelihood (Dev et al 2017).

Silvopasture

Silvopasture incorporates livestock into an intentional mixture of trees and pasture. Silvopasture is different from just "grazing the woods", because the spacing of the trees is carefully planned to allow enough sunlight for the forages below, and the livestock are kept from damaging the trees. The trees offer protection for livestock through shade during the heat of the summer and wind reduction in the cold winter (Brandle et al 2004, Kallenbach et al 2006).

Windbreaks

Windbreaks, also known as shelterbelts, were recognized early on as a useful agroforestry practice. Windbreaks prevent wind erosion, provide habitat for wildlife, and can increase water availability to nearby crops due to lower evapo-transpiration and the effects of catching snow (Brandle et al 2004). More water can mean higher production, leading to important economic benefits to farmers (Kort 1988). On a farmstead, windbreaks can decrease the heating and cooling needs for living and working spaces by reducing indoor air exchange caused by wind (Mize et al 2008).

Forest Farming

Forest farming includes practices such as raising mushrooms, harvesting medicinal herbs like ginseng and goldenseal, and marketing woody ornamental material (Gold et al 2015). This agroforestry approach usually occurs in established forests that are grown for timber and allows for income generation without major disturbance (Valdivia and Poulos 2008).

These practices therefore tend to complement, rather than compete with, existing production systems and may provide ways to contribute to food security by using resources that are otherwise underutilized.

Figure 1: *Examples of different agroforestry systems – alley cropping, windbreaks, silvopastures, traditional forest farming (Lin 2007)*

Trees Species Uses and Functions that Enhance Farmer's Adaptive Capacity

In the study area, agroforestry technologies involve agrosilvicultural (growing trees with crops), agrosilvopastoral (growing trees with pastures), agro horticultural, shifting cultivation and home garden (management of trees, crops and animals) were the main types of on-farm tree growing in the form of traditional agroforestry systems.

Table 5: *List of tree species, uses and functions that enhance farmer resilience*

Species name	Uses/Functions
Grevellia robusta	Timber, shade, firewood, add organic matter
Cordia africana	Timber, shade, firewood, soil improvement, fodder
Tamarindus indica	Timber, Fruits, firewood, organic matter
Syzigium cordatum	Timber, fruits, firewood, organic matter, shade
Balanites aegyptica	Timber, firewood, organic matter
Markamia obtusifolia	Firewood, Fodder, Nitrogen fixation, shade
Croton Macrostachyus	Firewood, Fodder, Nitrogen fixation, shade
Kigeria africana	Fruits, shade, Nitrogen fixation
Eucalyptus saligna	Poles, firewood, wind break, erosion control
Acacia species	Firewood, Fodder, Nitrogen fixation, shade
Salvadora persca	Firewood, Fodder, Nitrogen fixation, shade
Cordia sinensis	Firewood, organic matter, shade
Ficus species	Firewood, shade, windbreak, control soil erosion
Albizia schemperana	Timber, firewood, nitrogen fixation, fodder ,rain indicator
Faidherbia albida	Firewood, nitrogen fixation, fodder, shade
Commiphora eminii	Firewood, fodder, shade, organic matter, support climber
Artocarpus heterophyllus	Fruits, shade, control soil erosion

Persea americana	Fruits, Firewood, shade, soil improvement
Mangifera indica	Fruit, erosion control, windbreak, firewood
Anona muricata	Fruits, shade, live fence
Anona squamosa	Fruit, shade
Croton megarocarpus	Firewood, shade, organic matter, control soil erosion
Syzigium guineense	Timber, firewood, shade, soil improvement
Azadirachta indica	Firewood, shade, windbreak, control soil erosion
Psidium guajava	fruits, support climbers, firewood
Cocos nucifera	Fruits, windbreak
Rauvolfa caffra	Shade

Source: Charles et al 2013

The most frequent method of growing trees (except exotic trees and Coconut) was through deliberate retention and management of naturally regenerating tree seedlings. A total of 93 respondents were able to respond about types of tree species most preferred and planted or retained on their farms, and the answers are presented in Fig.1. Also result revealed that most of trees regenerated naturally, such as *Cordia africana, Croton macrostachyus, Markamia* spp., *Acacia* spp. and *Albizia chimerana. Eucalyptus saligna* were reported as among the tree species which have higher return and low cost of management due to its sprouting characteristics. Also it was observed to be grown along steep slope, unfertile land and away from water source. Also Eucalyptus trees were reported to take a few years to be due to a number of its advantages and their adaptations to a wide range of ecological conditions. It was observed that farmers' willingness to grow trees on their farms was a function of their socio-economical, environmental and cultural factors. For example, farmers ignored trees which are incompatible with arable crops such as *Cedrella odorata* and *Acrocarpus fraxinifolius.* Farmers' perceptions have showed a strong relation to the positive outcomes of tree planting (Table 5).

BENEFITS OF AGROFORESTRY

Agroforestry has positive effects on soil and water quality. Soil quality is improved by increased levels of organic matter, more diverse microbial populations, and improved nutrient cycling, which may increase crop productivity and the ability to cope with drought (Baah-Acheamfour et al 2014; Udawatta et al 2014 and Rivest et al 2013). The water quality benefits occur as non-point source pollution from row crops is reduced by incorporating agroforestry vegetative buffer strips (Udawatta et al 2002, Jose 2009, Garrett et al 2009).The increase in soil organic matter in the form of carbon not only improves the health of the soil, but it can also help to reduce atmospheric carbon dioxide that is implicated in climate change (Amundson et al 2015). Compared to a monoculture of crops or pasture, adding trees and shrubs to an agricultural landscape increases the level of carbon sequestration (Baah-Acheamfour et al 2014, Jose et al 2009).The mechanisms for increased carbon sequestration include better erosion control, more carbon being stored in woody perennials, reduced organic matter decomposition, and the fact that crop biomass is not harvested in agroforestry to the degree that it is in conventional systems (Kim et al

2016). The link between perennial systems and climate change may be an important one. Robertson et al 2000 studied the global warming potential of several annual and perennial systems. They found that none of the annual cropping systems reduced global warming potential, whether conventional, no-till, reduced input, or organic. Although the cropping systems did accumulate carbon in the soil, the gains were offset by nitrous oxide emissions. However, the perennial and early successional forest treatments including alfalfa, hybrid poplar, and abandoned early successional sites all reduced global warming potential. Reducing fossil fuel use is another important strategy for climate change mitigation (Reddy 2015). Bio-energy is one avenue to reduce fossil fuel dependence, but there are concerns about using valuable cropland to grow crops for energy instead of food (Foley 2016). By producing biomass from trees in combination with food on the same land, agroforestry may be one way to contribute to a secure energy future without compromising food production capabilities (Jose and Bardhan 2012, Holzmueller and Jose 2012).

Adding trees, shrubs, and other perennial vegetation to an agricultural landscape provides habitat for greater numbers and more diverse populations of wildlife (Jose et al 2012, Garrett et al 2009). Livestock can benefit from agroforestry as well. Windbreaks protect animals from harsh winds, while shade provided by trees can increase comfort in the heat of the summer and may encourage more even grazing over a paddock (Fagerholm et al 2016). Compared to annual monocultures, perennial polycultures like agroforestry are inherently more stable in the face of global market volatility and extreme climatic events (Foley 2016). In the event of fossil fuel scarcity, mature fruit and nut trees would continue to produce their products with relatively little interruption, though labor may have to be substituted for other inputs. Not only do agroforests sequester greenhouse gases that are driving global climate change, they are also more resilient to its likely effects. Deeper rooting systems and improved infiltration and water storage lessen the impact of drought, while trees' abilities to pump excess water out of the soil as well as withstanding inundation better than field crops means they are also more resilient to floods (Verchot et al 2007).

ADOPTION OF AGROFORESTRY SYSTEMS

Agroforestry systems provide the unique opportunity to utilize the resources to achieve maximum return from a unit of available soil, water, nutrient, and sunlight in an economically and ecologically sustainable manner. As highlighted earlier, the environmental benefits from these systems substantially outweigh the economic gains. At present the trade-offs between crop productivity and environmental functions are not well quantified. During the past decade, environmental services such as carbon sequestration, watershed protection and biodiversity conservation have been recognized as global environmental services and to some extent are being financed.

The available literature suggests that there are opportunities to make the agroforestry systems more attractive and profitable. There is a need to survey existing agroforestry systems to determine the interaction between component species, to classify the trees used, and then to refine the systems in view of soil, climate, and socio-economic limitations. Detailed studies on the competition and complementarity between trees and understorey agricultural crops for solar radiation, space, and soil factors are needed. Future systems should integrate trees that are tolerant to strong light and high evaporative demand and crops that are adapted to shade and relatively high humidity.

Table 6: *Examples of positive or negative implications of agroforestry practices for adaptation or mitigation to climate change*

		Mitigation	
		Positive	**Negative**
Adaptation	**Positive**	Soil carbon sequestration, improved water holding capacities, use of manure instead , mixed agroforestry for commercial products, income diversification with trees, reduced nitrogen fertilizer, fire management	Dependence on biomass energy, overuse of ecosystem services, Increased use of mineral fertilizers Poor management of nitrogen and manure, over extraction of non-timber products, timber extraction
	Negative	Integral protection of forest reserves, limited rights to agroforestry trees, Forest Plantation excluding harvest	Use of forest fires for pastoral and land management, tree exclusion in farming lands

Source: Mbow et al 2014

Agroforestry may nevertheless involve practices that raise GHG emissions, such as shifting cultivation, pasture maintenance by burning, nitrogen fertilization and animal production (Wilson et al 2016). In order to optimize agroforestry for adaptation and mitigation to climate change, there is a need for more integrated management to increase benefits and reduce negative impacts on climate (Table 6).

CHALLENGES TO AGROFORESTRY ADOPTION

The opportunities for agroforestry are exciting, but not without challenges. Agroforestry adoption has been surprisingly low, considering the well-documented benefits (Trozzo et al 2014, Sereke et al 2015, Faulkner et al 2014). Barriers have included the expense of establishment (Thevathasan et al 2012), landowner's lack of experience with trees (Strong and Jacobson 2006, Faulkner et al 2014), and the time and knowledge required for management (Valdivia et al 2012). For agroforestry systems that produce edible products such as fruits and nuts, the logistics of harvest can be challenging. For agroforestry systems to be economically competitive, mechanization may be required for larger plantings (Chestnuts-Harvesting 2016). This can be complicated if multiple fruit or nut species are grown. Non-traditional markets and delayed profits may be another deterrent (Strong and Jacobson 2006). The economic feasibility of some agroforestry systems such as silvopasture have been shown to be profitable, whereas other practices such as biomass plantings or riparian buffers may need the development of markets that offer compensation for the ecosystem services provided in order to make financial sense (Winans et al 2015, Holzmueller and Jose 2012, Grado and Husak 2004). Social change and networking will also play a role as mindsets evolve to include alternatives to the norm (Sereke et al 2015, Thevathasan et al 2012).

THE VALUE OF PROPERTY RIGHTS AND LAND USE POLICIES

As shown in the previous section, there is evidence that agroforestry is practiced around the world as a sustainable mode of land use that enhances the livelihoods of smallholder farmers and advances resilience and adaptation to the impacts of climate change. Further, agroforestry's ability to mitigate the causes of climate change through the sequestration and storage of carbon is significantly stronger than other forms of agricultural land use, making agroforestry a particularly promising form of climate smart agriculture. For all of these reasons one could say that climate smart agriculture, and agroforestry in particular, is sustainable development in action. Yet many smallholder farmers in the developing world still practice forms of agriculture that are vulnerable to drought and other climatological impacts, and which seem to be unsustainable in the face of inevitable climate variability and change caused by anthropogenic global warming. If agroforestry and other forms of climate smart agriculture hold so much promise for a resilient future, why are smallholder farmers everywhere not already taking advantage of the benefits? There are a number of reasons, including unequal distribution of knowledge and institutional support for CSA practices, state policies and funding that favor largescale conventional agriculture or other unsustainable practices, and the problem of insecure property rights (tenure). Agroforestry is practiced around the world in different capacities. Some forms are indigenous in nature and rely on traditional, local knowledge. Other forms of agroforestry are not necessarily native to a region but are promoted through institutional funding and support. This paper argues not that agroforestry is a silver bullet for climate change, but that it is a highly effective tool in the climate smart agriculture toolbox that should be promoted and supported for smallholder farmers around the developing world (McCabe 2013).

CONCLUSION

This study described the potential of agroforestry in increasing resilience of subsistence farmers who are different in altitudinal range and climatic conditions. Agroforestry systems reflected diversity in terms of the multiple benefits from trees, crops and livestock integrated in agriculture systems. Agroforestry products improved resilience of smallholder farmers against climate changes, particularly by improving farm production (food, fodder, timber, fuel wood, and manure), ecosystem services (soil improvement, climate amelioration, wind break, erosion control, and disease and pest control) and household income. Agroforestry practitioners were economic resilient that enabled to resist against stress, disturbance, shock and perturbation than those who depended on conventional agriculture. Agroforestry systems offer a win-win opportunity by acting as sinks for atmospheric carbon while helping to attain food security, increase farm income, improve soil health and discourage deforestation. While agroforestry systems clearly offer economic and ecological advantages, the management of these systems is complicated by the tradeoffs between subsistence requirements, acceptable risks, and costs involved. A critical and scientific assessment of these tradeoffs requires quantitative data and tools to make a realistic assessment of the complex interactions involved in natural resource use and economic and environmental benefits. Because of the long lead time required to study agroforestry systems thorough complete rotations and high research costs associated with such studies, much of the current assessment is based on incomplete data. Well calibrated and validated system simulation models have the potential to contribute significantly to the understanding and quantify the benefits.

REFERENCES

Altieri MA 1999 The ecological role of biodiversity in agroecosystems, *Agriculture Ecosystem Environment* **74** 19–31

Amundson R, Berhe AA, Hopmans JW, Olson C, Sztein AE and Sparks DL 2015 Soil and human security in the 21st century, *Science* **348** 1261-1271

Assogbadjo AE, Kakai RG, Vodouhe FG, Djagoun CAMS, Codjia JTC, and Sinsin B 2012 Biodiversity and socioeconomic factors supporting farmers' choice of wild edible trees in the agroforestry systems of Benin (West Africa), *Forest Policy Economics* **14** 41-49

Baah-Acheamfour M, Carlyle CN, Bork EW, and Chang SX 2014 Trees increase soil carbon and its stability in three agroforestry systems in central Alberta, Canada, *Forest Ecology and Management* **328** 131–139

Boesch D and Brinsfield R 2000 Coastal Eutrophication and Agriculture Contributions and Solutions, in *Biological Resource Management Connecting Science and Policy* (eds) Balázs E, Galante E, Lynch JM, Schepers JS, Toutant J-P, Werner D, Werry PATJ, *Springer Berlin*, Heidelberg, Germany, 93–115

Brandle JR, Hodges L, and Zhou XH 2004 Windbreaks in North American agricultural systems, *Agroforestry System* **61** 65–78

Brandle JR and Schoeneberger MM 2014 Working Trees: Supporting Agriculture and Healthy Landscapes, *Journal of Tropical Forest Science* **26** 305–308

Campbell GE, Lottes GJ, and Dawson JO 1991 Design and development of agroforestry systems for Illinois, USA: Silvicultural and economic considerations, *Agroforestry System* **13** 203–224

Cardinael R, Mao Z, Prieto I, Stokes A, Dupraz C, Kim JH, and Jourdan C 2015 Competition with winter crops induces deeper rooting of walnut trees in a Mediterranean alley cropping agroforestry system, *Plant and Soil* **391** 219–235

Caudill SA, DeClerck FJA, and Husband TP 2015 Connecting sustainable agriculture and wildlife conservation: Does shade coffee provide habitat for mammals? *Agriculture Ecosystem Environment* **199** 85–93

Charles Richard L, Munishi PKT, and Nzunda Emmanuel F 2013 Agroforestry as Adaptation Strategy under Climate Change in Mwanga District, Kilimanjaro, Tanzania, *International Journal of Environmental Protection* **1**(11) 29-38

Chestnuts Harvesting 2016 Available online http://msue.anr.msu.edu/topic/chestnuts harvest storage/harvesting (accessed on 6 June)

Daleszczyk K, Eycott AE, and Tillmann JE 2016 Mammal Species Extinction and Decline: Some Current and Past Case Studies of the Detrimental Influence of Man, in *Problematic Wildlife* (eds) Angelici FM, Springer International Publishing: New York, NY, USA, 21–44

Das HP 2003 Agrometeorology related to extreme events; CAgM-XI Working Group on Agro meteorology Related to Extreme Events, Geneva, Switzerland Secretariat of the World Meteorological Organization

Davis AS, Hill JD, Chase CA, Johanns AM and Liebman M 2012 Increasing Cropping System Diversity Balances Productivity, Profitability and Environmental Health, PLoS ONE **7** 147-149

DeSouza HN, DeGoede RGM, Brussaard L, Cardoso IM, Duarte EMG, Fernandes RBA, Gomes LC, and Pulleman MM 2012 Protective shade, tree diversity and soil properties in coffee agroforestry systems in the Atlantic Rainforest biome, *Agriculture Ecosystem and Environment* **146** 179-196

Dev I, Ahlawat SP, Palsaniya DR, Ram A, Newaj R, Tewari RK, Singh R, Sridhar KB, Dwivedi RP, Srivastava M, Chaturvedi OP, Kumar RV and Yadav RS 2017 Bamboo (*Dendrocalamus strictus*) + sesame (*Sesamum indicum*) based agroforestry model: A sustainable livelihood option for farmers of semi-arid region, *Indian Journal of Agricultural Sciences* **87**(11) 1528-34

Dixon RK, Winjum JK, Andrasko KJ, Lee JJ, and Schroeder PE 1994 Integrated land-use systems: Assessment of promising agroforest and alternative land-use practices to enhance carbon conservation and sequestration, *Climate Change* **27** 71–92

Fagerholm N, Torralba M, Burgess PJ and Plieninger T 2016 A systematic map of ecosystem services assessments around European agroforestry, *Ecological Indicators* **62** 47–65

FAO 2013 Climate smart agriculture source book, Retrieved from: http://www.fao.org/climatechange/climate smart/en/

FAO 2016 Dimensions of Need—Staple Foods: What do People Eat? Available online: http://www.fao.org/docrep/u8480e/u8480e07 htm (accessed on 12 April)

Faulkner PE, Owooh B, and Idassi J 2014 Assessment of the Adoption of Agroforestry Technologies by Limited-Resource Farmers in North Carolina, The *Journal of Extension* **52**(5) Available online: http://www.joe.org/joe/2014october/rb7.php (accessed on 16 October).

Foley JA 2011 Can We Feed the World and Sustain the Planet? *Science American* **305** 60–65

Foley J 2016 It's Time to Rethink America's Corn System. Available online: http://www.scientificamerican.com/article/time-to-rethink-corn/ (accessed on 24 April)

Frison EA, Cherfas J, and Hodgkin T 2011 Agricultural Biodiversity Is Essential for a Sustainable Improvement in Food and Nutrition Security, *Sustainability* **3** 238–253

Garrett HE, McGraw RL, and Walter WD 2009 Alley Cropping Practices, *North American Agroforestry Integrated Science Practice 2nd Ed.*, 133–162

Garrity DP, Akinnifesi FK , Ajayi OC, Weldesemayat SG, Mowo JG, Kalinganire A, Larwanou M, and Bayala J 2010 Evergreen agriculture: a robust approach to sustainable food security in Africa, *Food Security* **2** 1970-2214

Gold M, Cernusca M, and Hall M 2015 *Training Manual for Applied Agroforestry Practices*; University of Missouri Center for Agroforestry New Franklin, MO, USA, 2006

Grado SC and Husak AL 2004 Economic Analyses of a Sustainable Agroforestry System in the Southeastern United States, in *Valuing Agroforestry Systems*; Advances in Agroforestry; *Springer* Dordrecht, The Netherlands 39–57

Hansen M, Sato R, Ruedy P, Kharecha A, Lacis R, Miller L, Nazarenko K, Lo GA, Schmidt, G Russell, I Aleinov, S Bauer, E Baum, B Cairns, V Canuto, M Chandler, Y Cheng, A Cohen, A Del Genio, G Faluvegi, E Fleming, A Friend, T Hall, C Jackman, J Jonas, M Kelley, NY Kiang, D Koch, G Labow, J Lerner S, Menon T, Novakov V, OinasJa, Perlwitz, Ju, Perlwitz D, Rind A, Romanou, R Schmunk, D Shindell, P Stone, S Sun, D Streets, N Tausnev, D Thresher, N Unger, M Yao, and Zhang S 2007 Dangerous human-made interference with climate a GISS model E Atmos, *Chemical Physics* **7** 2287–2312

Holzmueller EJ and Jose S 2012 Biomass production for biofuels using agroforestry: Potential for the North Central Region of the United States, *Agroforestry System* **85** 305–314

IPCC 2001 Climate Change 2001 The Scientific Basis. Contribution of the Working Group 1 to the Third Assessment Report of the IPCC Cambridge, Cambridge University Press

IPCC 2007 Climate Change 2007 The Physical Science Basis. Contribution of Working Group I to the Fourth Assessment Report of the Intergovernmental Panel on Climate Change [Solomon S, Qin D, Manning M, Chen Z, Marquis M, Averyt K B, Tignor M and Miller H L (eds)]. Cambridge University Press, Cambridge, United Kingdom and New York, NY, USA, 996

IPCC 2014 The Physical Science Basis. Contribution of Working Group I to the Fifth Assessment Report of the Intergovernmental Panel on Climate Change

Jacobson M and Kar S 2013 Extent of Agroforestry Extension Programs in the United States, The *Journal of Extension* **51**(4)

Jacobsen SE, Sorensen M, Pedersen SM, and Weiner J 2013 Feeding the world: Genetically modified crops versus agricultural biodiversity, *Agronomy for Sustainable Development* **33** 651–662

Johnson RJ, Jedlicka JA, Quinn JE, and Brandle, JR 2011 Global Perspectives on Birds in Agricultural Landscape, in *Integrating Agriculture, Conservation and Ecotourism: Examples from the Field* (eds) Campbell WB, Ortiz SL, Springer: Dordrecht, The Netherlands, 55–140

Jordan NR and Davis AS 2015 Middle-way strategies for sustainable intensification of agriculture, *Bio Science* **65** 513–519

Jose S 2009 Agroforestry for ecosystem services and environmental benefits: An overview, *Agroforestry Syst*ems **76** 1–10

Jose S and Bardhan S 2012 Agroforestry for biomass production and carbon sequestration: An overview, *Agroforestry System* **86** 105–111

Jose S, Gold MA, and Garrett HE 2012 The Future of Temperate Agroforestry in the United States, in *Agroforestry—The Future of Global Land Use* (eds) Nair PKR, Garrity D, *Springer* Dordrecht, The Netherlands, 217–245

Jose S, Holzmueller EJ, Gillespie AR, and Garrett HEG 2009 Tree–Crop Interactions in Temperate Agroforestry. In *ACSESS Publications*; American Society of Agronomy Madison, WI, USA

Kallenbach RL, Kerley MS and Bishop-Hurley GJ 2006 Cumulative Forage Production, Forage Quality and Livestock Performance from an Annual Ryegrass and Cereal Rye Mixture in a Pine Walnut Silvopasture, *Agroforesrty System* **66** 43–53

Kim DG, Kirschbaum MUF, and Beedy TL 2016 Carbon sequestration and net emissions of CH_4 and N_2O under agroforestry: Synthesizing available data and suggestions for future studie, *Agriculture Ecosystem Environment* **226** 65–78

Kort J 1988 Benefits of windbreaks to field and forage crops, *Agriculture Ecosystem and Environment* **22** 165–190

Kumar BM and Nair PKR 2012 Carbon Sequestration Potential of Agroforestry Systems, Opportunities and Challenges, *Springer1-299 pages*

Kursten E 2000 Fuelwood production in agroforestry systems for sustainable land use and CO_2-mitigation, *Ecology Engineering* **16** S69-S72

Lin BB 2007 Agroforestry management as an adaptive strategy against potential microclimate extremes in coffee agriculture, *Agricultural and Forest Meteorology* **144** 85-94

Lusiana B, Noordwijk MV, and Cadisch G 2012 Land sparing or sharing? Exploring livestock fodder options in combination with land use zoning and consequences for livelihoods and net carbon stocks using the FALLOW model, *Agriculture Ecosystem and Environment* **159** 145-160

Malezieux E, Crozat Y, Dupraz C, Laurans M, Makowski D, Ozier-Lafontaine H, Rapidel B, de Tourdonnet S, and Valantin-Morison M 2009 Mixing plant species in cropping systems: Concepts, tools and models—A review, *Agronomy Sustainable Development* **29** 43–62

MbowCheikh, NeufeldtHenry, Minang, Peter Akong, Luedeling Eike, and Kowero Godwin 2014 Achieving mitigation and adaptation to climate change through sustainable agroforestry practices in Africa, *Current Opinion in Environmental Sustainability* **6** 8–14

McCabe Colin 2013 Agroforestry and Smallholder Farmers: Climate Change Adaptation through Sustainable Land Use, *Capstone Collection* 2612

McIsaac GF, David MB, Gertner GZ, and Goolsby DA 2001 Eutrophication: Nitrate flux in the Mississippi River, *Nature* **414** 166–167

Mendelsohn R and Neumann JE 1999 The Impact of Climate Change on the United States Economy. Cambridge University Press, Cambridge, UK

Mize CW, Brandle JR, Schoeneberger MM and Bentrup G 2008 Ecological Development and function of Shelterbelts in Temperate North America, in *Toward Agroforestry Design* (eds) Jose S, Gordon AM Springer Dordrecht, The Netherlands, 27–54

Nair PKR and Garrity D 2012 *Agroforestry—The Future of Global Land Use, Advances in Agroforestry*; *Springer,* Dordrecht, The Netherlands; 9

Norris K 2008 Agriculture and biodiversity conservation: Opportunity knocks, *Conservation Letters* **1** 2–11

OBoyle EJ 2006 Classical economics and the Great Irish Famine: A study in limits, *Forum Social Economics* **35** 21–53

Rao KPC, Verchot Louis V, and Laarman Jan 2007 Adaptation to Climate Change through Sustainable Management and Development of Agroforestry Systems, *SAT eJournal | ejournal.icrisat.org* **4**(1) 1-30

Reddy PP 2015 Impacts of Climate Change on Agriculture, in *Climate Resilient Agriculture for Ensuring Food Security*; *Springer* New Delhi, India 43–90

Reilly J, Tubiello F, McCarl B, and Melillo J 2001 The potential consequences of climate variability and change. Chapter 13 in Climate Change Impacts on the United States, Report for the U.S. Global Change Research Program US, National Assessment Synthesis Team. Cambridge University Press, Cambridge, UK

Rivest D, Lorente M, Olivier A, and Messier C 2013 Soil biochemical properties and microbial resilience in agroforestry systems: Effects on wheat growth under controlled drought and flooding conditions, *Science of Total Environment* **463–464** 51–60

Robertson GP, Paul EA, and Harwood RR 2000 Greenhouse Gases in Intensive Agriculture: Contributions of Individual Gases to the Radiative Forcing of the Atmosphere, *Science* **289** 1922–1925

Rosenzweig C and Hillel D 1998 Climate Change and The Global Harvest: Potential Impacts of the Greenhouse Effect on Agriculture, Oxford University Press, New York

Schoene D, Killmann W, Von Lupke H, and LoycheWilkie M 2007 Definition issues related to reducing emissions from deforestation in developing countries, *Forests and climate change working* FAO, Rome, Italy

Sereke F, Dobricki M, Wilkes J, Kaeser A, Graves AR, Szerencsits E, and Herzog F 2015 Swiss farmers don't adopt agroforestry because they fear for their reputation, *Agroforestry Systems* **90**(3) 385-394

Spracklen DV, Arnold SR, and Taylor CM 2012 Observations of increased tropical rainfall preceded by air passage over forests, *Nature Research Letter* **489** 282–285

Strong N and Jacobson MG 2006 A case for consumer-driven extension programming: agroforestry adoption potential in Pennsylvania, *Agroforestry Systems* **68** 43–52

Thevathasan NV, Gordon AM, Bradley R, Cogliastro A, Folkard P, Grant R, Kort J, Liggins L, Njenga F, and Olivier A 2012 Agroforestry Research and Development in Canada: The Way Forward, in *Agroforestry—The Future of Global Land Use* (eds) Nair PKR, Garrity D, *Springer* Dordrecht, The Netherlands 247–283

Thomas JA, Telfer MG, Roy DB, Preston CD, Greenwood JJD, Asher J, Fox R, Clarke RT, and Lawton JH 2004 Comparative Losses of British Butterflies, Birds, and Plants and the Global Extinction Crisis, *Science* **303** 1879–1881

Trozzo KE, Munsell JF, and Chamberlain JL 2014 Landowner interest in multifunctional agroforestry Riparian buffers, *Agroforestry Systems*; Springer: Dordrecht, The Netherlands; **88**(4) 619–629

Udawatta RP, Kremer RJ, Nelson KA, Jose S, and Bardhan S 2014 Soil Quality of a Mature Alley Cropping Agroforestry System in Temperate North America, *Communication of Soil Science Plant Analysis* **45** 2531–2551

Udawatta RP, Krstansky JJ, Henderson GS, and Garrett HE 2002 Agroforestry practices, runoff, and nutrient loss: A paired watershed comparison, *Journal of Environment Quality* **31** 1214–1225

Valdivia C, Barbieri C, and Gold MA 2012 Between Forestry and Farming: Policy and Environmental Implications of the Barriers to Agroforestry Adoption, *Canadian Journal of Agriculture Economy* **60** 155–175

Valdivia C and Poulos C 2008 Factors affecting farm operators' interest in incorporating riparian buffers and forest farming practices in northeast and southeast Missouri, *Agroforestry System* **75** 61–71

Verchot LV, Noordwijk MV, Kandji S, Tomich T, Ong C, Albrecht A, Mackensen J, Bantilan C, Anupama KV, and Palm C 2007 Climate change: Linking adaptation and mitigation through agroforestry, *Mitigation and Adaptation Strategies for Global Change* **12** 901–918

Wilson Matthew Heron and Lovell Sarah Taylor 2016 Agroforestry—The Next Step in Sustainable and Resilient Agriculture, *Sustainability* **574**(8) 1-15

Winans KS, Tardif AS, Lteif AE, and Whalen JK 2015 Carbon sequestration potential and cost-benefit analysis of hybrid poplar, grain corn and hay cultivation in southern Quebec, Canada, *Agroforestry System* **89** 421–433

CHAPTER 19

Growth, Biomass Production and CO_2 Sequestration of Some Important Multipurpose Trees Under Rainfed Condition

KN Patel, RR Shakhela and J R Jat

C. P. College of Agriculture, Sardarkrushinagar Dantiwada Agricultural University, Sardarkrushinagar - 385 506, Gujarat, India

**Corresponding author's Email: kpsoilscience@gmail.com*

ABSTRACT

An experiment was conducted at Agroforestry Research Station, Sardarkrushinagar Dantiwada Agricultural University, Sardarkrushinagar, Gujarat, India during 2014-15. The experiment was designed in randomized block design with four replications on 30 years old agroforestry tree plantation which consisted of four multipurpose tree species (MPTs) *viz., Azadirachta indica, Prosopis cineraria, Prosopis juliflora* and *Acacia tortolis*. Significantly highest tree canopy N-S (7.18 m) and E-W (7.32 m) was noted under *A. tortolis*. Above ground green and dry biomass and below ground green and dry biomass values did not differed significantly due to different tree species. The highest C content (41.55%) and CO_2 sequestration (1847.11 kg tree^{-1} and 2052.35 t ha^{-1}) was recorded under *P. juliflora* tree species. The descending order of carbon content in different tree species was *P. juliflora*>*A. indica*>*P. cineraria*>*A. tortolis*. Maximum SOC (0.544 Mg m^{-3}) was noted under *A. tortolis* which was 242 per cent higher over control.

Keywords: CO_2 sequestration, Multipurpose tree species, Soil organic carbon

INTRODUCTION

Agroforestry plays a great role in maintaining the natural resources and increases the productivity in the rainfed area of arid and semiarid region. Trees provide food, fuel wood, fodder, fertilizer and timber, reduction in incidence of total crop failure and sustained productivity. Trees also facilitates more efficient recycling of nutrients, reduction of surface run-off, nutrient leaching and soil erosion through impeding effect of tree roots. They also improve the microclimate, such as lowering of soil surface temperature, reduction of evaporation of soil moisture through a combination of mulching and shading, increment in soil nutrients through addition and decomposition of litter fall and improvement of soil structure through the constant addition of organic matter from decomposed litter. Carbon sequestration issues in tree has gained enormous momentum in the recent time. Forests store and release carbon dioxide through natural processes. As a tree grows, it takes in CO_2 from the atmosphere and releases O_2 in the process of photosynthesis. The carbon that is taken from the air is incorporated into sugars (such as glucose), that become the building blocks for production of wood. About one-half the weight of dry wood is carbon and that carbon is stored or sequestered as long as the wood is in existence. When trees die, decay or burn they release carbon stored in the soils and biomass (organic matter such as stems, stumps and slash) as CO_2 into the atmosphere. Carbon is also released as CO_2 when trees are harvested, although considerable carbon is stored in wood put into long-term use such as in houses, furniture and books. Hence tree constitutes a major 'C' sink owing to the photosynthesis and storage of CO_2 in live and dead organic matter.

Trees improve land cover in agricultural fields in addition to providing 'C' inputs to the soil. The most significant increase in 'C' stock occurs in fine textured soils, where 'C' is better protected through soil aggregation. Trees in degraded lands offer a great opportunity to sequester more carbon and seek carbon credits. There is constant addition of organic matter to the soil through decaying dead roots, which lead to improvement in the 'C' status of the soil. The overall fertility build-up in tree system may be attributed by the nature and diversity of species and due to heavy litter fall and their subsequent decomposing in the soil profile. Planting of multipurpose tree species in the field serves a dual purpose *i.e.,* promotion of biodiversity and carbon sequestration. There is variation in the carbon sequestration potential among different plantation species and variation in environmental conditions can also affect carbon sequestration. Therefore, this study was carried out to identify the trees which can sequester maximum carbon under given environment condition.

METHODOLOGY

The field experiment was carried out at Agroforestry Research Station, Sardarkrushinagar Dantiwada Agricultural University, Sardarkrushinagar, Gujarat during 2014-15. The experiment was laid out in randomized block design with a spacing of 3 m X 3 m and four replications. The experiment consisted of four multipurpose tree species (MPTs) viz *Azadirachta indica*, *Prosopis cineraria*, *Prosopis juliflora* and *Acacia tortolis*. The biometric observations of the tree species were recorded during February, 2016 from four plants of each tree species and with a total of 16 plants of each tree species were measured randomly. Height of trees was recorded with the help of Ravi's altimeter and collar girth (CG) and girth at breast height (GBH) was with the help of measuring tape. Biomass in the tree was estimated as per the formula suggested by Ahmed et al 2009 ($W = 0.25D^2H$) where, W =above ground weight of the tree in kg, D=Diameter of the trunk in cm (CD), and H=Height of the tree in meter. The root system weigh about 20 per cent, as much as the above ground weight of the tree. Therefore, the total green weight of the tree = W X 120%.The average tree is 72.5 per cent dry matter and 27.5 per cent moisture. Therefore, the total dry matter weight of the tree = W X 72.5%.

To study the carbon content in trees sample were collected from the different parts of the trees such as branches, leaves and barks. The carbon content in all collected plant samples were estimated by dry ashing method (Prasad et al 2010). The known quantity of oven dried sample (4.0 g) was placed in silica crucible and burnt in an electronic furnace

(GEI model-850) at 550°C for four hours. The ash content, inorganic elements in the form of oxide left after burning was weighted and carbon content was calculated by using following equation

Carbon content (%) = 100 – [Ash (%) + molecular weight of O_2 (53.3%) in $C_6H_{12}O_6$]

The weight of carbon dioxide sequestration in the tree was determined as

		Carbon content (%) X total DM (kg ha^{-1})
Carbon content (kg ha^{-1})	=	
		100

Soil organic carbon stock (SOC) was calculated as per the formula suggested by Batjes 1996. $Q_i = C_i D_i E_i \ (1 - G_i)$Where,

	Q	=	Soil Organic Carbon Stock (Mg m^{-3})
	E_i	=	Soil depth (m)
	C_i	=	Carbon content in soil (g C/g)
	D_i	=	Bulk Density (Mg m^{-3})
	G_i	=	Coarse fragments

RESULTS AND DISCUSSION

Growth Parameters of MPTS

Data on growth parameters (Table 1) revealed that among different tree species a significant variation existed in plant height at the end of 30 years of age.

Table 1: *Growth parameter of various MPTs under rainfed condition after thirty years of planting*

S.No.	Particular of treatment	Tree Height (m)	CD (cm)	DBH (cm) N-S	Tree Canopy (m)	
					E-W	
T_1	*A. indica*	9.29	20.08	17.63	5.06	4.80
T_2	*P. cineraria*	7.31	25.55	23.76	4.62	4.57
T_3	*P. juliflora*	8.02	29.16	19.55	6.94	6.78
T_4	*A. tortolis*	7.63	26.75	31.36	7.18	7.32
T_5	Control	-	-	-	-	-
	S.Em.(±)	0.30	1.65	1.14	0.48	0.45
	C.D. at 5 %	0.97	5.29	3.66	1.53	1.43
	C.V. (%)	7.53	13.04	9.92	16.04	15.19

Significantly the highest plant height was noted under *A. indica* (9.29 m) while the lowest plant height (7.31 m) was observed in *P. cineraria* whereas the highest CD (29.16 cm) was noticed in *P. juliflora* (T_3) and followed by *A. tortolis* (26.75 cm), *P. cineraria* (25.55 cm) and *A. indica* recorded the lowest CD (20.08 cm). The results are in agreement with those of Nandeshwar et al (2006) and Gill and Gupta (2005). Among the tree species significantly the highest DBH was noticed in *A. tortolis* (31.36 cm) and in *A. indica* recorded the lowest DBH (17.63 cm). The results are in agreement with those of Giri

Rao et al (2000) and Devevaranavadgi et al (2000). The canopy spread in N-S and E-W directions was significantly higher in *A. tortolis* (T_4) and followed by *P. juliflora* (T_3) *A. indica* and (T_1). Results are in conformity with the findings of Nandeshwar et al (2006) and Devevaranavadgi et al (2000).

BIOMASS PRODUCTION

The data presented in table 2 revealed that maximum above ground green (1822.91 kg tree^{-1}) and dry biomass (1321.61 kg tree^{-1}), below ground green (364.58 kg tree^{-1}) and dry biomass (264.32 kg tree^{-1}) was recorded under *P. juliflora* and followed by *A. tortolis. P. cineraria* and *A. indica.* Similar results were reported by Miria and Khan (2012).

Table 2: *Biomass allocation in above and below ground green biomass and dry biomass of MPTs after thirty years of plantation*

S.No.	Particular of treatment	AGGB (kg tree^{-1})	AGDB (kg tree^{-1})	BGGB (kg tree^{-1})	BGDB (kg tree^{-1})	Total dry biomass (t tree^{-1})
T_1	A. indica	1072.73	777.73	214.55	155.55	1036.97
T_2	P. cineraria	1208.02	875.82	241.60	175.16	1167.75
T_3	P. juliflora	1822.91	1321.61	364.58	264.32	1762.15
T_4	A.tortolis	1500.92	1088.16	300.18	217.63	1450.88
T_5	Control	-	-	-	-	-
	S.Em.(±)	188.33	136.54	37.67	27.31	182.06
	C.D. at 5 %	NS	NS	NS	NS	NS
	C.V. (%)	26.88	26.88	26.88	26.88	26.88

AGGB = Above ground green biomass, AGDB = Above ground dry biomass, BGGB = Below ground green biomass and BGDB =Below ground dry biomass

EFFECT OF DIFFERENT MPTS ON CO_2 SEQUESTRATION

The data pertaining to carbon content (percent) in tree indicated in table 3 showed significantly higher values in *P. juliflora* (41.55%). Whereas the *A. tortolis* (38.40%) have the lowest value as compare to other tree species. Total composite carbon content in tree was not influenced significantly due to different tree species (Table 3). However, the maximum total carbon content was recorded under *P. juliflora* (663.87 kg tree^{-1}) and minimum total carbon content was noted under *A. indica* (366.45 kg tree^{-1}) similar results were also reported by Dey (2005).

Table 3: *Carbon content and CO_2 sequestration in different tree species (30 years age)*

S. No.	Particular of treatment	Carbon content (%)	Total carbon content (kg tree^{-1})	CO_2 sequestration in tree (kg tree^{-1})
T_1	A. indica	39.24	366.45	1343.53
T_2	P. cineraria	38.91	408.54	1497.82

T_3	P. juliflora	41.55	663.87	1847.11
T_4	A. tortolis	38.40	500.25	1834.05
T_5	Control	-	-	-
	S.Em.(±)	0.20	68.46	223.22
	C.D. at 5 %	0.64	NS	NS
	C.V. (%)	1.02	28.25	27.38

The CO_2 sequestration of different MPTs did not exhibit significant difference under different trees species (Table 3). However the maximum value (1847.11 kg tree^{-1}) was noted under *P. juliflora* species and minimum (1343.53 kg tree^{-1}) was recorded under *A. indica* tree species. Similar results were also reported by Banerjee and Prakasam (2013).

SOIL ORGANIC CARBON STOCK (SOC)

A perusal of data presented in table 4 revealed that significantly higher SOC stock was recorded under *A. indica* at 0-30 cm soil depth (0.397 Mg m^{-3}). At 30-60 cm and 60-90 cm soil depth (0.574 and 0.669 Mg m^{-3}, respectively) higher SOC was recorded under *A. tortolis* tree species. Among the all treatments the lowest SOC stock (0.094, 0.133 and 0.250 Mg m^{-3}, respectively) at 0-30, 30-60 and 60-90 cm soil depth was recorded under open field. *A. tortolis* gave 242 per cent higher SOC over control (open field). The decreasing trend of SOC in different tree species was as *A. tortolis>P. juliflora>P. cineraria>A. indica*. Tree plays an important role in soil C sequestration. Tree helps in sequestering carbon in their biomass, soil and in wood products. Since a large land area is suitable for tree plantation, there is potential for large scale carbon sequestration (Subba Rao and Saha 2014). The similar results were also reported by Banerjee and Prakasam (2013)

Table 4: *Soil organic carbon stock (Mg m^{-3}) under different MPTs after thirty years of plantation at different soil depth*

S.No.	Particular of treatment	Soil organic carbon stock (Mg m^{-3})			Mean
		0-30 cm	30-60 cm	60-90 cm	
T_1	A. indica	0.397	0.331	0.448	0.392 (146.54)*
T_2	P. cineraria	0.300	0.341	0.625	0.422 (165.41)*
T_3	P. juliflora	0.263	0.556	0.503	0.441 (177.36)*
T_4	A. tortolis	0.392	0.574	0.669	0.544 (242.14)*
T_5	Control	0.094	0.133	0.250	0.159
	S Em(±)	0.02	0.02	0.03	-
	C.D. at 5 %	0.05	0.07	0.10	-
	C.V. (%)	11.72	11.71	12.69	-

* Data in parenthesis indicates per cent higher over control (T_5)

CONCLUSION

Among the tree growth parameters of 30 years of stand age, significant variation in plant height, CD, DBH and plant canopy has been observed. *A. tortolis* tree species stored 242.14 per cent higher organic carbon and SOC stock in soils over control. Maximum carbon content and CO_2 sequestration through tree was recorded under *P. juliflora*. Block based agroforestry system (trees in block) has the potential to sequester carbon from the atmosphere.

REFERENCES

Ahmed H, Yola S and Keles 2009 Changes in carbon and oxygen production in forest timber of Balci forest management unit in Turkey, *African Journal of Biotechnology* **8**(19) 4872-4883

Ahmedin AM, Bam S, Siraj KT, and Raju AJS 2013Assessment of biomass and carbon sequestration potentials of standing *Pongamia pinnata* in Andhra University, Visakhapatnam, India, *Bioscience Discovery* **4**(2) 143-148

Banerjee SK and Prakasam U 2013 Biomass carbon pool and soil organic carbon sequestration in *Tectona grandis* plantations, *Indian Forester* **139**(9) 797-802

Buvaneswaran C, Velumani R, and Jayaraj RSC 2012 Carbon Sequestration potential of *Tectona grandis L.* under agroforestry systems in Tamil Nadu, *Agroforestry for Mitigating Global Warming and Genetic Resource Conservation* 212-218

Devevaranavadgi SB, Guggari AK, Kalaghatagi SB, Wali SY, and Patil MB 2000 Performance of different tree species and their effect on physical and chemical properties of soil in northern dry zone of Karnataka, *Crop Research* **20**(3) 525-533

Dey SK 2005 A preliminary estimation of carbon stock sequestrated through rubber (*Hevea brasiliensis*) plantation in North Eastern Region of India, *Indian Forester* **131**(11) 1429-1436

Gill AS and Gupta SK 2005 Biomass production of trees under semi-arid rainfed agroforestry system, *Indian Forester* **131**(4) 591-594

Giri Rao LG, Joseph B, and Sreemannarayana B 2000 Growth and biomass production of some important multipurpose tree species on rainfed sandy loam soils, *Indian forester* **126**(7) 772-781

Jackson ML 1962 Soil Chemical Analysis Prentice Hall, Englowood Cliffs, NJ

Miria A and Khan AB 2012 Biomass estimation and carbon storage in selected multipurpose trees and common wild grasses, *Indian Forester* **138**(9) 853-856

Nandeshwar DL, Vijayaraghavan A, and Meshram M 2006 Performance of different multipurpose tree species in degraded land of Satpura region of Madhya Pradesh, *Indian Forester* **132(2)** 205-210

Prasad R, Saroj NK, Newaj R, Venkatesh A, and Dhyani CS 2010 Atmospheric carbon capturing potential of some agroforestry trees for mitigation of warming effect and climate change, *Indian Journal of Agroforestry* **12**(2)37-41.

Subba Rao A and Saha R 2014 Agroforestry for soil quality maintenance, climate change mitigation and ecosystem services, *Indian Farming* **63**(11) 26-29

CHAPTER 20

Biomass Production and CO_2 Mitigation Potential Under Different Land Use Systems

CP Rahangdale*, LD Koshta, Neelam Bisen[1], K Patle[2] and Rahul Dongre

Department of Forestry, J.N.K.V.V. Jabalpur, M.P., India

[1]Department of Agronomy, IAS, B.H.U. Varanasi, U.P., India

[2]Deputy Commissioner (Horticulture), Ministry of Agriculture, Govt. of India

**Corresponding author's Email: rchhatrapal87@gmail.com*

ABSTRACT

Carbon sequestration refers to the provision of long term storage of carbon in the terrestrial biosphere so that the buildup of carbon dioxide concentration in the atmosphere will be reduced or slowed in order to improve environmental conditions and check the processes of environmental degradation. In this concern, an experiment was carried out during June 2014 to April 2015 on farmers' field at Muskara village, Jabalpur, Madhya Pradesh, consisting of nine different land use systems in completely randomized block design with four replication. Results revealed that, the magnitude of total biomass production (20.0 t ha^{-1} yr^{-1}), carbon stock (10.99 t ha^{-1} yr^{-1}) and CO_2 mitigation potential (40.34 t ha^{-1} yr^{-1}) was significantly highest in eucalyptus + pigeonpea agrisilviculture practice as compared to other land use systems. However, the least total biomass production (9.42 t ha^{-1} yr^{-1}), carbon stock (4.91 t ha^{-1} yr^{-1}) and CO_2 mitigation potential (18.04 t ha^{-1} yr^{-1}) was found in moong-wheat conventional cropping land use. Agrisilviculture practice i.e., eucalyptus + pigeonpea seem to be better land use practice as compared to other land uses in term of expected value of carbon credit (₹ 26218 ha^{-1} yr^{-1}). From the results, it was concluded that, the integration of eucalyptus with pigeonpea under agrisilviculture system of agroforestry has more potential for mitigating the CO_2 from the atmosphere seems to be a best option for climate change mitigation

Keywords: Biomass production, Carbon stock, CO_2 mitigation, Eucalyptus clone

INTRODUCTION

Emission of greenhouse gases (GHG) has become a matter of great concern because of the future projection of the global warming and its related effects on biological life. Among the GHG, CO_2 is of major concern because it is accumulating in the atmosphere in its elemental form (C) at a very high rate (about 3.5 billion metric t annum^{-1}) as a consequence of mainly fossil and forest fuel combustion and tropical deforestation. Under such circumstances, agroforestry systems can be better climate change mitigation option than ocean and other terrestrial options because it is linked with the secondary environmental benefits such as food security and secured land tenure, increasing farm income, restoring and maintaining above and below ground biodiversity, maintaining watershed hydrology and soil conservation in addition to its high carbon mitigation potential which is primary environmental benefits.

Ever since the Kyoto protocol, agroforestry gained increased attention as a strategy to mitigate C in both developed and developing nations. The available estimates of C stored in agroforestry range from 0.29 to 15.21 Mg C ha^{-1} yr^{-1} aboveground and 30- 300 Mg C ha^{-1} upto 1.0 m depth in soil (Nair et al 2010). In one of the studies in Uttar Pradesh, 20 million tonnes of carbon has been mitigated by agroforestry plantations (Singh et al 2000). In small holding agroforestry systems in India, potential C mitigation ranges from 1.5 to 3.5 Mg C ha^{-1} (Dhyani et al 2009) having said this, it is also known that mitigation capability will certainly differ with the kind of land use, species, edaphic and climatic factors of the specific area. It is thus imperative, before framing region specific policy to identify location specific suitable land use systems which on one hand full fills the objective of carbon mitigation and on the other fit-in within the social-economic framework of society.

Eucalypts in India has been estimated to be about 107 million outside forest (FSI 2009). This genus has spread over most parts of the country primarily because of its versatile nature and amenability for harvesting on short rotations. Eucalypts and small timber played a vital role in meeting the requirement of pulp and plywood for industries and firewood needs of ever growing population. The productivity of eucalypts in India is about 10-15 m^3 ha^{-1} yr^{-1} (Venkatesh et al 1986). Technology based farm forestry plantations with genetically improved, high yielding and fast growing clonal planting stock of Eucalypts have tremendous potential for diversification of agriculture and meeting demand of industrial timber on sustainable basis. During the study six different agriculture crops viz., paddy, wheat, soybean, mustard, moong and pigeonpea were selected under different eucalyptus based agrisilviculture practices of agroforestry because little work has systematically been done so far on tree- crop interaction. The plenty of works on carbon mitigation/sequestration potential have been done in India and abroad both but all based on estimation rather than determination. Information on dynamics of CO_2 mitigation potential and expected value of carbon credit under conventional agriculture and alternate land use systems is still meagre. Therefore, keeping above facts in view, the present investigation was carried out.

METHODOLOGY

The study was carried out at the farmers' fields of Muskura village, Jabalpur District, Madhya Pradesh during June 2014 to April 2015. Geographically, the experimental site is located between 22° 49′ to 24° 8′ North Latitude and 78° 21′ to 80° 58′ East Longitude with an average altitude of 381.30 meters above the mean sea level. During the study total nine land uses systems were studied consist five conventional cropping viz., paddy-wheat (T_1), soybean-mustard (T_2), moong-wheat (T_3) in sequence, pigeonpea (T_4) and eucalyptus sole

(T_5) and four agrisilviculture practices viz., eucalyptus + paddy-wheat (T_6), eucalyptus + soybean-mustard (T_7), eucalyptus + moong-wheat (T_8) and eucalyptus + pigeonpea (T_9) in completely randomized block design with four replication. Plot size of 20 x 8 m for different land use systems were taken for study.

Aboveground biomass estimation of tree was carried out by harvesting one tree from each replication by using mean dbh of 40 trees in each tree based land use systems (total 5 trees in each tree based land use system were felled) during April 2015 (at the age of 4 year old). The trees were cut at ground level. Each harvested tree was partitioned into leaf, bark, twigs and bole without bark and fresh weight was recorded for each components and sample were collected. For determination of oven dry biomass, 500 g sample of each component were took in laboratory and kept in oven at 65 °C for 24 hrs and oven dry weight is recorded. The oven dry weight is used for determination of oven dry mass on hectare basis. The total aboveground tree biomass was worked out by summing of leaf, bark, twigs and bole biomass. For estimation of belowground biomass of tree, the non-destructive method were followed and calculated by multiplying aboveground biomass with a factor of 0.25 by using the guidelines of IPCC (1996). In crops, the biomass production was measured manually by harvesting the aboveground biomass by cutting at the ground level and below ground biomass by excavation method. Three random quadrates (1m x 1m) were laid out in each plot and harvested. Dry biomass is determined by drying the freshly harvested crops in hot air oven at 65 **°C**. Randomly collected plant sample of leaf, bark, twigs and bole of eucalyptus and shoot, grain and root sample of crops were dried and ground to pass through 0.2 mm sieve. The carbon content in plant biomass was determined by ash method. The oven dried sample of 5 gm for each part of the tree and crops were taken in pre-weighed crucible. The crucibles were placed in the muffle furnace adjusted at 400 **°C** for 2.30 hrs. The crucibles were cooled slowly inside the desiccators. After cooling the crucible with ash were weighed and percentage of organic carbon was calculated as formula given by Allen et al (1986). The carbon stock in dry biomass was worked out by multiplication of dry biomass with carbon content as per the formulae used by Rajput (2010). To determine carbon dioxide mitigation/sequestration by plants, the biomass carbon stock was multiplied with a factor of 3.67.

Carbon credit is a tradable currency showing that a government or company will pay for certain amount of carbon dioxide removed from the environment. One ton of net sequestered or mitigated carbon dioxide in the form of plant biomass is equal to one carbon credit. The expected earned value of carbon credit was calculated from the net sequestered or mitigated carbon dioxide by using the guidelines of IPCC (1996). (Standard Unit: 1 carbon credit = 1 t CO_2, Carbon credit value = ₹ 650 t^{-1} CO_2. Note: 1 Carbon credit @ US$ 10).

RESULTS AND DISCUSSION

Biomass Production

The aboveground, belowground and total biomass production was significantly influenced by different land use systems (Table 1). Among the nine land use systems, the aboveground biomass production was varied from 7.61 to 16.35 t ha^{-1} yr^{-1} and belowground biomass production was varied from 1.44 to 3.71 t ha^{-1} yr^{-1}. The magnitude of aboveground biomass production was found significantly highest (16.35 t ha^{-1} yr^{-1}) under eucalyptus + pigeonpea agrisilviculture land use, which was followed by the other land use systems. However, the significantly lowest aboveground biomass production was found under eucalyptus sole plantation (7.61 t ha^{-1} yr^{-1}). The belowground biomass production was found significantly highest under eucalyptus + pigeonpea agrisilviculture practice (3.71 t ha^{-1} yr^{-1}), which was followed by the other land use systems. However, the significantly lowest belowground biomass production was found under moong–wheat conventional cropping (0.81 t ha^{-1} yr^{-1}). A perusal of data in table 1, further showed that, the total biomass production (aboveground + belowground) by plant was significantly highest in eucalyptus + pigeonpea agrisilviculture (20.06 t ha^{-1} yr^{-1}) as compared to other land use systems, which was significantly superior to eucalyptus + paddy-wheat agrisilviculture (18.56 t ha^{-1} yr^{-1}), eucalyptus + moong-wheat agrisilviculture (16.80 t ha^{-1} yr^{-1}), eucalyptus + soybean-mustard agrisilviculture (14.61 t ha^{-1} yr^{-1}), paddy-wheat conventional cropping (11.77 t ha^{-1} yr^{-1}), soybean-mustard conventional cropping (10.39 t ha^{-1} yr^{-1}), pigeonpea conventional cropping (9.73 t ha^{-1} yr^{-1}), and eucalyptus sole plantation (9.51 t ha^{-1} yr^{-1}). However, the significantly lowest was observed under moong–wheat conventional cropping (9.42 t ha^{-1} yr^{-1}). The maximum biomass production under eucalyptus + pigeonpea agrisilviculture land use may be attributed due to complementary interaction between the eucalyptus and pigeonpea for sharing resources viz., light, water and nutrient and growth habit of pigeonpea. The biomass production in an agroforestry land use system depends on number of factors including the growth habit of species, site quality age, management practices and the interaction between tree and understory crops as also were reported by many researchers viz., Niu and Duiker (2006), Jena et al (2009), Kamine et al (2013) and Goswami (2014).

Carbon Stock

Similarly, the aboveground, belowground and total carbon stock significantly influenced by different land use systems (Table 1). Among the nine land use systems, the aboveground carbon stock was varied from 4.08 to 9.02 t ha^{-1} yr^{-1} and belowground carbon stock was varied from 0.38 to 1.97 t ha^{-1} yr^{-1}. The magnitude of aboveground carbon stock was found significantly highest (9.02 t ha^{-1} yr^{-1}) under eucalyptus + pigeonpea agrisilviculture land use, which was followed by the other land use systems. However, the significantly lowest aboveground carbon stock was found under moong-wheat conventional cropping (4.08 t ha^{-1} yr^{-1}). The belowground carbon stock was found significantly highest under eucalyptus + pigeonpea agrisilviculture practice (1.97 t ha^{-1} yr^{-1}), which was followed by the other land use systems. However, the significantly lowest belowground carbon stock was found under moong–wheat conventional cropping (0.38 t ha^{-1} yr^{-1}). Results further revealed that, the total carbon stock (aboveground + belowground) by plant was significantly highest in eucalyptus + pigeonpea agrisilviculture (10.99 t ha^{-1} yr^{-1}) as compared to other land use systems, which was significantly superior to eucalyptus + paddy-wheat agrisilviculture (10.27 t ha^{-1} yr^{-1}), eucalyptus + moong-wheat agrisilviculture (9.02 t ha^{-1} yr^{-1}), eucalyptus + soybean-mustard agrisilviculture (7.84 t ha^{-1} yr^{-1}), paddy-wheat conventional cropping (6.15 t ha^{-1} yr^{-1}), soybean-mustard conventional cropping land use (5.46 t ha^{-1} yr^{-1}), pigeonpea conventional cropping (5.28 t ha^{-1} yr^{-1}), eucalyptus sole plantation (5.10 t ha^{-1}

yr^{-1}). However, the significantly lowest was found under moong–wheat conventional cropping (4.91 t ha^{-1} yr^{-1}). It is deduced from the above discussion that, the variability in carbon stock under different land uses systems depends primarily on its complexity. Albrecht and Kandji (2003) have also reported that, the carbon variability in plant biomass can be high within complex system and productivity depends on several factors and the management of the system. Similar finding of carbon storage in plants were also observed by the Beer et al (1990), Kursten and Burschel 1993, Jensen (1993) and Kumar et al (2012).

CO_2 Mitigation Potential

The CO_2 mitigation potential in aboveground, belowground and total (above + belowground) significantly influenced by different land use systems (Table 1). Among the nine land use systems, the aboveground CO_2 mitigation potential was varied from 16.26 to 33.12 t ha^{-1} yr^{-1} and belowground CO_2 mitigation potential was varied from 1.38 to 7.22 t ha^{-1} yr^{-1}. The magnitude of aboveground CO_2 mitigation potential was found significantly highest (33.12 t ha^{-1} yr^{-1}) under eucalyptus + pigeonpea agrisilviculture practice, which was followed by the other land use systems. However, the significantly lowest aboveground CO_2 mitigation potential was found under moong–wheat conventional cropping (16.26 t ha^{-1} yr^{-1}). The belowground CO_2 mitigation potential was found significantly highest under eucalyptus + pigeonpea agrisilviculture practice (7.22 t ha^{-1} yr^{-1}), which was followed by the other land use systems. However, the significantly lowest belowground CO_2 sequestration was found under moong–wheat conventional cropping (1.38 t ha^{-1} yr^{-1}).

A perusal of data in table 1, further showed that, the total CO_2 mitigation potential (aboveground + belowground) by plant biomass was significantly highest in eucalyptus + pigeonpea agrisilviculture (40.34 t ha^{-1} yr^{-1}) as compared to other land use systems, which was significantly superior to eucalyptus + paddy-wheat agrisilviculture (37.69 t ha^{-1} yr^{-1}), eucalyptus + moong-wheat agrisilviculture (33.11 t ha^{-1} yr^{-1}), eucalyptus + soybean-mustard agrisilviculture (28.77 t ha^{-1} yr^{-1}), paddy-wheat conventional cropping (22.56 t ha^{-1} yr^{-1}), soybean-mustard conventional cropping (20.05 t ha^{-1} yr^{-1}), pigeonpea conventional cropping (19.38 t ha^{-1} yr^{-1}), eucalyptus sole plantation (18.71 t ha^{-1} yr^{-1}). However, the significantly lowest was observed under moong-wheat conventional cropping (18.04 t ha^{-1} yr^{-1}) land use system. The CO_2 mitigation potential by plant is directly related to biomass production of the different plant components. The highest CO_2 mitigation by eucalyptus + pigeonpea agrisilviculture land use can be attributed due to more biomass and more carbon stock level as compared to other land use systems. With adequate management of tree and incorporation of suitable agriculture crops under agroforestry systems as significant fraction of the atmospheric carbon could be captured and store in plant biomass. The results obtained are similar to as reported by Maikhuri et al (1999), Puri and Swami (2001), Montagini and Nair (2004) and Yadava (2011).

Table 1: *Biomass production potential, Carbon stock and CO_2 mitigation potential ($t\ ha^{-1}\ yr^{-1}$) under different land use systems.*

Land Use Systems	Biomass production ($t\ ha^{-1}\ yr^{-1}$)			Carbon stock ($t\ ha^{-1}\ yr^{-1}$)			CO_2 mitigation potential ($t\ ha^{-1}\ yr^{-1}$)		
	Above-ground	Below-ground	Total	Above-ground	Below-ground	Total	Above-ground	Below-ground	Total
T1-Paddy-wheat conventional cropping	10.85	0.92	11.77	5.72	0.43	6.15	20.98	1.58	22.56
T2-Soybean-mustard conventional cropping	8.95	1.44	10.39	4.74	0.72	5.46	17.40	2.65	20.05
T3-Moong-wheat conventional cropping	8.61	0.81	9.42	4.54	0.38	4.91	16.66	1.38	18.04
T4-Pigeonpea conventional cropping	8.08	1.66	9.73	4.43	0.85	5.28	16.26	3.13	19.38
T5-Eucalyptus sole plantation	7.61	1.90	9.51	4.08	1.02	5.10	14.97	3.74	18.71
T6-Eucalyptus+paddy-wheat agrisilviculture	15.61	2.94	18.56	8.74	1.53	10.27	32.05	5.64	37.69
T7-Eucalyptus + soybean-mustard agrisilviculture	11.96	2.66	14.61	6.44	1.40	7.84	23.62	5.15	28.77
T8-Eucalyptus + moong- wheat agrisilviculture	14.02	2.78	16.80	7.54	1.48	9.02	27.70	5.41	33.11
T9-Eucalyptus + pigeonpea agrisilviculture	16.35	3.71	20.06	9.02	1.97	10.99	33.12	7.22	40.34
SEm±	0.22	0.06	0.25	0.16	0.03	0.18	0.59	0.12	0.66
CD at 5%	0.65	0.18	0.74	0.47	0.10	0.52	1.72	0.35	1.93

Expected Value of C Credit

The carbon credit likely to be earned from the different land use systems (Fig. 1). The significantly highest value of C credit can be earned under T_9 i.e., eucalyptus + pigeonpea agrisilviculture (₹ 26, 218 $ha^{-1}\ yr^{-1}$) land use practice as compared to other land use systems, which was followed by T_6 i.e., eucalyptus + paddy-wheat agrisilviculture (₹ 24, 495 $ha^{-1}\ yr^{-1}$), T_8 i.e., eucalyptus + moong-wheat agrisilviculture (₹ 21,522 $ha^{-1}yr^{-1}$), T_7 i.e., eucalyptus + soybean-mustard agrisilviculture (₹ 18, 699 $ha^{-1}\ yr^{-1}$), T_1 i.e., paddy-wheat conventional cropping (₹ 14, 662 $ha^{-1}\ yr^{-1}$), T_2i.e., soybean-mustard conventional cropping (₹ 13,034 $ha^{-1}\ yr^{-1}$), T_4 i.e., pigeonpea sole conventional cropping (₹ 12, 599 $ha^{-1}\ yr^{-1}$), T_5 i.e., eucalyptus sole plantation (₹ 12,158 $ha^{-1}\ yr^{-1}$). However, the significantly lowest value of C credit may be earn under T_3 i.e., moong-wheat conventional cropping (₹ 11,724 $ha^{-1}\ yr^{-1}$) land use system. The maximum value of C credit under eucalyptus + pigeonpea agrisilviculture land use practice may be attributed due to higher biomass production and CO_2 sequestration potential (Rajput 2010 and Goswami 2014).

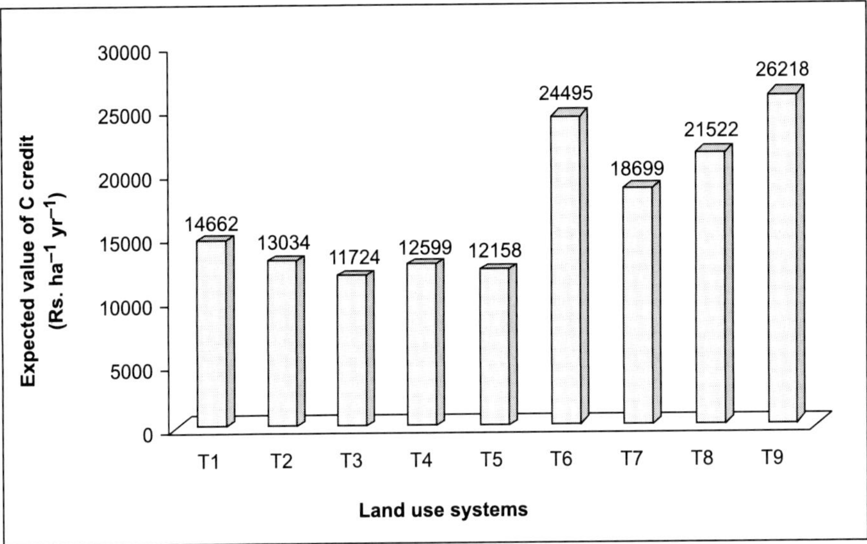

Figure 1: *Expected value of C credit (Rs. ha^{-1} yr^{-1}) under different land use systems*

CONCLUSION

The magnitude of total biomass production, carbon stock and CO_2 mitigation potential was significantly highest in eucalyptus + pigeonpea agrisilviculture practice as compared to other land use systems. However, the least total biomass production, carbon stock and CO_2 mitigation potential was found in moong-wheat conventional cropping land use. Agrisilviculture practice i.e., eucalyptus + pigeonpea seem to be best land use practice as compared to other land uses in term of expected value of carbon credit. Thus, the integration of eucalyptus with pigeonpea under agrisilviculture system of agroforestry has more potential for mitigating the CO_2 from the atmosphere and might be a best option for climate change mitigation.

REFERENCES

Albrecht A and Kandji ST 2003 Carbon sequestration in tropical agroforestry systems, *Agriculture, Ecosystems and Environment* **99** 15–27

Allen SE, Grimshaw HM, and Rowland AP 1986 Chemical Analysis, in *Method in Plant Ecology* (eds) PD. Moore and CHapman America: Blackwell Scientific Publications 285-344

Beer J, Bonnemann A, Chavez W, Fassbender HW, Imbach AC and Martel I 1990 Modelling Agroforestry systems of Cocoa (*Theobroma cocoa*) with laurel (*Cordia alliadora*) and poro (*Erythrina poeppigiana*) in Costa Rica V-Productivity indices, organic material models and sustainability over a period of ten years, *Agroforestry Systems* **12** 229-249

Dhyani SK, Newaj R, and Sharma AP 2009 Agroforestry: its relation with agronomy, challenges and opportunities, *Indian Journal of Agronomy* **54** 249–266

FSI2009 Indian State of Forest Report, 2009 Ministry of Environment, Forests and Climate Change. *Forest survey of India*, Government of India, Dehradun

Goswami S 2008 Appraisal of agroforestry land use systems for their carbon sequestration potential. *Ph.D. Thesis*, Dr.Y.S. Parmar University of Horticulture and forestry, Solan, H.P. India.134 p.

Goswami S, Verma KS and Kaushal R 2014 Biomass and carbon sequestration in different agroforestry systems of a Western Himalayan watershed, *Biological Agriculture & Horticulture* **30**(2) 88–96

IPCC 1996 Landuse Change and Forestry, In: Revised 1996 IPCC guidelines for National Greenhouse Gas Inventories, Cambridge University Press, Cambridge, U.K.

Jana BK, Biswas S, Majumder M, Roy PK, and Mazumdar A 2009 Carbon sequestration rate and aboveground biomass carbon potential of four young species, *Journal of Ecology and The Natural Environment* 115–24

Jenson M 1993 Productivity of Nutrient cycling of Javanese home gardens, *Agroforestry Systems* **24** 187-201

Kanime N, Kaushal R, Tewari SK, Raverkar KP, Chaturvedi S and Chaturvedi OP 2011 Biomass production and carbon sequestration in different tree based systems of central Himalayan Tarai region, *Forests, Trees and Livelihoods* **22** (1)38-50

Kumar S, Verma KS, and Kumar, N 2012 Aboveground Carbon Sequestration Potential of Agroforestry Systems of Western Himalaya, *Journal of Tree Sciences* **31** (1&2) 20-28

Kurston E and Burshel P 1993 CO_2 mitigation by agroforestry, *Water air and Soil Pollution* **70** (14) 533-544

Maikhuri RK, Rao KS, Saxena KG, and Semwal RL 1999 Traditional crop diversity based nutrition and prospects for sustainable development in central Himalaya, *Himalayan Paryavaran* 636-44

Montagnini F and Nair PKR 2004 Carbon sequestration: an under exploited environmental benefit of agroforestry systems, *Agroforestry Systems* **61** 281-295

Nair PKR, Nair VD, Kumar BM and Showalter JM 2010 Carbon sequestration in agroforestry systems, *Advances in Agronomy* **108** 237-307

Niu X and Duiker SW 2006 Carbon sequestration potential by afforestation of marginal agricultural land in the Midwestern US, *Forest Ecology and Management* **223** 415–427

Puri S and Swami SL 2001 Growth and biomass production in *Azadirachta indica* seedling in response to nutrient and moisture stress, *Agroforestry Systems* **51** 57-68

Rajput BS 2010 Bio-economic appraisal and carbon sequestration potential of different land use systems in temperate north-western Himalayas, Ph.D. Thesis, Dr Y.S. Parmar University of Horticulture and Forestry, Nauni, Solan (H.P.). India.

Singh TP, Vara Lakshmi V, and Ahluwalia SK 2000 Carbon sequestration through farm forestry: case from India, *Indian Forester* **25** 1257-1264

Venkatesan KR, Kumarvelu G. and Somasundaram K 1986 Tree improvement of Eucalyptus species in Tamilnadu, in *Eucalyptus in India: past present and future* (eds) Sharma JK, Nayar TS, Kedharnath S, Kondas SS. Kerala Forest Research Institute, Peechi pages 290–296

Yadava AK 2011 Potential of agroforestry systems in carbon sequestration for mitigating climate changes in Tarai region of central Himalaya, *Nature and Science* 972-80

CHAPTER 21

Improved Livelihood and Eco-System Services Through Agroforestry Based Watershed Interventions in Bundelkhand Region of Central India

Ramesh Singh*, Inder Dev, RK Tewari, RH Rizvi, Kaushal K Garg[1], Anand Kumar Singh, RP Dwivedi, KB Sridhar, Mahendra Singh, Dhiraj Kumar, PK Sarkar[2] and OP Chaturvedi

ICAR-Central Agroforestry Research Institute, Jhansi, U.P., India

[1]International Crop Research Institute for Semi Arid Tropics, Patancheru, Hyderabad, India

[2]ICAR Research Complex, Patna, Bihar, India

**Corresponding authors's Email: rameshsinghnrcaf@gmail.com*

ABSTRACT

The Bundelkhand region of central India is prone to drought and is the hotspot of water scarcity, land degradation, poor productivity, poor permanent vegetal cover and miserable socio-economic status. Natural resource management through watershed development interventions with major emphasis on agroforestry appears an effective tool to make the region drought resilient. To improve the natural resources in Parasai-Sindh watershed (1246 ha) about 1,25,000 m^3 rainwater harvesting facility was created through different rainwater harvesting structures including earthen bunding. Besides, 114 ha brought under various agroforestry interventions with majority under boundary plantation of teak. To improve the crop productivity, 265 participatory demonstrations were conducted with good package of practices. About 1,37,000 rooted slips of Napier bajra hybrid, guinea grass and TSH were transplanted on bunds, near checkdams and around ponds during 2013-14 & 2014-15. Ensuing study is an attempt to analyze the impact of agroforestry based watershed interventions on livelihood and eco-system services. There was no problem of drinking and irrigation water in spite of 30-35 per cent deficit in rainfall. Productivity of various crops has been enhanced by 20-70 per cent. Area under fodder cultivation has been increased by more than seven folds. Net groundwater recharge during monsoon season for pre and post interventions scenario was estimated at 53.7 and 92.3 mm, respectively, which is 71.8 per cent higher than pre intervention phase. Improved yield of open wells reduced *rabi* fallow by about 70 ha besides, ensuring irrigation water through dug wells to rabi crops predominantly wheat in entire treated area of the watershed. The study has proved that agroforestry based watershed interventions are the effective measures to change drought prone region into climate resilient region.

Keywords: Agroforestry interventions, crop productivity, groundwater recharge, *haveli* cultivation, rainwater harvesting structures

INTRODUCTION

Bundelkhand region (23° 8′- 26° 31′ N, 78° 11′-81° 30′ E) comprising 7 districts of UP (Jhansi, Jalaun, Lalitpur, Hamirpur, Mahoba, Banda and Chitrakoot) and 6 districts of MP (Sagar, Tikamgarh, Chhatarpur, Panna, Damoh and Datia) is spread over 7.16 million ha in Central India. The region is largely fed by rivers originating from central India. River Yamuna forms north-eastern boundary. Two major rivers that drain the region into Yamuna are Betwa and Ken. River Ken originates in Kaimour hills near Jabalpur (MP) and travels north east about 427 km to join Yamuna in Banda district (UP), most of its journey is through hills except a small part in plains of Banda. River Betwa originates in Vindhyan range near Hoshangabad and joins Yamuna in Hamirpur distt (UP). It travels about 610 km. About half of its course which is non navigable runs over Malwa plateaux before entering into Bundelkhand uplands. Dhasan and Jamni are main tributaries of Betwa. Thus it is clear that uplands of Bundelkhand do not get water from ice fed rivers of Himalaya and they solely depend upon rainfall received in central India. The region receives rainfall between 750-1200 mm in different parts and river flow is almost seasonal.

The region has experienced severe drought during 2004 to 2007 and 2014 to 2015. More than 80 per cent open wells dried up soon after monsoon season due to deficit rainfall and poor ground water recharge. During both the period, in the absence of drinking water and livelihood opportunities, a large proportion of rural population migrated to metros in search of livelihood. Cattles were abandoned due to water shortage and less fodder availability. In such situation, natural resource management through watershed development interventions with major emphasis on agroforestry appears an effective tool to make the region drought resilient. The ICAR-Central Agroforestry Research Institute, Jhansi (CAFRI) and ICRISAT Hyderabad in association with farming community identified Parasai-Sindh watershed in Babina block of Jhansi district during 2011 as a pilot site for improving groundwater recharge, water use efficiency, permanent vegetal cover through agroforestry interventions and strengthening eco-system services.

The general characteristics, land use and resources of watershed are presented in Table 1. The watershed is located about 30 km West of the district headquarter. The project comprises three villages namely Parasai, Chhatpur and Bachauni (Fig. 1).

METHODOLOGY

Experimental Site

The Parasai-Sindh watershed as part of Sindh river catchment is located at 25° 23′ 47.6″ - 25° 27′ 05.1″ N and 78° 20′ 06.5″ - 78° 22′ 33.0″ E, and about 270-315 m above mean sea level in Jhansi district of Uttar Pradesh (Fig. 1). The geographical area of watershed is

1246 ha, and has independent hydrology. The rainfall is highly erratic and uncertain. Long term weather data shows that the average rainfall in study region is 877 mm (standard deviation, σ = 251 mm), about 85% falling from June to September (Singh et al 2014). The watershed experiences semi-arid sub-tropical climate and characterized by dry and hot summer, warm and moist rainy season and cool winter with occasional rain showers. Mean annual temperature ranges from 24 to 25 °C. The mean summer (April-May-June) temperature is 34 °C which may rise to a maximum of 46 to 49 °C during the month of May and June. The mean winter temperature (December-January-February) is 16 °C. Further, the diurnal variation in temperature is quite high.

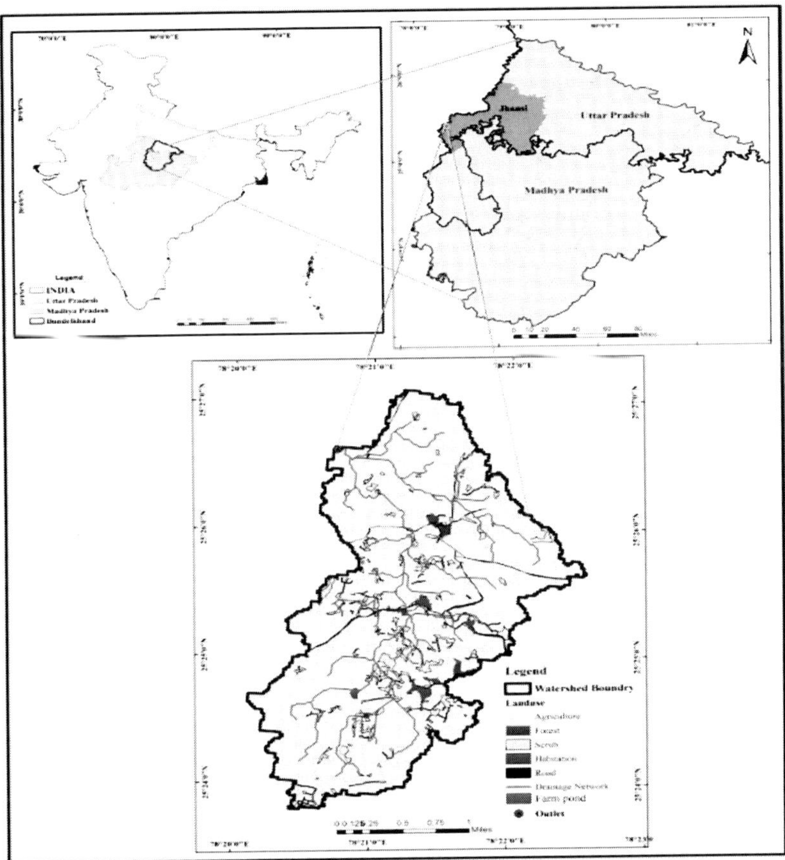

Figure 1: *Location map of study Parasai – Sindh watershed*

INTERVENTIONS IN THE WATERSHED

To augment water availability and groundwater recharge, several *in-situ* and *ex-situ* interventions were taken up. Till now, three *nallah* plugs (small checkdam), nine checkdams, one haveli (traditional rainwater harvesting structure), one community pond and one farm pond were constructed. To measure runoff and soil loss, twelve gauging stations were also constructed. In totality, about one lakh twenty five thousand cubic meter surface water storage has been created in the watershed. To improve *in-situ* moisture condition, field / contour bunding was done in 50 ha. Farmers were advised to cultivate across the slope. Besides, about 114 ha brought under various agroforestry interventions with majority under boundary plantation of teak. To improve the crop productivity, 265 participatory demonstrations were conducted with good package of practices. About 1, 37,000 rooted slips of Napier bajra hybrid, guinea grass and TSH were transplanted on bunds, near checkdams and around ponds during 2013-14 & 2014-15. On an average biomass yield of 7.2 DMY t ha^{-1} could be obtained from the mixed pasture. Several exposure visits and trainings were also organized to improve the skill of farming community and landless.

Rejuvenation of Traditional Rainwater Harvesting System

In this watershed, *Haveli* that was historically constructed at very upstream site was defunct due to the damaged outlet and washing out of nearly 50 m of the embankment. Mr. Kalyan Singh, a resident of Parasai village shared that this structure is nearly 300 years old. Masonry drop spillway (rectangular weir) and embankment along with core wall was constructed in 50 m breached area. Water harvested in *Haveli* during the monsoon period. The scientific team of ICAR-CAFRI along with the local community conducted the inspection of the structure and decided to rejuvenate the traditional system of rainwater harvesting. A drop spillway (rectangular weir) outlet was constructed to drain excess runoff during the rainy season. The weir was constructed at a height of 1.45 m from the bed level. The earthen embankment along with core wall was constructed in 50 m breached area. To control seepage, 147 m stone masonry wall was constructed along the embankment of *Haveli*. Submergence area of the structure is about 8.0 ha with harvesting capacity of 73000 cum. Due to the rejuvenation of *Haveli,* nearly 1-1.5 lakh cum. runoff are now being harvested every monsoon season that completely resolved the water scarcity issues of the villages.

RESULTS AND DISCUSSION

Impact of IWD Interventions on Provisioning Services

Water Availability

Rainwater harvesting capacity of about 1,25,000 cum. were created through different rainwater harvesting structures along with contour bunding and it actually harvested 2 to 3 times of its created capacity depending on annual rainfall. Groundwater recharge has been doubled. This ensured water availability for irrigation and drinking even in situation of 30-35 per cent deficit rainfall. During 2017, sufficient water is available to cater needs of drinking and irrigation in spite of 36 percent deficit rainfall.

Crop Productivity

To enhance the productivity, 265 participatory crop demonstrations with balance fertilization were conducted. New varieties of wheat, barley and soybean were introduced. Efforts resulted into improved productivity of different crop during both the season in the range of 16 to 70 per cent. Farmers are switched over to recommended seed rate of wheat.

Before interventions it was double the recommendation. Crop demonstrations and per cent age increase in yield over local variety were depicted in Fig. 2.

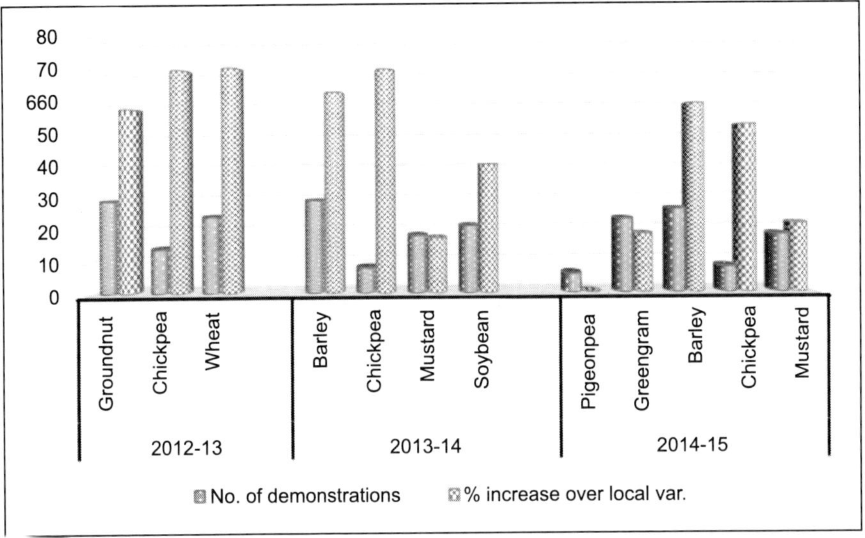

Figure 2: *Crop demonstrations and percentage Increase in crop yields (over local var.)*

IMPACT OF IWD INTERVENTIONS ON REGULATING SERVICES

Groundwater Recharge

The net groundwater recharge during monsoon was 3.54 and 5.98 m during pre and post-interventions, respectively. Considering 1per cent specific yield for the region, recharge of 35.4 and 59.8 mm rainfall is required to get respective water column during pre and post scenario. During monsoon season 2011, life saving irrigation was given in 828 ha. However, watering was done in 979 ha (18.2 per cent higher than pre intervention scenario) during 2013 due to enhanced availability of water in the watershed. During estimation it was assumed that irrigation was done at the rate of 70 per cent of required amount during pre-interventions, however, full amount of required irrigation was supplied during post scenario. Total groundwater recharge including irrigation, evaporation and water table buildup during pre and post interventions was estimated as 53.7 and 92.3 mm, respectively. It was 6.12 and 10.52 per cent of long term average of annual rainfall (877 mm) during respective scenario (Kumari et al 2014).

Base Flow

During pre-interventions only 1 to 1.5 percent of annual rainfall was partitioned as base flow, which disappears by end of September. Now, it has increased up to 5 to 6 percent of annual rainfall and it was throughout the year in case of normal rainfall year.

Impact of Trees on Temperature and Humidity

It was revealed from the study that, there was a significant difference in temperature and humidity from the outside extremes to the tree bases. For that, three different tree species were chosen based on crown density. Tree like Babool (*Acacia nilotica*) as light crown density, Neem (*Azadirachta indica*) as moderate crown density and Mahua (*Madhuca integrifolia*) as dense crown density tree were selected. The temperature reduced maximum at the tree bases in all the selected trees. It was found that temperature got reduced to 37.6 °C (Fig. 3) under Neem tree base from outside extreme (39.5 °C) but humidity got increased from 21.4 percent (outside) to 33.4 per cent (tree base), for Babool tree, temperature was recorded lower under tree as 38.8 °C and maximum was at outside extreme (40.2) but humidity was recorded higher at tree base (28%) and lower at outside extreme (22 percent) and for Mahua tree, temperature was recorded low as (38.1 °C) as compared to outside extreme (40.4°C) and humidity was measured as high at tree base (35 percent) as compared to outside extreme (22%).

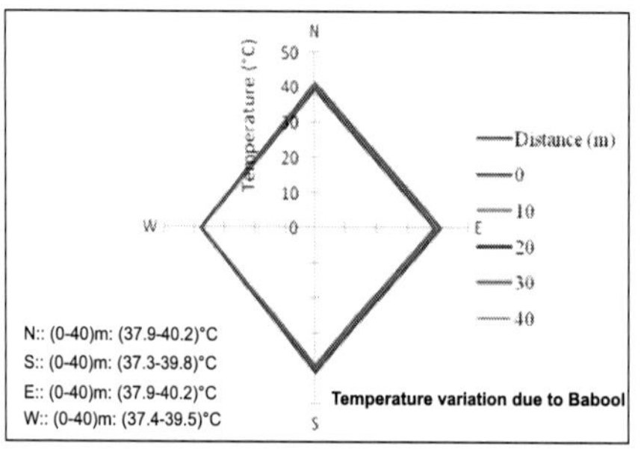

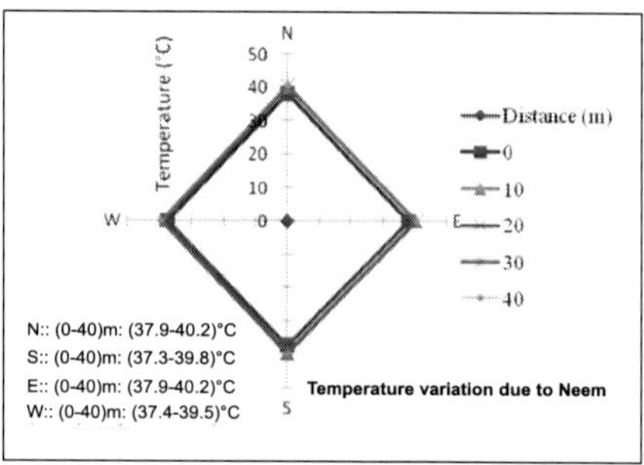

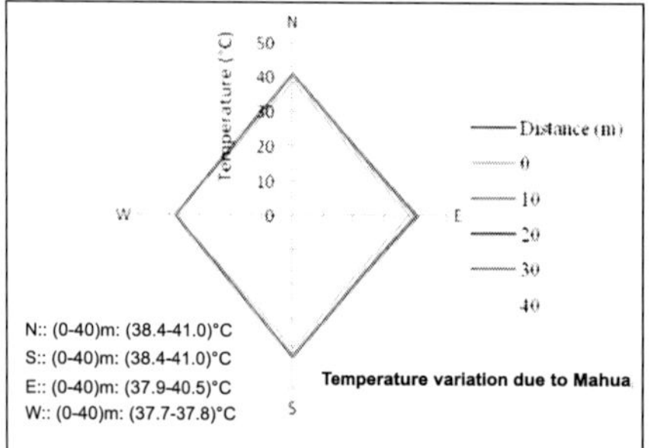

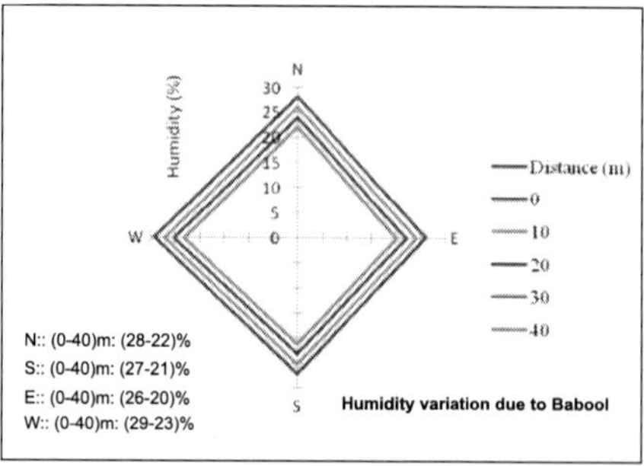

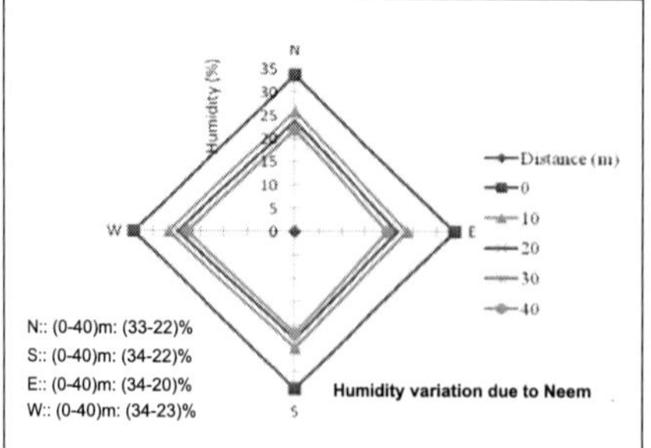

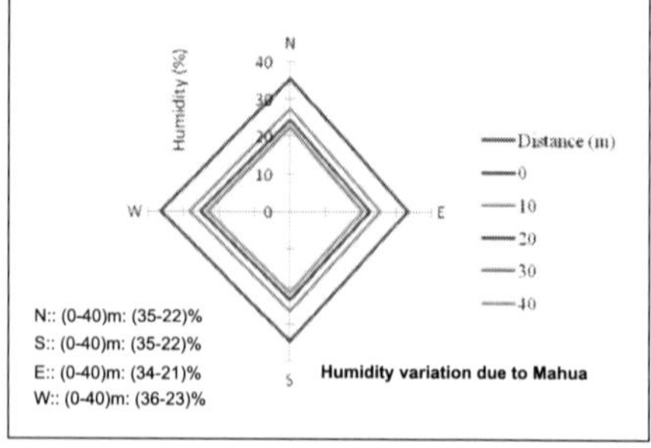

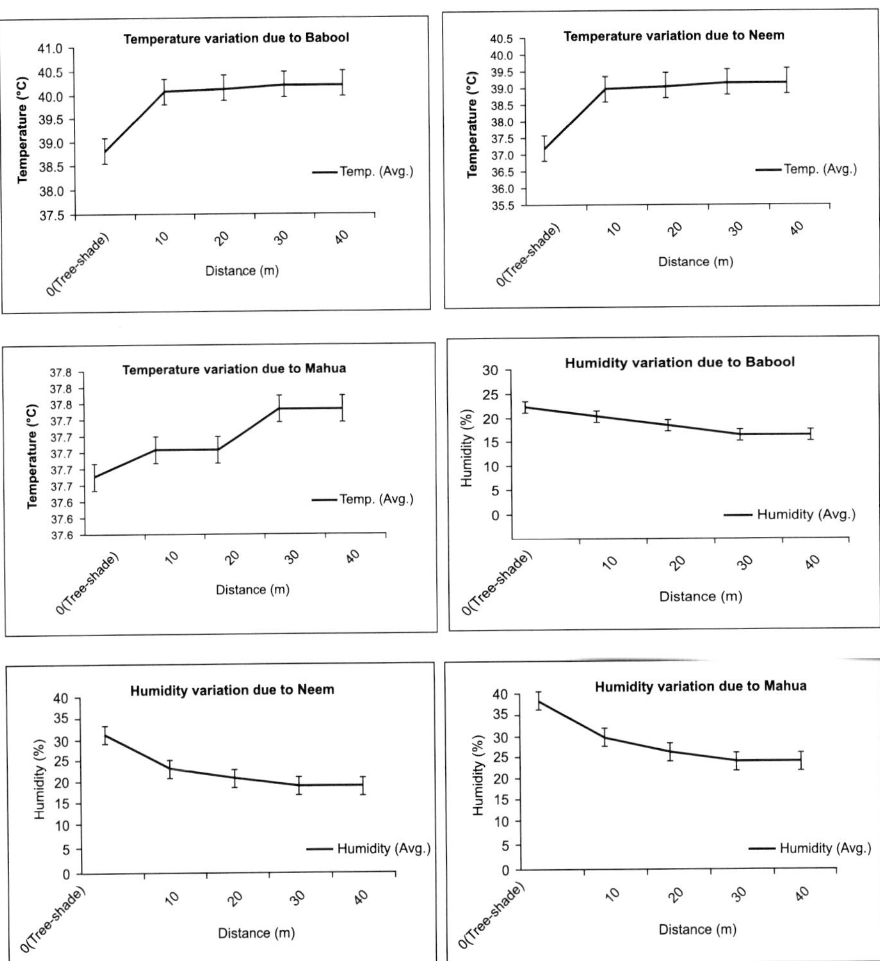

Figure 3: *Effect of trees on temperature and along and across the distances from tree base*

Impact of IWD Interventions on Supporting Services

Reduction in Soil Erosion

There is significant reduction in soil loss from the watershed after watershed interventions. It has been reduced by 60 to 75 per cent as compared to pre watershed interventions. In turn, it has improved the life of water reservoirs and enhanced the crop productivity.

Reduction of Drudgery and Cost of Cultivation

In Irrigating Wheat Crops

Before interventions, open wells were hardly supported for 1-2 hrs during rabi season due to low water column. Again, the pumping was done next day after well got recharged. Therefore, farmers generally took 10 to 15 days to irrigate one ha wheat crop. Mainly women and children were engaged in irrigation to field crops and generally they spent 40-50 hrs to complete irrigation in one ha. Now, majority of the wells are supporting round the clock and they can complete irrigation within a day (15-20 hrs). Now, the farmers, especially women have more time to care their children and other households' activities. It also reduces cost of cultivation by Rs. 6000.00 to 8000.00 per ha on account of reduced labour requirement for irrigation.

CASE STUDY OF PARASAI SINDH WATERSHED

Agroforestry Based Diversified Farming Brought Prosperity

Consequent upon water storage in *haweli*, the open well of Shri Rajveer Singh, marginal farmer of village Parasai, block-Babina, District-Jhansi which is located one km away from poundage area and lies parallel to pond, showed its reflection. There are 5 adults and 4 children in his family. He owns 5 acre land at a place. He had constructed one hut and keeps 2 cows and 1 buffalo at his holding. Farming and dairying are only means of his livelihood. He had one well in his holding.

Rejuvanation of tradition rainwater harvesting system (*haveli cultivation*) in Bundelkhand region	Rainwater harvesting at ephemeral drains in Parasai-Sindh watershed alongwith gauging station
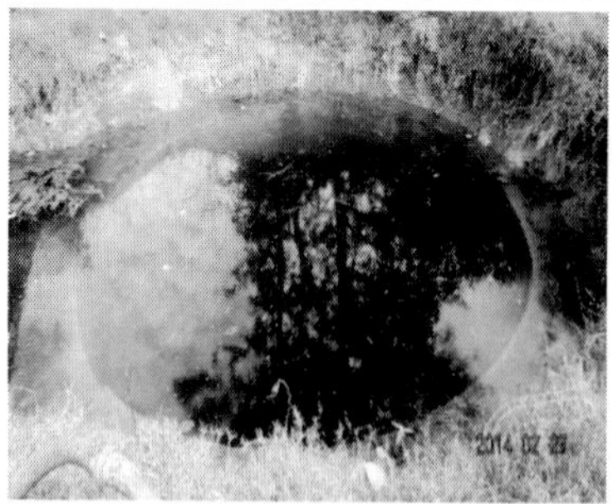	
Recharged open well due to rejuvanation of *haveli* and checkdams at *nallah*	Cultivation of fodder crops during summer and teak on field bund

Gradually water level in his well went down. Due to slow recharge in well, he could hardly earn his livelihood. Therefore, he decided to dig another well in his holding. He contacted local experts who predict availability of ground water at given location. He invested his all saving and further borrowed money from relatives and well-wishers to dig the well in 2010 at a cost of ₹ 3.0 lakh. He was happy as there was plenty of water in his 2nd well. With his hard work he was trying his best to repay loan. Meanwhile, the first well (old one) with rejuvenation of *haweli*, also had enough water to support 12 hours water lifting with 5 hp diesel engine in the month of October i.e. when farmers were giving pre-sowing irrigation to wheat. The second one, newly constructed well was full to its brim and the farmer did not feel need to operate it. Even in the month of April after wheat harvest, his first well was yielding adequate water. He was not using 2nd well as first well was meeting his requirement. He was lamenting that ₹ 3.0 lakh expenditure on digging 2nd well has truly gone waste.

Before, rejuvenation of *haweli*, Shri Rajveer was cultivating groundnut in 4.0 acre and blackgram in 1.0 acre of land during *kharif*, however during *rabi* he sowed 2.5 acre wheat and kept remaining fifty per cent area fallow due to lack of irrigation water. He was also selling 8 liter milk on daily basis. The net income was ₹ 49923.00 per year from all sources before rejuvenation of traditional rainwater harvesting. After rejuvenation of *haweli*, his both the open wells were yielding sufficient water to take *rabi* crops in whole areas. He diversified crops (groundnut, black gram, wheat, barley, vegetables and fodder) and also invested ₹ 3.0 lakh to purchase 6 buffaloes during 2014 and started cultivating fodder and vegetable crops during summer. The productivity of various crops had gone up by 21 to 35 per cent and he is earning regularly Rs. 900 per day from dairying enterprise. Now, his annual net income has gone up from ₹ 50000.00 to ₹ 283000.00 from agriculture and livestock.

Shri Rajveer Singh has planted 28 guava and 60 teak along field bunds as long term investment for future. He also initiated lac cultivation on his farm on *Butea monosperma* trees. Five *ber* plants on his field were top worked with *Banarasi Karaka* variety in 2013 and they are producing fruits. So for, He has not marketed them but considers them as priced possession. He obliges farm visitors by offering quality *ber* fruits. After seeing the quality *ber* fruits, he top worked his all 95 desi *ber* trees during 2015 rainy season. His success is an example to other farmers of similar land holding and status. Now, several families are following his path for prosperity, risk proofing and sustainability. Similar efforts in other parts of the region if adopted can ensure sustainable development even in changing climate scenario.

CONCLUSION

The livelihood can be ensured in Bundelkhand region by improving different eco-system services through integrated watershed development interventions with major emphasis on woody perennials.

ACKNOWLEDGEMENT

The help rendered by team members and stakeholders in implementation of the programme is gratefully acknowledged.

REFERENCES

Dhyani SK, Samra JS, Ajit, Handa AK and Uma 2007 Forestry to support increased agricultural production: focus on employment generation and rural development, *Agricultural Economics Research Review* **26** (2) 179–202

Kumari Reena, Singh Ramesh, Singh, RM, Tewari RK, Dhyani SK, Dev Inder, Sharma Babloo, and Singh AK 2014 Impact of rainwater harvesting structures on water table behavior and groundwater recharge in Parasai-Sindh watershed of Central India, *Indian Journal of Agroforestry* **16**(2) 47-52

Singh Ramesh, Grag KK, Wani SP, Tewari RK, and Dhyani SK 2014 Impact of water management interventions on hydrology and ecosystem services in Garhkundar- Dabar watershed of Bundelkhand region, Central India, *Journal of Hydrology* **509** 132-149

CHAPTER 22

Groundwater Drought Study of Bearma Basin in Bundelkhand Region Using Groundwater Drought Index

Dinesh Kumar*, T Thomas[1] and RM Singh

Department of Farm Engineering, Institute of Agricultural Sciences, Banaras Hindu University, Varanasi-221005, Uttar Pradesh, India

[1]National Institute of Hydrology, Ganga Plains South Regional Centre, WALMI CAMPUS, Bhopal-462016, Madhya Pradesh, India

**Corresponding author's Email: dineshcae10@gmail.com*

ABSTRACT

Drought is a natural phenomenon characterized by a significant decrease of water availability during crucial ~~a~~ significant period of time. In recent years which caused severe problems for most of areas in Bundelkhand region in Central India. Bearma basin is one of the considerable groundwater resource in South portion of Bundelkhand in Madhya Pradesh, which is Sub-basin of Ken River. The 26 years (1984-2010) quarterly groundwater levels of 76 observation wells falling in and around of Bearma basin has been used. The groundwater drought characteristics in Bearma basin have been studied based on the drought severity classification of Groundwater Drought Index (GDI) for the selected 76 observation wells in the basin falling in Sagar and Damoh districts during various months of the identified drought years 2002-03 and 2007-08. The study reveals that each groundwater drought event is unique in its characteristics and has varying degree of influence in different zones of the basin. The groundwater drought distribution and pattern is completely different in two of the meteorological drought years of 2002-03 and 2007-08 with the southern portion of the basin facing worse drought scenario during 2002-03 whereas in 2007-08 the northern portions of the basin faced more severe drought conditions as compared to southern part.

Keywords: Bearma basin, Bundelkhand, Damoh and Sagar, Groundwater drought, Groundwater Drought Index

INTRODUCTION

A drought is an extended period when a region notes a deficiency in its water supply (Beran and Rodier 1985). Generally, this occurs when a region receives consistently below average precipitation. Drought and its consequences have long plagued ecosystems and society (Le Roy Ladurie 1971). The consequences of drought vary greatly depending on its location, timing, extent and the type of society or social sector impacted by the drought (Gleick 1993). The different types of droughts have their own specific spatio-temporal characteristics (Peters et al 2006, Tallaksen et al 2009). Different types of drought are meteorological, hydrological, agricultural and socio-economic (Hisdal and Tallaksen 2000). Meteorological drought simply refers to the atmospheric conditions that result in the absence or reduction of precipitation. Hydrological drought is defined as a significant reduction in the availability of water in all its forms appearing in the land phase of the hydrological cycle which includes stream flow, groundwater and lake and reservoir storages (Nalbantis 2009). The surface water and groundwater droughts will occur more or less simultaneously if the surface water is mainly fed by groundwater. The lag between a meteorological drought and a hydrological drought may amount to months or even years. Also the groundwater storage also recovers slowly, which implies that the effects of a groundwater drought may be felt long even after the meteorological drought has ended. Because of the slow reactions of rainfall on groundwater levels, only the major meteorological droughts finally appear as groundwater drought. (Calow et al 1999) states that, groundwater drought describes a situation where groundwater sources fail as a direct consequence of drought. This definition of a groundwater drought apparently also incorporates human demand of water. (van Lanen and Peters 2000) defined that a groundwater drought occurs if the groundwater heads in an aquifer have fallen below a critical level over a certain period of time, which results in adverse effects. It is the natural decline in the ground water levels that may result in dewatering of the aquifer completely or partly, or to a point where it could cause serious water supply problems. In this study, an attempt has been made to evaluate the groundwater drought characteristics by developing Groundwater Drought Index (GDI) for Bearma basin in Bundelkhand region. The Bearma basin has been selected as a pilot basin and the study has been carried out to mainly characterize the temporal variation of groundwater drought. The groundwater drought characteristics evaluated using the percentile approach or an appropriate drought index.

MATERIALS AND METHODS

Study Area

The river Bearma is one of the important tributaries of Ken river passing through the heartland of Bundelkhand in the State of Madhya Pradesh and is located between north latitudes of 23° 07′ and 24° 18′ and east longitudes of 78° 54′ and 80° 00′. The Bearma basin is a leaf shaped elongated basin with a total catchment area of 5890 km^2. The average rainfall for the basin is about 1186.85 mm. The climatic conditions prevailing in the basin vary from semi-arid to dry. The five rain gauge stations at *Deori, Rehli, Damoh, Jabera* and *Hatta* represent the rainfall pattern in the basin.

Hydrological Data

To investigate the groundwater drought, there are 26 years (1984-2010) quarterly ground water levels of 76 observation wells falling in Sagar and Damoh districts have been used for identifying the groundwater drought characteristics in Bearma basin. The ground water levels are being monitored by the State Ground Water Survey, Govt. of Madhya Pradesh.

Groundwater Levels

The reduced level of the ground (RL_G) as well as the height of measuring point (H_{MP}) from where the measurements for the groundwater levels are always carried out have also been collected for each observation well along with the time series of groundwater levels in all four quarters of the water year. The reduced level of the groundwater table has been computed from the equation as given below:

$$RL_{GWL} = R_{LG} + H_{MP} - D_{GWL} \qquad \text{...(1)}$$

Where, RL_{GWL} = Reduced level of groundwater level (m)

RL_G = Reduced level of ground (m)

H_{MP} = Height of measuring point above ground level (m)

D_{GWL} = Depth of groundwater level below measuring point (m)

After obtaining the time series of reduced level of groundwater levels, it was compared with the time series of adjacent locations for finding out the outliers, if any. The processed data have been subsequently used in the development of the groundwater drought index and evaluation of groundwater drought characteristics.

Groundwater Drought Index (GDI)

The most well-known methods used in groundwater drought analysis from ground water level data are the threshold level approach and the Sequent Peak Algorithm (Tallaksen and van Lanen 2004). The GDI is computed by normalizing quarterly/seasonal groundwater levels and dividing the difference between the quarterly/seasonal water level and its long-term seasonal mean by its standard deviation. For normalization, an incomplete gamma function was used for water level data before using them for calculating GDI. The GDI is an indicator of water-table decline and an indirect measure of recharge, and thus an indirect reference to drought. The GDI is computed as per the following equation given below:

$$GDI = \left\{ \frac{GWL_{ij} - GWL_{im}}{\sigma} \right\} \qquad \text{...(2)}$$

Where, GWL_{ij} = seasonal water level for the ith well and j^{th} observation,

GWL_{im} = seasonal mean,

σ = is the standard deviation.

Groundwater Drought Characteristics

The classification used for identifying the groundwater drought characteristics based on GDI (van Lanen and Peters 2000) is given in Table 1.

Table 1: *Standard ranges of GDI values and their classification*

S.No.	GDI range	Classification
1.	0.0 to –0.99	Mild drought
2.	–1.0 to –1.49	Moderate drought
3.	–1.5 to –1.99	Severe drought
4.	–2.0 ≤	Extreme drought

RESULTS AND DISCUSSION

Groundwater Drought Charactersitics in Sagar District

The groundwater drought characteristics evalauted at the 17 observation wells in Sagar district given in Table 2. The maximum number of 10 drought events occurred at *Kesli* with a total severity of -26.82 with a average drought intensity of -0.31 followed by 9 drought events at *Copra* observation well having an average drought intensity of -0.34. The cumulative groundwater drought duration of 130 months has been observed at *Chhula* with an average drought intensity of -0.20. The maximum average drought intensity of -0.60 has been observed at *Gourjhamar* followed by -0.59 at *Chandpur* and -0.57 at *Patha.* The overall drought intensity in the observation wells in Bearma basin falling in Sagar district is -0.35.

Table 2: *Groundwater drought charactersitics in wells of Sagar district in Bearma basin*

Name of Village	Onset of drought	Termination of drought	Duration (months)	Severity	Intensity
Chheola	2/5/1986	5/18/1987	15	-7.24	-0.48
	5/25/2000	8/29/2003	15	-7.29	-0.48
	5/21/2007	5/6/2008	12	-1.40	-0.12
Reechai	11/19/1984	10/31/1987	35	-5.98	-0.17
	11/13/1988	11/5/1990	12	-7.27	-0.60
	5/26/2000	1/4/2002	19	-4.23	-0.22
Titarpani	11/22/1984	6/15/1985	6	-3.71	-0.62
	12/2/1991	11/17/1993	23	-5.00	-0.21
	5/21/1996	11/1/1996	5	-1.60	-0.32
	5/21/2000	5/27/2001	12	-1.76	-0.14
	11/1/2001	5/27/2003	19	-2.41	-0.13
Patha	12/6/1991	5/27/1992	6	-2.24	-0.37
	2/26/1993	8/19/1993	5	-2.24	-0.40
	5/21/2000	8/28/2003	39	-8.07	-0.20
	11/17/2004	5/27/2005	6	-1.20	-0.18
	1/25/2007	5/27/2008	16	-4.04	-0.25
	11/4/2008	5/25/2009	7	-1.52	-0.22

Kachhi-pipariya	2/24/1992	6/8/1992	3	-1.40	-0.46
	5/29/2000	5/6/2004	47	-4.16	-0.08
	11/6/2004	5/31/2005	7	-2.94	-0.42
	11/8/2006	5/30/2008	19	-4.72	-0.25
	1/18/2009	8/20/2009	7	-5.09	-0.72
Chhirari	11/6/1989	6/12/1990	7	-1.79	-0.25
	6/3/2000	5/5/2004	48	-2.49	-0.05
	11/5/2004	5/25/2005	7	-1.86	-0.27
	11/6/2005	5/28/2008	31	-6.40	-0.20
	1/10/2009	8/8/2009	7	-5.19	-0.74
Chhula	10/24/4986	5/4/1987	6	-1.31	-0.21
	5/22/1988	5/23/1990	24	-3.98	-0.16
	5/26/1991	6/17/1992	11	-2.20	-0.20
	5/30/2000	1/10/2004	41	-5.10	-0.12
	11/8/2004	5/31/2005	7	-2.09	-0.29
	11/6/2005	5/31/2008	31	-2.72	-0.08
	11/6/2008	8/26/2009	10	-3.49	-0.35
Maharajpur	10/15/1992	5/16/1994	19	-7.34	-0.38
	11/11/1994	11/14/1995	12	-4.77	-0.39
	5/26/2000	5/26/2001	12	-4.09	-0.34
	11/30/2001	5/25/2002	6	-2.23	-0.37
	11/30/2002	8/27/2003	9	-3.71	-0.41
	11/4/2004	5/31/2005	7	-2.23	-0.32
Sahajpur	5/24/2000	5/24/2001	9	-2.55	-0.28
	11/3/2003	5/31/2004	7	-0.45	-0.06
	11/10/2004	5/30/2005	6	-0.76	-0.12
	11/16/2005	5/28/2006	6	-2.08	-0.34
	11/17/2006	5/30/2007	6	-1.88	-0.31
	11/4/2007	5/14/2008	6	-5.95	-1.00
	11/4/2008	8/3/2009	9	-12.60	-1.40
Ghana	11/16/1984	6/8/1985	7	-2.19	-0.32
	2/6/1989	11/8/1990	20	-3.34	-0.17
	5/29/1991	11/18/1994	41	-2.03	-0.04
	11/18/1999	8/31/2002	33	-5.30	-0.16
	1/31/2008	8/2/2008	7	-2.63	-0.37

Kesli	11/15/1984	6/6/1985	5	-2.29	-0.45
	6/27/1986	5/19/1987	11	-3.02	-0.27
	2/6/1989	3/7/1990	13	-2.71	-0.20
	12/6/1991	5/20/1992	6	-3.17	-0.52
	2/28/1993	11/28/1993	9	-3.17	-0.35
	5/30/1995	5/30/1996	12	-2.28	-0.19
	11/17/1996	5/30/1997	5	-1.77	-0.35
	11/17/1999	8/30/2000	9	-2.02	-0.22
	1/24/2001	11/4/2002	21	-4.36	-0.20
	11/412007	5/14/2008	6	-2.03	-0.33
Tada	11/20/1999	11/7/2003	48	-7.79	-0.16
	11/17/2004	11/8/2005	11	-2.08	-0.18
	5/25/2007	5/27/2008	12	-1.79	-0.15
	1/11/2009	8/9/2009	7	-4.29	-0.61
Copra	11/7/1989	5/13/1990	6	-2.43	-0.40
	6/25/1992	5/25/1993	11	-3.23	-0.29
	11/12/1995	5/11/1997	18	-5.20	-0.28
	11/12/1999	5/28/2000	6	-2.28	-0.37
	11/11/2000	5/28/2001	6	-3.50	-0.58
	11/11/2001	8/26/2003	21	-3.98	-0.19
	11/7/2004	5/31/2005	7	-2.50	-0.35
	5/26/2007	5/4/2008	11	-3.03	-0.27
	11/19/2008	8/27/2009	9	-2.68	-0.29
Chandpur	11/29/1984	6/10/1985	7	-6.93	-0.99
	6/12/1986	3/4/1987	8	-2.67	-0.33
	11/7/1989	5/13/1990	6	-5.70	-0.94
	11/11/2000	5/28/2001	6	-2.60	-0.43
	11/11/2001	11/11/2002	12	-3.41	-0.28
Gunjora	11/10/1986	5/4/1987	7	-2.63	-0.37
	2/12/1988	2/25/1989	12	-1.89	-0.15
	12/5/1991	3/9/1993	15	-4.57	-0.30
	5/20/1995	5/20/1996	12	-1.90	-0.15
	11/7/1996	5/20/1997	7	-1.46	-0.20
	11/7/1999	8/21/2003	46	-4.15	-0.09
Deori	12/3/1991	5/14/1994	30	-5.93	-0.19
	5/4/1995	5/4/1996	12	-3.33	-0.27
	11/25/1996	5/4/1997	6	-1.61	-0.26
	1/2/2000	1/10/2003	36	-5.94	-0.16

Gourjhamar	8/15/1993	1/17/1994	5	-1.81	-0.36
	11/17/1999	5/24/2000	6	-2.14	-0.35
	11/13/2000	5/24/2001	5	-4.70	-0.93
	11/12/2001	8/28/2002	9	-7.02	-0.77

Groundwater Drought Charactersitics in Damoh District

The groundwater drought characteristics evalauted for 59 observation wells in Damoh district given in Table 3. The maximum number of 9 drought events occurred at *Fatehpur, Hatta* and *Bari-kalan* with total durations varying between 83 to 105 months and total severity varying between -23.47 to -26.81 and average drought intensity varying between -0.20 to -0.35. The cumulative groundwater drought duration of 127 months has been observed at *Devron* followed by 116 months at *Kusumi* with an average drought intensity of -0.21 and -0.28 respectively. The maximum average drought intensity of -1.28 has been observed at Sanga followed by -0.88 at Tori and -0.77 at *Kudai*. The overall drought intensity in the observation wells in Bearma basin falling in Damoh district is -0.38.

Table 3: *Groundwater drought characteristics in wells of Damoh district in Bearma basin*

Name of Village	Onset of drought	Termination of drought	Duration (months)	Severity	Intensity (Severity/ Duration)
Luharra	5/29/1995	11/6/1995	6	-1.32	-0.22
	5/23/1996	11/7/1996	6	-1.20	-0.20
	5/24/1997	11/5/1997	6	-1.86	-0.30
	1/7/2002	8/31/2002	8	-3.98	-0.49
	11//10/2003	5/27/2004	6	-1.58	-0.26
	1/8/2009	8/31/2009	8	-4.99	-0.62
Khejara	10/28/1989	5/28/1990	6	-2.03	-0.33
	1/1/2000	5/20/2001	17	-5.13	-0.30
	11/5/2001	5/25/2003	18	-4.22	-0.23
	11/3/2004	5/23/2005	6	-1.80	-0.30
	11/9/2006	5/23/2008	18	-6.11	-0.33
	5/31/2009	11/12/2009	6	-3.15	-0.52
Khajari	8/30/1998	5/20/1999	8	-4.55	-0.56
	5/9/2004	11/6/2004	5	-3.69	-0.73
	1/21/2007	8/30/2008	19	-4.55	-0.23
Hathani	3/9/1984	10/15/1985	20	-6.78	-0.33
	10/22/1986	6/2/1987	8	-2.35	-0.29
	2/28/1988	5/28/1990	27	-4.37	-0.16
	11/3/1991	6/5/1993	19	-5.31	-0.27
	8/31/2009	1/9/2010	4	-4.34	-1.08

	6/3/1989	6/5/1990	12	-2.01	-0.16
	2/27/1992	11/2/1993	11	-2.17	-0.19
	5/20/1994	11/2/1995	7	-2.76	-0.39
	5/21/1996	8/20/1997	14	-2.01	-0.14
Bilai	5/21/1998	11/2/1998	5	-1.42	-0.28
	5/26/2000	11/2/2000	5	-1.39	-0.27
	5/25/2001	11/2/2001	5	-1.40	-0.27
	11/4/2007	5/31/2008	7	-6.18	-0.88
	10/30/1984	5/12/1985	7	-2.54	-0.36
	10/15/1989	5/27/1990	7	-1.54	-0.21
Bandakpur	11/2/2000	5/25/2001	7	-1.18	-0.16
	1/1/2004	5/22/2008	53	-7.10	-0.13
	8/31/2009	8/31/2010	12	-10.90	-0.90
	9/29/1995	1/1/1996	3	-3.66	-1.21
	11/8/2001	11/10/2002	11	-1.40	-0.12
Abhana	11/5/2007	5/23/2008	6	-3.84	-0.63
	11/1/2008	8/23/2010	22	-10.84	-0.49
	5/20/1989	2/15/1990	8	-0.60	-0.07
Deodongra	5/23/2006	5/20/2008	23	-13.18	-0.57
	8/29/2009	1/7/2010	9	-6.72	-0.74
	5/26/1996	11/4/1996	5	-1.68	-0.33
	1/5/1999	8/27/1999	7	-2.96	-0.42
	1/9/2002	8/4/2002	7	-1.35	-0.19
Ghatera	5/30/2004	11/10/2004	5	-2.09	-0.41
	5/18/2007	11/6/2007	5	-2.45	-0.48
	5/20/2009	8/23/2010	14	-5.26	-0.37
	10/1/1984	3/6/1985	5	-7.80	-1.55
	3/5/1989	3/5/1991	23	-2.93	-0.12
Banwar	5/17/2007	5/27/2008	11	-3.34	-0.30
	1/4/2009	1/1/2010	12	-2.86	-0.23
	11/6/1986	5/9/1987	7	-1.36	-0.19
Kundalpur	11/2/2007	5/26/2009	19	-13.49	-0.71
	11/10/2009	5/31/2010	6	-5.49	-0.91
	11/4/2000	5/21/2001	7	-4.16	-0.59
	11/7/2001	5/20/2002	7	-4.60	-0.65
Hardua-sadak	1/1/2003	8/22/2006	44	-3.31	-0.07
	1/1/2007	8/24/2010	44	-7.02	-0.15

Bamhori	11/26/1984	5/15/1985	6	-2.87	-0.47
	2/20/1993	5/29/1995	27	-4.70	-0.17
	8/23/2008	1/4/2009	4	-1.28	-0.32
	8/23/2009	1/2/2010	4	-5.65	-1.41
Kusmi	10/27/1984	11/5/1985	12	-2.20	-0.18
	2/28/1989	11/15/1990	20	-3.58	-0.17
	6/8/1991	11/9/1993	28	-5.56	-0.19
	5/22/2000	11/3/2002	29	-3.33	-0.11
	5/30/2004	8/25/2005	15	-1.81	-0.12
	1/4/2007	8/25/2007	8	-0.82	-0.10
	8/24/2009	1/2/2010	4	-4.50	-1.12
Jhalon	1/5/1995	11/3/1995	10	-3.09	-0.30
	8/30/2001	1/5/2002	4	-2.38	-0.59
	8/26/2002	1/2/2004	16	-3.80	-0.23
	8/25/2004	1/7/2005	4	-1.96	-0.49
	8/26/2007	1/7/2008	4	-2.95	-0.73
	8/27/2008	8/22/2010	24	-7.03	-0.29
Tardehi	6/7/1991	6/2/1992	11	-3.48	-0.31
	11/9/2000	11/7/2002	23	-7.07	-0.30
	5/21/2004	11/3/2004	5	-1.46	-0.29
	11/2/2006	8/27/2007	9	-0.80	-0.08
Bamhori(T)	11/8/1984	5/14/1985	5	-1.93	-0.33
	5/20/1994	5/21/1995	12	-3.74	-0.31
	5/30/2000	11/5/2002	29	-5.93	-0.20
	5/24/2009	11/7/2009	6	-3.90	-0.64
Jharoli	5/26/1989	11/10/1990	17	-2.45	-0.14
	1/2/1995	8/29/1995	8	-1.42	-0.17
	1/9/2005	11/4/2005	10	-1.23	-0.12
	5/25/2006	5/27/2007	12	-2.96	-0.24
	11/13/2007	5/24/2009	18	-4.27	-0.23
Purena	3/12/1986	11/6/1986	20	-0.95	-0.04
	10/23/1989	5/25/1990	7	-2.28	-0.32
	6/6/1991	5/24/1992	11	-0.99	-0.09
	8/23/1998	5/26/1999	9	-1.53	-0.16
	11/8/2000	5/27/2001	7	-0.60	-0.08
	11/8/2001	8/27/2002	10	-0.96	-0.09
	11/2/2006	1/1/2009	26	-9.76	-0.37
	8/20/2009	1/5/2010	4	-1.73	-0.43

	10/31/1991	3/2/1993	16	-1.79	-0.11
	8/29/1996	1/5/1997	4	-1.84	-0.45
	1/1/1999	8/26/1999	8	-0.95	-0.11
	11/1/2000	5/20/2003	30	-2.61	-0.08
Palar	11/1/2003	5/9/2004	5	-0.96	-0.19
	1/19/2007	5/27/2008	11	-11.71	-1.06
	5/31/2009	5/31/2010	12	-5.22	-0.43
	3/12/1984	3/5/1986	23	-4.21	-0.18
	11/7/1986	6/12/1987	7	-3.82	-0.54
	10/19/1988	5/19/1990	19	-4.42	-0.23
	1/10/1995	8/21/1995	11	-2.24	-0.20
Sagoni	1/2/1996	5/22/1997	17	-3.28	-0.19
(Kundalpur)	5/27/2000	5/30/2001	12	-2.86	-0.23
	1/20/2007	5/29/2008	16	-5.16	-0.32
	8/29/2009	1/8/2010	4	-2.88	-0.72
Amkhera	2/20/2007	11/4/2007	5	-3.52	-0.70
	11/26/1984	3/10/1988	9	-4.33	-0.48
	10/24/1989	3/6/1991	16	-5.80	-0.36
	11/11/1991	11/6/1993	23	-2.72	-0.11
Ghatpiparia	5/22/1994	11/10/1994	5	-1.58	-0.31
	11/2/2000	5/22/2001	6	-1.99	-0.33
	11/3/2001	5/25/2002	6	-1.24	-0.20
	11/5/2000	5/20/2001	6	-3.22	-0.53
	11/5/2001	5/24/2003	18	-5.04	-0.27
Pura	11/5/2004	5/22/2006	18	-2.22	-0.12
	11/2/2006	5/29/2008	13	-11.82	-0.90
	2/27/1989	3/5/1990	12	-1.66	-0.13
	5/30/2000	5/27/2001	12	-3.21	-0.26
Parsoria	5/29/2002	5/20/2003	12	-5.36	-0.44
	11/2/2005	5/15/2006	6	-1.90	-0.31
	11/15/2006	5/28/2008	17	-11.07	-0.63
Karraiya	2/23/1989	2/23/1990	12	-6.53	-0.54
	6/7/1991	6/5/1993	24	-5.27	-0.21
	11/5/1988	5/27/1990	18	-6.38	-0.35
	11/8/2000	5/28/2001	7	-1.99	-0.28
Hinota	11/2/2006	1/2/2009	25	-7.45	-0.29
	8/21/2009	1/6/2010	16	-4.85	-0.30

Mariadeo	11/5/1998	5/24/1999	5	-3.39	-0.67
	11/6/1999	5/28/2000	7	-1.74	-0.24
	11/5/2000	5/1/2001	7	-3.23	-0.46
	11/9/2002	5/30/2003	7	-2.24	-0.31
	11/5/2003	5/29/2004	7	-1.31	-0.18
	11/8/2004	5/22/2005	7	-1.94	-0.27
	11/6/2005	5/28/2006	7	-2.04	-0.29
	11/9/2006	8/30/2010	58	-6.39	-0.11
Muhari	10/28/1989	5/28/1990	7	-2.79	-0.39
	11/9/1998	5/29/1999	7	-1.53	-0.21
	11/10/2000	5/29/2001	7	-3.45	-0.49
	11/9/2001	5/22/2002	7	-3.91	-0.55
	11/9/2002	5/27/2003	7	-2.78	-0.39
	11/3/2006	11/3/2009	37	-5.38	-0.14
Fatehpur	5/20/1987	11/20/1987	5	-0.41	-0.08
	2/20/1989	5/20/1990	14	-1.52	-0.10
	5/25/1991	5/19/1992	12	-1.62	-0.13
	11/3/2000	5/22/2001	7	-2.92	-0.41
	11/8/2002	5/28/2003	7	-2.38	-0.33
	11/6/2004	5/21/2005	7	-2.25	-0.32
	11/6/2005	5/27/2006	7	-2.08	-0.29
	11/7/2006	1/7/2009	26	-6.40	-0.24
	8/27/2009	1/3/2010	3	-3.89	-1.29
Luhari	3/11/1989	5/29/1990	14	-1.70	-0.12
	11/2/2000	5/21/2001	7	-1.62	-0.23
	11/7/2001	8/21/2002	9	-3.71	-0.41
	5/20/2006	1/4/2009	30	-6.84	-0.22
	5/27/2009	8/31/2010	14	-5.43	-0.38
Hatta	2/26/1985	11/21/1985	10	-2.40	-0.24
	11/3/1986	6/14/1987	7	-3.15	-0.44
	3/11/1989	2/23/1990	12	-3.76	-0.31
	10/30/1991	6/8/1992	7	-2.89	-0.41
	5/20/1998	8/21/1999	15	-3.56	-0.23
	11/7/2000	5/27/2001	7	-1.84	-0.26
	11/8/2001	5/20/2002	7	-1.77	-0.25
	11/1/2006	1/4/2008	14	-5.54	-0.39
	8/20/2008	1/1/2009	4	-1.90	-0.47

Kudai	11/10/2000	5/28/2001	7	-2.41	-0.34
	11/2/2002	5/25/2003	7	-2.73	-0.39
	5/26/2004	8/24/2005	15	-4.89	-0.32
	1/1/2006	5/20/2008	29	-10.12	-0.34
	11/4/2008	1/9/2009	2	-4.93	-2.46
Majhaguwa-patol	3/5/1989	5/17/1992	39	-4.02	-0.10
Ghutaria	10/27/1991	6/11/1992	7	-0.87	-0.12
	11/5/1996	5/29/1997	7	-1.72	-0.24
	5/23/2000	8/26/2002	27	-4.25	-0.15
	1/9/2003	8/27/2003	8	-1.60	-0.20
	1/6/2004	8/24/2004	8	-0.71	-0.08
	1/7/2006	5/21/2008	17	-7.73	-0.45
Barkhera-chain	10/31/1984	6/11/1985	7	-1.27	-0.18
	5/25/1989	5/20/1990	12	-2.01	-0.16
	5/28/1995	11/5/1995	5	-0.45	-0.09
	11/4/2001	5/31/2002	7	-1.47	-0.20
	1/10/2007	5/24/2008	17	-10.68	-0.62
	8/23/2009	1/2/2010	4	-4.18	-1.04
Kachanari	8/24/1998	1/9/1999	4	-1.38	-0.34
	1/3/2001	8/26/2001	8	-1.44	-0.17
	1/4/2002	8/29/2002	8	-2.54	-0.31
	11/3/2006	8/26/2010	46	-6.94	-0.15
Hanmatpatti	10/23/1988	2/21/1990	16	-3.34	-0.20
	2/20/1991	5/19/1992	15	-4.20	-0.27
	11/7/2001	5/29/2002	7	-1.18	-0.16
	11/5/2006	5/24/2008	19	-7.29	-0.38
	11/4/2008	8/28/2010	22	-5.25	-0.23
Gaisabad	11/9/1984	6/11/1985	7	-1.97	-0.28
	11/6/1991	6/7/1991	7	-3.33	-0.47
Barkhera	10/31/1986	2/18/1990	40	-3.15	-0.07
	11/12/1992	5/26/1993	7	-0.66	-0.09
	11/9/2000	5/29/2001	7	-3.47	-0.49
	11/2/2001	5/25/2003	7	-3.42	-0.48
	11/1/2004	5/31/2005	7	-1.24	-0.17
	1/17/2007	5/20/2008	17	-5.71	-0.33

Nohata	3/8/1984	3/1/1985	12	-2.82	-0.23
	10/17/1991	5/20/1992	7	-1.99	-0.28
	11/1/1995	5/24/1996	7	-1.75	-0.25
	8/28/2004	1/1/2005	4	-1.37	-0.34
	1/3/2009	8/21/2010	20	-8.67	-0.43
Bari-kanora	6/19/1985	11/11/1986	17	-3.50	-0.20
	5/25/1987	6/15/1988	11	-2.33	-0.21
	2/21/1989	5/20/1990	15	-2.94	-0.19
	10/18/1991	5/19/1992	7	-1.26	-0.17
	5/28/2000	5/23/2001	12	-3.05	-0.25
	11/4/2001	5/27/2002	7	-1.64	-0.23
	11/7/2002	5/28/2003	7	-0.78	-0.11
	11/4/2005	5/26/2006	7	-0.62	-0.08
	11/7/2006	8/23/2008	22	-8.80	-0.40
Bhainsa	3/4/1989	5/21/1990	15	-2.33	-0.15
	2/21/1991	5/18/1992	15	-0.65	-0.04
	11/6/2000	5/27/2001	7	-2.12	-0.30
	11/7/2001	5/29/2003	19	-3.11	-0.16
	11/8/2003	5/22/2004	7	-1.59	-0.22
	11/4/2006	5/25/2008	19	-10.12	-0.53
	8/22/2009	1/8/2010	4	-3.14	-0.78
Raneh	3/14/1984	3/13/1986	24	-7.18	-0.29
	11/7/1986	6/9/1987	7	-1.43	-0.20
	5/21/2004	11/4/2004	4	-1.06	-0.26
	5/24/2006	5/25/2008	36	-7.82	-0.21
	11/6/2008	8/28/2010	22	-5.15	-0.23
Lidhora(harat)	10/28/1989	5/28/1990	7	-2.81	-0.40
	2/16/1991	5/18/1992	15	-4.00	-0.26
	5/22/2001	11/3/2001	7	-1.31	-0.18
	5/27/2007	5/27/2010	7	-8.42	-1.20
Devron	11/7/1991	11/5/1993	24	-3.62	-0.15
	11/9/1994	5/29/1995	7	-0.25	-0.03
	11/9/1995	5/30/1996	7	-0.61	-0.08
	11/6/1996	5/20/1997	7	-1.61	-0.22
	11/7/2001	11/10/2003	36	-4.28	-0.11
	5/27/2004	11/2/2004	4	-1.70	-0.42
	5/27/2005	11/2/2005	4	-2.36	-0.58
	11/15/2006	1/19/2010	38	-5.11	-0.13

	11/6/1984	2/23/1986	16	-5.45	-0.34
	11/11/1986	2/18/1988	16	-3.00	-0.18
	10/30/1988	5/26/1992	43	-5.07	-0.11
Kumhari	5/28/2000	11/5/2001	7	-2.28	-0.32
	5/25/2002	11/9/2002	4	-1.77	-0.44
	5/23/2003	11/4/2003	4	-1.36	-0.38
	5/31/2004	8/29/2004	3	-1.14	-0.38
Balakota	11/7/1988	11/7/1989	12	-2.47	-0.20
	3/12/1991	11/10/1991	8	-2.58	-0.32
	10/23/1986	6/3/1987	7	-1.14	-0.16
	6/11/1988	3/8/1990	20	-4.68	-0.23
Patana	11/7/2001	11/10/2002	12	-1.40	-0.12
	5/23/2004	11//2004	6	-0.78	-0.13
	1/26/2007	5/22/2008	16	-9.69	-0.60
	8/30/2009	1/12/2010	4	-6.51	-1.62
	11/1/1985	5/18/1986	7	-0.96	-0.13
	5/20/1995	11/7/1995	5	-1.55	-0.31
Khamaria	5/25/2006	11/5/2006	5	-0.94	-0.18
	5/21/2008	11/1/2008	5	-2.24	-0.44
	5/22/2009	8/28/2010	15	-12.40	-0.82
	10/22/1986	6/5/1987	7	-0.99	-0.14
Tori	5/22/2007	5/25/2008	12	-6.93	-0.57
	8/31/2009	1/13/2010	4	-7.72	-1.93
	11/4/2006	5/26/2007	7	-4.03	-0.57
Tejgarh(105)	11/7/2007	5/28/2008	7	-4.01	-0.57
	8/22/2009	8/23/2010	12	-7.11	-0.60
	8/27/2002	1/4/2003	4	-1.17	-0.29
Pura(106)	8/25/2004	1/7/2005	4	-0.73	-0.18
	1/2/2006	1/6/2009	36	-4.42	-0.12
	8/28/2009	8/27/2010	12	-8.23	-0.68
	5/27/2000	1/10/2001	7	-2.69	-0.38
	8/21/2001	1/5/2002	4	-1.54	-0.38
Dhaneta	8/27/2002	1/7/2003	4	-1.90	-0.47
	8/25/2004	1/7/2005	4	-1.62	-0.40
	1/2/2006	1/6/2010	48	-8.47	-0.17
	1/2/2000	5/25/2003	29	-5.85	-0.20
Bakan	5/24/2004	5/27/2005	12	-1.99	-0.16
	1/6/2006	11/6/2008	34	-5.14	-0.15

Batiagarh	11/6/2000	8/21/2001	10	-1.91	-0.19
	1/2/2002	5/26/2003	17	-4.67	-0.27
	1/4/2004	5/30/2005	17	-2.51	-0.14
	5/27/2007	5/29/2008	12	-5.21	-0.43
	8/28/2009	8/20/2010	12	-3.01	-0.25
Majhagaon-patol	11/13/2003	5/25/2004	7	-1.21	-0.17
	11/1/2004	5/30/2005	7	-2.41	-0.34
	11/2/2005	5/20/2007	19	-3.80	-0.20
	11/8/2007	5/24/2008	7	-2.59	-0.36
	8/30/2009	1/2/2010	4	-5.88	-1.47
Sanga	8/24/2008	1/10/2009	4	-4.44	-1.11
	8/28/2009	1/10/2010	4	-5.87	-1.46
Sarra	5/28/1989	6/14/1992	48	-4.13	-0.08
	5/22/1994	11/9/1995	17	-4.84	-0.28

CONCLUSION

The study found that the northern and southern part of the basin was more affected by extreme groundwater drought condition during meteorological drought years 2002-03 and 2007-08, whereas the central portion of basin was affected by mild to moderate groundwater drought condition and very less parts were free from the drought event. Groundwater drought characteristics indicating the imbalance variation between the amount of ground water recharge with its discharge. So, this situation threatens that there should be proper management of surface and groundwater for sustainable development through augmenting groundwater recharge. This analysis would help to understand the variation of groundwater drought and its areal influence during various seasons of a drought years.

REFERENCES

Beran MA and Rodier JA 1985 Hydrological Aspects of Drought: A Contribution to the International Hydrological Programme, *World Meteorological Organization Studies and Reports on Hydrology* Paris 39

Gleick PH 1993 *Water in Crisis: A Guide to the World's Fresh Water Resources*. Oxford University Press, New York

Hisdal H and Tallaksen LM 2003 Estimation of Regional Meteorological and Hydrological Drought Characteristics: A Case Study for Denmark, *Journal of Hydrology* **281** 230-247

Nalbantis and G Tsakiris 2009 Assessment of Hydrological Drought Revisited, *Water Resource Manage* **23** 881–897

Le Roy Ladurie E 1971 Times of Feast, Times of Famine: A History of Climate since the Year 1000 (Barbara Bray Trans), Doubleday *and Co,* New York, NY

Peters E and van Lanen HAJ 2000 Definition, Effects and Assessment of Groundwater Droughts: Drought and Drought Mitigation in Europe, *Advances in Natural and Technological Hazards Research* **14** 44-61

Tallaksen LM, van Lanen HAJ, van Loon AF and Wanders N 2013 Hydrological Drought across the World: Impact of Climate and Physical Catchment Structure, *Hydrology and Earth System Sciences* **17** 1715–1732

Van Lanen HAJ and Peters E 2000 Definition Effects and Assessment of Groundwater Droughts, in Drought and Drought Mitigation in Europe (eds) Vogt J Vand Somma F, Kluwer, Dordrecht pp 49-61.

CHAPTER 23

DNRA in Agroforest Soils: A Tool for Cleaner Environment

C. B. Pandey

ICAR-Central Arid Zone Research Institute, Jodhpur - 342 003, Rajasthan, India

Corresponding author's Email: cbpandey5@rediffmail.com

ABSTRACT

Humid tropics covering 10% of geographical area of the world is poorly studied for biogeochemical cycling of different ecosystems. Dissimilatory nitrate reduction to ammonium (DNRA) is a forgotten pathway of nitrogen cycling invented recently in terrestrial ecosystems, it reduces nitrate to ammonium N in soils, which without being incorporated into microbial biomass goes to mineral N pool and serves readily available nitrogen (N) to primary producers. Simultaneously, it reduces size of nitrate N in soils substantially and results in no gaseous losses (N_2O/N_2) of nitrogen, protects nitrogen (NO_3^-) from the leaching loss and enriches soils with mineral N (NH_4^+). Therefore, DNRA is an important part of nitrogen cycling. But, for a long time it was known to be mediated by anaerobic bacteria, and was reported from animal rumen, marine sediments, marshy places, sludge, etc. Of late with turn of 21st century it was found to occur in humid tropical forest soils experiencing rainfall 3000 to 4000 mm yr^{-1}. Later it is now reported to occur in grassland and rice fields. With these reports now it is well established that DNRA is an integral part of nitrogen cycle in terrestrial ecosystems. Three factors were known to regulate the DNRA which are: labile carbon, nitrate pool, and redox potential of soils, and high ratio of labile C/NO_3 was known to favour DNRA to occur. It was argued that labile carbon provided electron to nitrate. Recently agroforestry trees are also reported to reduce oxygen concentration in soils through root respiration and stimulate nitrate reductase enzymes, which intern triggers DNRA to occur, but this finding is not studied much so far. This paper describes DNRA, its implications, microorganisms and factors involved to regulate this process of nitrogen cycling in terrestrial ecosystems.

Keywords: Agroforestry, DNRA, Environment, Microorganisms

INTRODUCTION

Biogeochemistry of humid tropic ecosystems is relatively poorly studied though it covers 10% of total geographical area of the world (Pandey et al 2010). Nitrogen (N) is a major nutrient required in bulk in all ecosystems in the humid tropics, hence N-cycling is important from systems productivity view point. In general, nitrogen is found in soils in organic form which is oxidized to ammonium N that is quickly converted to nitrate N under aerobic conditions. But, in humid tropics where rainfall is high, even a small area of a landscape having different ecosystems is characterized with a mosaic of anaerobic and aerobic microsites / microhabitats which inhabit both aerobic and anaerobic microorganisms that mediate nitrification and methanogenesis on the same piece of land (Silver et al 2001). During aerobic conditions organic N is mineralized and its rate of mineralization is very high due to high temperature as microbial activity is found to increase with temperature up to 35 °C (Pandey et al 2010). The mineralized N is prone to heavy leaching losses, and denitrification is another process, which is known to occur in humid tropic soils. These two processes together weaken nitrogen cycling in humid tropical ecosystems. There are some studies, which report that high rainfall makes humid tropic trees to decline in stature and productivity owing to poor availability of mineral N (Schuur and Matson 2001).

Assimilatory cycling of nitrogen in different ecosystems is well known. It makes mineral N available to primary producers in a steady state in dry tropical conditions. Microbes are known to immobilize mineral N in soils during summer season when plants either do not require or they need it in a very little amount as most of the vegetation is dried. The immobilized N is, however, re-mineralized during following rainy season, when plants demand for it. This tight coupling between immobilization and mineralization is known to protect mineral N (NO_3^-) from leaching losses. This N conservation mechanism, however, works well in dry tropics (Singh et al 1989), but in humid tropics where clear cut season does not prevail, this mechanism does not hold good (Pandey et al 2009). In the humid tropics, unlike the dry tropics, microbial biomass (MB-C) is a steady state function of net soil N-mineralization rather than its conservation as it is evident from positive correlation between MB-C and net soil N-mineralization (Pandey et al 2009). Of late, with turn of 21st century, dissimilatory nitrate reduction to ammonium (DNRA) is postulated to be a mechanism, which protects mineral N (NO_3^-) from leaching losses in the humid tropical soils (Silver et al 2001, Pandey et al 2009). The DNRA is a newly recognized pathway of N cycling in terrestrial ecosystems, and found to occur in forests, grasslands and rice fields (Muller et al 2009, Pandey et al 2009, Souridis et al 2011, Silver et al 2001, Yin et al 2002). It converts rapidly NO_3^- to NH_4^+ in soils, without incorporation in tissues (Tiedje 1988) and provides additional NH_4^+ for uptake and assimilation by primary producers (Kraft et al 2014, Silver et al 2001). Plants prefer NH_4^+ over NO_3^- as they save energy of conversion of NO_3^- in tissues (Puri and Ashman 1999). The DNRA is known to consume almost all NO_3^- in Chilean temperate forests (Huygenes et al 2007, Rutting et al 2008) and 82% in the Andaman's tropical rainforests (Pandey et al 2009).

DNRA and denitrification is found to occur in same environmental conditions. Both are regulated by redox potential of soils, labile carbon, and labile carbon / NO_3^- ratio. For a long time DNRA was known to occur in highly reduced conditions like anaerobic sediments, rumen, sludge and other similar habitats. Silver et al (2001) reported it for the first time to occur in humid tropical forest soils and labile carbon provided electrons to reduce the NO_3^-. Other studies reported no relation between DNRA and labile carbon. Minick et al (2016) recently found that it is NO_3^- pool size rather than labile carbon, which regulated DNRA to occur. All these observations indicate that it is still inconclusive what factors regulate DNRA to occur.

MICROORGANISMS

Long ago DNRA was reported to be catalyzed by bacteria through fermentative rather than oxidative metabolism which occurs in denitrifiers. *Clostridia* and rumen bacteria (i.e. *Wolinella* and *Selenomonas*) were recognized in early nineteen eighties to catalyze DNRA. Later *Veillonella* and *Desulfovibrio* were added to the list. Now it is well known that DNRA microorganisms are obligate anaerobes, facultative anaerobes and aerobes. Obligate anaerobes include primarily *Clostridia* and rumen bacteria (i.e. *Wollinella* and *Selanominas*). Obligate anaerobes are some species of Bacteriodes, which reduce nitrate (Krieg and Holt 1984), but it is not yet known if some dissimilates nitrite to ammonium (Tidje 1988). Among the facultative anaerobes the capacity to dissimilate nitrate to ammonium is rather common, it is reported in seven genera of Enterobacteriaceae, photo bacterium and Bacillus species. Virtually all members of Enterobacteriaceae and Vibrionaceae reduce nitrate. Aeromonas and Vibrio are common anaerobic nitrate reducing microorganisms in estuarine sediments (Cole and Brown 1980, MacFarlane and Herbert 1982). Pseudomonads and Neisseria subflava are reported to dissimilate nitrate to ammonium under certain conditions (Grant and Payne 1981, Samuelsson 1985). Denitrifiers are basically aerobic bacteria and they have capacity to reduce nitrogen oxides when O_2 becomes limiting, hence they are facultative. *Pseudomonads* and *Alcaligens* are well known denitrifers.

A fungal denitrifier *Fusarium oxysporum* is the first example of denitrifying eukaryotes and can reduce nitrate up to N_2O (Shoun and Tanimoto 1991). This denitrifier fungus functions as nitrate respiration under O_2 limited conditions (Kobayashi et al 1996) like the bacterial one, and it occurs in mitochondria (Kobayashi et al 1996, Takaya et al 1999). Fates of nitrate in culture of *Fusarium oxysporum* depends on O_2 supply (Zhou et al 2001). When O_2 supply is enough, cell growth occurs through O_2 respiration where nitrate is reduced to ammonium and assimilated. In anaerobic conditions, however, it is reduced to N_2O through denitrification process. Under the aerobic conditions the nitrate assimilatory genes are induced as a Nir A-dependent manner, and at the same time denitrifying genes are repressed by Rox1p, and thus the fungus assimilates nitrate. However, under anaerobic conditions, Rox1p does not repress expression of the denitrifying genes and Nir A-dependent induction of the gene expression occur (Takaya et al 2002).

PHYSIOLOGY

DNRA is a two step reaction, in the first step nitrate is reduced to nitrite which is known as nitrate respiration. This step is well studied in *E. coli* and other enteric microorganisms. In most organisms this step produces ATP through electron transport phosphorylation (ETP). This step, however does not differentiate DNRA that from denitrification. The second step is reduction of nitrite to ammonium. This step is rate limiting because denitrifiers coexisting in a community may convert it (NO_2^-) to N_2O / N_2 gas. Therefore, critical step in DNRA is from nitrite-to-ammonium conversion (Tidje 1988). *Clostridium* catalyzes reduction of nitrate to ammonium but does not couple ETP with the reduction. Molar growth yield of *Clostridium* is found to increase 15.7% on glucose using this dissimilatory pathway (Caskey and Tiedje 1980), this could be due to substrate-level phoshorylation by alternative electron acceptors (Tiedje 1988).

Campylobactor sputorum (deVries et al 1974), *Desulfovibrio gigas* (Barton et al 1983) and *Wolinella* (Bokranz et al 1983) are found to conserve energy during nitrate reduction to NH_4^+, in this reaction H_2 or formate is oxidized (Tidje 1988). *E. coli* has also been found to couple energy production from reduction of nitrate to ammonium, but with formate-coupled nitrate reductase (Motteram et al 1981), this pathway accounts, however, only 20% of the nitrate reduction to ammonium in *E. coli*. These aforesaid organisms use a membrane bound nitrite reductase (Tiedje 1988). In DNRA organisms where nitrite reductase is soluble, they can not conserve energy (Tiedje 1988). The non-energy-linked nitrite reducers seem to carry out nitrite reduction to ammonium by different soluble nitrite reductases (Coleman et al 1978). NADH is one of them found in *E. coli*. Sulfite reductase is another enzyme. DNRA is reported high in marine sediments, which are rich in sulfate, this indicates that nitrite reduction to ammonium may be enhanced because it is an alternative substrate for sulfite reduction pathway. Some of the nitrite reductases were speculated to have assimilatory functions. Yordy and Delwiche (1979) found that synthesis of nitrite reductase in Veillonella required nitrite or nitrate and was repressed by ammonium and tryptone, but the organisms excreted ammonium from nitrite reduction. A number of organisms with assimilatory nitrite-reducing pathway probably use this pathway during the oxygen stress period to accept electrons for their survival. Smith and Zimmerman (1981) and Tiedje et al (1982) found higher number of DNRA organisms in aerobic and speculated that DNRA activity could be due to other organisms like *Pseudomonads* rather than obligate anaerobes and the Enterobacteriaceae.

ENVIRONMENT OF DNRA

The soil oxidation state is a principal factor that influences the importance of DNRA compared to denitrification (Matheson et al 2002) with DNRA by bacteria and fungi occurring under more reducing (anoxic) conditions (Takaya 2002, Yin et al 2002). On the other hand, other studies showed that DNRA is less sensitive to variable redox conditions (Pett-Ridge et al 2006) and O_2 than denitrification (Fazzolari et al 1998). In the latter study soil aggregates were incubated under various O_2 levels with same NO_3^- concentrations combined with different levels of glucose C additions. The studies showed that the effect of variable O_2 on DNRA was dependant on the C/NO_3^- ratio and concluded that C rather than O_2 was the key factor regulating NO_3^- partitioning between denitrification and DNRA (Fazzolari et al 1998). Tiedje et al (1982) hypothesized that C/ NO_3^- ratio decides partitioning NO_3^- reduction between DNRA and denitrification. Yin et al 1998 showed that significant DNRA occurred only at a C/NO_3^- ratio above 12. However, Matheson et al 2002 argued that the effect of C/ NO_3^- ratio on DNRA may be an artifact of experimental approaches. Experiments investigating DNRA under different C/ NO_3^- ratios usually artificially alter either the organic C or NO_3^- content that can result in enhanced O_2 consumption due to stimulation of microbial activity or in altered soil redox potential due to that NO_3^- is an oxidizing agent itself (Matheson et al 2002). Therefore, Matheson et al (2002) concluded that experimental evidence of the effect of C/ NO_3^- ratio are most likely due to altered soil oxidation state, which is hence the key partitioning factor.

Tiedje et al 1988 pointed out that it is the availability of a suitable organic C source, supporting respiration or fermentation that regulates the population of DNRA bacteria. Several studies showed that the addition of glucose, carbohydrate that support respiration as well as fermentation, stimulated DNRA (Buresh and Patrick 1978, Yin et al 2002, Caskey and Tiedje 1980, Yin et al 1998). In a ^{15}N labeling study with soil slurries Morley and Baggs (2010) reported that DNRA appeared to be stimulated more by carbohydrates (glucose and mannitol) than amino acids and butyrate, but Yin et al (1998) reported that the carbohydrates glycerol and succinate do not support DNRA. Moreover, in two anaerobic soils, addition of glucose did not influence DNRA (Chen et al 1995). deCatanzaro et al (1987) also did not find an effect of glucose addition on DNRA, but observed that DNRA was stimulated by alfalfa addition, probably because it released sulphur (-SH group) upon mineralization, which served as reducing agent. Under anaerobic conditions, sulphide stimulates DNRA, by serving as an electron donor, and depresses denitrification, by repressing NO and N_2O reductase.

Other C sources, like straw, glycerol, methanol and succinate did not promote DNRA (Buresh and Patrick 1978, deCatanzaro et al 1987, Yin et al 1998). The reasons for this are not fully understood and needs further studies. But, Buresh and Patrick (1978) and Yin et al (1998) attributed this finding to the fact that the mentioned C sources are poor substrates for fermentative process (Cole and Brown, 1980), hence did not promote DNRA. However, as two distinct pathways of DNRA exist, one fermentative and other respiratory (Simon 2002, Mohan et al 2004), this cannot be the sole explanation. Simon (2002) for the respiratory DNRA listed formate, H_2 and sulphide as substrates. Possibly the above mentioned C sources neither favour the fermentative nor the respiratory DNRA pathway. This may also explain why in some cases addition of glucose does not support DNRA and some of the above mentioned contradictions in the response to C/NO_3^- as this may depend on whether respiratory or fermentative DNRA bacteria are present. This is supported by deCatanzaro et al (1987) in respect to sulphide.

Yin et al (2002) studied DNRA in two soils, one from China and other from Australia and found greater DNRA in Australian soils. They attributed it to more population of nitrate-ammonifers in the Australian soils. They manipulated ambient conditions of the soils to know the factors which favour DNRA. They observed that addition of NO_3 at 100 mg N kg^{-1} soils increased DNRA, but it was more in Australian soils. When reducing agent L-cystein was added in the soils 56.4% added NO_3 was converted to NH_4^+ in the Australian soils. They found that DNRA occurred faster than denitrification in an experiment where ^{15}N-lebeled NH_4^+ increased much before 36-108 hr of incubation whereas denitrification produced N_2O / N_2 after 180 hr.

COMPETITION BETWEEN DENITRIFIERS AND NITRATE AMMONIFIERS

As known, denitrification losses fixed nitrogen in the form of green house gas, whereas nitrate-ammonification reduces size of nitrate and either reduce substantially or stops denitrification, therefore it is interesting to find out factors, which help nitrate-ammonification to outcompete denitrification. These two microbial processes i.e. denitrification and nitrate-ammonification, compete for nitrate as electron acceptors. The C/N ratio, nitrite vs. nitrate concentrations, soil sand content, availability of fermentable carbon compounds, temperature and sulfide concentration are speculated to be important environmental controls on this competition, but, it is still not clear which factors are more important (Kraft et al 2014). They did a long term incubation experiment with natural communities known to contribute denitrification and ammonification with the objective to know factors which select nitrate-ammonification over denitrification and found that supply of nitrite relative to nitrate and C/N ratio are key environmental factors that determine whether nitrite will be reduced to N_2O/N_2 or ammonium.

CONCLUSION

DNRA is speculated to detoxify accumulated NO_2^-. It acts as an electron sink which reoxidize NADH which is reduced during the nitrate respiration. It produces ATP through ETP during nitrite reduction. All above processes benefit to DNRA organisms. Ecologically, however, it makes environment N_2O free by reducing NO_3^- to NH_4^+, otherwise NO_3^- is lost as N_2O gas via denitrification. NH_4^+ being a readily available mineral N, increases availability of nitrogen to plants, and thereby it increases soil fertility. NH_4^+ resists leaching losses and makes groundwater nitrate pollution free.

REFERENCES

Barton L L, LeGall J, Odom JM, and Peck HD Jr 1983 Energy coupling to nitrite respiration in the sulfate-reducing bacterium Desulfovibrio gigas, *Journal of Bacteriology* **153** 867-871

Bokranz M, Katz J, Schroder I, Roberton AM and Kroger A 1983 Energy metabolism and biosynthesis of Vibrio succinogenes growing with nitrate or nitrite as terminal electron acceptor *Archives of Microbiology* **135** 36-41

Buresh RJ and Patrick WH 1978 Nitrate reduction to ammonium in anaerobic soils, *Soil Science Society of America Journal* 42 913-918

Caskey WH, and Tiedje JM 1980 The reduction of nitrate to ammonium by a Clostridium species isolated from soil, *Journal of general microbiology* **119** 217-223

Chen DL, Chalk PM and Freney JR 1995 Distribution of reduced products of 15N-labelled nitrate in anaerobic soils, *Soil Biology and Biochemistry* 27 1539-1545

Cole JA and Brown CM 1980 Nitrate reductase to ammonia by fermentative bacteria: a short circuit in the biological nitrogen cycle, *FEMS Microbiology Letters* **7** 65-72

Coleman KJ, Comish-Bowden A and Cole JA 1978 Purification of properties of nitrite reductase from Escherichia coli K12, *Biochemistry Journal* **175** 483-493

deCatanzaro JB, Beauchamp EG and Drury C F 1987 Denitrification vs. dissimilatory nitrate reduction in soil with alfa, straw, glucose and sulfide treatments, *Soil Biology and Biochemistry* **19** 583-587

deVries W, van Wyck-Kapteyn WMC and Oosterhuis S K H 1974 The presence and function of cytochromes in *Selenomonas ruminantium*, *Anaerovibrio lipolytica* and *Veillonella alcalescens*, *Journal of General Microbiology* **81** 69-78

Fazzolari E, Nicolardot B, and German JC 1998 Simultaneous effects of increasing levels of glucose and oxygen partial pressure on denitrofication and dissimilatory nitrate reduction to ammonium in repacked soil cores, *European Journal of Soil Biology* **34** 47-52

Grant MA, and Payne WJ 1981 Denitrification by strains of *Neisseria, Kingella,* and *Chromobacterium International journal of* systematic *bacteriology* **31** 276-279

Huygens D, Rutting T, Boeckx P, Van Cleemput O, Godoy R, and Muller C 2007 Soil nitrogen conservation mechanisms in a pristine south Chilean Nothofagus forest ecosystem, *Soil Biology and Biochemistry* **39** 2448-2458

Kobayashi M, Matson Y, Tanimoto T, Suzuki S, Maruo F, and Shoun H 1996 Denitrification, a novel type of respiratory metabolism in fungal mitochondrion, *Journal of Biological Chemistry* **271** 16263-16267

Kraft B, Halina E, Tegetmeyer, Sharma R, Kolotz M G, Ferdelman T G, Hettich R L, Geelhoed JS, and Stous M 2014 The environmental controls that govern the end product of bacterial nitrate respiration, *Science* **345** 676-679

Krieg NR, Holt JG (eds.) 1984 *Bergey's manual of systematic bacteriology*, Vol. I. Williams & Wilkins, Baltimore, 964

MacFarlane GT, and Herbert RA 1982 Nitrate dissimilation by Vibrio spp. Isolated from estuarine sediments, *Journal of general microbiology* **128** 2463-2468

Matheson FE, Nguyen MC, Cooper AB, and Burt TP 2003 Short-term nitrogen transformation rates in riparian wetland soil determined with nitrogen-15, *Biology and Fertility of Soils* **38**(3) 129-136

Minick KJ, Pandey CB, Fox TR and Subedi S 2016 Dissimilatory nitrate reduction to ammonium and N_2O flux: effect of soil redox potential and N fertilization in loblolly pine forests, *Biology and Fertility of Soils* **52** 601-614

Mohan SB, Schmid M, Jetten M, and Cole J 2004 Detection and widespread distribution of the nrfA gene encoding nitrite reduction to ammonia, a short circuit in the biological nitrogen cycle that competes with denitrification, *FEMS Microbiology Ecology* **22** 57-63

Morley N, and Baggs EM 2010 Carbon and oxygen controls on N_2O and N_2 production during nitrate reduction, *Soil Biology and Biochemistry* **42** 1864-1871

Motteram PA, McCarthy JEG, Ferguson SJ, Jackson JB and Cole JA 1981 Energy conservation during the formate dependant reduction of nitrite by Escherichia coli, *FEMS Microbiology Letters* **12** 317-320

Muller C, Rutting T, Abbasi MK, Laughlin RJ, Kammann C, Clough TJ, Sherlock RR, Kattge J, Jagar H, and Watson CJ 2009 Effect of elevated CO_2 on soil N dynamics in a temperate grassland, *Soil Biology & Biochemistry* **41** 1996-2001

Pandey CB, Singh GB, Singh SK, and Singh RK 2010 Soil nitrogen and microbial biomass carbon dynamics in native forests and derived agricultural land uses in a humid tropical climate of India, *Plant Soil* **333** 453-467

Pandey CB, Srivastava RC, and Singh RK 2009 Soil nitrogen mineralization and microbial biomass relation, and nitrogen conservation in humid tropics, *Soil Science Society of America Journal* **73** 1142-1149

Pett-Ridge J, Silver WC, and Firestone M K 2006 Redox fluctuations frame microbial community impacts on N-cycling rates in a humid tropical forest soil, *Biogeochemistry* **81** 95-110

Puri G, and Ashman MR 1999 Microbial immobilization of 15N-labelled ammonium and nitrate in a temperate woodland soil, *Soil Biology & Biochemistry* **31** 929-931

Rutting T, Huygens D, Muller C, Van Cleemput O, Godoy R, and Boeckx P 2008 Functional role of DNRA and nitrite reduction in a prestine south Chilean Nothofagus forest, *Biogeochemistry* **90** 243-258

Samuelsson MO 1985 Dissimilatory nitrate reduction to nitrite, nitrous oxide, and ammonium by Pseudomonas putrefaciens, *Applied and Environmental Biology* **50** 812-815

Schhur EAG, Matson PA 2001 Net primary productivity and nutrient cycling across a mesic to wet precipitation gradient in Hawaiian montane forest, *Oecologia* **128** 431-442

Sgouridis F, Heppell CM, Wharton G, Lansdown K, and Trimmer M 2011 Denitrification and dissimilatory nitrate reduction to ammonium (DNRA) in a temperate reconnected floodplain, *Water Research* **45** 4909-4922

Shoun H, and Tanimoto T 1991 Denitrification by the fungus *Fusarium oxysporum* and involvement of cytochrome P-450 in the respiratory nitrite reduction, *Journal of Biological Chemistry* **266** 11078-11082

Silver WL, Herman DL, and Firestone MK 2001 Dissimilatory nitrate reduction to ammonium in upland tropical forest soils, *Ecology* **82** 2410-2416

Simon J 2002 Enzymology and bioenergetics of respiratory nitrite ammonification, **FEMS Microbiology** Reviews **26** 285-309

Singh JS, Raghuvanshi AS, Singh RS, Srivastava S C 1989 Microbial biomass acts as a source of plant nutrients in dry tropical forest and savanna, *Nature* **338** 499-500

Smith, M.S., and Zimmerman, K. 1981 Nitrous oxide production by non-denitrifying soil nitrate reducers, *Soil Science Society of America Journal* **45** 865-871

Takaya N, Suzuki S, Kuwazaki S, Shoun H, Maruo F, Yamaguchi M, and Takeo K 1999 Cytochrome P450 involved in nitrate respiration in the fungus Fusarium oxysporum, *Archives of Biochemistry and Biophysics* **372** 340-346

Takaya N, Uchimura H, Lai Y, and Shoun H 2002 Transcriptional control of nitrite oxide reductase gene (CYP55) in the fungal denitrifier Fusarium oxysporum, *Bioscience Biotechnology* and *Biochemistry* **66**(5)1039-1045

Tiedje JM 1988 Ecology of denitrification and dissimilatory nitrate reduction to ammonium, in: A. J. B., *Environmental Microbiology of Anaerobes,* Zehnder (ed.). John Wiley and Sons, NY

Tiedje JM, Sexstone AJ, Myrold DD, Robinson JA 1982 Denitrification: ecological niches, competition and survival. Antonie Leeuwenhoek, *Journal of Microbiology* **48** 569-583

Yin S X, Chen D, Chen LM, and Edis R 2002 Dissimilatory nitrate reduction to ammonium and responsible microorganisms in two Chinese and Australian paddy soils, *Soil Biology & Biochemistry* **34** 1131-1137

Yin S X, Shen Q, Tang Y, and Cheng L M 1998 Reduction of nitrate to ammonium in selected paddy soils of China, *Pedosphere* **8** 221-228

Yordy DM, and Delwiche EA 1979 Nitrite reductase in Veillonella alcalescens, *Jornal of Bacteriololgy* **137** 905-911

Zhou Z, Takaya N, Sakairi MAC, Shoun H 2001 Oxygen requirement for the denitrification by filamentous fungus *Fusaroum oxysporum*, *Archives of Microbiology* **175** 19-25

CHAPTER 24

Agroforestry and Soil Health: An Overview

S Sarvade*, DS Gautam, VB Upadhyay, RK Sahu, AK Shrivastava, Rajesh Kaushal[1], Rahul Singh[2] and AG Yewale[3]

College of Agriculture, Balaghat, JNKVV, Jabalpur (MP), India

[1]ICAR-Central Soil and Water Conservation Research and Training Institute, Dehradun (UK), India

[2]Department of Agriculture, Mata Gujri College, Fategarh Sahib, Punjab University Patiala, Punjab, India

[3]KVK, Ranichauri, Uttarakhand University of Horticulture and Forestry, Tehri Garhwal, Uttarakhand, India

**Corresponding author's Email: somanath553@gmail.com*

ABSTRACT

Soil health/quality is continued capacity of soil to function as a living system, which sustain biological productivity, environmental quality, and plant and animal health. But increasing human and livestock population along with unsustainable use of natural resources, soil health is degrading day by day. Many researchers reported unplanned and unscientific agricultural practices, deforestation, over grazing, over-construction and alignment of roads soil health degrading causes. Along with these causes climate change catalyse the soil health/quality degradation processes and made researchers/scientists to rethink about these problems and their solutions. Perennial woody plant based systems had potential to improve soil health/quality through different processes. In case of soil health, soil physical, chemical and biological properties were improved by tree species as they add high amount of leaf, root, twig, flowers and fruits biomass to the soil. Decomposition, nutrient release and return to the soil (nutrient cycling), biological nitrogen fixation, nutrient pumping and reduction of soil erosion improves soil properties under agroforestry systems.

Keywords: Agroforestry, Biological nitrogen fixation, Nutrient cycling and pumping, Soil conservation, Soil health

INTRODUCTION

Soil is the basis of production in agriculture and forestry, the nourisher of mankind and an important component of the human environment. Health/quality of soil is considered to be an important factor for the success of agricultural production. Soil health has been defined as the continued capacity of soil to function as a living system, which sustained biological productivity, environmental quality, and plant and animal health (Karlen et al 1997, Oliver et al 2013, FAO 2015). Sustainable agriculture production depends on the scientific management of soil health to meet people's present and future needs. But now a days, the health of soil facing high degradation problems due to increased pressure of human and livestock population on inadequate natural resources. Semwal et al (2009) and Araujo et al (2012) reported the possible reasons of the soil degradation such as unplanned and unscientific agricultural practices, deforestation, over grazing, over-construction and alignment of roads. Carter et al (2004) reported scientists and researchers interest in soil and land quality and understanding changes in soil fertility studies.

The population of human is increasing with an alarming rate. Due to rapid increase in human population and decrease in land holding, people moved towards intensive agriculture instead of adopting sustainable farm producing practices. Intensification of agriculture results in depletion of soil health/quality through overexploitation of resources and nutrient losses through open nutrient cycling (Sarvade et al 2014a). The degradation of soil health is observed in all agro-climatic regions on all continents (Anonymous 2010). The degradation of soil is more unspoken but grim environmental problems that our world faces. An uncontrolled grazing is also one of the reason for depletion of soil health through reducing ground vegetation (by grazing and trampling) and increasing surface runoff. Soil of India are deficient in major plant nutrients like nitrogen, phosphorus and potash in some cases, secondary nutrients, like sulphur, calcium, and magnesium. In case of Micronutrients, zinc, boron and a limited extent of iron, manganese, copper and molybdenum have also been reported to be deficient.

Along with rising human population, climate change, unsustainable use of natural resources affects soil nutrient status of the country (Sarvade et al 2014b). Climate change is the disturbance in temperature and rainfall patterns for a certain period of time due to the greenhouse gasses emission in the atmosphere. High intensity rain fall in small time period causes soil erosion problems. The soil erosion through runoff water is the most serious degradation problem during monsoon season, which results in loss of essential topsoil nutrients (Joshi 2004, Lakaria et al 2011, Grogan et al 2012, Bhattacharyya et al 2015). This process not only affects the land/soil but also cause loss of bio- diversity including base resource itself, and human life. About 126 M ha area of the country has been found suffering from various degrees of water erosion (Anonymous 2010). An annual loss of 8 M t plant nutrients through 5.3 billion t of soil lost by water erosion reported by Prasad and Biswas (1999). The increase of 2.5° C triggered total organic carbon depletion in different layers of soil due to accelerating decomposition rate of organic matter (Anonymous 2012).

Soil health is an important factor for controlling yields of the crops. In future, sustainable soil conservation efforts would be needed for tackling problems such as soil health depletion, climate change and food insecurity.

Agroforestry is a collective name for land-use systems in which woody perennials (trees, shrubs, etc.) are grown in association with herbaceous plants (crops, pastures) and/or livestock in a spatial arrangement, a rotation or both, and in which there are both ecological and economic interactions between the tree and non-tree components of the system (Lundgren and Raintree 1982). Figure 1 explains the soil quality/health indicators, which were influenced by different soil improving practices in agroforestry systems. Agroforestry is sustainable land use system, many scientists and researchers from the different parts of the country and world were reported their potential for soil conservation and improvement of soil physical, chemical and biological properties through multifunctional approach (Pandey 2007, FAO 2010, Das et al 2011, Anju and Koppad 2013, Chauhan et al 2013, Mandal et al 2013, Sarvade 2014e, Guleria et al 2014, Nair 2014, Sarvade et al 2014a, Sarvade and Singh 2014d, Berhe and Retta 2015). The combination of trees, crops and livestock mitigates environmental risk, creates a permanent soil cover against erosion, minimises damage from flooding and acts as water storage, benefitting crops and pastures (FAO 2014). Agroforestry systems can reverse land degradation problem in the tropics through their beneficial effects on soil resource (Tornquist et al 1999). The amount of litter addition, litter quality, their decomposition rate and an improvement of soil physico-bio-chemical properties is controlled by kind of tree species, age and density of tree species (Isaac and Nair 2006, Nair 2007, Xiong et al 2008, Sarvade et al 2014c).

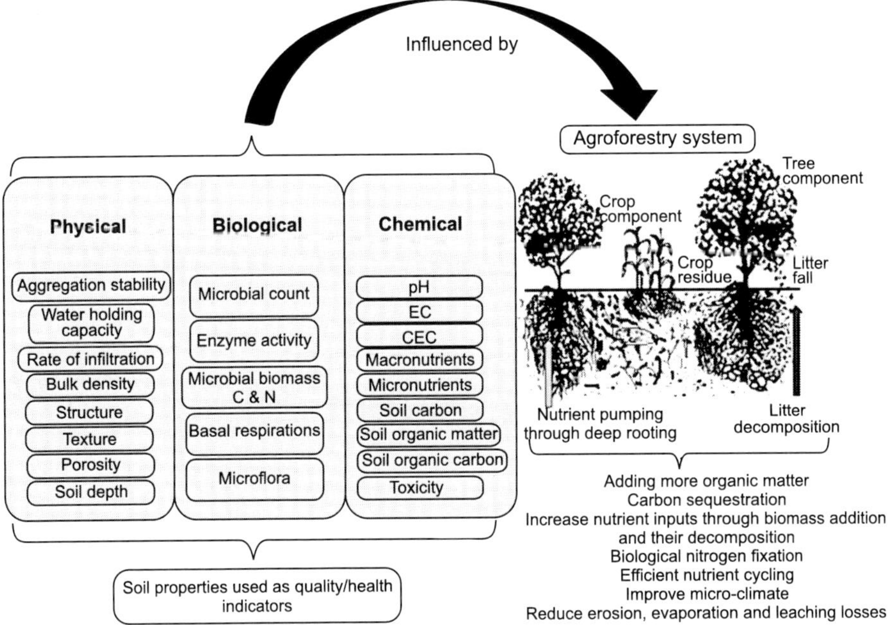

Figure 1: *Soil quality/ health indicators influenced through different soil improving processes in agroforestry systems*

SOIL PHYSICAL PROPERTIES

The soil physical properties are important for improving soil chemical and biological properties, which helps to healthy growth of plants (Sarvade et al 2014c, Berhe and Retta 2015, Uthappa et al 2015). The addition of organic matter and decomposition after coming in contact with soil and microbial fauna improves soil physical properties such as structure, bulk density, particle density, porosity and moisture holding capacity (Hillel 1982, Singh and Sharma 2007). Litter added by tree species act as protective layer for maintaining soil physical properties such as soil moisture conservation, shielding against soil compaction and reducing soil losses from erosion or leaching (Xiong et al 2008).

Soil Aggregate

Structural stability of the soil depends on the stability of soil aggregate and decline in soil aggregate stability is a powerful indicator of land degradation. Perennial plant species can improve soil aggregate stability because of adding organic matter through litter and plant root turnover. The grass land system recorded greater proportion of larger aggregate (>5mm) as compared to cultivated fields, where lowest aggregates was recorded by Somasundaram et al (2012) in Chambal region, Rajasthan, India. Mostafa et al (2008) and Lawal et al (2009) reported that mean weight diameter (MWD) and water stable aggregate were greater in soils of perennial plant species based systems as compared to soils of cultivated annual crops. Gupta et al (2009) studied soil aggregation under poplar based agroforestry system in relation to tree age and reported that the water stable aggregates (WSA >0.25 mm) were increased by 14.4, 32.6 and 56.9 times in soils with 1, 3 and 6 years poplar plantation, respectively, from that in sole crop. Singh et al (2004) reported that an individual tree species can result in better soil structure through various processes such as increasing organic matter and reducing erosion losses. Better soil aggregation improves water transmissivity and infiltrability by increasing soil organic carbon and reducing soil erosion (Saha et al 2010). The agroforestry systems of northwest Himalayan region improved soil physical properties (soil depth, texture) as compared to irrigated and rain fed cropland (Mandal et al 2013). Odewumi et al (2013) investigate the effects of teak (*Tectona grandis*) cultivation on soil physical properties in Ajibode community, Ibadan, Oyo state and reported that considerably higher concentration of clay content in soil under teak canopy than in soil outside teak canopy. Soils under MPTs significantly increased 65 per cent mean weight diameter (MWD) of aggregates and 21 per cent moisture under hilly ecosystems of Northeast India (Ramesh et al 2013). Paudel et al (2012) reported higher water stable aggregates in surface soils under grass buffer and agroforestry buffer than grazed pasture and row crop management treatments (Fig. 2).

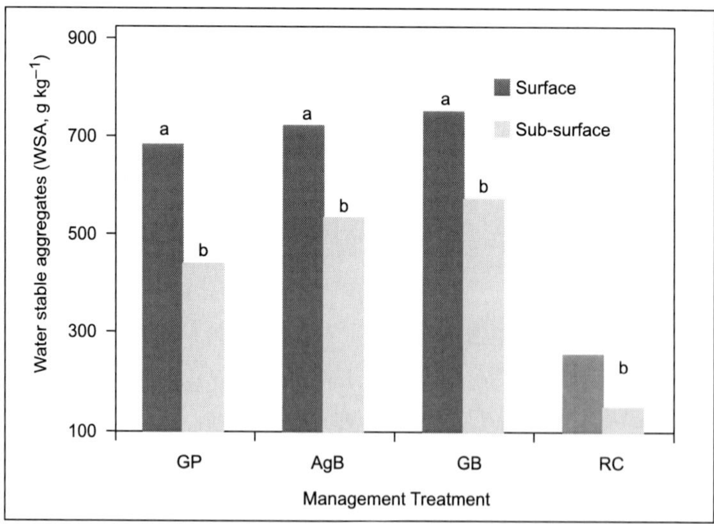

Figure 2: *Water stable aggregate proportions (WSA) for the grazed pasture (GP), agroforestry buffer (AgB), grass buffer (GB) and row crop (RC) management treatments (Paudel et al 2012).*

Soil Water Holding Capacity

An improved soil structure in tree based systems also helps to conserve more soil moisture through adding organic matter and reducing moisture losses (evaporation and erosion). The perennial tree species from natural forest and agroforestry systems improves soil moisture content in each soil profile depth (Saha et al 2004) due to increasing organic matter and reducing erosion and evaporation losses. Tripathi et al (2009) reported higher water holding capacity (WHC) under Alder (*Alnus nepalensis*) based agroforestry system as compared to Khasi mandarin (*Citrus reticulata*) and Khasi pine (*Pinus kesiya* Royle ex Gordon) based agroforestry systems. They also reported reverse trend between soil porosity and bulk density; values were greater in surface layer than subsurface layer. Saha et al (2007) reported the potential of MPTs in improving soil hydro-physical characteristics with decreasing bulk density and erosion ratio by 15.9 and 39.5 per cent respectively and increasing soil porosity by 10.9 per cent, aggregate stability by 24.0 per cent and available soil moisture by 33.2 per cent in agroforestry systems of north-east India. Mishra (2011) reported high soil moisture content (74 per cent) under broad-leaved tree species compared to *Pine* trees at Upper Shillong.

Bulk Density

The soil physical properties also affected by kind, age of tree and density of tree species. Bulk density is an indicator of soil health and affect plant rooting depth, soil water holding capacity, soil porosity, infiltration rate and other soil bio-chemical properties which influence soil productivity (Ernest et al 2015). The soil bulk density decreased under high density (3m×1m spacing) *Poplar, Leucaena* and *Melia* based agroforestry systems due to high rate of litter addition than Eucalypt based agroforestry systems (Sarvade et al 2014c). Araujo et al (2004) recorded lower soil bulk density in rubber tree plantation as compared to annual cropping system. The slightly higher bulk density was reported by Devi et al (2013) in plantations as compared to natural forest due to lower soil organic carbon in plantation. The 12.6 per cent increase in bulk density and 64.89 per cent decrease in hydraulic conductivity in scrub land and terraced agricultural lands as compared to tree based system in north-western hilly region was reported by Dadhwal et al (2011). The lower mean bulk density under agroforestry systems, however slight but not significant increase in bulk density was recorded in grazed land by Tornquist et al (1999). Ramesh et al (2013) observed 7 per cent reduction in bulk density as compared to the control plot (without tree plantation) under hilly ecosystems of Northeast India. The bulk density and spore space has inverse relation; the lower bulk density and higher porosity in soil under teak canopy than in soil outside teak canopy reported by Odewumi et al (2013). Paudel et al (2012) recorded lower soil bulk density (1.31g per cm^3) under agroforestry systems as compared to grass buffer, grazed pasture and row crop at the Horticulture and Agroforestry Research Centre (HARC) of the University of Missouri in New Franklin.

SOIL CHEMICAL PROPERTIES

Soil Organic Matter and Organic Carbon

The 38 per cent higher soil organic matter (SOM) and organic carbon (SOC) was recorded in poplar + sugarcane compared to poplar + fodder-wheat rotation, this may be ascribed to the addition of sugarcane trash in the fields in addition to litterfall of poplar plants (Singh and Sharma 2007). They also reported higher SOM and SOC status of soils under the older plantations resulted from higher quantities of litterfall addition and root necromass.

Soil PH, EC and CEC

The availability of many nutrients are largely controlled by soil pH, which also influences their uptake and tree growth (Londo et al 2006). Soil pH is good determiner of soil microbial composition and activity, and is very important in determining the mineralization and availability of nitrogen (Tian et al 2013). Soils with a pH of 6.0-7.0 typically have high concentrations of available nutrients. Tree species significantly influence soil pH; lower pH under agroforestry ascribed to the increased accumulation of aboveground biomass, associated cation uptake and production of organic acids by the tree component of agroforestry systems (Tornquist et al 1999, Gupta and Sharma 2009, Sarvade et al 2014c). Yan et al (2003) reported lowering trend in soil pH with going towards the citrus trees from wheat rows due to increase in leaf litter addition in citrus + wheat agroforestry. Rathore et al (2013) reported reduction in soil pH (12.12–15.62 per cent) in fruit based agri-horticultural model (mango + cowpea - toria) compared with the initial value recorded in 1995 before the establishment of orchard. The significantly lower pH values were reported by Dong et al (2009) under shrubs than between shrubs ($P < 0.01$) in Northeast China.

Electrical conductivity (EC) is a measure of the amount of salts in soil (salinity of soil). It is an important indicator of soil health and commonly expressed in units of deci-siemens per meter (dS/m), which is equal to the reading in milli Siemen per meter divided by 100. The significantly higher soil EC was reported by Dong et al (2009) under sand-fixing shrubs on sand dune than between shrubs ($P < 0.01$) at the 0-20 cm depth in Northeast China.

Exchangeable cations are a short-term storage pool that can rapidly replenish nutrient ions in soil solutions. Cation exchange is the major nutrient reservoir of K^+, Ca^{2+}, and Mg^{2+} and important for holding N in the ammonium (NH_4^+) form, and to some extent supplies micronutrient trace metals like Zn^{2+} and Mn^{2+}. Soil organic matter is a very important factor in soil fertility under agroforestry systems. It acts as a reservoir of plant nutrients, has a high CEC, buffers soil pH, and chelate micronutrients. Cation exchange helps the soils to resist changes in pH in addition to retaining plant nutrients. Soil cation exchange capacity (CEC) increased with increase in plantation age of *Caragana microphylla* plantation and reported positive correlated with soil pH (Zhang et al 2013).

Soil Macro and Micro Nutrients

Nutrient return to soil varies greatly depending on tree species, spacing, plantation age, intercrops and management techniques (Singh and Sharma 2007), because of variation in the litter fall addition. They reported 1.5, 2.7, 3.8 and 4.5 Mg ha^{-1} litter production in 2, 3, 4 and 5 year old poplar, respectively; the return of 4.72-7.32 mg kg^{-1} N, 1.96-3.75 mg kg^{-1} P and 3.71-5.49 mg mg kg^{-1} K. The higher concentration of Fe (15.5 per cent), Mn (13.0 per cent) and Cu (10.5 per cent) at successive sampling times was reported in 4 years poplar plantations with sugarcane initially and fodder-wheat rotation afterwards. Yadav and Bisht (2014) studied litter fall and nutrient returns from pecan nut (*Carya illinoinensis*) in agroforestry system in Indian Himalaya and reported 2143.3 kg ha^{-1} yr^{-1} litterfall production and returns 901.91 kg ha^{-1} yr^{-1} carbon, 57.44 kg ha^{-1} yr^{-1} nitrogen, 3.21 kg ha^{-1} yr^{-1} phosphorus and 43.29 kg ha^{-1} yr^{-1} potassium through break down of litter in agri-horticultural system. They also reported 42.1 per cent C, 2.68 per cent N, 0.15 per cent P and 2.02 per cent K from the fall of pecan nut tree. Semwal et al (2003) reported highest rate of N and P release through decomposition of added leaf litter by six multipurpose tree species, viz. *Alnus nepalensis, Albizzia lebbek, Boehmeria rugulosa, Dalbergia sissoo, Ficus glomerata* and *F. roxburghii* in a mixed plantation established on an abandoned agricultural land in Central Himalaya. Singh et al (2007) observed that 0.473, 7.77, 1.05 and 0.116 kg per ha Zn, Fe, Mn and Cu, respectively, were returned to soil through poplar litterfall in 5 year old plantation with trees spaced at 5 × 4 m. The significantly increase in exchangeable calcium (Ca), magnesium (Mg), potassium (K), saturations of Mg and K, whereas Ca saturation tended to decrease with the increasing age of *Caragana microphylla* (Zang et al 2013). Yadav et al (2008) studied litter dynamics and soil properties under different multipurpose tree species and reported higher total litterfall production under *P. cineraria* followed by *A. leucophloea*, *A. nilotica* and *D. sissoo* based agroforestry systems of Semi-Arid Eastern Plain Agroclimatic Zone of Rajasthan, India. Litterfall from *P. cineraria* reported highest NPK concentration. Macro and micronutrients return through poplar litterfall followed the order: Ca > N > K > S > P and Fe > Mn > Zn > Cu, respectively. The improvement in total nitrogen (6.25–56.25 per cent), phosphorus (10.27–48.63 per cent) and potash (20.7–58.46 per cent) under fruit based agri-horticultural model under rainfed situation of Western Himalaya, India (Rathore et al 2013). Soils under MPTs (*Michelia oblonga* Wall, *Parkia roxburghii* G. Don, *Alnus nepalensis* D. Don, and *Pinus kesiya* Royle ex- Gordon) recorded significant increases of 17 per cent soil organic carbon, 26 per cent available nitrogen (AN), 28 per cent phosphorus (AP) and 50 per cent potassium (AK) (Ramesh et al 2013). Yan et al (2003) reported higher total C in the terraced soils under citrus trees than the wheat, tea bush and forestry on unterraced slopes. A similar trend was observed for total N, reactive C and readily available nutrients (N, P and K).

Table 1: *Chemical characteristics of litter of four MPTs under agri-silviculture system*

Parameter	P. cineraria	D. sissoo	A. leucophloea	A. nilotica
C (%)	48.7dc	49.3ab	51.2a	48.9cb
N (%)	2.68a	2.31b	1.87d	2.12b
P (%)	0.22a	0.19b	0.23a	0.12c
K (%)	0.89a	0.72b	0.33d	0.68b
C/N	18.2c	21.3b	27.4a	23.1b
C/P	224.4c	254.1b	226.5c	404.1a
Lignin (%)	21.6a	7.94c	14.3b	7.9c
Lignin/N	8.06a	3.47b	7.65a	3.73b
ADF (%)	38.7b	21.8c	47.7a	19.5c
NDF (%)	49.5b	35.2c	56.5a	29.2d

Source: Adapted from Yadav et al (2008)

BIOLOGICAL NITROGEN FIXATION

Increasing use of N fertilizers in intensive agriculture decreases its use efficiency and causes environmental problems. Many of the marginal farmers cannot afford to buy nitrogen fertilizers and therefore yield of the annual crop suffer. The atmospheric biological nitrogen fixation has nutrient cycling potential in agroforestry systems and play important role in resolving these problems. In case of legume, most of the nitrogen fixing tree species are from Mimosoideae and Caesalpinioideae, with fewer in Papilionoideae subfamilies. Within these sub-families, 98, 60 and 30 per cent of the tested mimosoids, papilionoids and caesalpiniods, respectively showed potential to fix atmospheric nitrogen. The non-legume families, such as Betulaceae, Casuarinaceae, Chrysobalanaceae, Coriariaceae, Eleagnaceae, Myricaceae, Rhamnaceae, Rosaceae, Ulmaceae and Zmiaceae also showed potential for N_2 fixation. The important nitrogen fixing plant species with fixed atmospheric nitrogen quantity at kg per ha per year is given in Table 2.

Substantial amounts of N input from biological N_2 fixation by leguminous trees in symbiosis with rhizobial bacteria can yield positive effects for soil improvement and productivity of agroforestry. The fixed N is released into the soil through litter fall, root and nodule decay. However, such inputs are subject to a number of variables, including soil, climate, and legume-Rhizobium interaction management conditions (Mafongoya et al 2004). On the other hand, the influence of the age of leguminous trees on N_2 fixation is not well documented. Such information is important for assessment of the sustainability of agroforestry in circumstances where N is exported from the field in timber trees. Export of timber trees can potentially produce a negative N balance in the timber agroforestry system. Biological nitrogen fixation mediated by nitrogenase enzymes in soil and depends on ecological conditions in association with the specific nitrogen fixation capabilities of certain microorganisms and plant genotypes at different climatic conditions (Shridhar 2012).

Table 2: *Important N_2 fixing plant species with associated bacterial strains/ microorganisms*

Plant	Scientific name	Family	Strains/ microorganism	Nitrogen fixed (kg N ha^{-1} yr^{-1})	Reference
Silver wattle	*Acacia dealbata*	Mimosoideae	Rhizobia	12-32	Adams and Attiwill (1984), Leary et al (2006)
Black wattle	*Acacia mearnsii* De Wild.		Bradyrhizobium	200	Nair (1993), Silva and Uchida (2000)
Gum Arabic tree	*Acacia nilotica*		Rhizobium	40-100	Sharma and Kapoor (2005), Shetta et al (2011)
Koko; Woman's tongues tree	*Albizia lebbeck*		Bradyrhizobium	94	Liya et al (1990)
Indian alder	*Alnus nepalensis* D. Don	Betulaceae	Frankia	29-117	Sharma and Kapoor (2005), Mishra and Rai (2013)
European alder	*Alnus glutinosa*		Frankia	43-60	Sharma and Kapoor (2005)
Beef wood	*Casuarina equisetifolia* L.	Casuarinaceae	Frankia	60-110	Nair (1993), Silva and Uchida (2000)

Erythrina	*Erythrina poeppigiana (Walp.) Cook*	Papilionoideae	Bradyrhizobium	60	Nair (1993), Silva and Uchida (2000)
Ana Tree	*Faidherbia albida* (Delile) A.Chev.	Mimosoideae	Bradyrhizobium	20	Nair (1993), Silva and Uchida (2000)
Gliricidia	*Gliricidia sepium* (Jacq.) Kunth ex Walp.	Fabaceae	Rhizobium	13-108	Nair (1993), Silva and Uchida (2000)
Inga	*Inga jinicuil*	Mimosoideae	Rhizobium	35-40	Roskoski (1981)
Subabool, Ipil-Ipil	*Leucaena leucocephala* (Lam.) de Wit.		Rhizobium; occasionally with Bradyrhizobium	100-500	Nair (1993), Silva and Uchida (2000)
Mesquite	*Prosopis juliflora*		Rhizobia	30-80	Sharma and Kapoor (2005), Leary et al (2006)
New dhaincha	*Sesbania rostrate*	Fabaceae	Sinorhizobium	70–458	Nair (1993), Silva and Uchida (2000), Leary et al (2006)
Dhaincha	*Sesbania sesban* (Jacq.) W. Wight		Sinorhizobium; Rhizobium	43–600	Ndoye and Dreyfus (1988), Shaheen et al (2004), Sharma and Kapoor (2005), Leary et al (2006), Degefu et al (2011)
True indigo	*Indigofera tinctoria* L.		Rhizobium	79	Date (1991), Nair (1993), Castro et al (1999), Silva and Uchida (2000), Zang et al (2001), Workalemahu (2009), Madhaiyan et al (2009) Kumari et al (2010), Date and Eagles (2010), Dubey et al (2010), Solomon et al (2012), Zahran et al (2013), Parthiban and Mahesh (2015), Reeve et al (2015)
Stylo (Brazilian Lucerne)	*Stylosanthes* spp.		Bradyrhizobium	20-263	
Horse bean	*Vicia faba* L.		Rhizobium	45–552	
Pigeon pea	*Cajanus cajan* L. Millsp.		Rhizobium	68-88	
Cowpea	*Vigna sinensis* L. Walp.		Rhizobium	73–354	
Black gram, Mung bean	*Vigna mungo* L. Hepper		Rhizobium	63–342	
Soybean	*Glycine max* L. Merr.		Bradyrhizobium	60–168	
Chickpea	*Cicer arietinum* L.		Rhizobium	103	
Lentil	*Lens esculenta* Moench		Rhizobium	88–114	
Peanut, Groundnut	*Arachis hypogaea* L.		Bradyrhizobium	72–124	
Pea	*Pisum sativum* L.		Rhizobium	55–77	
Bean	*Phaseolus vulgaris* L.		Rhizobium	40–70	
Alfalfa, Lucerne	*Medicago sativa* L.		Sinorhizobium	229–290	
Clover	*Trifolium* spp.		Rhizobium	128–207	
Sunhemp	*Crotalaria juncea* L.		Methylobacterium	199–223	
Wild tantan	*Desmanthus virgatus* (L.) Willd.		Rhizobium	196–226	

Nutrient Pumping

The growth and development of root system of multipurpose tree governs the nutrient pumping in agroforestry systems. It is considered that the deep root system act as 'safety net' and absorb leached nutrients, and pumped it to the aboveground growing part of the tree (Suprayogo et al 2010, Kumar 2011). In further process such as leaf litter addition, decomposition, release and return to the soil system (Figure 3). Toky and Bisht (1992) reported length (cm) and basal diameter (cm) of main and lateral roots of some important agroforestry multipurpose tree species (Table 3). They reported *Prosopis cineraria* with highest main root length (288 cm) followed by *Acacia nilotica* (215 cm) and *Eucalyptus tereticornis* (179 cm); whereas *Populus deltoides* with highest lateral root (271 cm) system.

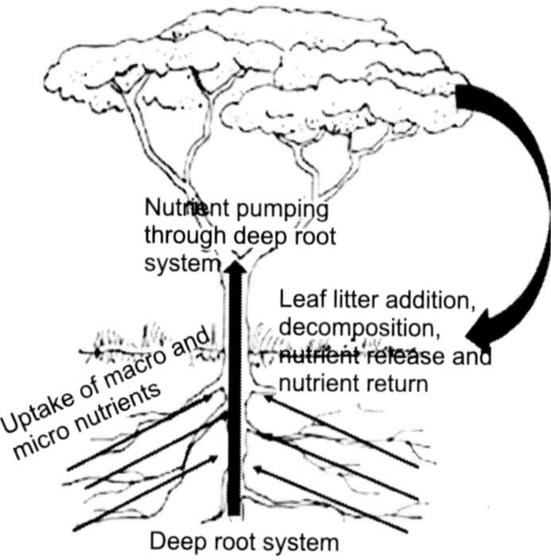

Figure 3: *Nutrient pumping through deep root system of multipurpose tree species in agroforestry systems.*

Schroth (1999) and Gowda and Kumar (2008) reported that the rooting depth of extent plants (e.g. trees) determines their nutrients uptake ability from subsoil and make them available in topsoil which may be available for associated plants (e.g. crops) with shallower root systems through nutrient pumping. The closer associated tree components in coconut+dicot tree system had high subsoil root activity and recorded greater nutrient capturing potential from the lower leaching nutrients (Gowda and Kumar 2008). Montagnini and Nair (2004) reported grevillea may utilize (capture/pump) nutrients from deeper soil layers and made available to the maize crop which capture nutrients from surface soil layer in grevillea and maize intercropping system.

Table 3: *Length and diameter of nutrient pumping main and lateral roots (Adopted from Toky and Bisht 1992)*

Species	Main roots		Lateral roots	
	Length (cm)	Basal diameter (cm)	Length (cm)	Basal diameter (cm)
Acacia catechu	143 ± 7.3	8.6 ± 3.1	99 ± 7.9	2.4 ± 0.3
Acacia nilotica	215 ± 7.8	15.4 ± 1.1	139 ± 10.8	2.3 ± 0.2
Albizzia lebbek	118 ± 36.8	11.4 ± 2.1	141 ± 16.7	3.7 ± 0.3
Azadirachta indica	117 ± 22.3	12.2 ± 1.4	82 ± 7.6	2.5 ± 0.2
Dalbergia sissoo	130 ± 17.0	10.1 ± 0.5	134 ± 11.9	2.5 ± 0.2
Melia azedarach	92 ± 6.7	12.4 ± 0.5	66 ± 3.50	3.0 ± 0.3
Morus alba	72 ± 6.2	9.5 ± 1.9	160 ± 13.6	3.0 ± 0.3
Prosopis cineraria	288 ± 43.3	9.1 ± 4.3	157 ± 20.2	2.7 ± 0.3
Zizyphus mauritiana	102 ± 1.8	9.2 ± 1.2	112 ± 14.0	2.4 ± 0.4
Populus deltoides	85 ± 10.4	26.4 ± 2.4	271 ± 60.9	6.8 ± 1.0
Eucalyptus tereticornis	179 ± 12.4	17.9 ± 1.0	167 ± 23.3	3.5 ± 0.3
Leucaena leucocephala	148 ± 32.7	11.7 ± 0.6	72 ± 6.3	2.7 ± 0.3

SOIL BIOLOGICAL PROPERTIES

The decomposition of plant litter and most soil activities such as nutrient cycling, energy flow are influenced by soil microflora such as fungi and bacteria (Tian et al 2012). Soil biological indicators such as microbial count, microbial biomass C and N, microflora and basal respiration that are commonly used to measure soil health and quality are playing very important role in improving soil physico-chemical properties; described by several researchers (Bastida et al 2008, Panwar et al 2013, Berhe and Retta 2015, Zornoza et al 2015). Whereas the soil biological properties are influenced by soil organic matter and organic carbon added through litter fall, plant residue, decomposition of litter and plant species composition under different agroforestry systems (Tangjang and Arunachalam 2009, Tian et al 2012).

Microbial Count

Tangjang and Arunachalam (2009) reported higher bacterial and fungal count under traditional agroforestry systems in Northeast India. They also reported the strong positive correlation of bacterial count with clay and moisture content in soil, whereas fungal count showed strong positive correlation with total nitrogen content in soil. The soil pH and organic carbon are deciding factors for microbial population; neutral soil pH and high soil organic carbon increases microbial population under agroforestry systems which is influenced by increased soil toxicity and high fluctuations in soil pH (Hombegowda et al 2015). Unger et al (2013) observed that the soils under agroforestry vegetation filter strip treatment increased by 15 per cent more total bacteria (Fig. 4), 21 per cent more gram-negative bacteria, 23 per cent more anaerobic bacteria, and 35 per cent more mycorrhizal fungi than the cropland soil (Fig. 5) in north-central MO, USA and adding organic matter inputs in variable amount were the most likely reasons for similar results. Whereas, grass VFS supported larger communities of mycorrhizal fungi and protozoa than the cropland (44 and 43 per cent greater, respectively) (Fig. 5). Hailemariam et al (2013) reported higher level of root colonization by Arbuscular Mycorrhiza (AM) for *Croton macrostachyus* (45 per cent) tree than *Albizia gummifera* (41 per cent). They also observed significantly higher root colonization of maize crops grown under the canopy of *A. gummifera* and *C. macrostachyus* trees by AM than outside the canopy (P < 0.001).

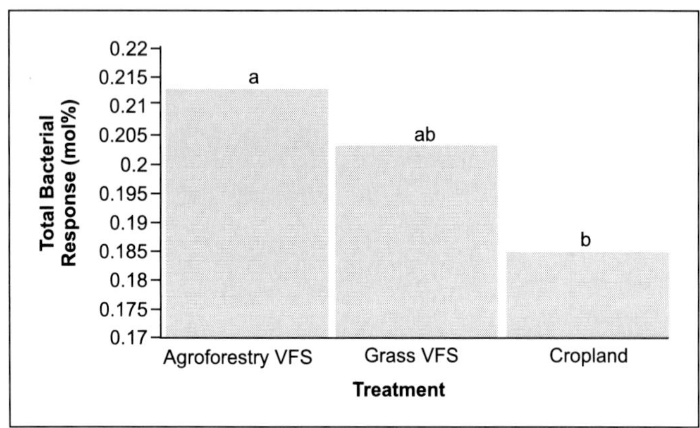

Figure 4: *Total bacteria response of the soil microbial community in the vegetative filter strips (VFS) and cropland. Bars with same letter are not significantly different (a < 0.05) (adopted from Unger et al 2013)*

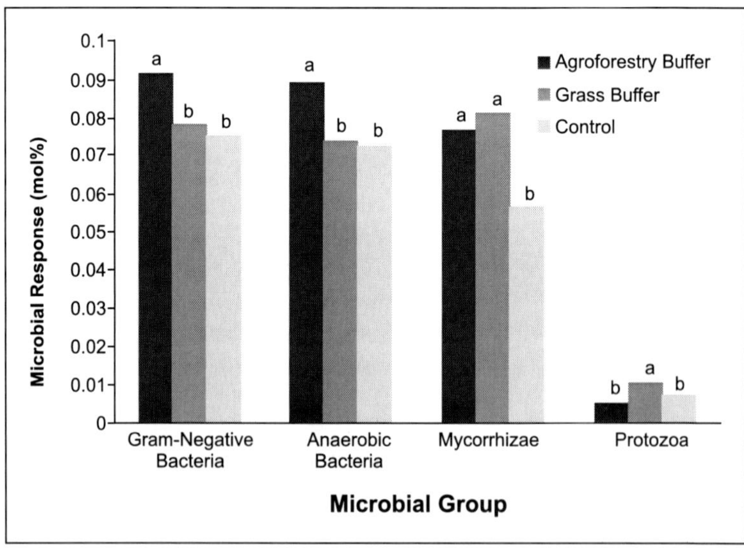

Figure 5: *Response of the soil microbial community in the vegetative filter strips (VFS) and cropland. Bars with same letter are not significantly different (a < 0.05) (adopted from Unger et al 2013)*

Microbial Biomass Carbon and Nitrogen

Panwar et al (2013) reported higher microbial biomass carbon (MBC) from surface soil of *Terminalia arjuna* (526 mg per kg) followed by *Gmelina arborea* (408.33 mg per kg) and least in subsurface soil of *Salix tetrosperma* (280 mg per kg) based agri-silviculture systems in West Bengal, India. Rodrigues et al (2015) assessed soil microbial biomass under babassu palm-*Brachiaria brizantha* associations (with different palm densities) and reported higher soil microbial biomass C (MBC) and N (MBN) in high density of babassu palm than the straight grass pasture. They also reported higher rate of plant litter addition (15 t per ha), that supply microbial biomass with C sources, which favoured the accumulation of soil microbial biomass. The quality and quantity of litter addition, root exudations also influence soil microbial biomass. Seasonal changes in microbial activity was reported by Kumar et al (2008); they recorded higher microbial activity in autumn and lower in winter season. The higher microbial biomass carbon (MBC) was observed under *Terminalia arjuna* which was followed by *Eucalyptus tereticornis* and *Swetenia mahogany*. Whereas, higher microbial quotient (MQ) was recorded from soils under *Pongamia pinnata* as compared to *E. tereticornis, S. mahogany* and *T. arjuna*. Tornquist et al (1999) investigate the effects of agroforestry systems on soil characteristics of the Sarapiqui region of Costa Rica and reported higher litter production in agroforestry and pasture land was the responsible factor for higher microbial activity, soil microbial biomass C (SMBC) and soil microbial biomass N (SMBN). Kaur et al (2000) studied tree based systems, *Acacia* + rice-berseem, *Eucalyptus* + rice-berseem and *Populus* + rice-berseem and compered with sole cropping; they reported 42 per cent increase in microbial C and 13 per cent microbial N in tree-based systems as compared to mono-cropping. In comparison if different agroforestry (tree based) systems *Acacia* + rice-berseem showed higher soil microbial biomass carbon (153.40 μg per g soil) and microbial biomass nitrogen (32.57 μg per g soil), whereas higher microbial C: N (6.92) from soils under *Eucalyptus* sole planting. Tian et al (2012) reported higher microbial biomass C, N in *Ginkgo-Tea* agroforestry compared to monoculture; 1.25-1.31 times higher basal respirations and 1.15-1.2 times higher metabolic quotient also observed in their study on soil microbiological properties. The 34 per cent increase in soil MBC was recorded under tree species compared to control by Ramesh et al (2013); they also reported positive correlation between amount of net carbon stored in the soil and MBC ($r = 0.706^{**}$) under hilly ecosystems of Northeast India (Figure 6). Some studies from India and other country have been conducted to know the relations between soil microbial biomass C, N and P with soil organic matter and soil fertility in agroforestry systems, and reported positive correlation (Kaur et al 2000, Sharma et al 2004, Zaia et al 2008).

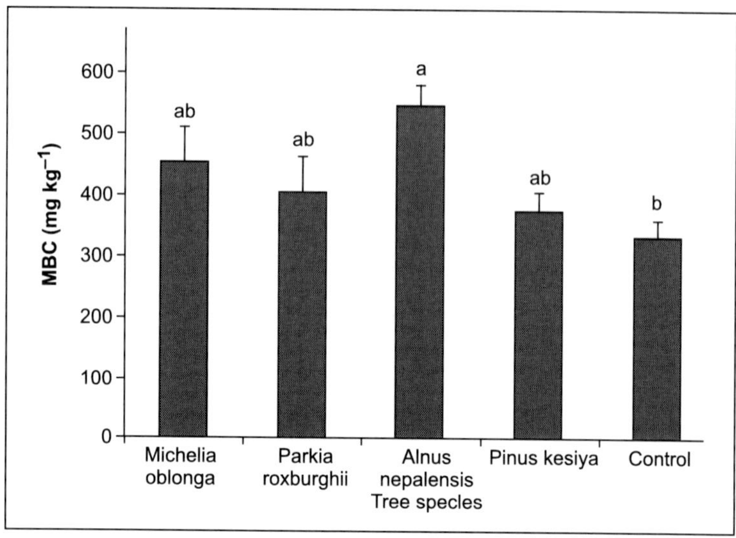

Figure 6: *Soil microbial biomass carbon (mg per kg) under various multipurpose tree species at Umiam in NEH region of India*

Soil Basal Respiration (SBR)

Microbial activity in the soil indicated by rate of soil respiration and influence organic matter decomposition (Santos et al 2012, Tian et al 2012); soil respiration might be indicated by either a disturbance of the soil or a high level of productivity in the ecosystem (Islan and Weil 2000). A large pool of labile C substrates under agroforestry system may considered a responsible factor for higher rate of soil basal respiration (Wang et al 2005). They also reported positive correlation between SBR and total organic carbon.

Micro-Flora and Micro-Fauna

This is one most important soil health indicator and respond delicately to land management practices and correlate well with beneficial soil functions including water storage, decomposition and nutrient cycling through improving soil physico-chemical properties (Akinnifesi et al 2010). Soil microflora such as fungi and bacteria are responsible for the decomposition of plant litter. Maurya et al (2012) reported higher *Azospirillum* and *Azotobacter* population in agroforestry and lower in rice-wheat as compared to vegetables and grassland based crop rotations in Eastern Uttar Pradesh of India.

Enzyme Activities

Soil enzyme activity are important indicator for soil quality through changes in metabolic capacity and nutrient cycling. Wang et al (2005) reported dehydrogenase activity reflects the total range of oxidative activity of soil microflora and a good indicator of microbiological activity, which helps in cycles of some important nutrients such as N, P and C, so it generally used as an early indicators of soil health/quality changes. Their activity may change form soil type, vegetation types and different seasons. Significantly correlated between soil enzyme (β-glucosidase, β-glucosaminidase, dehydrogenase and β-glucosaminidase) activities and soil organic carbon content ($r > 0.94$, $P < 0.0001$) reported by Paudel et al (2012). The 1.1-1.13 times higher catalase and 1.56 and 1.40 times higher protease activity; whereas higher in some extent of polyphenoloxidase, dehydrogenase, urease and invertase activities also recorded under *Ginkgo-Tea* agroforestry systems compared to monoculture (Tian et al 2012). They also reported strong positive correlated between organic C content and catalase, dehydrogenase, urease, protease and invertase enzyme activities ($r = 0.95$, 0.95, 0.87, 0.87, and 0.93, respectively) under agroforestry systems. Such results were experienced by different researchers and they hypothesised that the perennial vegetation provide environmental conditions suitable for greater accumulation of organic C and total N through adding tree leaf litter and crop residue. The enzyme activities were influenced by replacing meadows with pine plantations under a temperate climate Mijangos et al (2014), however enzyme activities are highly sensitive to human-induced alterations. Hydrolysis of fluorescein diacetate (FDA) and dehydrogenase activity (DHA) was reported higher in agroforestry systems with high densities of babassu palm in Brachiaria brizantha pasture (Rodrigues et al 2015). Mungai et al (2005) reported from their studies that enzyme activities are greater in agroforestry alley cropping practices due to differences in litter quantity and quality and, root exudates. Yadav et al (2010) investigated the effect of traditionally grown trees (*Prosopis cineraria, Dalbergia sissoo, Acacia leucophloea* and *Acacia nilotica*) on soil biological characteristics and reported higher dehydrogenase and alkaline phosphatase activity under different tree based agroforestry systems as compared to mono-cropping system.

TREES AND CONTROL OF SOIL EROSION

In last four decades sustainability of natural resources had threatened due to misuse and mismanagement of natural resources for human greed. The 57.08 per cent (187.8 m ha) of the total geographical area (329 m ha) is suffering from various forms of degradation like water erosion (79 per cent of 187.8 m ha), wind erosion, chemical and physical degradation. Agroforestry systems play a great role in conservation of natural resources, especially soil. The long root system of tree species also acts as safety net for controlling soil erosion losses caused due to water and wind actions. Udawatta et al (2002) reported 23 and 10 per cent runoff reductions, and 20 per cent N loss reduction by agroforestry and contour strip treatments based on calibrations. The potential of large cardamom based agroforestry system in reducing soil erosion in Sikkim Himalaya was reported by Mishra and Rai (2013).

Table 4: *In-situ retention of rainfall under different land use systems*

Land use	Mean Slope (%)	Conservation measures	Annual rainfall Retained (%)
Agriculture-Fodder	32.0	Contour bunds + bench terraces	98.53-99.77
Forestry	38.0	Vegetation	92.79-98.27
Agro-forestry	33.0	Tree+Crop/herb/grass	96.83-99.52
Agriculture-Food	32.1	Contour trench + bench terraces	0.00-98.68

Agri-horti-silvi pastoral	32.4	Contour bunds, bench & half-moon terraces	98.27-99.99
Horticulture	41.8	Contour bunds, bench & half-moon terraces	96.18-99.75
Natural fallow	32.4	Natural vegetation	Trace to 98.46

Source: Adopted from Bundela 2007

Bundela (2007) reported potential of agroforestry system in soil and water conservation through performing as vegetation barrios for soil erosion processes, and helping in retaining annual rainfall in the field for increasing soil moisture and improving other soil properties (Table 4).

CONCLUSION

The perennial woody plant based systems assisted for improving soil health/ quality (soil physico-bio-chemical properties) and help in achieving sustainable development goal. In that, leaf litter addition, decomposition, release and return of nutrients to the soil system makes very closed nutrient cycling in agroforestry system. The biological atmospheric nitrogen fixation and carbon sequestration (soil) helps to improve soil nutrient status which helps in eradication of poverty from the world. Not only soil improvement benefits but soil and water conservation to avoid erosion losses also helps to increase productive capacity of the soil through improving soil health.

ACKNOWLEDGEMENT

Authors would like to thanks Dean, College of Agriculture, Balaghat for encouraging and providing assistance for writing the manuscript.

REFERENCES

Adams MA and Attiwill PM 1984 Role of *Acacia* spp. in nutrient balance and cycling in regenerating *Eucalyptus regnans* forests. II. Field studies of acetylene reduction, *Australian Journal of Botany* **32** 217–223

Akinnifesi FK, Ajayi OC, Sileshi G, Chirwa PW, and Chianu J 2010 Fertiliser trees for sustainable food security in the maize-based production systems of East and Southern Africa. A review, *Agronomy for Sustainable De*velopment **30** 615-629 DOI: 10.1051/agro/2009058

Anju SV and Koppad AG 2013 Influence of in situ soil moisture conservation measures on growth and productivity of *Acacia auriculiformis*, *Karnataka Journal of Agricultural Sciences* **26**(1) 170-171

Anonymous 2010 Degraded and wastelands of India: status and spatial distribution. Indian Council of Agricultural Research, New Delhi pp 155

Anonymous 2012 Climate change and Indian agriculture: impact, adaptation and vulnerability. Division of Environmental Sciences, Indian Agricultural Research Institute, New Delhi pp 26

Araujo ASF, Leite LFC, Iwata BF, Jr M de AL, Xavier GR, and Figueiredo MVB 2012 Microbiological process in agroforestry systems. A review, *Agronomy for Sustainable Development* **32** 215-226

Araujo QR, Comerford NB, Ogram AV, Al-Agely, A, Filho LPS, and Santos JG 2004 Soil carbon and physical property changes in Brazilian Coastal Tableland soils with land use following deforestation, *Agroforestry Systems* **63** 193-198

Bastida F, Zsolnay A, Hernandez T, and Garcia C 2008 Past, present and future of soil quality indices: a biological perspective, *Geoderma* **147** 159-171

Berhe DH and Retta AN 2015 Soil improvement by trees and crop production under tropical agroforestry systems: a review, *Merit Research Journal of Agricultural Science and Soil Sciences* **3**(2) 18-28

Bhattacharyya R, Ghosh BN, Mishra PK, Mandal B, Rao CS, Sarkar D, Das K, Anil KS, Lalitha M, Hati KM, and Franzluebbers AJ 2015 Soil degradation in India: challenges and potential solutions, *Sustainability* **7** 3528-3570 DOI: 10.3390/su7043528

Bundela DS 2007 *Water Management in Northeast India - Some Case Studies*. Kaledhonkar et al (Eds). CSSRI, Karnal, India. Pp. 40-46

Carter MR, Andrews SS, and Drinkwater LE 2004 Systems approaches for improving soil quality, in *Managing Soil Quality: Challenges in Modern Agriculture* (eds) Schjonning P, S Elmholt and BT Christensen, CABI International, Wallingford, UK, 261-281pages

Castro S, Permigiani M, Vinocur M, and Fabra A 1999 Nodulation in peanut (*Arachis hypogaea* L.) roots in the presence of native and inoculated rhizobia strains, *Applied Soil Ecology* **13** 39-44

Chauhan SK, Dhillon WS, Singh N, and Sharma R 2013 Physiological behaviour and yield evaluation of agronomic crops under agri-horti-silviculture system, *International Journal of Plant Research* **3**(1) 1-8

Dadhwal KS, Mandal D, Srimali SS, Dhyani SK, Mohanand SC, and Raizada A 2011 Landscape-scale soil quality assessment under different land use systems in north-western hilly region, *Indian Journal of Soil Conservation* **39**(2) 128-135

Das K, Deb S, and Arunachalam A 2011 Effect of earthworm and plant residues for soil enrichment under agroforestry systems for north- east India, *Indian Journal of Soil Conservation* **39**(1) 73-79

Date RA 1991 Nitrogen fixation in *Desmanthus*: strain specificity of *Rhizobium* and responses to inoculation in acidic and alkaline soil *Tropical Grasslands* **25** 47-55

Date RA and Eagles DA 2010 *Bradyrhizobium* strain effectiveness for *Stylosanthes macrocephala*, *Tropical Grasslands* **44** 158-164

Degefu T, Wolde-meskel E, and Frostegard A 2011 Multilocus sequence analyses reveal several unnamed *Mesorhizobium genospecies* nodulating *Acacia* species and *Sesbania sesban* trees in Southern regions of Ethiopia, *Systematic and Applied Microbiology* **34** 216-226 DOI:10.1016/j.syapm.2010.09.006

Devi B, Bhardwaj DR, Panwar P, Pal S, Gupta NK, and Thakur CL 2013 Long term effects of natural and plantation forests on carbon sequestration and soil properties in mid-hill sub-humid condition of Himachal Pradesh, India, *Range Mgmt. and Agroforestry* **34** (1) 19-25

Dong XW, Zhang XK, Bao XL, and Wang JK 2009 Spatial distribution of soil nutrients after the establishment of sand-fixing shrubs on sand dune, *Plant Soil Environ* **55**(7) 288-294

Dubey RC, Maheshwari DK, Kumar H, and Choure K 2010 Assessment of diversity and plant growth promoting attributes of rhizobia isolated from *Cajanus cajan* L, *African Journal of Biotechnology* **9**(50) 8619-8629 DOI: 10.5897/AJB10.1127

Ernest H, Kumar A, Bhople BS, Srivastava PK, Singh RD, and Sahu SK 2015 Soil quality and health management through farming practices: to ensure agricultural sustainability, *Research Journal of Pharmaceutical, Biological and Chemical Sciences* **6**(4) 1983-1992

FAO 2015 Healthy soils are the basis for healthy food production. Rome, Italy pp 4

Gowda HBS and Kumar BM 2008 Root competition for phosphorus between coconut palms and inter-planted dicot trees along a soil fertility gradient, in *Towards Agroforestry Design: An Ecological Approach* (eds) Jose S and Gordon A, Springer, The Netherlands, 175-193 pages

Grogan P, Lalnunmawia F, and Tripathi SK 2012 Shifting cultivation in steeply sloped regions: a review of management options and research priorities for Mizoram state, Northeast India, *Agroforestry Systems* **84** 163-177

Guleria V, Vashisht A, Gupta A, Salven T, Thakur C, and Kumar P 2014 Carbon stocks and soil properties under nitrogen-fixing trees on degraded site in subtropical Himalayan region, *Indian Journal of Soil Conservation* **42**(3) 293-297

Gupta MK and Sharma SD 2009 Effect of tree plantation on soil properties, profile morphology and productivity index: poplar in Yamunanagar district of Haryana, *Annals of Forestry* **17**(1) 43-70

Gupta N, Kukal SS, Bawa SS, and Dhaliwal GS 2009 Soil organic carbon and aggregation under poplar based agroforestry system in relation to tree age and soil type, *Agroforestry Systems* **76** 27-35

Hailemariam M, Birhane E, Asfaw Z, and Zewdie S 2013 Arbuscular mycorrhizal association of indigenous agroforestry tree species and their infective potential with maize in the rift valley, Ethiopia, *Agroforestry Systems* **87** 1261-1272 DOI 10.1007/s10457-013-9634-9

Hillel D 1982 Introduction to soil physics, Academic Press, San Diego, CA

Hombegowda HC, Straaten OV, Kohler M and Holscher D 2015 On the rebound: soil organic carbon stocks can bounce back to near forest levels when agroforests replace agriculture in southern India *Soil Discuss* **2** 871-902

Isaac SR and Nair MA 2006 Litter dynamics of six multipurpose trees in a homegarden in Southern Kerala, India, *Agroforestry Systems* **67** 203-213

Islan KR and Weil RR 2000 Soil quality indicator properties in mid-Atlantic soils as influenced by conservation management, *Journal of Soil Water Conservation* **55** 69-78

Joshi BK 2004 Soil, water and nutrient conservation in different landuse system in Mid Hill of Indian Central Himalaya- a case study from Bhetagad watershed, *Journal of Soil and Water Conservation* **3**(3&4) 113-122

Karlen DL, Mausbach, MJ, Doran JW, Cline RG, Harris RF, and Schuman GE 1997 Soil quality: a concept, definition, and framework for evaluation, *Soil Science Society of America Journal* **61** 4-10

Kaur B, Gupta SR and Singh G 2000 Soil carbon, microbial activity and nitrogen availability in agroforestry systems on moderately alkaline soils in northern India, *Applied Soil Ecology* **15** 283-294

Kumar BM 2011 Quarter century of agroforestry research in Kerala: an overview, *Journal of Tropical Agriculture* **49** (1-2) 1-18

Kumar K, Laik R, Das DK, and Chaturvedi OP 2008 Soil microbial biomass and respiration in afforested calciorthent, *Indian Journal of Agroforestry* **10**(2) 75-83

Kumari BS, Ram MR, and Mallaiah KV 2010 Studies on nodulation, biochemical analysis and protein profiles of *Rhizobium* isolated from *Indigofera* species, *Malaysian Journal of Microbiology* **6**(2) 133-139

Lakaria BL, Narayan D, Katiyar VS, Biswas H, Jha P, and Somasundaram J 2011 Runoff and soil loss under varying rainfall magnitudes and its prediction for different vegetative covers in Bundelkhand region, *Indian Journal of Soil Conservation* **9**(1) 1-8

Lawal HM, Ogunwole JO, and Uyovbisere EO 2009 Changes in soil aggregate stability and carbon sequestration mediated by land use practices in a degraded dry savanna Alfisol, *Tropical and Subtropical Agroecosyst*ems **10** 423-429

Leary JK, Singleton PW, Scowcroft PG, and Borthakur D 2006 Symbiotic diversity in the cosmopolitan genus *Acacia*, *Symbiosis* **41** 107-117

Liya SM, Odu CTI, Agboola AA, and Mulongoy K 1990 Estimation of N_2 fixation by nitrogen fixing trees in the subhumid tropics using 15-N dilution and difference methods, Paper presented at the fourth African Association for Biological Nitrogen Fixation, HTA. Ibadan, 24–28 Sept 1990

Londo AJ, Kushla JD and Carter RC 2006 Soil pH and tree species suitability in the south, *Southern Regional Extension Forestry* **2** 1-4

Lundgreen BO and Raintree JB 1982 Sustained agroforestry, in *Agricultural research for development: potentials and challenges in Asia* (eds) Nestel B, ISNAR. Hague Netherlands 37-49 pages

Madhaiyan M, Poonguzhali S, Senthilkumar M, Sundaram SP, and Sa T 2009 Nodulation and plant-growth promotion by methylotrophic bacteria isolated from tropical legumes, *Microbiological Research* **164** (1) 114-120 DOI:10.1016/j.micres.2006.08.009

Mafongoya P, Giller KE, Odee D, Gathumbi S, Ndufa SK, and Sitompul SM 2004 Benefiting from N2-fixation and managing rhizobia, in *Below-ground interactions in tropical agroecosystems. Concepts and models with multiple plant components* (eds) Van Noordwijk, Cadisch G and Ong CK CABI Wallingford, UK. 227-242 pages

Mandal D, Dhyani BL, Kumar A, Singh C, Bihari B, Muruganandam M, and Madhu M 2013 Impact of different land use systems on soil quality in northwest Himalayan region, *Indian Journal of Soil Conservation* **41**(2) 200-205

Maurya BR, Kumar A, Raghuwanshi R, and Singh V 2012 Diversity of *Azotobacter* and *Azospirillum* in rhizosphere of different crop rotations in Eastern Uttar Pradesh of India, *Research Journal of Microbiology* **7** 123-130 DOI: 10.3923/jm.2012.123.130

Mijangos I, Epelde L, Garbisu C, and Gonzalez-Oreja JA 2014 Modification of soil enzyme activities as a consequence of replacing meadows by pine plantations under temperate climate, *Pedobiologia* **57** 61-66

Mishra BP 2011 Vegetation composition and soil nutrients status from polyculture to monoculture, *African Journal of Environmental Science and Technology* **5**(5) 363-366

Mishra PK and Rai SC 2013 Use of indigenous soil and water conservation practices among farmers in Sikkim Himalaya *Indian Journal of Traditional Knowledge* **12**(3) 454-464

Montagnini F and Nair PKR 2004 Carbon sequestration: an underexploited environmental benefit of agroforestry systems, *Agroforestry Systems* **61** 281-295

Mostafa E, Mehdi E, Majid B, Hamed F, and Mehboub S 2008 Effect of land use change on selected soil physical and chemical properties in North Highlands of Iran *Journal of Applied Sciences* **8** (3) 496-502 DOI: 10.3923/jas.2008.496.502

Mungai NW, Motavalli PP, Kremer RJ, and Nelson KA 2005 Spatial variation of soil enzyme activities and microbial functional diversity in temperate alley cropping systems, *Biology and Fertility of Soils* **42** (2) 129-136 DOI: 10.1007/s00374-005-0005-1

Nair PKR 1993 *An Introduction to Agroforestry*, Kluwer Academic Publishers. Dordrecht, Boston, London 491p

Nair PKR 2007 The coming of age of agroforestry, *Journal of the Science of Food and Agriculture* **87** 1613-1619 DOI: 10.1002/jsfa.2897

Nair PKR 2014 Grand challenges in agroecology and land use systems, *Frontiers in Environmental Science* **2** 1-4 DOI: 10.3389/fenvs.2014.00001

Ndoye I and Dreyfus B 1988 N_2 fixation by Sesbania rostrata and Sesbania sesban estimated using N15 and total N difference methods, *Soil Biology and Biochemistry* 20 209-213

Odewumi SG, Iwara AI, and Ogundele FO 2013 Effects of teak (*Tectona grandis*) cultivation on soil physical and chemical properties in Ajibode community, Ibadan, Oyo state, *Wudpecker Journal of Agricultural Research* **2**(2) 49-54

Oliver K, Njira W, and Nabwami J 2013 Soil management practices that improve soil health: Elucidating their implications on biological indicators, *Journal of Animal and Plant Sciences* **18**(2) 2750-2760

Pandey DN 2007 Multifunctional agroforestry systems in India *Current Science* **92**(4) 455-463

Panwar P, Pal S, Chakravarty S, and Alam M 2013 Soil quality and production of low land paddy under agri-silviculture systems in acid soil of West Bengal, India, *Range Management and Agroforestry* **34** (1) 51-57

Parthiban P and Mahesh V 2015 Isolation and characterization of *Rhizobium* and their effect on *Vigna Mungo* L, *International Journal of Scientific Research* **4**(7) 764-766

Paudel BR, Udawatta RP, Kremer, RJ, and Anderson SH 2012 Soil quality indicator responses to row crop, grazed pasture, and agroforestry buffer management, *Agroforestry Systems* **84** 311-323 DOI 10.1007/s10457-011-9454-8

Prasad RN and Biswas PP 1999 *Soil Resource of India*, in *50 years of natural resource management research* (eds) GB Singh and BR Sharma, Indian Council of Agricultural research ICAR, New Delhi, 644 pages

Ramesh T, Manjaiah, KM, Tomar Tomar, and Ngachan SV 2013 Effect of multipurpose tree species on soil fertility and CO_2 efflux under hilly ecosystems of Northeast India, *Agroforestry Systems* **87** 1377-1388

Rathore AC, Saroj PL, Lal H, Sharma NK, Jayaprakash J, Chaturvedi OP, Raizada A, Tomar JMS, and Dogra, P 2013 Performance of mango based agri-horticultural models under rainfed situation of Western Himalaya, India, *Agroforestry Systems* **87** 1389-1404

Reeve W, Ardley J, Tian R, Eshragi L, Yoon JW, Ngamwisetkun P, Seshadri R, Ivanova NN, and Kyrpides NC 2015 A genomic encyclopedia of the root nodule bacteria: assessing genetic diversity through a systematic biogeographic survey, *Standards in Genomic Sciences* **10**(14) 1-8

Rodrigues RC, Araujo RA, Costa CS, Lima, AJT, Oliveira, ME, Cutrim jr JAA, Santos FNS, Araujo JS, Santos VM, and Araujo ASF 2015 Soil microbial biomass in an agroforestry system of Northeast Brazil, *Tropical Grasslands - Forrajes Tropicales* **3** 41-48 DOI: 10.17138/TGFT(3)41-48

Roskoski JP 1981 Nodulation and N_2 fixation by Inga jinicuil, a woody legume in coffee plantations. I. Measurements of nodule biomass and field C_2H_2 reduction rates, *Plant and Soil* **59**(2) 201-206

Saha R, Ghosh PK, Mishra VK, Majumdar B, and Tomar JMS 2010 Can agroforestry be a resource conservation tool to maintain soil health in the fragile ecosystem of north-east India? *Outlook on Agriculture* **39**(3) 191-196 DOI: 10.5367/oa.2010.0004

Saha R, Mishra VK, and Tomar JMS 2004 Soil water retention-transmission characteristics of various agro-forestry systems under hilly agriculture in Meghalaya, *Indian Journal of Hill Farming* **17** 106-110

Saha R, Tomar JMS, and Ghosh PK 2007 Evaluation and selection of multipurpose tree for improving soil hydro-physical behaviour under hilly eco-system of northeast India, *Agroforestry Systems* **69** 239-247

Santos VB, Leite LFC, Nunes LAPL, and Melo WJ 2012 Soil microbial biomass and organic matter fractions during transition from conventional to organic farming systems, *Geoderma* **170** 227-231 DOI: 10.1016/j.geoderma.2011. 11.007

Sarvade S 2014c Agroforestry: refuge for biodiversity conservation, *International Journal of Innovative Research in Science and Engineering* **2**(5) 424-429

Sarvade S and Singh R 2014d Role of agroforestry in food security, *Popular Kheti* **2**(2) 25-29

Sarvade S, Mishra HS, Kaushal R, Chaturvedi S, and Tewari S 2014c Wheat (*Triticum aestivum* L.) yield and soil properties as influenced by different agri-silviculture systems of Terai Region, Northern India, *International Journal of Bio-resource and Stress Management* **5**(3) 350-355

Sarvade S, Singh R, Prasad H, and Prasad D 2014a Agroforestry practices for improving soil nutrient status, *Popular Kheti* **2**(1) 60-64

Sarvade S, Singh R, Vikas G, Kachawaya DS, and Khachi B 2014b Agroforestry: an approach for food security *Indian Journal of Ecology* **41**(1) 95-98

Schroth G 1999 A review of belowground interactions in agroforestry, focussing on mechanisms and management options, *Agroforestry Systems* **43** 5-34

Semwal DP, Uniyal PL, Bahuguna YM, and Bhatt AB 2009 Soil nutrient storage under different forest types in a part of Central Himalayas, India, *Annals of Forestry* **17**(1) 43-52

Semwal RL, Maikhuri Rk, Rao KS, Sen, KK, and Saxena KG 2003 Leaf litter decomposition and nutrient release patterns of six multipurpose tree species of central Himalaya, India, *Biomass Bioenergy* **24** 3-11

Shaheen N, Hussain M, Yousaf F, Qureshi MS, and Idrees S 2004 Effect of rhizobium strains on growth of two *Sesbania* species, *Pakistan Journal of Life and Social Sciences* **2**(1) 79-81

Sharma P, Rai SC, Sharma R, and Sharma E 2004 Effects of land use change on soil microbial C, N and P in a Himalyan watershed, *Pedobiologia* **48** 83-92

Sharma PK and Kapoor KK 2005 Problems and perspectives of nitrogen fixation in agroforestry system, *Proceedings of Indian National Science Aca*demy **71**(3&4) 145-161

Shetta ND, Al-Shaharani TS, and Abdel-Aal M 2011 Identification and characterization of rhizobium associated with woody legume trees grown under Saudi Arabia condition, *American-Eurasian Journal of Agriculture & Environmental Sciences* **10**(3) 371-379

Shridhar BS 2012 Review: nitrogen fixing microorganisms, *International Journal of Microbiological Research* **3(1)** 46-52

Silva JA and Uchida R 2000 Biological nitrogen fixation: nature's partnership for sustainable agricultural production, in *Plant nutrient management in hawaii's soils, approaches for tropical and subtropical agriculture*, University of Hawaii at Manoa, 121-126 pages

Singh AN, Raghubanshi AS, and Singh JS 2004 Comparative performance and restoration potential of two *Albizia* species planted on mine spoil in a dry tropical region, India, *Ecological Engineering* **22** 123-140

Singh B and Sharma KN 2007 Tree growth and nutrient status of soil in a poplar (*Populus deltoides* Bartr.)-based agroforestry system in Punjab, India, *Agroforestry Systems* **70** 125-134

Singh B, Gill RIS, and Kaur N 2007 Litter fall and nutrient return in poplar plantation varying in row directions and spacings, *Indian Journal of Agroforestry* **9**(1) 33-37

Solomon T, Pant LM, and Angaw T 2012 Effects of Inoculation by *Bradyrhizobium japonicum* strains on nodulation, nitrogen fixation, and yield of soybean (*Glycine max* L. Merill) varieties on Nitisols of Bako, Western Ethiopia, International Scholarly Research Network ISRN Agronomy, 1-8 pages DOI: 10.5402/2012/261475

Somasundaram J, Singh RK, Ali S, Sethy BK, Singh D, Lakaria BL, Chaudhary RS, Singh RK, and Sinha NK 2012 Soil aggregates and other properties as influenced by different long term land uses under table landscape topography of Chambal region, Rajasthan, India, *Indian Journal of Soil Conservation* **40**(3) 212-217

Suprayogo D, Hairiah K, Noordwijk M, and Cadisch G 2010 Agroforestry interactions in rainfed agriculture: can hedgerow intercropping systems sustain crop yield on an ultisol in Lampung (Indonesia)? *Agrivita* **32**(3) 205-215

Tangjang S and Arunachalam K 2009 Microbial population dynamics of soil under traditional agroforestry systems in Northeast India, *Research Journal of Soil Biology* **1**(1) 1-7

Tian Y, Cao F, Wang G, Zhang W, and Yu W 2012 Soil Microbiological properties and enzyme activities in Ginkgo-Tea agroforestry compared to monoculture, *Forest Research* **1**(2) 1-6 DOI: 10.4172/2168-9776.1000107

Tian Y, Takanashi K, Toda H, Haibara K, and Ding F 2013 pH and substrate regulation of nitrogen and carbon dynamics in forest soils in a karst region of the upper Yangtze River basin, China, *Journal of Forest Research* **18** 228-237 DOI 10.1007/s10310-012-0341-6

Toky OP and Bisht P 1992 Observations on the rooting patterns of some agro forestry trees in an arid region of north-western India, *Agroforestry systems* **18** 249-261

Tornquista CG, Hons FM, Feagley SE, and Haggar J 1999 Agroforestry system effects on soil characteristics of the Sarapiqui region of Costa Rica, *Agriculture, Ecosystems and Environment* **73** 19-28

Tripathi OP, Pandey HN, and Tripathi RS 2009 Litter production, decomposition and physico-chemical properties of soil in 3 developed agroforestry systems of Meghalaya, Northeast India, *African Journal of Plant Science* **3**(8) 160-167

Udawatta RP, Krstansky JJ, Henderson GS, and Garrett HE 2002 Agroforestry practices, runoff, and nutrient loss: a paired watershed comparison, *Journal of Environmental Quality* **31** 1214–1225

Unger IM, Goyne KW, Kremer RJ, and Kennedy AC 2013 Microbial community diversity in agroforestry and grass vegetative filter strips, *Agroforestry Systems* **87** 395-402 DOI 10.1007/s10457-012-9559-8

Uthappa AR, Chavan SB, and Handa AK 2015 Trees for soil amelioration- a review, *Global Journal for Research Analysis* **4**(6) 320-322

Wang H, Huang Y, Huang H, Wang KM, and Zhou SY 2005 Soil properties under young Chinese fir-based agroforestry system in mid-subtropical China, *Agroforestry Systems* **64** 131-141

Workalemahu A 2009 The effect of indigenous root-nodulating bacteria on nodulation and growth of faba bean (*Vicia Faba*) in the low-input agricultural systems of Tigray Highlands, Northern Ethiopia, *Alemayehu* **1**(2) 30-43

Xiong Y, Xia H, Li Z, Cai X, and Fu S 2008 Impacts of litter and understory removal on soil properties in a subtropical *Acacia mangium* plantation in China, *Plant Soil* **304** 179-188

Yadav RP and Bisht JK 2014 Litter fall and potential nutrient returns from pecan nut (*Carya illinoinensis*) in agroforestry system in Indian Himalaya, *International Journal of Herbal Medicine* **2** (1) 51-52

Yadav RS, Yadav BL, and Chhipa BR 2008 Litter dynamics and soil properties under different tree species in a semi-arid region of Rajasthan, India, *Agroforestry Systems* **73** 1-12 DOI 10.1007/s10457-008-9106-9

Yadav RS, Yadav BL, Chhipa BR, Dhyani SK, and Ram M 2010 Soil biological properties under different tree based traditional agroforestry systems in a semi-arid region of Rajasthan, India, *Agroforestry Systems* **81**(3) 195-202 DOI: 10.1007/s10457-010-9277-z

Yan T, Yang L, and Campbell CD 2003 Microbial biomass and metabolic quotient of soils under different land use in the Three Gorges Reservoir area, *Geoderma* **115** 129-138 DOI: 10.1016/S0016-7061(03)00082-X

Zahran HH, Chahboune R, Moreno S, Bedmar EJ, Abdel-Fattah M, Yasser MM, and Mahmoud AM 2013 Identification of rhizobial strains nodulating Egyptian grain legumes, *International Microbiology* **16** 157-163 DOI: 10.2436/20.1501.01.190

Zaia, FC, Gama-Rodrigues AC, Gama-Rodrigues EF, and Machado RCR 2008 Fo´sforo orgaˆnico em solos sob agrossistemas de cacau, *Revista Brasiliera De Ciencia Do Solo* **32** 1987-1995

Zhang H, Zhang H, Li X, and Yang S 2001 A *sinorhizobium* fredll strain that effectively nodulate *Medicago sativa*, *Wei Sheng Wu Xue Bao* **41**(2) 127-32

Zhang Y, Xu Z, Jiang D, and Jiang Y 2013 Soil exchangeable base cations along a chronosequence of *Caragana microphylla* plantation in a semi-arid sandy land, China, *Journal of Arid Land* **5**(1) 42-50 DOI: 10.1007/s40333-013-0140-8

Zornoza R, Acosta JA, Bastida F, Dominguez SG, Toledo DM, and Faz A 2015 Identification of sensitive indicators to assess the interrelationship between soil quality, management practices and human health, *Soil* **1** 173-185

CHAPTER 25

Potential of Bamboos for Livelihood and Income Generation

Salil Tewari

Agroforestry Research Centre, GB Pant University of Agriculture & Technology, Pantnagar (Uttarakhand), India

Corresponding author's Email: saliltewari@gmail.com

ABSTRACT

Bamboos are versatile group of plants with multiple utility commonly found in moist deciduous, semi evergreen, tropical, subtropical and temperate forest. India is 2nd largest country in bamboo cultivation after China. In India, there are 23 genera and 125 indigenous and exotic species of bamboo which constitutes 12.8 per cent of the total forest area in India. Bamboo is an important economical and ecological resource, and can be harvested at 4 years rotation. Low gestation period, multiple uses and easy marketability makes bamboo excellent component for agroforestry system. In India bamboo used as paper pulp, housing, non-residential uses, rural uses, fuel, packing, transport, furniture and other wood industries uses. Besides its different uses its potentiality has also been reported for environment protection, income and employment generation.

Keywords: Agroforestry, Bamboo distribution, Livelihood security

INTRODUCTION

Agroforestry is the cultivation of woody perennial trees along with cereals, vegetables, flowers and other crops with the objective of reducing the risk of crop failure, increasing yield, enhancing soil fertility hence earning more profit with positively affecting the ecological balance. In this practice, farmers cultivate annual crops with the woody perennial forest trees, which prove to be far more lucrative than the individual cultivation of any crop as mono-crop. One such important upcoming agro-forestry tree is Bamboo, which is a fast growing perennial woody grass. There are about 1250 species across 75 generas of this giant grass whereas in India- there are 130 species of 18 generas known. After China, India is 2nd largest country in bamboo cultivation. Current demand of bamboo in total is 26.69 million tonnes while supply is 13.12 million tonnes. Biggest use of bamboo in India is in building and construction purposes, which is approximately 3.4 million tonnes, whereas Bamboo demand in small and cottage industry is 3.32 million tonnes. It is usually taken as wood substitute for pulp, which is about 2.5 million tonnes, and in paper industry, demand is about 1 million tonnes. Bamboos are versatile group of plants of multiple utility from rich belt of vegetation in moist deciduous, semi evergreen, tropical, subtropical and temperate forest. About 14 million ha of the earth surface is covered by bamboo forest with 80 per cent in Asia. Bamboos are fairly well represented in all the continents except Europe. India is endowed with rich diversity of bamboo in different bioclimatic zones of country. In India, there are 23 genera and 125 indigenous and exotic species of bamboo which constitutes 12.8 per cent of the total forest area in India. The total annual production of bamboo is estimated to be 4.5 million tonnes. India is one of the leading countries of the world, second only to China in bamboo production. About 50 per cent of the bamboo produced in India grows in North Eastern region and West Bengal.

BAMBOO DISTRIBUTION AND STATUS

Bamboo grows from sea level in the tropics to 3700 meters above sea level in the temperate region (Varmah and Bahadur 1980). The geographical distribution of bamboo is governed largely by the conditions of rainfall, temperature, altitude and soil. Most of the bamboo requires a temperature from 8 °C to 36 °C with a minimum annual rainfall of 1000mm and high atmosphere humidity for good growth.

Generally the genera *Bambusa* and *Dendrocalamus* are found under tropical condition whereas *Arundinaria* and its allies occur in the temperate region and on high elevation in the western and eastern Himalayas. *Dendrocalamus strictus* is dominant in dry deciduous forest while *Bambusa bambos* in moist deciduous forest. *Gigantochloa rostrata* is most important bamboo of semi evergreen forest of Andaman. In eastern region i.e. West Bengal, Assam and North Eastern Himalayas the commercially important bamboos are *B. tulda, D. hamiltonii* and *Melocana baccifera*.

Principal Bamboo Genera are:

Arundinaria	*Phyllostachys*
Bambusa	*Pleioblatsus*
Chimonobambusa	*Pseudosasa*
Dendrocalamus	*Pseudoxytenanthera*
Dinochloa	*Racemobambos*
Gigantochloa	*Schizostachyum*
Melocanna	*Sinarundinaria*
Ochlandra	*Thamnocalamus*
Oxytenanthera	*Thyrsostachys*

Of all the genera mentioned, the genus *Bambusa* is the widely distributed in India and *Dendrocalamus* is next widely available genus. These are found in the understorey of forest of almost all the states except Kashmir valley.

Table 1: *Bamboo area and its potential availability*

Name of state	Forest Area (km^2)	Bamboo bearing forest area (km^2)	Total bamboo bearing area (km^2)	Main bamboo Species	Potential availability of bamboo (mt)
Andhra Pradesh	43,290	9882.57	11401.95	*B. arundinacea, D. strictus*	0.255
Arunanchal Pradesh	68,602	1200.00	1218.75	*Melocanna baccifera, D. hamiltonii, Arundinaria baccifera, B. polymorpha*, others	0.20
Assam	23,824	-----	498.00	Others	1.21
Bihar	26,524	551.00	2895.06	*D. strictus, D. hamiltonii*	0.046
Gujrat	12,578	-----	845.63	*D. strictus*	-----
Haryana	604	25.69	413.50	*D. strictus*	0.003
Himanchal Pradesh	12,521	91.45	238.95	*D. hamiltonii*	0.48
Karnataka	32,403	6000.00	6970.94	*Bambusa arudinaria, D. strictus*	1.05
Madhya Pradesh	131,195	22261.13	23849.88	*D. strictus, D. hamiltonii*	0.30
Meghalaya	15,657	3992.00	4028.00	*D. hamiltonii*	0.49
Orissa	46,941	15309.00	16165.88	*D. strictus, D. hamiltonii*	0.49
Punjab	1,387	-----	446.567	*D. strictus*	0.009
Rajasthan	13,353	793.82	1854.451	*D. strictus*	---
TamilNadu	17,064	84.51	1233.89	*D. strictus*	7.20
Tripura	5,546	812.250	870.75	*Melocana baccifera*	0.215
Uttar Pradesh	33,994	647.15	4131.84	*D. strictus*	0.041
West Bengal	8,349	13.50	630.62	*D. strictus*, others	0.02

Source: National Data Base of Bamboo (INDIA), 1995

The total geographical area of the country is 3287263 sq km. The Table-1 shows that the maximum forest area is of M.P. (131,195 km^2) and total maximum bamboo bearing area is also of M.P. (23849.88 km^2) and minimum is in H.P. (238.95 km^2). However the potential availability of bamboo is maximum in undivided Assam (1.21 million tonnes) and minimum in Punjab (0.009 million tonnes). In Uttarakhand region bamboo grows on steep mountain slopes of Garhwal and Kumaon hills. Major bamboo area of the state is in Lansdowne forest division where *D. stictus* occur as predominant 9 Table 2). Other spp. in Uttarakhand are *B. bambos, B. nutans and D. hamiltonii.*

IMPORTANCE OF BAMBOO CULTIVATION FOR FARMERS

Bamboo is an important economic and ecological resource, which is a fast growing and can grow up to 3-30 metres within 3-4 months after planting. The crop can be harvested at 4 years rotation. It can flourish well, even on degraded land or non-cropped land where no other kind of cultivation is feasible. It is a drought tolerant tree and can tolerate diverse soil and moisture regime. Bamboo sparsely requires 1270mm of rainfall to 6350mm or sometimes even more, whereas others occur in dry deciduous forest with very low rainfall. It has the capacity to reclaim the degraded land area by helping in effective binding of the soil in top 1 metre. It also contributes through carbon sequestration by providing healthier environment. One-hectare bamboo plantation can sequester up to 12 tonnes of carbon dioxide from the atmosphere. Its leaf foliage acts as effective soil mulch and leaves are used as fodder. Bamboo is a crop, which should be grown in a well-drained sandy loam to loamy clay soil. Although it prefers well drained soil but some species can flourish well along the riverbanks, beds of stream and other moist sites. Wide range of Bamboo species are there with different preferences.

Bamboo Plantation on Farmland Under Agroforestry

Bamboo is basically a house hold plant spp. Grown around houses and compounds.

Bamboo advantage over perennial cash crops:

- Low gestation period
- Fast growth
- Diversified uses
- Easier marketability

Bamboo can be grown as farm crop:

- Exclusively for their edible shoots
- Dual purpose
- Exclusively for mature culms

All ensuring its potential as an excellent agroforestry crop. The scope for bamboo as a component of agroforestry system is much there under uncertain weather conditions, increase cost of labour involved in agriculture and larger area of uncultivated lands. Traditionally 1-5 clumps of bamboo in homestead are existing with every household in N-E hill region. It has soil erosion control capacity and ameliorate immediate environment in humid tropical and subtropical condition.

Different type of farm models

- Bamboo + Tea
- Bamboo + Some of viable timber trees or crops
- Bamboo + Fish pond
- Bamboo + Poultry farm
- Bamboo + Edible fungi
- Bamboo + Medicinal plant

Bamboo is a long term cropping plant and it cannot be replaced easily even after several rotations or harvestings. Therefore, bamboo should be planted in the fields where it can be placed on long-term basis. Bamboo can be grown in the field as per following ways:

Bund Plantation

Bamboo is sown in furrows or bunds either to limit the boundary of the plots or to act as a windbreak. For bund purpose the cultivars used should be thorny or shrubby in nature like *Bambusa bambos, Dendrocalamus strictus.* In this kind of plantation, the distance maintained between two plants is 1-3 m and within 1-2 years, plants form a thick fence through vegetable propagation. Timely and regular pruning of this crop is required otherwise it can spread in the nearby field.

Block Plantation

In this method of planting, Bamboo is planted in degraded, less fertile part of the field and there is no direct contact between the planted Bamboo and other crops grown. This provides congenial environment for easy maintenance of both the crops separately.

Bamboo as an Intercrop

In this method, bamboo can be grown in direct contact with other crops hence competition for growth is more. Distance between rows and plants should not be kept less than 8-10m or even more (Dev et al 2016; 2017). Another technique is to plant two rows at a distance of 1-2m, which are further, kept at a distance of 10-20m with other two rows. Distance between plants to plant should be 5-8m. The pit where bamboo is to be planted should be filled with 5 kg compost, 100g urea, 100 g SSP and 50g of MOP and 5 g of Chlorpyriphos powder should also applied in the soil for protection of plant from insect pest.

Table 2: *Various important genus of bamboos grown in different regions*

Region	Species
North East India	*Bambusa spp, Dendrocalamus, Schizostachym, Arudinaria, Gigantodoa*
North West India	*Bambusa, Dinochlos, Dendrocalamus, Pseudoxythenanthera*
Indogangetic plains	*Dendrocalamus, Bambusa*
Peninsular India	*Ochlandra, Bambusa, Dinochlos, Schizostachyum*
Andaman and Nicobar Islands	*Dendrocalamus, Dinochlos, Schizostachyum*

BAMBOO UTILISATION

Bamboo has been put to use for various applications ranging from construction to household utilities and have more than 1500 documented uses including important industrial uses. Due to Plethora of essential uses, bamboo has been aptly described as 'Poor man's timber', 'Green gold', 'Friend of people', 'The cradle to coffin timber', 'Green Gasoline' etc. The bamboo main uses can be grouped as follows:

Edible Item

Bamboo shoots are used for vegetable and pickles and beverage like bamboo juice and bamboo bear. Bamboo shoots are a good potential source of human food as high in protein, amino acid and fibres. In India *D. asper, Bambusa spp. and Melocana spp.* are used for shoot production.

Building & Construction Material

Bamboo is structurally stronger than steel and yet it is light weight, easily workable and have vibration dumping and heat insulation properties. It finds application in scaffolding, housing and roads. Bamboo professed as scaffolding material because of its flexibility in variety of length. In housing its potential usage of mats is bamboo roofing and reinforced bamboo structure. Uses of bamboo grids in road making are another potential area.

Paper Pulp

Bamboo is considered to be most important raw material for paper industry in India. At present 50 to 60 per cent of total fibrous raw material is bamboo component consumed by paper mills. *D. strictus* is widely used for this purpose. Other spp. are *B. bambos, B. tulda, D. hamiltonii, G. rostrata* etc.

Handicrafts

There are numerous design and varieties of handicrafts made out of bamboo. India is rich in bamboo handicrafts.

Fodder

Bamboo leaves are much valued fodder for cattle especially during the period of scarcity. Generally *B. bambos, D. hamiltonii, Arundinaria spp, Ochlandra spp* is used as fodder.

Medicine

Tabashir is a siliceous secretion found in the internodes of the clump of various spp. of bamboos. It is used as a cooling tonic and aphrodisiac also in the treatment of asthma and cough and other debilitating diseases.

Charcoal

Bamboo charcoal is superior to many other for use in electric batteries. In India it is used for Pharmaceutical purposes. Jewelers prefer bamboo charcoal.

Along with these there are so many other uses of bamboos.

Table 3: *Major uses of bamboo*

S.No.	Uses	Consumption (%)
1	Paper Pulp	35.0
2	Housing	20.0
3	Non-residential	5.0
4	Rural uses	20.0

5	Fuel	8.5
6	Packing, Including basket	5.0
7	Transport	1.5
8	Furniture	1.0
9	Other wood industries	1.0
10	Others, including ladders, staff etc.	3.0

Source: Solanki 2003

POTENTIALITY OF BAMBOO

In Environment Protection

Bamboo is an ideal spp. for soil and moisture conservation. It found suitable in rehabilitation of degraded lands. It controls landslide, soil erosion and flood, due to its vast underground rhizome network. It has very important role in retardation of climate change. Bamboo provided re-greening in Hiroshima after the blast in 1945.

Wood Substitute

As the forest reserve is depleting globally and with concern of environment growing, timber are getting scare day by day. This is due to long period of time taken by even softwood to attain maturity, so bamboo is found as an alternative.

In Tiny and Cottage Industry

Use of bamboo in making agarbattis and woven mats has been traditional and would continue to grow. The current size of bamboo in agarbatti market is approximately ₹ 135 crores. Close to 1 m tonnes of bamboo is utilized by ice-cream, kites and cracker industries and other miscellaneous items, which are valued at ₹ 186 crores at present.

In Handicrafts

With 23 million people involved in crafts it forms second largest employment sector in India next to agriculture and especially in times of drought, lean harvest, floods or famine. Bamboo based handicrafts have been promoted during last 2 years in N-E region, Jharkhand, Chattishgarh, M.P., U.P. and Maharasthra.

In Large Bamboo Based Industry

Bamboo constitute an important raw material for manufacturing of various types of papers. It has been estimated that current annual consumption of bamboo by paper mills in India is approximately 1.9 million tonnes.

The main population dependent on the bamboo cultivation in India is:

- Tribal and forest dwellers.
- Bamboo craftsman and artisans.

- Bamboo based small scale industries.
- Large scale industries.
- Population dependent on bamboo shoots and Seeds as food material.

Table 4: *Current and expected market of bamboo*

Product/Application	Current market (₹ Cr.)	Expected market (₹ Cr.)
Bamboo shoots	4.8	300 (growing at 25%)
Bamboo as wood substitute	10,000 (import value)	30,000 (in next 20 yrs)
Bamboo plyboard	200	500
Bamboo plyboard for use in trucks and railways	1000	3,408 (2015)
Bamboo flooring	100 (domestic) 100 (export)	1,950 (2015)
Bamboo pulp	100	2,088 (2015)
Bamboo furniture	380	3,265 (2015)
Building and construction material	-	-
Scaffolding	-	861 (2015)
Housing	-	1,163 (2015)
Roads	-	274
Bamboo girds	-	1,000 (2005)
Tiny and cottage industry	-	-
Agarbatti	-	-
Miscellaneous industry (ice cream sticks, fire crackers, lathis, ladders etc.)	394	600 (2004)
Pencil industry	-	-
Match industry	-	-

Source: National Mission on Bamboo Technology and Trade Development, Planning Commission 2001

EMPLOYMENT GENERATION

- One hectare of bamboo plantation can produce 200 man days of employment and 900 man days in cottage industry with the raw material from this plantation (Lakshman 1999).
- From the total area 108306 ha plantation, 21.6 million man days equivalent to ₹ 1080 million wage earning has already been generated. If corresponding cottage industry can develop, these plantations can generate 97.2 million man days equal to ₹ 4860 million wage earning in the cottage industry sector annually.
- Harvesting of bamboo from natural forests generate every year 2.25 million man days equal to ₹ 112.5 million wage earning.

This amply proves the significance of bamboo not only as a produce to meet the demand of the consumers but also as a production process to provide employment opportunities to the impoverished rural poor. Bamboo stands as an ideal plant capable of achieving eco-restoration of degraded lands beside ensuring bio diversity, conservation, livelihood security and economic empowerment due to its manifold uses and industrial application. Bamboos are fast, widespread, renewable, versatile, low or no cost, environment enhancing resources with a potential to improve livelihood security in both rural and urban areas. An intensive awareness programme should be launched about bamboos, people and environment. Considering ecological significance and vast economic potential of bamboo, its scientific management should be promoted. Thus we can say that bamboo is currently in the process of being rediscovered in India.

REFERENCES

Anonymous 2004 The Bamboo Book, National Mission on Bamboo Applications 7 - 11

Anonymous 2004 Understanding Bamboo Craftsmen in Uttaranchal, Uttaranchal Bamboo & Fiber Development Board 29 – 34

Anonymous 2004 VIIth World Bamboo Congress, New Delhi 1 – 59

Dev I, Ahlawat SP, Palsaniya DR, Ram A, Newaj R, Tewari RK, Singh R, Sridhar KB, Dwivedi RP, Srivastava M, Chaturvedi OP, Kumar RV, and Yadav RS 2016 A sustainable livelihood option for farmers' of semi-arid region. Bamboo+Chickpea based Agroforestry model, *Indian Journal of Agroforestry* **18**(1) 84-89

Dev I, Ahlawat SP, Palsaniya DR, Ram A, Newaj R, Tewari RK, Singh R, Sridhar KB, Dwivedi RP, Srivastava M, Chaturvedi OP, Kumar RV, and Yadav RS 2017 Bamboo (*Dendrocalamus strictus*) + sesame (*Sesamum indicum*) based agroforestry model: A sustainable livelihood option for farmers of semi-arid region, *Indian Journal of Agricultural Sciences* **87**(11) 1528-34

Gaur RC 1986 Bamboo Research in India. Systematic Botany Branch Forest Research & Education, Dehradun 26 - 31

Haridasan K and Tewari S 2008 Bamboo for PAN India Plantation Programmes – Preliminary Observations from Recent Trials, *The Indian Forester* **134**(3) 314-324

Singhal RM and Gangopadhyay PB 1999 Bamboo in India & Data Base. Indian Council of Forestry Research & education, Dehradun 1 -143

Pande VC, Kurothe RS, Rao BK, Kumar G, Parandiyal AK, Singh AK, and Kumar A 2012 Economic Analysis of Bamboo Plantation in Three Major Ravine Systems of India, *Agricultural Economics Research Review* **25**(1) 49-59

Solanki KR 2003 Bamboo: A dependable source of woody raw material. Indian Council of Agriculture Research, Krishi Bhavan, New Delhi 31 - 34

Tewari DN 1996 Monograph of Bamboo, Indian Council of Forest Research and Education, Dehradun 1 - 23

Tewari DN 1998 Bamboos in India, Indian Council of Forest Research and Education, Dehradun 18–22

Tewari S, Kaushal R, Tewari L and Banik RL 2014 Genotype x environmental interaction in bamboo species in pan India: Results from a multi-location trial, *International Journal of Innovative Horticulture* **3**(1)71-77

Tewari S, Kaushal R, Banik RL, Tewari L, and Chaturvedi S 2014 Evaluation of Bamboo species in India: Results from a multi-location trial, *Indian Journal of Agroforestry* **16**(1) 68-73

CHAPTER 26

Agroforestry for Diversification Through Livestock

M Ramachandran

Institute of Animal Nutrition, Tamil Nadu Veterinary and Animal Sciences University, Kattupakkam, Kancheepuram District– 603 203, Tamil Nadu, India

Corresponding author's Email: ian@tanuvas.org.in

ABSTRACT

India has a large livestock population and projections indicate that to feed this large livestock population by the year 2020, India would require 526 million tonnes (mt) of dry fodder, 855 mt of green fodder and 56 mt of concentrate feed. On the other hand land allocation for cultivation of fodder crops in India is only 5 per cent of the gross cropped area and India has a long history of shortage of fodder for livestock. Livestock production in India faces many constraints, the chief of which is shortage of quality feed / fodder. Dwindling pasture lands and land allocated for fodder production aggravates the situation. Therefore, in this context, newer feed resources are being explored as alternative to conventional feeds. Though tree fodder has been used for feeding livestock traditionally, with ban on lopping fodder in forests / reserve forests, this practice is slowly vanishing. Agroforestry which allows growing of trees in agricultural land ensures year around availability of tree fodder for livestock feeding. The critical role of trees and shrubs in livestock production has been well recognized in throughout the world. Tree leaves, fruits, pods, twigs etc., are traditionally used for feeding livestock especially small ruminants. In arid and semi arid regions of the country, tree fodder is an important feed resource or the only feed resource available for livestock. It is in this scenario that to alleviate the existing fodder shortage, to create fodder banks for the future, to overcome problems associated with deforestation and mitigate climate change, the concept of integrating fodder trees in agricultural land, wasteland, community land and pastures is gaining momentum.

***Keywords*:** Agroforestry, Fodder shortage, Livestock, Pasture, Silvipasture

INTRODUCTION

In Indian agriculture, livestock plays a pivotal role in the development and progress of mankind with crop production program as a complementary enterprise. However, livestock productivity is constrained by an acute shortage of feed and fodder. A general agreement is that there is a shortage of 40.4% and 24.7% dry fodder and green fodder, respectively. In India, there is a deficit of green fodder, particularly during the summer season. In India, only 4.4% of the cultivated area is under fodder crops with annual total forage production of 846 million t (Valli et al 2016). Besides providing various ecosystem services, agroforestry systems are the also considered is an important source of fodder. Diversification of land use systems with agroforestry is a necessary strategy for providing a variety of products for meeting requirements on a long term basis (Dhyani et al 2016). The current area under agroforestry in India is estimated as 25.32 m ha or 8.2% of the total geographical area of the country. There is further scope of increasing the area under agroforestry in future by another 28.0 m ha. The major share of the land to be brought under agroforestry will come from fallows, cultivable fallows, pastures, groves, and rehabilitation of problem soils. Thus, a total of 53.32 m ha, representing about 17.5% of the total reported geographical area of the country, could potentially be brought under agroforestry in the near future, which will make agroforestry a major land-use activity, after agriculture and forestry in India. Agroforestry answers many problems that are faced by today's agriculture in terms of stability in production, regular returns, restoration of fertility, indiscriminate deforestation, drought mitigation and environmental pollution.

One of the major constraints for increasing the production from Indian livestock is their poor nutritional status. It is evident from the chronic feed and fodder deficiency. About a one third of total feed intake by Indian livestock is met from grazing which are nutritionally very poor and are severely eroded. In a comprehensive study at Salem district of Tamil Nadu (India), it was concluded that the grass cover was deficient in protein at least for six months and both protein and energy for rest of the year. It is also reported that production potential of these grazing lands was only 0.55 t ha^{-1}. The average cost of livestock feed works out to about 45% in cows and 41% in buffaloes. In small farms the corresponding figures were 38 and 41%. In our country in one hand there has been serious fodder deficiency and on the other valuable land resource is wasted. Therefore changing the land use through agroforestry can solve the multiple problems related to natural resources, fodder shortage and livestock. Agroforestry, the integration of trees and shrubs with annual crops production, is an age old management system practiced by farmers to provide shade, a steady supply of food and/or income throughout the year, arrest degradation and maintain soil fertility, diversify income sources, increase and stabilize income, enhance use efficiency of soil nutrients, water and radiation, and provide regular employment.

UTILITY OF TREE FODDER

The trees are integral component of any agroforestry system. The tree fodders are in general rich in protein, low in fibre and digestible energy. They are rich in calcium and poor in phosphorus. The fruits and pods of legume trees are also have high protein and feeding value. The tree fodders which have tolerant level of tannins provide the much needed undegradable protein to the high producing animals and digestible protein to average yielders. Thus the tree fodder can be used as source of supplementary protein to livestock. Thus integrating fodder production programmes with existing Agriculture, Horticulture, Sericulture etc, and utilizing the uncultivable waste lands for fodder production and improving the quality of so called pasture lands in to multi-tire silvipasture systems or legume pasture and providing complementary grazing to the livestock in the systems will not only increase the land utilization but also increase fodder availably and *vis-a-vis* the production potential of livestock.

VALUE ADDITION IN TREE FODDER

When tree fodder is available in surplus it is lopped, dried, ground and incorporated in concentrate mixture at varying levels in replacement of conventional feed ingredients and fed to livestock. This helps to reduce cost of concentrate feed and maintaining the desired nutritive value.

Pic.1 (a) Tree fodder based concentrate feeding

Pic.1 (b) Neem Silage

Pic.1 (c) Gliricidia Hay

Pic.1 (d) Complete feed block incorporating tree fodder

Pic.1 (e) Extruded feed incorporating tree fodder

Pic.1 (f) Tree leaf meal

High quality silage could also be prepared from tree fodder using additives such as 0.5% salt, 1% molasses and non protein nitrogen sources such as urea at 0.5 to 1%. Tree fodder can also be lopped, dried and stored as hay. From dried tree fodder complete feed block, extruded feed, tree leaf meal and tree leaf meal incorporated anthelmintic salt licks can be prepared (Pic. 1).

BIOMASS YIELD AND NUTRITIVE VALUE OF FODDER TREES

Regardless of the production systems of fodder trees, biomass / foliage yield is affected by factors such as tree species, density of planting and harvesting management *i.e.,* age at first harvest, height and frequency of cutting and season of harvest. Similarly a high variability in the nutritive value of fodder trees exists and this could be attributed to species variability, plant age and part, harvesting regimen, season and location. The digestibility of tree fodder depends on the lignifications of plant tissue which in turn depends on the age of the plant. As plant age increases the lignin content increases and digestibility decreases. Table 1 provides details on common fodder trees of southern India with their biomass yield and nutritive values.

Table 1: *Common fodder trees with their biomass yield and nutritive value*

S.No.	Botanical name	Biomass yield (tones dry matter ha^{-1})	Crude protein (%)	Crude fibre (%)	TDN (%)
1.	*Acacia nilotica*	0.5 – 2.0	12	28	45
2.	*Acacia leucophloea*	0.5 – 2.0	14	25	45
3.	*Acacia planifrons*	0.5 – 2.0	9.6	22	45
4.	*Ailanthes exelsa*	3.0 - 4.5	6.0	23	63
5.	*Albizia lebbeck*	1.0 – 2.0	16	18	50
6.	*Albizia amara*	2.0 – 3.0	15	20	50
7.	*Azardirachta indica*	-	6	27	55
8.	*Bahunia spp.*	1.5 – 2.5	17	22	55
9.	*Dalbergia sissoo*	-	17	25	50
10	*Erythrina indica*	1.0 – 2.0	26	12	60
11.	*Ficus bengalensis*	-	12	27	52
12	*Hardwickia binata*	-	10	26	50
13.	*Gliricidia sepium*	2.0-15.0	22	25	65
14.	*Inga dulcis*	2.0 – 4.0	20	25	55
15.	*Milligtonia hortensis*	-	6	20	62
16.	*Leucaena leucocephala*	2.0-20.0	21	24	68
17.	*Lannea coromandalica*	-	4	18	65
18.	*Sesbania sesban*	2.0 -10.0	24	28	65
19.	*Sesbania grandiflora*	2.0 -10.0	22	27	65
20.	*Thespesia populnea*	-	11	20	60

Source: Valli and Murugan 1998

Tree fodders are rich in minerals compared to fodder crops. The mineral content of tree fodders are variable depending upon tree species, age of the tree, stage of lopping, season of lopping, soil mineral profile, soil pH, etc. The calcium to phosphorus ratio is wide in tree fodders hence necessitates supplementation of phosphorus while being fed to animals by way of supplementing with bran. Table 2 gives the mineral content of some common fodder trees.

Table 2: *Mineral content of common fodder trees*

S.No.	Botanical name	Calcium (%)	Phosphorus (%)	Magnesium (%)	Iron (ppm)	Zinc (ppm)	Copper ppm
1.	*Acacia nilotica*	2.80	0.52	11.28	–	58.68	12.64
2.	*Acacia leucophloea*	2.29	0.41	12.08	257.44	15.41	10.79
3.	*Acacia planifrons*	2.45	0.04	11.28	335.42	58.68	12.64
4.	*Albizia lebbeck*	3.04	0.04	13.50	887.76	6.97	11.70
5.	*Azardirachta indica*	2.13	0.04	–	–	16.28	6.71
6.	*Bahunia sp*	2.90	0.03	10.55	–	–	–
7.	*Dalbergia sissoo*	2.65	0.03	13.39	323.37	95.76	11.73
8.	*Erythrina indica*	2.40	0.04	16.18	–	38.3	8.17
9.	*Ficus bengalensis*	2.33	0.04	10.78	300.24	18.72	11.24
10.	*Hardwickia binata*	9.75	0.02	9.81	1162.00	56.73	10.66
11.	*Gliricidia sepium*	2.47	0.27	14.98	–	26.64	8.84
12.	*Inga dulcis*	3.23	0.06	13.30	–	30.37	16.84
13.	*Milligtonia hortensis*	2.35	–	15.15	–	–	–
14.	*Leucaena leucocephala*	2.29	0.06	12.33	253.38	11.71	32.61
15.	*Lannea coromandalica*	2.25	–	14.86	–	17.3	6.36
16.	*Sesbania grandiflora*	2.20	0.05	13.87	226.71	23.55	9.57
17.	*Thespesia populnea*	4.03	0.06	14.01	260.02	53.44	16.13

Source: Valli and Murugan 1998

Though tree fodders are nutritious and provide year long fodder to Indian livestock, some of them contain anti nutritional factors that make them unpalatable or harmful to livestock when consumed in large amounts. Some of the common antinutritional factors in tree fodders include mimosine, or its metabolite 3-hydroxy-pyridone (DHP), tannins and saponins. A number of methods can be used to overcome toxicity problems. Wilting, drying and dilution with other feeds are some of the methods. In the past decade, researchers have also proved that tannins and saponins inspite of their antinutritional property also have methane mitigation potential in ruminant livestock. Table 3 lists out the probable anti nutritional factors that could be prevalent in some tree fodders.

Table 3: *Anti nutritional factors present in some common tree fodders.*

S.No.	Botanical name	Anti-nutritional factor	Concentration (%)
1.	*Acacia nilotica*	Tannin	9 - 10
2.	*Acacia leucophloea*	Tannin	8 - 9
3.	*Acacia planifrons*	Tannin	9 - 10
4.	*Ailanthes exelsa*	Tannin	2 - 8
5.	*Albizia lebbeck*	Tannin Saponins	7 - 9 0.5 - 1
6.	*Albizia amara*	Tannin Saponins	7 - 9 0.5 - 1
7.	*Azardirachta indica*	Azadirachtin	0.02
8.	*Bahunia sp*	Tannin	10 - 14
9.	*Dalbergia sissoo*	Tannin	6 - 7
10.	*Erythrina indica*	-	-
11.	*Ficus bengalensis*	Tannin	3 - 4
12.	*Hardwickia binata*	Tannin	11
13.	*Gliricidia sepium*	Tannin	2 - 3
14.	*Inga dulci*	Tannin	5 - 8
15.	*Milligtonia hortensis*	-	-
16.	*Leucaena leucocephala*	Tannin Mimosine	9 - 11 2 - 6
17.	*Lannea coromandalica*	Tannin	3 - 5
18.	*Sesbania sesban*	Tannin Saponin	8 2- 11 0.5-1
19.	*Sesbania grandiflora*	Tannin Saponin	8-11 0.5-1
20.	*Thespesia populnea*	Tannin	8-9

Source: Valli and Murugan, 1998

AGROFORESTRY MODELS FOR LIVESTOCK INTEGRATION

During the past two decades at Institute of Animal Nutrition developed several agroforestry models for livestock integration have been studied. The following are the developed agroforestry models (Pic 2).

Pic 2(a) Hortipasture: *Fodder crops are cultivated in the understorey of fruit trees*

Pic 2(b) Agrisilvipasture: *Cash crops are cultivated in the understorey of trees.*

Pic 2(c) Silvipasture: *Fodder trees are grown in pastures for livestock.*

Pic 2(d) Three tier agroforestry model: *Fodder trees are grown with grasses and legumes in three tiers so that year round fodder is ensured.*

Pic 2(e) Agroforestry model in degraded wasteland: *Fodder trees are grown in degraded wastelands/ barren wastelands and they are converted into fodder bank for livestock.*

Pic 2(f) Boundary plantations: *Fodder trees are grown in field bunds or boundaries ensuring shade and supply of fodder for livestock*

CARRYING CAPACITY OF DIFFERENT AGROFORESTRY MODELS

Carrying capacity of an agroforestry model is the number of livestock that it can sustain. The carrying capacity of agroforestry models differ based on the soil status, fodder and tree components involved in the agroforestry models. Based on the trials conducted in this institute, carrying capacity of livestock has been depicted in the following table.

Table 4: *Carrying capacity of different agroforestry models*

Agroforestry model	Tree and fodder component	Carrying capacity
Silvipasture	*Leucaena leucocephala* *Gliricidia sepium* *Cenchrus ciliaris*	*Leucaena leucocephala* can replace 20 per cent of green fodder and can hold 6 -7 cows per year per hectare without adverse effect on its milk yield and composition in dairy animals *Gliricidia sepium* can replace 15 per cent of green fodder and can hold 14 -15 cows per year per hectare without adverse effect on its milk yield and composition in dairy animals.
Hortipasture	*Cocos nucifera,* *Psidium guajava* *Desmanthus virgatus* *Stylosanthes hamata* *Crotolaria juncea* *Cenchrus ciliaris*	*Desmanthus virgatus* can hold 70 calves ha^{-1} fed at 2 kg $calf^{-1}$ day^{-1} apart from reducing the requirement of concentrate feed by 37.5 per cent. *Stylosanthes hamata* can hold 60 adult sheep / goats per year per ha fed at 1kg day^{-1} $animal^{-1}$. *Crotolaria juncea* can hold 12 -14 adult sheep / goats per year per ha fed at 1kg day^{-1} $animal^{-1}$ or 50 -55 rabbits per year fed at 0.25kg day^{-1} animal. *Cenchrus ciliaris* can hold 4 cattle or 16 sheep / goats per year per ha.
Hortisilvipasture	*Leucaena leucocephala* and *Mangifera indica* Fodder sorghum (CoFS 29)	The fodder produced in this model could support 36 calves for a year per ha when fed at 8 kg day^{-1} $animal^{-1}$.

Source: Valli et al 2016

Several studies indicated that the Agrisilvi system with cultivated fodder in irrigated condition could hold 80-90 adult sheep or 10-12 cows per ha, depending up on the manner of integration. The Horti-silvipasture in coconut orchard in combination with irrigated fodder can support 25-30 sheep. The dry land Hortipasture with *Stylosanthes hamata* can maintain 7 sheep or one cow. The Silvipasture with *Leucaena leucocephala* and *Cenchrus ciliaris* can hold 9 sheep or 1 cow.

ROLE OF AGROFORESTRY FOR IMPROVING SOIL FERTILITY

The mixed silvipasture system in red soil with *Leucaena leucocephala, Gliricidia sepium*, Teak and Eucalyptus having understorey of *Sehima nervosum* after four years of establishment was found to reduce soil pH from 8.8 to 7.7 and increase organic matter by 280 %. There was substantial increase in Total nitrogen and Potassium. Thus it is obvious that the Silvipasture systems are one of the best tools for improving the soil quality in degraded waste lands. Nutrient mining from continuous cropping without adequately fertilizing or fallowing the land is often cited as the main constraint to increase in productivity. Agroforestry systems have attracted considerable attention as an attractive and sustainable pathway to improve soil fertility. World Agroforestry Center made substantial progress in the identification and promoting of agroforestry systems aimed at improving soil fertility.

According to Sanchez et al (1997), there are four ways through which trees can contribute to the improved nutrient supply - increase nutrient inputs to the soil, enhance internal cycling, decrease nutrient losses from the soil, and provide environmental benefits. After extensive experimentation with a wide range of soil fertility replenishment practices, the World Agroforestry Center has developed a system with three components that can be used in combination or separately: (i) nitrogen-fixing leguminous tree fallows, (ii) indigenous rock phosphates in phosphorus deficient soils, and (iii) biomass transfer of leaves of nutrient-accumulating shrubs (Sanchez 2002).

AGROFORESTRY AND CLIMATE CHANGE

The greatest role of Agroforestry in relation to climate change is perhaps in mitigating the emissions of CO_2 by productively sequestering carbon from the atmosphere. The tree component of the agroforestry systems can be a significant sink for carbon in lands devoted to agriculture. The three major paths through which tree can help reduce atmospheric carbon are: conservation of existing carbon pools through practices such as avoided deforestation and alternatives to slash and burn; sequestration through improved fallows and integration with trees, and substitution through bio-fuel and bio-energy plantations to replace fossil fuel use (Montagnini and Nair 2004). Cultivated field crops results in nitrogen emission and carbon depletion, leading to reduction of carbon in sink i.e. soil. Whereas, agroforestry plays an important role in capturing the atmospheric carbon dioxide and deposit in soil.

CONSTRAINTS

- Though silvipasture is the age old practice it has not percolated into the farming systems in spite of backing with new scientific interventions and trainings given to the farmers.
- The farmers training should have more field visits and interaction with successful farmers.
- More data and success stories need to be presented to the farmers.
- Motivation on nursery production, management and planting techniques are essential.
- Farmers prefer horticultural trees rather than fodder or multipurpose trees. If the horticultural components are added and silvi-horti pasture or horti-silvi models are presented to the farmers, the adoption rate may increase.
- More awareness on lopping management to be created among farmers.
- Timely availability of quality fodder seeds with good germination rate at nominal cost is the biggest problem.
- Most of the farmers are not aware of the seed treatment techniques for different fodder seeds.
- Most of the farmers who opt for agroforestry system immediately increase the number of livestock once they start harvesting fodder from the agroforestry system which once again results in fodder deficiency in the farm.

CONCLUSION

Agroforestry is a means of sustainable agriculture, is a tool to mitigate climate change, source of additional revenue to farmers, protects environment and also provides livestock with superior feed / fodder augmenting their productivity. In todays consumerist society, production of value added livestock products or designer livestock products production is possible using tree fodder. Methane mitigation in ruminants, protected form of nutrient delivery, health / growth promoting effects is possible by using tree fodder. Availability of fodder in drought, natural calamities is possible only through agroforestry systems.

REFERENCES

Dhyani SK, Ram A and Dev I 2016 Potential of agroforestry systems in carbon sequestration in India, *Indian Journal of Agricultural Sciences* 86 (9) 1103–1112

Montagnini F and Nair PKR 2004 Carbon sequestration: An underexploited environmental benefit of agroforestry systems. *Agroforestry System* 61 281–295

Sanchez PA, Izac AMN and Buresh RJ 1997 Soil fertility replenishment in Africa as an investment in natural resource capital in *Replenishing soil fertility in Africa*. ASA-SSSA Special Publication.

Schroeder P 1994 Carbon storage benefits of agroforestry systems. *Agroforestry System* 27 89–97

Sanchez PA 2002 Soil fertility and hunger in Africa. *Science* 295 2019–2020

Valli C, Murugan M, Gunasekaran S, Mynavathi VS, Karu Pasupathi, Murugeswari R, Babu M 2015 Handbook on Agroforestry models in Tamil Nadu for livestock integration ISBN 978-81-924045-5-4

Valli C, Karu Pasupathi, Mynavathi VS, and Manobhavan M 2016 Workshop compendium on Agroforestry systems for livestock integration

Valli C and Murugan M 1998 Chemical composition of certain tree fodders of Tamil Nadu in national seminar on Integration of livestock and agroforestry systems in wasteland development, Chennai 75-77

CHAPTER 27

Performance of Ginger and Chittaratha Under Varying Spacings of Seven Year Old Bamboo (*Dendrocalamus strictus* (Roxb.) Nees) in Central Kerala, India

BH Kittur*, K Sudhakara[1], BM Kumar[2] and TK Kunhamu[1]

Department of Farm Forestry, University of Agricultural Sciences, Dharwad, 5800051, India

[1]College of Forestry, Kerala Agricultural University, Vellanikkara, Thrissur, Kerala, 680656, India

[2]School of Ecology and Environment Studies, Nalanda University, Rajgir, District Nalanda, Bihar, 803116, India

**Correspondence author's Email:* kittur412@gmail.com

ABSTRACT

Bamboo-based agroforestry is a promising land management option for humid tropics of India. Optimal management of bamboo-based mixed species production systems requires an understanding of stand density, soil nutrients and understorey behavior. We evaluated the growth and productivity of ginger and chittaratha as influenced by varying spacings of seven year bamboo (*Dendrocalamus strictus* (Roxb.) Nees). The growth observations of bamboo revealed that closest spacings (4x4 m) recorded largest clump height (9.11 m) and due to increasing spacing to 12x12 m clump height decreased by 19 per cent. The leaf area index and spacings of bamboo are inversely related. About 57:48:33 per cent of higher total N: avail. P: avail. K accumulation in closest spacings of bamboo were observed as compared to bambooless plot. The rhizome yield of ginger was high (19.55 Mg ha^{-1}) at 10x10 m spacings of bamboo which decreased to 3.19 Mg ha^{-1} when the spacings of bamboo decreased to 4x4 m. However, Chittaratha responded highest (18.28 Mg ha^{-1}) in bambooless open plot, Nevertheless it was found better under widest spacings (12x12 m, 16.34 Mg ha^{-1}) of bamboo. Study revealed that managing bamboo based agroforestry systems to optimize the intercrop productivity without compromising the bamboo yields. Probably intermediary bamboo stocking such as 8x8 m and 10x10 m would be ideal.

Keywords: Chittaratha, *Dendrocalamus strictus*, Ginger, Intercropping spacing

INTRODUCTION

Bamboo based agroforestry can play an important role in enhancing productivity, sustainability and resource conservation. In the traditional land use systems of India and elsewhere, bamboos frequently occur on farmlands as scattered clumps and/or as hedgerows on boundaries. There is plausible informations available on understorey productivity in intercropping system is largely a function of the competitive interactions for light and space, both belowground and aboveground (Kiaer et al 2013, Ehrmann and Ritz 2014). However, the intricacies associated with these interactions have not been unraveled so far in bamboo based agri-silviculture systems. Regulation of planting spacing in bamboo might considerably influence the understorey PAR, which in turn affects the growth of understorey crops (Wilkinson and Elevitch 2000). Incorporation of medicinal plants with bamboo forms an important land management aspect in Kerala and can be studied for optimizing understorey productivity. Zingiberaceae members are important medicinal and aromatic oil yielding herbs and have a great demand for ayurvedic and culinary use. The medicinal properties are innumerable and very ancient. Kirtikar and Basu (1988) stated that the rhizome is very pungent, laxative, anthelmentic, alexeteric and emollient. It is also having antioxidant (Chen *et al*., 2008), anti-inflammatory, anti-cancer (Wohlmuth 2008), antibacterial (Bhunia and Mondal 2012) and antipyretic (De Padua et al 1999) properties. It is very important to explore the resource utilization patterns in bamboo based intercropping systems, in order to provide management strategies for deriving optimum yield and productivity. Therefore, the present study was formulated to evaluate the growth and productivity of ginger and chittaratha as influenced by varying spacings of 7 year bamboo (*Dendrocalamus strictus* (Roxb.) Nees).

METHODOLOGY

The study was conducted in a seven year old bamboo stand at Vellanikkara, Thrissur, Kerala. This experimental stand was established in 2004 with five spacing treatments (4x4, 6x6, 8x8, 10x10 and 12x12 m with corresponding densities of 625, 277, 156, 100 and 69 clumps per ha) and a control plot (without bamboo). The precipitation pattern at the site is bimodal; both South-west (June to September) and North-east monsoons (October to December) supply moisture, with a mean annual rainfall of 3062 mm, most of which was received during the South-west monsoon. The mean maximum temperature during the experimental period ranged between 29.10 °C (July) and 35.49 °C (March). The mean minimum temperature varied between 22.19° and 24.83°C in the months of December and May respectively. The trial was laid out in a randomized block design with three replications and the individual plots were of size 30x30 m. The physico-chemical properties of soil (0-20 cm soil depth) were determined before understorey planting. Triplicate soil samples (0-20 cm layer) were collected from all plots (three replications and six treatments) randomly from between the bamboo rows before understorey planting (April 2012). Soil bulk density was determined using a steel cylinder and soil pH measured in 1:2.5 soil: water suspension (Jackson 1958). Total soil nitrogen was determined by Kjeldahl method (Jackson 1958), available phosphorus by Bray No.1 extraction followed by reduced molybdate blue colour estimation method (Watanabe and Olsen 1965) and exchangeable potassium by neutral normal ammonium acetate solution followed by flame photometry (Jackson 1958).

The leaf area index of bamboo stand was estimated using a Plant Canopy Analyzer (LAI 2000, LI-COR Inc., Lincoln, Nebraska, USA). The photosynthetically active radiation (PAR) was taken at mid-day (0200 hrs) in each treatment during September 2012 using a Line Quantum Indicator (LQI 2404, K131, Li-Cor, Lincoln, USA). Within each plot, the line quantum sensor was installed on wooden platforms in random bamboo alleys at 70 cm above the ground on two consecutive days. The line of the sensor was oriented east-west direction so as to receive maximum incident radiation. PAR incident above the canopy of each plot was simultaneously recorded by the data logger using a Point Quantum Sensor (LI-COR LI-191) mounted on a 10 m pole rising above the canopy. A battery powered data logger (LI 1000, LI-COR Inc) integrated the mean flux of PAR at hourly intervals (μ moles s^{-2} m^{-1}). PAR was then converted to canopy transmittance - the ratio of light below the canopy to light incidence on the top of the canopy expressed as per cent. The mean mid-day PAR above the bamboo canopy recorded 1220 μmol. per sec per m^2.

Ginger (*Zingiber officinale* Roscoe. var. Varade) and chittaratha or lesser galangal (*Alpinia calcarata* Roscoe. var. local) were grown in beds of size 3 mx1.2mx30cm in between the centrally located hedge-rows of bamboo plot during May 2012. The intercrops were planted at 25x25 cm spacing following recommended package of practices (KAU, 2011). In addition to these treatments, an absolute bambooless control was established where ginger and chittaratha were raised in open contiguous plots. Prior to planting, the ginger and chittaratha rhizomes were treated with Dithane M 45 @ 3 g $litre^{-1}$ and Ekalux @ 1 ml $litre^{-1}$ as prophylactic measure against pests and diseases. The rhizome of average size 15 g were planted in planting hole of 5 cm depth at spacing of 25x25 cm in a bed area of 300 x 120 x 30 cm between the two central rows of bamboo in each plot during May 2012. Immediately after sowing, farm yard manure (FYM) at the rate of 35 Mg ha^{-1} (wet weight basis) was broadcast-applied on the beds as recommended and mulched with leaves of *Macaranga peltata* and *Trema orientalis* at the rate of 15 Mg per ha fresh weight basis. Mulching was repeated at the same rate during the 2nd and 3rd month. Beds were weeded 60 and 120 days after planting. Fertilizers (50:50:50 kg N: P_2O_5: K_2O per ha) were applied at first weeding with light tillage and earthing up was done. To control rhizome rot, wilt and leaf spot diseases, *Pseudomonas fluorescens* P1/PGPR II was sprayed and soil drenched at 45 days after planting. The growth and yield parameters of ginger were recorded by destructive sampling at 230 days after planting (DAP) and chittaratha at 360 DAP. Finely ground samples of dried mature rhizomes were also analysed for oleoresin content by solvent extraction. For this, a 10 g finely powdered sample (one sample per replication) was placed in a filter paper pouch and distilled in a Soxhlet apparatus with 250 ml petroleum ether (boiling point 60–80° C) for 8 hrs. The extract was then transferred to a 250 ml flask, petroleum ether was evaporated and the difference in weight of flask recorded for estimating oleoresin content.

Statistical Analysis

The data on growth of bamboo, ginger and chittaratha were analysed following the SPSS statistics for Windows at $p < 0.05$ following randomized-block design, with spacing of bamboo as treatments. NPK content of soil was analyzed following the SPSS statistics for Windows at $p < 0.05$. Least significant difference (LSD) was used for comparing differences between means. Standard error between the replications was also calculated.

RESULTS AND DISCUSSION

As spacing of bamboo increased the clump height significantly decreased (Table 1). The closest spacing (4x4 m) recorded largest clump height (9.11 m) and due to increasing spacing to 12x12 m clump height decreased by 19 per cent. The decrease in clump height with increasing spacing of bamboo may be attributed to less crown competition. Similar observations were made by Kibwage et al (2008) in the bamboo-tobacco growing regions of Kenya. About 53 per cent and 73 per cent increase of clump DBH and crown width were recorded due to widest spacing (12x12 m). The variation in height and DBH with spacing revealed that trees may grow taller in denser stands and larger girth in less denser stands. The number of culms was also significantly less (47.66) in closest spacing and increased with increasing spacing of bamboo. In the present study as bamboo spacing increased, the stand leaf area index of bamboo decreased. It was lowest (0.45) in widest spacing (12x12 m) and highest (6.78) in closest spacing (4x4 m). In agroforestry practices, the factors like LAI and crown development are of outmost importance for standardization of spacing. The changes in mean mid-day PAR (0200 hrs) among varying spacing was significant. The over storey recorded maximum PAR (1220 μmol per sec. per m^2); at this time, the understory PAR was 89 per cent less in closest spacing and 0.8 per cent under widest spacing (12x12 m).

Table 1: *Growth parameters, LAI and PAR of differentially spaced 7 year old bamboo (Dendrocalamus strictus) in central Kerala, India.*

Spacing (m)	Height (m)	Clump DBH (m)	Crown width (m)	Number of culm clump^{-1}	LAI	Mean mid-day PAR (μ moles/s^2/m^1)
4 × 4	9.11(0.06)^d	1.03(0.01)^a	4.69(0.10)^a	47.66(6.69)^a	6.78(0.30)^d	132(21)^a
6 × 6	8.18(0.12)^c	1.28(0.02)^b	6.61(0.13)^b	75(12.50)^b	4.76(0.11)^c	242(25)^b
8 × 8	8.03(0.03)^c	1.44(0.01)^c	7.36(0.09)^c	103.66(5.69)^c	3.38(0.024)^b	321(7)^c
10 × 10	7.6(0.10)^b	1.55(0.004)^d	7.79(0.09)^d	111.33(1.85)^{cd}	1.11(0.011)^a	1059(42)^d
12 × 12	7.31(0.05)^a	1.58(0.003)^d	8.13(0.08)^e	130(5.36)^d	0.00	1210(9)^e

Values in the parenthesis are Standard Deviation of the Mean; Values followed by same superscript in a column do not differ significantly (LSD, $P<0.05$).

Among the spacing of bamboo, the bulk density increased significantly with spacing from 1.11 Mg per m^3 in 4x4 m to 1.54 Mg per m^3 in 12x12 m, implying that, changes in spacing of bamboo have significant influence on bulk density of soil (Table 2). About 57:48:33 per cent of higher total N: available P: available K accumulation in closest spacings of bamboo was observed as compared to bambooless plot. The total N content was higher (2197 kg ha^{-1}) in closest spacing and it was decreased by 56 per cent in 12x12 m spacing. This emphasizes the need to minimize competition between trees and crops and maximize the use of available resources for overall productivity in any agroforestry system (Malezieux et al 2009).

Table 2: *Soil physico-chemical properties of differentially spaced 7 year old bamboo (Dendrocalamus strictus) plantation prior to understorey ginger and chittaratha planting in central Kerala, India.*

Spacing (m)	Bulk Density (Mg m^{3})	Soil pH	Total N (kg ha^{-1})	Available P (kg ha^{-1})	Available K (kg ha^{-1})
4 × 4	1.11(0.01)^{a}	5.83(0.15)^{a}	2197.70(34.76)^{e}	21.32(0.29)^{e}	203.49(0.47)^{d}
6 × 6	1.18(0.01)^{b}	5.8 (0.05)^{a}	1807(17.38)^{d}	19.02(0.26)^{d}	202.39(0.78)^{d}
8 × 8	1.24(0.01)^{c}	5.93(0.12)^{a}	1556.50(5.77)^{c}	17.20(0.17)^{c}	192.77(1.79)^{c}
10 × 10	1.31(0.01)^{d}	6.0 (0.06)^{a}	1466.71(17.38)^{b}	15.99(0.22)^{b}	164.26(0.31)^{b}
12 × 12	1.44(0.01)^{e}	6.1 (0.10)^{a}	1404.97(11.69)^{a}	14.73(0.19)^{a}	153.26(0.26)^{a}
Bambooless control	1.54(0.01)^{f}	5.96 (0.09)^{a}	1396.41(9.22)^{a}	14.43(0.05)^{a}	152.86(0.77)^{a}

Values in the parenthesis are Standard Deviation of the Mean; Values followed by same superscript in a column do not differ significantly (LSD, $P<0.05$).

RESPONSE OF UNDERSTOREY GINGER AND CHITTARATHA

In general, ginger and chittaratha plant height, number of tillers and rhizome yield increased with time course (duration). About 131 per cent increase in plant height of ginger at 10 x 10 m spacings of bamboo was observed compared to closest spacing (4x4 m); however this height decreased (25 per cent) with further increase of spacings to 12 x 12 m and bambooless control plot (11 per cent). In case of chittaratha the number of tillers at 12x12 m spacing of bamboo was increased by 295 per cent at 360 DAP compared to closest spacing of bamboo (4x4 m). Biscoe and Gallagher (1975) reported that plant dry matter production and rhizome yield is generally directly related to the intercepted radiation. Wider spacings of bamboo intercepted less solar radiation and allowed more solar radiation to reach the ground which might have stimulated the understorey growth and resulted in more understorey dry matter production (Table 3). Kasanga and Monsi (1954) also revealed that, the understorey growth rate is proportional to the amount of radiation intercepted by the canopy. In a study elsewhere in China revealed that low understorey PAR levels resulting from high level of LAI significantly reduced yield of winter wheat near tree row (Chirko et al 1999).

Table 3: *Growth and yield of ginger and chittaratha as influenced by varying spacing for 7 year old bamboo (Dendrocalamus strictus) in central Kerala, India*

	Spacings	Plant height (cm)	Number of Tillers	Dry matter (M ha^{-1}) production	Rhizome yield (M ha^{-1})	Oleoresin content (%)
Ginger	4 × 4	32.89(2.08)^{a}	1.81(0.13)^{a}	0.63(0.02)^{a}	3.19(0.04)^{a}	3.24(0.31)^{a}
	6 × 6	36.26(1)^{b}	1.63(0.23)^{a}	1.22(0.05)^{b}	3.59(0.09)^{a}	3.60(0.30)^{a}
	8 × 8	45.2(1.01)^{c}	4.63(0.75)^{b}	3.76(0.12)^{d}	8.29(2.17)^{b}	5.90(0.93)^{b}
	10 × 10	76.23(2.07)^{f}	9.62(0.71)^{d}	5.00(0.15)^{f}	19.55(2.47)^{d}	5.74(0.36)^{b}
	12 × 12	56.91(0.57)^{d}	8.77(0.49)^{cd}	4.49(0.01)^{e}	18.02(0.15)^{cd}	5.57(0.65)^{b}
	Control	67.75(1.94)^{e}	8.03(0.76)^{c}	2.86(0.15)^{c}	16.69(1.51)^{c}	5.71(0.62)^{b}

Chittaratha	4 × 4	15.27(0.54)[a]	5.40(0.42)[a]	1.33(0.05)[a]	7.10(0.94)[a]	5.75(0.26)[a]
	6 × 6	17.66(0.16)[b]	6.92(0.06)[b]	1.74(0.06)[b]	9.82(0.58)[b]	6.56(0.09)[b]
	8 × 8	29.35(0.52)[c]	13.81(0.73)[c]	2.89(0.11)[c]	11.6(0.56)[c]	7.84(0.80)[c]
	10 × 10	47.53(0.86)[d]	17.14(0.61)[d]	4.06(0.12)[d]	15.78(0.58)[d]	8.32(0.16)[cd]
	12 × 12	57.46(0.50)[e]	21.44(1.30)[e]	4.21(0.24)[d]	16.34(0.90)[d]	8.74(0.14)[d]
	Control	64.43(0.17)[f]	21.81(0.97)[e]	4.77(0.11)[e]	18.28(0.96)[e]	8.77(0.03)[d]

Values in the parenthesis are Standard Deviation of the Mean

Values followed by same superscript in a column do not differ significantly (LSD, $P < 0.05$).

It was clearly found that rhizome yield of ginger was significantly high (19.55 Mg ha-1) at 10×10 m spacings of bamboo which decreased to 3.19 Mg ha-1 when the spacing of bamboo decreased to 4×4 m (Table 3). Kunhamu et al (2008) also reported that ginger rhizome yield increased linearly with decreasing stand density of *Acacia mangium* AFS. The oleoresin per cent in ginger and chittaratha significantly varied over the spacing, implying that, oleoresin content is perhaps dependent on spacing of bamboo. But the previous study conducted by Latha (1994) reported that oleoresin and shade were independent. However in general for better growth performance of understory ginger and chittaratha PAR of 62 to 82 per cent resulted to be ideal. The lesser growth response in close spacings might also be due to presence of phenolics in bamboo leaves which inhibited the growth and development of groundnut seedlings (Eyini et al 1989).

CONCLUSION

Bamboo forms an important component in the agroforestry systems in India. The practice of bamboo cultivation and management provides a better option for sustainable use of land. Present study revealed the height growth in bamboo decreased with increasing spacings of bamboo; whereas higher leaf area index and lesser light penetration in close spacings of bamboo decreased ginger and chittaratha yield; however both crops had positive growth response with increasing spacings. The NPK content of soil gradually decreased with proportional increase of spacings of bamboo. The ginger responded better for 10×10 m spacings of bamboo. Conversely, chittaratha is better under open condition; nevertheless chittaratha performed good under wider spacings (12x12 m) of bamboo. Despite the better intercrop performance at lower bamboo stocking, there could be possible reduction in the bamboo yield which may not be desirable from the economic stand point. Hence a possible trade off need to be evolved for bamboo based agroforestry systems to optimize the intercrop productivity without compromising the bamboo yields. Probably intermediary bamboo stocking such as 8x8 m and 10x10 m would be ideal.

ACKNOWLEDGMENT

This work forms part of the PhD dissertation project of the first author. The bamboo spacing trial was established using an AP Cess fund research Grant from the Indian Council of Agricultural Research, New Delhi to B. Mohan Kumar (2004–2007). The authors thank Dr. Asha K Raj and Dr. AV Santosh Kumar for the help in analyzing the data. We also thank Dr. PS John for valuable comments and suggestions during crop measurements and nutrient analysis. The financial assistance provided to the first author by the Kerala Agricultural University for undertaking this research project is gratefully acknowledged.

REFERENCES

Bhunia D and Mondal AK 2012 Antibacterial Activity of Alpinia L (Zingiberaceae) from Santal and Lodha tribal areas of Paschim Medinipur District in Eastern India, *Advance in Bioresearch* **3**(1) 54 – 63

Biscoe PV and Gallagher J N 1975 Weather, dry matter production and yield, in *Environmental effects on crop physiology* (eds) Landsberg JJ and Cutting, Academic Press, London, 75-100

Chen IN, Chang CC, Ng CC, Wang CY, Shyu YT and Chang TL 2008 Antioxidant and antimicrobial activity of Zingiberaceae plants in Taiwan, *Plant Foods Hum Nutrition* 63(1) 15-20

Chirko CP, Gold MA and Jiang JP 1999 Influence of direction and distance from trees on wheat yield and photosynthetic photon flux density in a Paulownia and wheat intercropping systems, *Forest Ecology Management* **83** 171-180

De Padua LS, Bunyapraphatsara N, and Lemmens RH 1999 Medicinal and poisonous plants, Plant resources of South-East Asia No.12 (1), Buckhuys Publishers, Leiden 96-101

Ehrmann J and Ritz K 2014 Plant soil interactions in temperate multi-cropping production systems, *Plant and Soil* **376** (1) 1-29

Eyini M, Jayakumar H, and Panniselvam S 1989 Allelopathic effect of bamboo leaf extract on the seedlings of groundnut, *Tropical Ecology* **30** 138–141

Jackson ML 1958 Soil chemical analysis, Asia Publishing House, New Delhi, India, 498

Kasanga K and Monsi M 1954 On the light transmission of leaves and its meaning for the production of matter in plant communities, *Journal of Japanes Botany* **14** 304-324

Kerala Agricultural University (KAU 2011) Package of practices recommendations crops, 14th edn Kerala Agricultural University, Thrissur, 360

Kiaer LP, Weisbach AN, and Weiner J 2013 Root and shoot competition: a meta-analysis, *Journal of Ecology* **101** 1298–1312

Kibwage JK, Netondo GW, Odondo AJ, Oindo BO, Momanyi GM, and Jinhe F 2008 Growth performance of bamboo in tobacco-growing regions in South Nyanza, Kenya, *African Journal of Agriculture Research* **3**(10) 716–724

Kirtikar KR and Basu BD 1988 Indian medicinal plants, Vol. II International Book Distributors, DehraDun, 2444-2449

Kunhamu TK, Kumar BM, and Viswanath S 2008 Does thinning affect litterfall, litter decomposition, and the associated nutrient release in *Acacia mangium* stands of Kerala in the peninsular India, *Canadian Journal of Forest Research* **39** 792-801

Latha EV 1994 Evaluation of kacholam (*Kaempferia galangal*) types for morphological variability and yield. M. Sc. (Agri.) Thesis submitted to Kerala Agricultural University India 93-96

Malezieux E, Crozat Y, Dupraz C, Laurans M, Makowski D, Ozier-Lafontaine H, Rapidel B, De Tourdonnet S and Valantin-Morison M 2009 Mixing plant species in cropping systems concepts, tools and models: A review, *Agronomy Sustainable Development* **29** 43–62

Watanabe FS and Olsen SR 1965 Test of an ascorbic acid method for determining phosphorus in water and $NaHCO_3$ extracts from soil, *Soil Science Society of American Proceeding* **29** 677–678

Wilkinson KM and Elevitch CR 2000 Integrating Understory Crops with Tree Crops: An Introductory Guide for Pacific Islands. Agroforestry Guides for Pacific Islands #4. Permanent Agriculture Resources, Holaloa, Hawaii, USA, Web site: http://www.agroforestry.net.

Wohlmuth H 2008 Phytochemistry and pharmacology of plants from the ginger family, Zingiberaceae. Ph.D Thesis, Department of Natural and Complementary Medicine Southern Cross University Lismore, Australia 168-169

CHAPTER 28

Performance of Wheat and Paddy Intercropped Under Poplar (*Populus deltoides* Bartr. ex Marsh) Based Agrisilviculture System

Pradyuman Singh*, OP Rao, KS Bangarwa[1] and Neeraj

N. D. University of Agriculture and Technology, Kumarganj, Faizabad -224229 U. P., India

[1]CCS Haryana Agriculture University Hisar, 125004 Haryana, India

**Corresponding author's Email:pradyuman12@gmail.com*

ABSTRACT

The performance of three wheat varieties (NW-1067, NW-1014 and HD -2643) and three paddy varieties (sarjoo-52, NDR-359 and swarna) were assessed under six year old *Populous deltoides* plantation at different distances from tree base (0.5 m, 1.0 m, 1.5 m and 2.5 m) and control (open condition). Maximum plant height (90.90 cm) was found for wheat variety HD-2643. In case of paddy, plant height (96.34 cm) and straw yield (2.47 t ha^{-1}) were found better for paddy variety NDR-359, when the crop was grown at 0.5 m away from the tree base. However, maximum number of shoots (m^{-2}), grain ear^{-1} -$panicle^{-1}$, test weight (1000 grain wt. g.), grain yield (t ha^{-1}) were recorded in case of wheat variety NW-1067 and paddy variety NDR- 359, when the crop was grown in control (open condition). Amongst different varieties, wheat variety NW-1014 and paddy variety NDR-359 were found most suitable.

Keywords: Agrisilviculture system, Poplar, Wheat

INTRODUCTION

Agroforestry as alternative land management system addresses many of the global challenges, including deforestation, unsustainable cropping practices, loss of biodiversity, increased risk of climate change, as well as rising hunger, poverty and malnutrition (Garrity 2004). Agroforestry systems are believed to arrest land degradation and sustain soil productivity (Jose 2009, Tsonkova et al 2012) and can re-establish some of the forests ecological goods and services like timber production, potentially improves agricultural productivity, soil fertility improvement and enhances biodiversity (Montagnini and Nair 2004, Sharma et al 2004, Chauhan and Ritu 2005). Poplar is the very prominent taxonomic group of tree species in plantation forestry as well as in natural forests north-western state of India. Its natural population is small and found only in the mountains with six indigenous species. As an exotic species *(Populous deltoides*), it acquire great role afforestation/reforestation programs, agroforestry and conservation activities. According to India country report on poplar and willow (Anonymous 2012), area under poplar outside the forest in India is estimated 312,000 ha. The annual return from poplar at current market rates are estimated to be around ₹ 60-80,000 acre^{-1} yr^{-1}, which is much higher than any other on-farm intervention (Dhiman 2012). Poplar is normally planted and grown under wide spacing accommodating 400 to 500 trees per ha. Its stand density permits enough sunlight and air circulation and therefore gives better crops yields (Tewari 2002). In many locations, farmers now plant more trees per unit area than recommended number of 400-500 trees ha^{-1} with the hope to get increased returns. This study was therefore, planned to study the effect of different distances from tree base on wheat and paddy performance in *Populus deltoides* based agri-silvicultural system. Poplar based agroforestry systems are economically viable and more profitable than many other crop rotations (Jain and Singh 2000, Chauhan and Mangat 2006). It is usually managed in 6-8 years rotation cycle under agroforestry system. Owing of its rapid growth, high biomass, adoptability, early economic returns and compatibility with crops prompted to introduce poplars in agroforestry systems. Wheat (*Triticum aestivum* L.) is generally intercropping with high density short rotation tree species is the best option to meet increasing food and industrials raw material requirement through sustainable utilization of natural resources (Sarvade et al 2014). Cropping with tree species is an ancient practice and very important tools to achieve goal of National Forest Policy (1988). It has been reported as an important component of the 'evergreen revolution' movement in the country (Puri and Nair 2004). Growth of the poplar depends upon various factors such as clone, quality of planting stock, spacing of the trees, intercrops, site quality, climate and management practices (Tewari 1995, Chauhan et al 2012a, 2015a).Its popularity with the farmer is mainly due to its fast growth, multi-utility wood, high market prices, less competition with associated crops and pruning tolerant nature. Poplar based agroforestry systems are economically viable and more profitable than many of the other crop rotations in the region (Jain and Singh 2000, Chauhan et al 2012b, 2015b). It benefits the farmers both economically and from soil physical fertility point of view.

METHODOLOGY

The present study was conducted during 2011-2012 in already established 6 years old poplar plantation spaced at 5 m x 4 m at N. D. University of Agriculture and Technology, Kumarganj, Faizabad (26^{0} 27' N latitude and 82^{0} 12' E longitudes at an elevation 113 m above mean sea level). The annual rainfall during 2011-2012 was 874.4 mm. The mean monthly maximum temperature ranged from 20.4 ^{0}C (January) to 41.6 ^{0}C (June) and mean monthly minimum temperature varied from 7.8 ^{0}C (December) to 28.0 ^{0}C (June) during the study period. The soil of experimental site is partially improved salt affected soil (pH=8.5). It is medium in available nutrient (NPK) and organic matter. The trees at random in 5 m × 4 m spacing was measured for their top height and girth at breast height (GBH). The height was measured from ground to top of the trees. The seeds of three varieties of wheat (NW-1067, NW-1014 and HD-2643) were sown during the 15th November, 2011. Keeping row to row distance of 22.5 cm with a seed rate of 100 kg ha^{-1}. One half of the nitrogen and whole of the phosphorus and potash was applied at the time of sowing and remaining N was applied to wheat crop under 5 m x 4 m spacing of poplar plantation and control (without trees) after the first irrigation. However, seedlings of the three paddy varieties (Sarjooo-52, NDR-359 and Swarna) were raised in well prepared nursery bed during 30th May to 29th June 2012, and transplanted on 30th June at the distance of 20 cm from row to row and 10 cm from hill to hill . Before transplanting of seedlings, half dose of nitrogen and full dose of phosphorus and potash were applied. Remaining N was used twice *i.e.* after 30 and 60 days of transplanting. Wheat and paddy crop was estimated in terms of growth and yield parameters by quadrate method at time of harvest. Five quadrates of 1 m^2 were selected per replication between rows of poplar. The yield of produce (grain) was extrapolated to be expressed in t ha^{-1}. The experiment was laid out in Factorial Randomized Block design with five replications. All cultural practices were adopted as per recommended for cultivation of wheat and paddy. Data were analysed statistically as per method given by Panse and Sukhatme (1989) and results were expressed at 5 per cent levels of significance. The critical difference (CD) was calculated using following formula:

$$\text{C.D.} = \sqrt{\frac{2MS_E}{r}} \times t \text{ value of error degree of freedom at 5 per cent of significance.}$$

RESULTS AND DISCUSSIONS

Tree Growth Parameters

Tree height, DBH, canopy width and canopy length were significantly affected by the wheat and paddy crop. Growth parameters recorded in poplar revealed that it gained 15.60 m height, 28.06 cm DBH, 6.24 m canopy width and 7.58 m canopy length at the of harvesting.

Crop Parameters

Table 1 indicated the variation in growth characters were significant in different wheat varieties. Results revealed that wheat variety HD-2643 grown 0.5 m away from tree base produced maximum plant height (90.90 cm), which was significantly higher than other varieties of wheat followed by NW-1067 (88.56 cm). However, significantly maximum number of shoots m^{-2}, number of grains ear^{-1} and test weight (1000-grain weight) was recorded in variety NW-1067, when the crop was grown in control (open condition). Wheat grown in alleys at 0.5 m,1.0 m,1.5 m and 2.5 m distance from tree base produced lower number of shoots m^{-2}, number of ears m^{-2}, number of grains per ear and test weight (1000-grain weight g) compared to control (open condition). Similar results have also been reported by Gandhi (2008) and Dhillon et al (2016).

Data presented in Table 2 indicated that the grain yield was significantly influenced by variety and distance and the interaction between variety and distance were also found significant. Significantly higher grain yield (2.32 t ha^{-1}) was recorded for wheat variety NW-1014, which was closely followed by NW-1067 (2.30 t ha^{-1}) in control (open condition). The maximum straw yield (2.89 t ha^{-1}) was found in case of wheat variety NW-1014 at 0.5 m away from tree base. The variation in yield of different varieties at varies distance from tree base might be due to differential competition for light, soil moisture and nutrients. Similar results have also been reported by Singh et al (2014).

Table 1: *Plant growth for wheat varieties under Populus deltoides based agri-silvicultural System*

Distance from tree base (m)	Varieties			Mean
	NW-1067	NW-1014	HD-2643	
Plant height (cm)				
0.5	88.56 (212.45)	87.56 (200.79)	90.90 (198.79)	89.00 (204.01)
1.0	86.90 (213.33)	86.56 (206.44)	88.67 (200.88)	87.38 (206.88)
1.5	86.20 (219.44)	85.23 (207.54)	87.24 (206.44)	86.23 (211.14)
2.5	83.90 (220.45)	83.56 (213.54)	86.23 (210.78)	84.56 (214.92)
Control	80.23 (224.45)	78.90 (221.43)	82.32 (220.12)	80.48 (222.00)
Mean	85.16 (218.02)	84.36 (209.95)	87.07 (207.40)	
	Variety (V)	Distance (D)	V × D	
CD at 5%	0.32** (1.81**)	0.41** (2.34**)	0.71** (4.05**)	
(Values in parentheses indicate No. of shoots m^{-2})				
Number of grains ear^{-1}				
0.5	25.01 (31.36)	24.67 (29.14)	24.01 (2802)	24.56 (29.51)
1.0	26.67 (31.39)	25.65 (29.31)	25.01 (28.10)	25.78 (29.60)
1.5	29.01 (31.71)	26.32 (29.35)	26.34 (29.05)	27.22 (30.03)
2.5	30.01 (32.08)	27.67 (30.64)	27.67 (30.44)	28.45 (31.05)
Control	32.34 (32.49)	28.99 (31.33)	30.32 (31.06)	30.55 (31.63)
Mean	28.61 (31.81)	26.66 (29.95)	26.67 (29.33)	
	Variety (V)	Distance (D)	V × D	
CD at 5%	0.53** (0.37**)	0.68** (0.48**)	1.18** (0.83**)	
(Values in parentheses indicate test weight(1000-grain wt. g)				

** Significant at 1% level

Table 2: *Grain yield of three wheat varieties under Populus deltoides based agri-silviculture System*

Distance from tree base (m)	Varieties			Mean
	NW-1067	NW-1014	HD-2643	
Grain yield t ha^{-1}				
0.5	1.14 (2.44)	1.15 (2.89)	1.14 (2.82)	1.14 (2.72)
1.0	1.44 (2.50)	1.32 (1.94)	1.28 (2.13)	1.35 (2.19)
1.5	1.99 (1.76)	1.46 (1.82)	1.37 (1.61)	1.61 (1.73)
2.5	2.27 (1.86)	1.74 (1.53)	1.82 (1.43)	1.95 (1.61)
Control	2.30 (1.48)	2.32 (1.42)	2.25 (1.26)	2.29 (1.13)
Mean	1.83 (2.01)	1.60 (1.92)	1.57 (1.85)	
	Variety (V)	Distance (D)	V × D	
CD at 5%	0.59** (0.66**)	0.76** (0.86**)	1.31**(1.49**)	
(Values in parentheses indicate straw yield)				

** Significant at 1% level

The variations in growth characters were significant in different paddy varieties. The maximum plant height (96.34 cm) was recorded for paddy variety NDR-359, which was significantly higher than all other paddy varieties followed by Swarna (92.01 cm), when the crop was grown at 0.5 m away from tree base. However, the maximum number of shoots m^{-2} (178.23), number of panicle m^{-2} (175.79) was found in Sarjoo-52, and the maximum number of grains $panicle^{-1}$ (204.45), 1000-test weight (26.67g) was found in NDR-359, when the crop was grown in control (open condition). It may be due to the fact that direct light intensity favors number of shoots (Table 3). Perusal of data in Table 4 indicated that variety and distance was significantly affected the yield of paddy crop. Among all the three varieties of paddy crop significantly higher grain yield (2.31 t ha^{-1}) was recorded in paddy variety NDR-359, which was closely followed by Sarjoo-52 (2.21 t ha^{-1}) variety in control (open condition). Whereas, the maximum straw yield (2.47 t ha^{-1}) was also observed in NDR-359 followed by Swarna (2.46 t ha^{-1}) when the crop was grown at 0.5m away from tree base. The reduction in the grain yield under tree canopies may be due to the fact that shade negatively affects grain yield. Similar findings have also been reported by Sirohi and Rao (2012). In case of maximum straw yield was found at 0.5m away from tree base and the minimum value are directly related to the plant population and plants getting quantum of light ,thus affecting the photosynthesis.

Table 3: *Plant growth for paddy varieties under Populus deltoides based agri-silvicultural System*

Distance from tree base (m)	Plant height (cm)			Mean
	Varieties			
	Sarjoo-52	NDR-359	Swarna	
0.5	79.01 (161.89)	96.34 (166.42)	92.01 (165.67)	89.12 (164.66)
1.0	76.34 (166.56)	93.34 (169.23)	90.32 (169.00)	86.66 (168.23)
1.5	76.01 (168.23)	91.01 (171.56)	88.34 (172.67)	85.12 (170.82)
2.5	70.34 (171.89)	91.34 (173.89)	88.67 (173.22)	83.45 (172.22)
Control	65.67 (178.23)	85.34 (176.52)	82.67 (177.10)	77.89 (177.10)
Mean	73.47 (169.36)	91.47 (171.52)	88.41 (170.81)	
	Variety (V)	Distance (D)	V × D	
CD at 5%	1.06** (1.10**)	1.37** (1.42**)	2.37** (2.47**)	
(Values in parentheses indicate number of shoots m^{-2})				
Number of grains $panicle^{-1}$				
0.5	179.23 (21.14)	184.00 (26.03)	171.23 (20.20)	178.15 (22.45)
1.0	180.12 (21.52)	185.33 (26.07)	172.22 (20.24)	179.22 (22.61)
1.5	182.34 (21.48)	186.68 (26.11)	173.67 (20.26)	180.89 (22.62)
2.5	182.67 (21.81)	188.34 (26.22)	174.34 (20.46)	181.78 (22.83)
Control	194.33 (22.50)	204.45 (26.67)	186.54 (20.96)	195.11 (23.38)
Mean	183.74 (21.69)	189.76 (26.22)	175.60 (20.42)	
	Variety (V)	Distance (D)	V × D	
CD at 5%	0.98**	1.27**	2.19**	
(Values in parentheses indicate test weight(1000-grain wt. g)				

** Significant at 1% level

Table 4: *Grain yield of three paddy varieties under Populus deltoides based agri-silvicultural system*

Distance from tree base (m)	Varieties			Mean
	Sarjoo-52	NDR-359	Swarna	
	Grain yield t ha^{-1}			
0.5	1.56 (2.33)	1.58 (2.47)	1.33 (2.36)	1.49 (2.36)
1.0	1.71 (1.98)	1.69 (1.97)	1.54 (1.82)	1.65 (1.65)
1.5	1.87 (1.77)	1.79 (1.84)	1.64 (1.76)	1.77 (1.74)
2.5	1.90 (1.70)	1.99 (1.76)	1.74 (1.63)	1.87 (1.82)
Control	2.21 (1.66)	2.31 (1.37)	2.18 (1.65)	2.23 (1.67)
Mean	1.85 (1.89)	1.87 (1.88)	1.69 (1.85)	
	Variety (V)	Distance (D)	V × D	
CD at 5%	0.43** (0.64**)	0.55** (0.82**)	0.96** (2.26**)	
(Values in parentheses indicate straw yield)				

** Significant at 1% level

CONCLUSION

Based on the result obtained from the present investigation, it may be concluded that there was a significant effect of wheat and paddy on the height and DBH of the perennial plants. The number of shoots m^{-2}, number of ears panicle^{-1} m^{-2}, grains panicle^{-1}, test weight and grain yield were found to be maximum in control (open condition). Whereas, the plant height and straw yield were maximum, when crop was grown in whole area including plant basin. On the basis of the yield potential, it may be concluded that NW-1067 was found to be the best variety of wheat under this system. In case of paddy, NDR-359 is a suitable variety in comparison to the other. Among all the three varieties of wheat and paddy, NW-1067 (wheat variety) and NDR-359 (paddy variety) gave better result under *Populus deltoides* based agri-silviculture system.

REFERENCES

Anonymous 2012 India country report on poplars and willow, National Poplar Commission of India by Indian Council of Forestry Research and Education, Dehradun Uttrakhand

Chauhan SK and Ritu 2005 Towards agro-ecological health: diversifying traditional crop rotation through agroforestry in Punjab, India, *Asia-Pacific Agroforestry News,* **27** 3-4

Chauhan SK and Mangat PS 2006 Poplar based agroforestry is ideal for Punjab, India, Asia-*Pacific Agroforestry News* **28** 7-8

Chauhan SK, Brar MS, and Sharma R 2012 Performance of poplar (*Populous deltoides* Bartr) and its effect on wheat yield under agroforestry system in irrigated agro-ecosystem, India, *Caspian Journal of Environmental Science,* **10** (1) 53-60

Chauhan SK, Sharma R, and Dhillon WS 2012 Status of intercropping in poplar based agroforestry in India, International Poplar Commission Workshop at ICFRE Dehradun (Oct.29-Nov.2, 2012) *Envis Forestry Bulletin* **12** 49-87

Chauhan SK, Sharma R, Singh B, and Sharma SC 2015 Biomass production, carbon sequestration and economic of on-farm poplar plantation in Punjab, India, *Journal of Applied and Natural Science* 7 (1) 452-458

Chauhan SK, Saini KS, Saralch H S, Rani S, and Verma A 2015 Wheat and barley crop performance under different sowing methods under poplar based agroforestry system, *Indian Journal of Ecology* **42** (2) 528-530

Dhillon RS, Bhardwaj KK, Beniwal RS, Bangarwa KS, Kumari S, Godara AS and Sheokand RN 2016 Performance of wheat as intercrop under different spacings of poplar plantations in semi-arid ecosystem of Northern India, *Indian Journal of Ecology* **43** 323-327

Dhiman RC 2012 An innovative partnership for poplar based agro-forestry in India, *Asia-Pacific Agroforestry News* **40** 4-6

Gandhi JN 2008 Study on yield potential of wheat crop under different spacing of Poplar plantation. M.Sc. Thesis, Department of Agronomy, HNB, Gharwal, University, Srinagar

Garrity DP 2004 Agroforestry and the achievement of the millennium development goals, *Agroforestry Systems* **61** 5-17

Jain SK and Singh P 2000 Economic analysis of industrial agroforestry: poplar (*Populous deltoides*) in Uttar Pradesh (India), *Agroforestry Systems* **49** 255-273

Jose S 2009 Agroforestry for ecosystem services and environmental benefits: an overview, *Agroforestry Systems* **76** 1-10

Montagnini F and Nair PKR 2004 Carbon sequestration: An under–exploited environmental benefit of agroforestry system, *Agroforestry Systems* **61** 281-295

Panse VG and Sukhatme PV 1989 Statistical Methods for Agricultural Workers, ICAR, New Delhi

Puri S and Nair PKR 2004 Agroforestry research for development in India: 25 years of experiences of a national program, *Agroforestry Systems* **61** 437-452

Sarvade S, Mishra HS, Kaushal R, Chaturvedi S, Tewari S, and Jadhav TA 2014 Performance of wheat (*Triticum aestivum* L.) crops under different spacings of trees and fertility levels, *African Journal of Agricultural Research* **9** (9) 866-873

Sharma P, Rai SC, Sharma R, and Sharma E 2004 Effects of land use change on soil microbial C, N and P in a Himalayan watershed, *Pedobiologia* **48** 83-92

Sirohi C and Rao OP 2012 Varietal comparison of wheat and paddy under *Populus deltoides* based silviculture system in sodic soil, *Indian Journal of Forestry* **35** (3) 291-294

Tewari DN 1995 Agroforestry for increased productivity, sustainability and poverty alleviation, International book distributors, Dehradun 799

Tewari DN 2002 A monograph on Poplar, Surya publication Dehradun

Tsonkova P, Böhm C, Quinkenstein A, and Freese D 2012 Ecological benefits provided by alley cropping systems for production of woody biomass in the temperate region: review, *Agroforestry Systems* **85** 133-152

CHAPTER 29

Growth Performance, Essential Oil Recovery and Financial Flows of *Ocimum* spp. Under *Pongamia pinnata* - *Ocimum* spp. Based Silvi-Medicinal Agroforestry Systems

NS Thakur*, AH Suvera, SK Jha and DB Patel

Navsari Agricultural University, Navsari, Gujarat – 396 450, India

**Corresponding author's Email:* drnsthakur74@gmail.com

ABSTRACT

Four *Ocimum* spp. viz., *O. sanctum*, *O. tenuiflorum*, *O. basilicum* and *O. gratissimum* were intercropped under 3 years old Karanja [*Pongamia pinnata* (L.) Panigrahi] based agroforestry systems named as silvi-medicinal (Pongamia+*Ocimum* spp.) and sole cropping systems in factorial randomized block design (2 x 4) with three replications. Significantly better growth, above and below ground and total herbage (panchang) yield of *Ocimum* spp. were recorded under silvi-medicinal as compared to sole cropping systems. Out of four *Ocimum* spp., maximum plant height, number of branches and leaves per plant and higher herbage yield were attained by *O. tenuiflorum*. There was no significant effect of land use systems on the oil recovery (%) of all the *Ocimum* spp. Among *Ocimum* spp., statistically maximum essential oil recovery was recorded from *O. gratissimum*. The interaction due to land use systems and *Ocimum* spp. had significant effect on the oil recovery. Among intercrops, *O. tenuiflorum* accrued highest net returns and benefit-cost ratio from herbage production. The investigation evinced that cultivation of *Ocimum* spp. can make karanja plantation ecologically and economically sustainable by providing early returns.

Keywords: Agroforestry, *Ocimum* spp., *Pongmia pinnata*, Silvi-medicinal, Tree borne oil seed

INTRODUCTION

Pongamia pinnata (L.) Panigrahi is a potential tree borne oil seed (TBOs) species, efficient nitrogen fixer, able to grow in problematic soils and can withstand drought conditions (Duke 1983). It is considered as suitable tree species for agroforestry (Beniwal and

Chauhan 2011). It is advocated to grow short rotation intercrops during initial years to make TBOs plantations economically sustainable. Shade loving medicinal and aromatic plants are suitable intercrops with different timber and fruit yielding woody perennials (Patil et al 2005, Singh et al 2008 a, b, Ravitchandirane and Haripriya 2011, Thakur et al 2014). Since majority of the medicinal plants are found in forest and also shade tolerant; therefore agroforestry system offers a convenient strategy for promoting their cultivation and conservation. The most important question is selection of suitable medicinal and aromatic plants, which have market demand, to intercrop with woody component having long gestation period. The present investigation, therefore, intended to integrate *Ocimum* spp. with *P. pinnata*. More than 150 basil species belonging to the genus *Ocimum* have been reported and many of them are used in perfumery and cosmetics industries (Javanmardi et al 2002). Sweet basil *(O. basilicum)* and holy basil *(O. sanctum)* are the most widely grown in the world either for the fresh herb market or for essential oil production (Zheljazkov et al 2008). *O. americanum*, *O. basilicum*, *O. sanctum* and *O. tenuiflorum* are among the 178 high trade species with estimated annual trade of 500-1000, 1000-2000, 2000-5000 and 2000-5000 tonne, respectively (www.nmpb.nic.in). Therefore, owing to the importance of both karanja and *Ocimum* spp., present investigation was carried to screen out the compatible species.

METHODOLOGY

The present investigation was carried out during 2011-2012, at College of Forestry, Navsari Agricultural University, Navsari, Gujarat situated at 20.95°N latitude, 75.90°E longitude at an altitude of 10 m above the mean sea level. The climate of the area is typically tropical. Four *Ocimum* spp. viz., *O. sanctum* (S_1), *O. tenuiflorum* (S_2), *O. basilicum* (S_3) and *O. gratissimum* (S_4) were intercropped under 3 year old Karanja plantations spaced at 3x3m. The cropping systems so formed were named as silvi-medicinal (Pongamia+*Ocimum* spp. =L_1) and sole cropping system or open land use (*Ocimum* spp. in open=L_0).The available N and P_2O_5 (kg ha^{-1}) were estimated following Subbiah and Asija (1956) and Olsen et al (1954) whereas, K_2O (kg ha^{-1}), organic carbon (%) and pH were estimated following Jackson (1973), under both the land use systems. Land was prepared before commencement of monsoon and one month old seedlings of each spp. were transplanted at 30x30cm spacing in both the land use systems. Farm yard manure @ 20 tonnes/ha and Nitrogen (in the form of urea) @ 20 kg/ha were applied to all the plots as basal dose. The growth parameters of *Ocimum* spp. collar diameter, number of branches per plant, leaf area (cm^2) were recorded as per standard procedures. Herbage yield (above ground, below ground and total) was estimated at maturity by uprooting the plants. The dry herbage yield was estimated by taking the representative fresh herb samples, which were dried in shade till constant weight was attained.

Essential oil recovery (%) was estimated by hydro distillation method using Clevenger Apparatus taking 100 g of fresh herbage sample, comprising of leaves, inflorescence and very small twigs. The chopped sample was put in 500 ml capacity flask half filled with water and distilled for about an hour. The oil being lighter than water was collected in the burette and reading was recorded. The gross realization in terms of Indian Rupees per hectare was worked out on the basis of total dry herbage yield (stem, root, leaves, inflorescence and seeds, '*Panchang*' referred herein). The prevailing market price of

Ocimum spp. was considered to calculate the gross returns. The total cost of cultivation was deducted from gross returns to estimate the net returns. The data generated were subjected to the statistical analysis using factorial randomized block design (FRBD).

RESULTS AND DISCUSSION

Average height and collar diameter of karanja trees were 2.88 m and 55.90 cm, respectively. The average crown spread was 2.19 m and 2.71 m in east-west and north-south directions, respectively. The available N, P_2O_5, K_2O (kg ha^{-1}), and organic carbon (%) under silvi-medicinal system (Pongamia + *Ocimum* spp.) were 233.76, 34.02, 321.71 and 0.82, respectively; and the respective values in sole *Ocimum* spp. plots were 206.00, 30.67, 301.89 and 0.68. The soil reaction was above neutral (7.4). The soil physico-chemical analysis evinced that, nutrient and organic matter status was better under karanja. Improved soil physico-chemical properties in karanja based agroforestry systems have been reported by Banerjee et al (2013). The light intensity was recorded in each treatment at 10.00, 13.00 and 16.00 h, during the investigation period, at five days interval. The mean monthly light intensity values [recorded from July to November (5 months, cropping period of *Ocimum* spp.)] were 354.64, 352.68, 362.29, 352.21,364.27 under silvi-medicinal agroforestry systems and 564.64, 572.88, 529.97, 545.54, 578.15 under sole cropping systems (Fig. 1). The light intensity was less under silvi-medicinal systems as compared to open land use during the course of study.

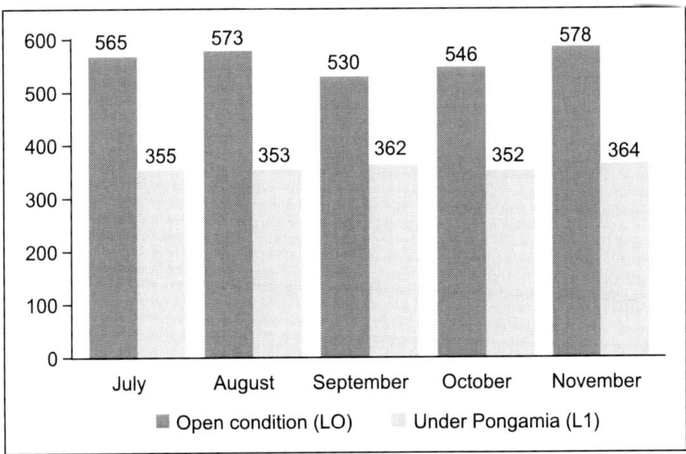

Figure 1: *Average light intensity in open and under silvi-medicinal systems during intercropping period*

GROWTH OF *OCIMUM*

The growth attributes of *Ocimum* spp. were significantly affected due to land use systems *i.e.* karanja based silvi-medicinal and sole cropping. These attributes also varied among *Ocimum* spp. as well (Table 1a and b). Significantly maximum plant height (0.90 m), collar diameter (8.41 mm), number of branches per plant (13.53) and number of leaves per plant

(168.74) were recorded under silvi-medicinal system for all the *Ocimum* spp. as compared to sole crop. Amongst *Ocimum* spp., superior plant height (1.11 m), maximum number of branches (15.10) and leaves per plant (232.47) were recorded for *O. tenuiflorum*. It also attained maximum leaf area (20.02 cm^2) as compared to all other intercrops. Results bore out that growth attributes of intercrops significantly differed due to interactions of land use systems and intercrops. *O. tenuiflorum* produced tallest plants (1.12 m) when grown as sole crop *i.e.* L_0S_2. Sole grown *O. basilicum* had higher collar diameter (9.09 mm). However, most number of branches (17.45) and leaves per plant (257.03) were recorded for *O. tenuiflorum* in interaction L_1S_2. Leaf area of intercrops was not affected whether grown under silvi-medicinal system or as sole crop. Better plant growth of *Ocimum* spp. under karanja may be due to partial shade (Fig. 1) provided by karanja trees. The shaded conditions under agroforestry systems have moderation effect on the ambient micro-climatic conditions *i.e.* air temperature, soil moisture, reduction in the rate of evaporation of water from soil and more nitrogen supply to plants through water absorption which enhance the vegetative growth (Singh et al 1997a, Ravindran and Kulandaivelu 1998). Intercrops under agroforestry, due to shade effect, produce larger leaves, leaf area and number of branches in order to capture more light because of shade avoidance mechanism and favoured the vegetative growth (Pooter and Oliver 2000). Favourable effect of shade in agroforestry systems on plant growth has also been reported by other workers (Pathiratna 2004, Patil et al 2005, Singh et al 2008 a, b, Khawlhring et al 2011, Ravitchandirane and Haripriya 2011). The leaf fall from trees add organic matter in soil (Anon 2008b) and improve soil fertility (Osman et al 2001, Mohsin 2005, Chaudhry et al 2007). The difference in soil nutrient status may be ascribed to more plant growth under silvi-medicinal as compared to sole cropping systems. The difference in growth attributed among *Ocimum* spp. may be attributed to genetic makeup of individual species.

Table 1a: *Growth parameter of Ocimum spp. under Pongamia pinnata based silvi-medicinal and sole cropping systems*

Land use systems	*Plant height (m)*				*Mean*	*Collar diameter (mm)*				*Mean*
	***Ocimum* spp.**					***Ocimum* spp.**				
	(S_1)	*(S_2)*	*(S_3)*	*(S_4)*		*(S_1)*	*(S_2)*	*(S_3)*	*(S_4)*	
(L_0)	0.76	1.12	0.70	0.69	0.82	7.26	6.57	9.09	7.62	7.64
(L_1)	0.92	1.09	0.75	0.85	0.90	8.34	8.88	8.76	7.65	8.41
Mean	0.84	1.11	0.73	0.77		7.80	7.73	8.92	7.64	
Sources	S. E. ±		C.D. at 5 %			S. E. +		C.D. at 5 %		
L	0.02		0.06			0.22		0.65		
S	0.03		0.08			0.30		0.92		
M x S	0.04		0.11			0.43		1.30		

(S_1) = *O. sanctum*, (S_2) = *O. tenuiflorum*, (S_3) = *O. basilicum*, (S_4) = *O. gratissimum*, (L_0) = Sole, (L_1) = Karanja + *Ocimum* spp.

Table 1b: *Growth parameter of Ocimum spp. under Pongamia pinnata based silvi-medicinal and sole cropping systems*

Land use systems	Number of branches plant^{-1}				Mean	Number of leaves plant^{-1}				Mean	Leaf area (cm^2)				Mean
	Ocimum spp.					*Ocimum* spp.					*Ocimum* spp.				
	(S_1)	(S_2)	(S_3)	(S_4)		(S_1)	(S_2)	(S_3)	(S_4)		(S_1)	(S_2)	(S_3)	(S_4)	
(L_0)	14.13	12.75	11.88	9.57	12.08	158.58	207.90	121.57	78.77	141.70	5.63	5.38	5.26	20.07	9.09
(L_1)	15.85	17.45	12.30	8.52	13.53	185.50	257.03	147.23	85.20	168.74	6.08	5.58	5.65	19.97	9.32
Mean	14.99	15.10	12.09	9.04		172.04	232.47	134.40	81.98		5.86	5.48	5.46	20.02	

Sources	S. E. +	C.D. at 5 %	S. E. +	C.D. at 5 %	S. E. +	C.D. at 5 %
L	0.33	1.01	2.80	8.50	0.25	NS
S	0.47	1.43	3.96	12.02	0.35	1.08
M x S	0.67	2.02	5.61	17.00	0.50	NS

(S_1) = *O. sanctum*, (S_2) = *O. tenuiflorum*, (S_3) = *O. basilicum*, (S_4) = *O. gratissimum*, (L_0) = Sole, (L_1) = Karanja + *Ocimum* spp.

HERBAGE YIELD *(PANCHANG)*

The land use systems significantly affected the yield of intercrops. Maximum above (2.66 t ha^{-1}) and below ground (1.00 t ha^{-1}) and total herbage yield (3.66 t ha^{-1}) of *Ocimum* spp. was recorded under silvi-medicinal as compared to sole cropping systems (Table 2). Among intercrops, *O. tenuiflorum* produced higher above ground (3.95 t ha^{-1}) and total herbage yield (4.77 t ha^{-1}). Herbage yield of *O. basilicum, O. gratissimum*, obtained in the present study, is within the range reported by Kumar et al (2015). The increased herb yield of *Ocimum* spp. under silvi-medicinal systems may be ascribed to positive agroforestry interface (above and below ground) *i.e.* partial shade provided by karanja trees and difference in nutrient status of land use systems. Higher yields of medicinal and aromatic, cereals and vegetables, under different agroforestry systems have been reported in various studies (Pooter and Oliver 2000, Vyas 2001, Karikalan et al 2002, Gangadharan and Menon 2003, Kurian et al 2003, Kumar et al 2005, Sharanabasappa et al 2007, Ravitchandirane and Haripriya 2011) are in conformity with present investigation.

Table 2: *Herbage yield (DM t ha^{-1}) of Ocimum spp. under Pongamia pinnata based silvi-medicinal and sole cropping systems*

Land use systems	Above ground (t ha^{-1})				Mean	Below ground (t ha^{-1})				Mean	Total (Panchang) (t ha^{-1})				Mean
	Ocimum spp.					*Ocimum spp.*					*Ocimum spp.*				
	(S_1)	(S_2)	(S_3)	(S_4)		(S_1)	(S_2)	(S_3)	(S_4)		(S_1)	(S_2)	(S_3)	(S_4)	
(L_0)	1.96	3.69	1.98	2.01	2.41	0.94	0.76	1.12	0.80	0.91	2.90	4.46	3.10	2.81	3.32
(L_1)	2.10	4.20	2.40	1.93	2.66	1.01	0.87	1.36	0.77	1.00	3.11	5.07	3.76	2.70	3.66
Mean	2.03	3.95	2.19	1.97		0.98	0.82	1.24	0.78		3.01	4.77	3.43	2.75	

Sources	S. E. +	C.D. at 5%	S. E. +	C.D. at 5 %	S. E. +	C.D. at 5 %
L	0.08	0.23	0.03	0.08	0.10	0.30
S	0.11	0.32	0.04	0.114	0.14	0.42
M x S	0.15	NS	0.05	NS	0.20	NS

(S_1) = *O. sanctum*, (S_2) = *O. tenuiflorum*, (S_3) = *O. basilicum*, (S_4) = *O. gratissimum*, (L_0) = Sole, (L_1) = Karanja + *Ocimum* spp., DM = Dry matter

ESSENTIAL OIL RECOVERY (%)

Land use systems (silvi-medicinal and sole cropping) failed to impose any effect on the oil recovery from *Ocimum* spp. (Fig. 2). Among *Ocimum* spp., significantly maximum essential oil recovery was from *O. gratissimum*. The interaction due to land use systems and *Ocimum* spp. had significant effect on the oil recovery.

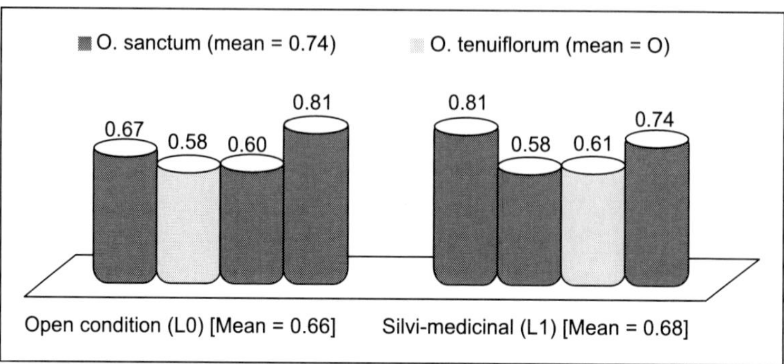

Figure 2: *Oil recovery (%) of Ocimum spp. under silvi-medicinal systems and open conditions [S.E.± (Land use=0.007; Species=0.009; Land use x species= 0.013), CD=Land use=NS; Species=0.03; Land use x species= 0.04)*

Maximum oil was recovered from *O. sanctum* grown under silvi-medicinal system, which was at par with oil recovered from *O. gratissimum* grown as sole crop. The oil recovery in the present investigation fall within the range obtained in earlier studies under agroforestry (Thakur et al 2009, Kumar et al 2016). Earlier studies have also reported significant effect on essential oil recovery from *O. sanctum* grown under Peach, *Grewia* and *Morus* based agroforestry system (Thakur et al 2009). The biosynthesis of secondary metabolites in plants is an adaptive mechanism to shaded conditions, and is controlled by environment. These secondary products and alkaloids result from the breakdown and re-synthesis from the primary products and this process is favoured by shaded conditions (Ram et al 1999, Maheswarappa et al 1998).

NET RETURNS FROM HERBAGE OF *OCIMUM* SPP.

O. tenuiflorum grown under silvi-medicinal system and sole crop system accrued highest net returns amounting to ₹ 79,428.00 ha^{-1} (BCR 3.61) and ₹ 67,228.00 ha^{-1} (BCR 3.06), respectively (Table 3 and 4). Kumar et al (2015) have reported net returns amounting to ₹ 38,018.60 ha^{-1} (BCR 1.85) and ₹ 25,418.60 ha^{-1} (BCR 1.24) from *O. gratissimum* under sole cropping system and silvi-medicinal system (Teak+*Ocimum* spp.), respectively.

Table 3: *Price (₹) of inputs and outputs and cost of cultivation of Ocimum spp.*

A.	*Price of inputs/outputs*	*Price (□)*
1.	Cost of *Ocimum* species seed per kg	1000.00
2.	Fertilizer (Urea) N per kg	5.60
3.	Farm Yard Manure (FYM) per tonnes	300.00

4.	Tractor hiring charges per hour	450.00
5.	Labour wages per day	100.00
6.	Dry herbage (Panchang) per kg**	20.00
B.	***Cost of cultivation (per ha)***	***Cost (ha^{-1})***
1.	Seed	200.00
2.	Land preparation and layout	2250.00
3.	Farm Yard Manure (FYM)	6000.00
4.	Nursery bed preparation, raising and maintenance	1520.00
5.	Transplanting	800.00
6.	Weeding and hoeing	1800.00
7.	Application of fertilizer 20 kg N/ ha	240.00
8.	Irrigation	600.00
9.	Plant protection measures, watch and ward	3400.00
10.	Harvesting	1200.00
11.	Managerial cost 10 % of total cost	1801.00
12.	Interest on working capital @ 12 %	2161.20
	Total cost	21972.20

Source: *Anonymous 2008a, ** Ved and Goraya 2007

Table 4: *Net returns (₹ ha^{-1}) from dry herbage of Ocimum spp. grown under Pongamia pinnata based silvi-medicinal and sole cropping systems*

Ocimum spp.	***Yield t ha^{-1}***	***Cost of Cultivation***	***Gross returns (□ ha^{-1})***	***Net returns (□ ha^{-1})***	***BCR***
Sole cropping					
O. sanctum	2.9	21972	58000	36028	1.64
O. tenuiflorum	4.46	21972	89200	67228	3.06
O. basilicum	3.1	21972	62000	40028	1.82
O. gratissimum	2.81	21972	56200	34228	1.56
Silvi-medicinal system					
O. sanctum	3.11	21972	62200	40228	1.83
O. tenuiflorum	5.07	21972	101400	79428	3.61
O. basilicum	3.76	21972	75200	53228	2.42
O. gratissimum	2.7	21972	54000	32028	1.46

CONCLUSION

Higher growth and herbage yield of *Ocimum* spp. were observed when grown with Karanja as compared to sole cropping. Among four *Ocimum* spp., *O. tenuiflorum* gave maximum herbage yield. The economic analysis underpinned that among intercrops, *O. tenuiflorum* irrespective of land use system accrued highest net returns and benefit-cost ratio from herbage. Thus, suggesting it as more suitable species to be integrated with karanja.

ACKNOWLEDGEMENT

The present study was carried out in Karanj plantations established under NOVOD Board, Gurgaon, sponsored project to Navsari Agricultural University Navsari, Gujarat. Therefore, authors are thankful to NOVOD Board.

REFERENCES

Anonymous 2008 Propagation, cultivation and value addition of herbal plants. Training literature part-II extension and advisory services unit IHBT (Institute of Himalayan Bioresource Technology) Palanpur, H. P. 66

Banerjee H, Dhara PK, Mazumdar D and Alipatra A 2013 Development of Tree Borne Oilseeds (TBOS) Plantation and Scope of Intercropping on Degraded Lands, *Indian Journal of Agricultural Research* **47**(5) 453-456

Beniwal RS and Chauhan R 2011 *Pongamia pinnata* as an alternative source of renewable energy, *Asia-Pacific Agroforestry Newsletter* **38** 14-15

Channabasappa KS, Kumar P and Madiwalar SL 2007 Influence of arecanut on economic yield and quality parameters of medicinal plants, *Karnataka Journal of Agricultural Science* **20**(4) 880-882

Chaudhry AK, Khan GS and Ahmad I 2007 Effect of poplar tree intercropping at various densities on the post-harvest soil nutrient contents, *Pakistan Journal of Agriculture Science* **44**(3) 468-472

Gangadharan H and Menon MV 2003 Performance of kacholam (*Kaempferia galanga*) ecotypes as influenced by variations in shade and preparatory cultivation, *Journal of Medicinal and Aromatic Plant Sciences* **25** 976-980

Jackson ML 1973 *"Soil Chemical Analysis"*. Prentice-Hall of India Pvt. Ltd., New Delhi, India, 39- 415

Javanmardi J, Khalighi A, Kashi A, Bais HP and Vivanio JM 2002 Chemical characterization of basil (*Ocimum basilicum* L.) found in local accessions and used in traditional medicines in Iran, *Journal of Agriculture and Food Chemistry* **50** 5878-5883

Karikalan TV, Yassin MM, Divya MP and Gopi D 2002 Effect of intercropping and nitrogen management on growth and yield of medicinal plants under kapok, *Indian Journal of Agroforestry* **4**(2) 88-93

Khawlhring N, Thanga JLT and Lalnunmawia F 2011 Plant performance of *Anthurium andreanum* as affected by shade conditions and different conventional nutrient sources, *Journal of Horticulture and Forestry* **4**(2) 22-26

Kumar BM, Kumar SS and Fisher RF 2005 Galangal growth and productivity related light transmission in single-strata, multistrata and no-over-canopy systems, *Journal of New Seeds* **7** 111-126

Kumar M, Thakur NS and Hegde HT 2015 Growth, herb yield and financial flows from *Ocimum* spp. intercropped under teak (*Tectona grandis* L. f.)-*Ocimum* spp. based silvi-medicinal system in Gujarat, India, *International Journal of Innovative Horticulture* **4**(2) 113-118

Kumar M, Thakur NS and Hegde HT 2016 Fresh herb, essential oil yield and net returns from *Ocimum* spp. grown under teak (*Tectona grandis* Lf) based silvi-medicinal systems in south Gujarat, India, *Indian Journal of Ecology* **43** 306-311

Kurian A, Augustin A and Nybe EV 2003 Economic analysis of resource-based cropping in selected medicinal species, in: *National Interactive meet on Medicinal and Aromatic Plants*, Central Institute of Medicinal and Aromatic Plants, Luknow 47-49

Maheswarappa H P, Hegde M R and Nanjappa M V 1998 Kacholum (*Kaempferia galanga*) a potential medicinal-cum- aromatic crop for coconut gardens, *Indian Coconut Journal (Cochin)* **29** (5)4-5

Mishra RK and Pandey VK 1998 Intercropping of turmeric under different tree species and their planting pattern in agroforestry systems, *Range Management & Agroforestry Journal* **19** 199-202

Mohsin F 2005 Effect of litterfall of short-rotation trees on herbage and oil yield of aromatic plants under Agroforestry System, *Indian Journal of Agroforestry* **7**(1) 25-31

Olsen SR, Cole CV, Watanabe FS and Dean LA 1954 Estimation of available phosphorus in soil by extraction with sodium bicarbonate, *USDA circular* 939

Osman KT, Rahman MM and Barua P 2001 Effects of some forest tree species on soil properties in Chittagong University Campus, Bangladesh, *Indian Forester* **127**(4) 431-442

Pathiratna LSS, Joseph KDSM and Perera MKP 2004 The effect of some cultural practices on the growth and yield of the medicinal plant *Aerva lanata* (L) Juss Ex, schult (Polpala) Cey, *Journal of Biological Science* **32** 67-74

Patil DR, Patil HB, Patil SN and Prashanth JM 2005 Studies on evaluation of intercrops in mango orchards, Abstract of National seminar on commercialization of Horticulture in non-traditional areas held at CIAH, Bikaner from 5-6 February 81

Pooter H and Oliver YH 2000 Effects of seasonal drought gap on under storey seedlings in a Bolivian moist forest, *Journal of Tropical Ecology* **16** 481-498

Ram M, Ram D, Singh S, Naqvi AA and Kumar S 1999 Studies on intercropping of patchouli (*Pogostemon patchouli*) with papaya (*Carica papaya*), *Indian Journal of Medicinal and Aromatic Plants* **21** 358-360

Ravindaran KC and Kulandaivelu G 1998 Effect of light intensity on growth and yield of cardamom (*Elettaria cardamomum*), *Journal of Plantation Crops* **26**(1) 87-88

Ravitchandirane V and Haripriya K 2011 Intercropping with medicinal plants in mango cv Alphonso, *Plant Archive,* **11**(1) 413-416

Sharanabasappa AP, Madiwalar SL, Channabasa KS and Kumar P 2007 Performance of medicinal and aromatic plants as intercrops with teak, *Karnataka Journal of Agriculture Science* **20**(1) 179-180

Singh B, Singh V, Singh RP and Srivastava BK 1997 Effect of young *Eucalyptus* trees on the growth, yield and quality of vegetable intercrops, *Indian Journal Horticulture* **54** (4) 320-326

Singh RA, Shamim M, Singh PV, Singh MK and Pandey RK 2008 Agri and vege- horti systems with mango in gangetic alluvial tract of U.P, *The Asian journal of Horticulture* **3**(2) 226-228

Singh RA, Shamim M, Singh PV, Singh MK and Pandey RK 2008 Cultivation of crops in association of mango under two tier system of agro forestry, *The Asian Journal of Horticulture* **3**(2) 270-272

Subbiah BV and Asija GL 1956 A rapid procedure for the estimation of available nitrogen in soils, *Current Science* **25** 259-260

Thakur NS, Verma KS and Rana RC 2014 Growth and yield performance of ashwagandha (*Withania somnifera*) under agroforestry, *Indian Journal of Agricultural Sciences* **84** (8) 937-941

Thakur NS, Verma KS and Rana RC 2009 Effect of tree-crop combinations and nitrogen levels on fresh herbage and oil yield of sacred basil (*Ocimum sanctum*) grown in agrihorti-silvi-pasture system in mid hill Himalayas, *Indian Perfumer* **53** 39-44

Ved DK and Goraya GS 2007 Demand and supply of medicinal plant in India, NMPB, New Delhi and FRLHI, Banglore India

Vyas SP 2001 Effect of *Prosopis cineraria* (L macbride) on growth and productivity of *Cassia angustifolia* (vahl) under medico- forestry system in aride region of Thar Desert, *Journal of Tropical Forestry* **17** (3)71-73

Zheljazkov VD, Callahan A, Cantrell CL 2008 Yield and oil composition of 38 Basil (*Ocimum basilicum* L) accessions grown in Mississippi, *Journal of Agriculture Food Chemistry* **56** 241-245

CHAPTER 30

Quality Fodder Production Through Silvo-pastoral System: A Review

S Sarvade*, VB Upadhyay and SB Agrawal

College of Agriculture, Balaghat, Jawaharlal Nehru Krishi Vishwa Vidhyalaya, Jabalpur (MP), India

** Corresponding author's Email: somanath553@gmail.com*

ABSTRACT

The production of food items from livestock with an intensive agriculture practices is the most important enterprise to meet out our basic needs. Country has world's 18 % livestock population. So that in India, many scientists and institutes are familiar with continuous increase in the demand of dry and green fodder (million tonnes) up to 2050. Therefore silvo-pastoral systems, by introducing trubs and grass/crop components under trees into natural pasture and degraded lands are important to provide nutritious green fodder throughout the year for higher livestock production. Tree species like subabul, sirish, shisham, mulberry, kachnar, agast, gliricidia, jamun, ber, anjan, peepal, neem etc. are the most important quality fodder yielding species. The fodder yielding tree species may be planted at 60 cm × 60 cm to 3 m ×3 m spacing for high quality fodder production. Suitable management practices may help in improving sustainable fodder production throughout the year. In such management practices, the pruning of fodder tree species or grazing in silvopastoral systems may not be permitted earlier than the third year since establishment for attaining an optimum combination of quantity and quality of fodder. In case of cut-and-carry module, labour cost will be increased but the losses generated due to the trampling of grazing/browsing animals are minimised, which may increases quality forage production. Cutting height should be 20cm to 150cm and 6-12 weeks cutting frequency favours quality fodder yield. Napier grass (250–300 t ha^{-1}) produces higher green fodder, which is followed by lucerne/alfalfa (140–150 t ha^{-1}) and berseem (70–110 t ha^{-1}), whereas, guinea grass yields higher (15-18 t ha^{-1}) dry fodder in silvo-pastoral systems. Fodder yield from tree species varies with species, locality factors and age of tree species.

Keywords: Cutting frequency, Fodder, Grazing, Livestock, Silvo-pastoral system

INTRODUCTION

Conservation of natural resources such as water, vegetation and land in sustainable manner is vital to the economic growth of any country or a region (Kapur, 2016). The 16.7 % of the world human population (Sarvade 2014a) and 18 per cent livestock population (Sarvade et al 2017) is directly or indirectly depends upon these natural resources which are continuously exhausted. Livestock is the most important source of income in India and has 512 million livestock population including cattle (37.28%), buffaloes (21.23%), sheep (12.71%), goats (26.4%) and pigs (2.01%) (Islam et al 2016). The requirement of dry and green fodder is 2050 million tonnes and increasing continuously. Though, production of forage is increasing, but not sufficient to meet total fodder demand (Fig. 1, Anonymous, 2013). Moreover, Gebremedhin (2015) reported that, India is deficient in 84% concentrate feed and 88% green fodder. Devendra and Leng (2011) reported the looming climate change, spiralling food costs and financial crises are crucial for creating crisis in fodder production and supply. Before 1979, grazing was permitted in the forest areas with free, concessional and commercial rates. Now, the forest (22.7%) and CPR other than forest (17.7%) are the two main grazing resources in India (Fig. 2). Now conservation of forest as well as cropped areas is needed for better livestock production, and its management. For this purpose stall feeding of nutritive fodder is one of better options (Misri 1986, Dwivedi 1992, Misri 1995, Puri 2007).

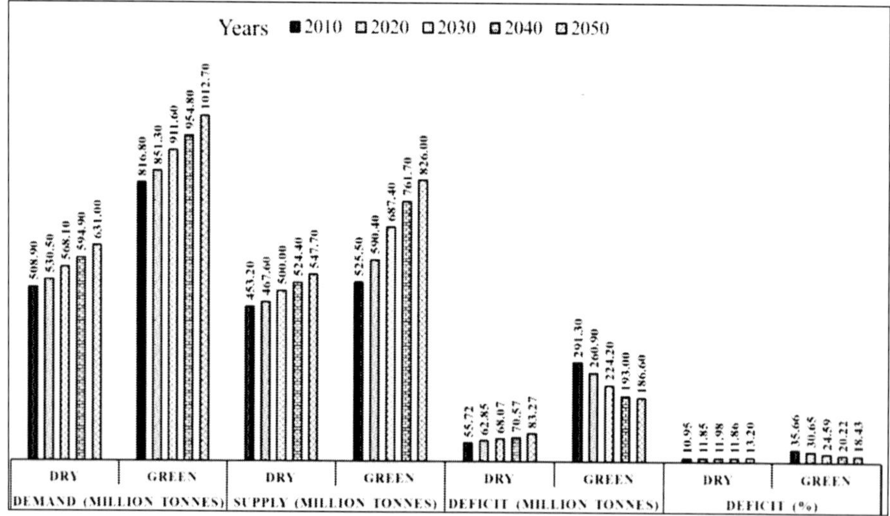

Figure 1: *Demand and supply estimates of dry and green forages (million tonnes) in India up to 2050 (Anonymous 2013).*

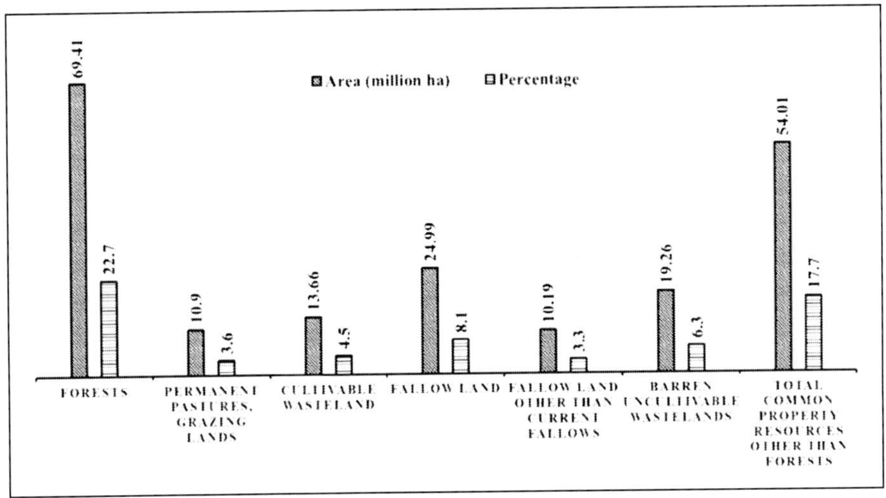

Figure 2: *Available grazing resources in India (Anonymous, 2013)*

In rural areas, nourishment of animal is an important component of farming system. It provides milk, meat, manure and other valuable products (Sushant 2013). Required nutritive fodder can be meet out from the palatable annual grass/crop residues, green forages and perennial tree species (Devendra et al 1997, Patidar et al 2014, Devendra 2014). Tree leaves and pods, traditionally used as a natural source of feed and fodder for domestic livestock. It serves as a protein banks to supplement grass or crop residues during summer season. Hence it is suggested to develop silvo-pastoral systems/models by introducing trees, shrubs and annual grasses as well as crop components under trees on natural pasture and degraded lands to get nutritious green forage throughout the year for higher livestock production (Nair 1984, Cameron et al 1991, Tewari et al 2007, Panda 2013).

The term 'silvi' means 'tree' and the 'pasture' means 'grasses' or 'grass + legume' crops. Silvo-pastoral systems are defined as growing of ideal combination of grasses/ crops, legumes and fodder trees for producing highly nutritious fodder and forage, fuel wood, timber and optimising land productivity, therefore conserving soil and nutrients on sustainable manner from the same piece of land. There are mainly four type of silvo-pastoral system i.e. Scattered trees and shrubs on pasture, Intensive silvi-pastoral (fodder trees with agri. fodder crops), live fence of fodder trees or boundary systems and hedge, and protein bank (cut and carry) (Nair 1984, Dwivedi 1992, Nair 1993, Puri 2007, Panda 2013, Soni et al 2016, Mathukia et al 2016).

In general, the silvi-pastoral system benefited through diversify farm enterprise, enhanced the growth of trees, reduced stress and improved animal productivity, increased nutrient cycling and enhanced wildlife habitat. Tree components of the system change the microclimate which boosts up the forage production under canopy. Healthy growth of understory forage can heal the eroded land which was prohibited for animal grazing. Similarly, planting of trees provides an opportunity to reduce runoff and soil erosion. At small farms/field in the tropics, diversification of animal and plant species is the most efficient utilization of the resources and provides a variety of products (Cameron et al 1991, Feldhake 2001, Feldhake 2002; Menezes et al 2002, Choocharoen et al 2014, Devendra 2014, Sarvade et al 2014c). Singh (2008) reported that the forage-based feeding systems have high production and economic efficiencies in ruminants. Silvo-pastoral system provides employment to the rural people in the activities of animal husbandry; collection, processing, and manufacturing of value added products from trees and grasses (Nair 1984, Cameron et al 1991, Devendra 2012, Yadav et al 2014, Devendra 2014). Apart from direct economic benefits, it also add the recreational and aesthetic value. Whereas, silvo-pastoral system have some limitations such as: producer should already be practicing some type of rotational grazing, distance and access to water, challenges establishing young trees, challenges introducing forages to existing woodlands, maintaining proper light levels and fencing issues (Cameron et al 1991).

Integrated Farming System (IFS) under Rainfed Area Development Program, covered 15851 ha of livestock based farming system, 7129 ha of dairy based farming system, 15069 ha of agro-forestry based farming system and 4133 ha of silvi-pasture based farming system (Ministry of Agriculture & Farmers' Welfare, 2016). Such programs may help in increasing the area under silvo-pastoral system and livestock production.

SYSTEM COMPONENTS

Grasses-Legumes-Crops

Fodder grasses, legumes and crops selection depends upon agro ecological situations such as edaphic, climatic, biotic and topographic factors. Based on AES, species may be selected for a particular place and requirement of the farmers. Napier grass produces higher green fodder (250–300 t ha^{-1}), followed by lucerne (140–150 t ha^{-1}) and berseem (70–100 t ha^{-1}) (Pandey and Roy 2011, Vaghela et al 2014). On the basis of production potential, farmers may go for selection of species for fodder under silvo-pastoral system. The detailed information on varieties, seed rate, sowing time and yield potential of fodder crops, grasses and legumes are presented in Table 1.

Perennial Tree Species

Fodder tree species may be planted on farm bunds and on cultivable-degraded lands in rows, strips or block planting. Number of fodder trees are available, which have an excellent potential to produce quality nutritive fodder (Table 2). In general, those species establish easily, grow fast, out-compete weeds, produce high-quality fodder, having high regeneration capacity after harvest and under harsh conditions respect to moisture, nutrients, sunlight needs to be selected. Bhatt et al (2003) reported biophysical aspects of tree- crop interactions in silvo-pastoral system, which affect selection of crops, grasses and legumes. The tree canopy which occupy strata producing leaf forage, fruits, and other products are key determinants for the choice of tree species for silvo-pastoral systems (Cajas-Giron and Sinclair 2001).

Table 1: *Fodder producing grasses/legumes/ crops cultivated in agri-silviculture system.*

S. No	Common Name	Botanical Name	Varieties	Seed Rate	Sowing time	Yield		Source
						Green	Dry	
1	Berseem	*Trifolium alexandrinum*	Berseem Ludhiana-1 (BL-1); Jawahar Berseem-1 (JB-1)	25-30 kg ha^{-1}	from last week of September to first week of December	70–100 t ha^{-1}	15–18 t ha^{-1}	Pandey and Roy 2011
2	Lucerne/ Alfalfa	*Medicago sativa*	Chetak (S-244)	25–30 kg ha^{-1}	Sept-December	140–150 t ha^{-1}	8-9 t ha^{-1}	Pandey and Roy 2011
3	Jowar (Sorghum)	*Sorghum bicolor* (L.) Moench	MP Chari, Jawahar Chari-6, Jawahar Chari-69	30 kg ha^{-1}	At onset of monsoon	53 t ha^{-1}	15 t ha^{-1}	Pandey and Roy 2011
4	Stylosanthes	*Stylosanthes* spp.	Stylosanthes Phule Kranti (RSS-2000-95)	5–6 kg ha^{-1}	June-July to Sept.-Oct	25–30 t ha^{-1}	10–12 t ha^{-1}	Pandey and Roy 2011, Trivedi 2002
5	Hedge Lucerne	*Desmanthes virgatus* (L.) Willd.	Desmenthus virgatus	2–3 kg ha^{-1}	It can be sown in any season	35-80 t ha^{-1}	10-15 t ha^{-1}	Vaghela *et al.*, 2014
6	Guinea Grass	*Panicum maximum* Jacq.	Bundel Guinea-1 (JHGG-96-5); Bundel Guinea-2 (JHGG 04 –01)	3–6 kg ha^{-1}	Jun – Sep / Oct - Nov	50–60 t ha^{-1}	15–18 t ha^{-1}	Vaghela *et al.*, 2014
7	Anjan Ghas	*Cenchrus ciliaris* Linn	Bundel Anjan-1; Bundel Anjan-3 (IGFRI- 727)	5 kg ha^{-1}	At onset of monsoon	35–40 t ha^{-1}	6-12 t ha^{-1}	Mishra et al 2010, Vaghela et al 2014
8	Saen Grass	*Sehima nervosum* Forssk.	Bundel sain Ghas-1 (IGS 9901)	33000-35000 seedlings ha^{-1}	Before on set of monsoon	18.3 t ha^{-1}	4.7 t ha^{-1}	Vaghela et al 2014
9	Dharaf Grass	*Chrysopogon fulvus* (Spreng.) Chioiv.	Bundel Dhawalu Ghas-1 (IGC 9903)	4 to 5 kg ha^{-1}	In June-July (on set of Monsoon)	25–30 t ha^{-1}	6-7 t ha^{-1}	Vaghela et al 2014
10	Napier Grass	*Pennisetum purpureum* Schumach	Pusa Giant Napier, IGFRI-10, CO-1; JP-1, JP-13	10,000 stem cutting or rooted slips for one hectare	Pre monsoon in the month of June-July	250–300 t ha^{-1}	14-15 t ha^{-1}	Vaghela et al 2014
11	Dinanath Grass	*Pennisetum pedicellatum* Trin.	Jawahar Pennisetum-12; Bundel-1; Bundel-2	2.5–5 kg ha^{-1}	Onset of monsoon	55–60 t ha^{-1}	14.0 t ha^{-1}	Pandey and Roy 2011

Table 2: *Fodder yielding perennial tree species cultivated in silvo-pasture system.*

Sl. No	Common Name	Botanical Name	Variety	Planting time	Spacing	Yield		Source
						Green	Dry	
1	Subabul	*Leucaena leucocephala* (Lam.) De Wit	K-6; K-8; K-102; K-23; K-28; K-29; K-341; K-585, K-601, S-22, S-24; E1 Salvador; Subabul CO-1 (P); FD 1423; K-156 (*L. diversifolia*) and K-743A (*L. leucocephala* x *L. diversifolia*)	July-August	3m×3m	55-85 t ha^{-1}	20 t ha^{-1}	Anonymous 2002, Pandey and Roy 2011, Singh et al 2012, Meena et al 2013, Khanna 2013, Gunasekaran et al 2014, Reddy 2016
2	Sirish	*Albizia procera* (Roxb.) Benth.		After monsoon rain sets in	3m×3m	40-70 t ha^{-1}	11-15 kg $tree^{-1}$	Prinsen 1986, Newaj et al 2005, Khanna 2013, Gunasekaran *et al.*, 2014, Reddy 2016
3	Shisham	*Dalbergia sissoo* Roxb. ex DC.		After monsoon rain sets in	2.5m×2.5m	40-70 t ha^{-1}	5-6 kg $tree^{-1}$	Gupta et al 2009, Khanna 2013, Singh et al 2013
4	Mulberry	*Morus alba*	L-146, RSS-175, RSS-135, S-54, TR-10, K-2	July-August	4m×4m	40-120 t ha^{-1}	43 t ha^{-1}	Dwivedi 1992, Datt et al 2008, Khanna 2013
5	Kachnar	*Bauhinia variegata*	Alboflava, Candida, Chinensis	Onset of monsoon	2.5m×2.5m or 3m×3m	7.7kg/tree	6.7 kg $tree^{-1}$	Khanna 2013, Singh et al 2013
6	Agast	*Sesbania grandiflora*		June-July	1.2m×1.2m	80 t ha^{-1}	20 t ha^{-1}	Khanna 2013, Sarvade 2014b; Mariswamy et al 2017
7	Jait/Jainti	*Sesbania sesban*	Nubica	June-July	1m×1m or 2m×2m	15-40 t ha^{-1}	4-20 t ha^{-1} (in 9-10 cuts)	Gupta et al 1983, Shukla et al 2002, Gomase et al 2012, Khanna, 2013, Mathukia et al 2016
8	Gliricidia	*Gliricidia sepium*		After onset of monsoon	1m×1m or 2m×2m	43 t ha^{-1}	5-6 t ha^{-1}	Khanna 2013, Gunasekaran et al 2014, Mathukia et al 2016
9	Ber	*Zizyphus nummularia*		July-August	3m×3m	46 t ha^{-1}	3.5 t ha^{-1}	Pandey et al 2010, Khanna 2013, Verma 2016
10	Anjan	*Hardwickia binata*		After onset of monsoon	3m×3m	24-26 t ha^{-1}	4-10 t ha^{-1}	Tewari et al 2007
11	Neem	*Azadirachta indica*		Onset of monsoon	3m×3m	40-70 t ha^{-1}	40-60 kg $tree^{-1}$	Khanna 2013, Reddy 2016, Bijalwan et al 2017

Suitable Traits of Fodder Tree Species Are:

- Easy to establish and adoptive under existing agro ecological situations.
- The orientation of tree row should be in an east-west direction for minimizing competition between tree and grass/legumes.
- Tolerance to side shading.
- Sparse branching and optimum light penetration to the ground.
- Suitable for fodder with an ability to tolerate lopping, pruning and coppicing as well as having resistant to pest and diseases.
- Plant MPTs (Multipurpose trees and shrubs) which are able to produce fodder, fuel, fruit, food and fiber.
- Leaf litters having rapid decomposition ability e.g. *Leucaena*, *Grevillea*, etc.
- Ability to fix atmospheric nitrogen: species should have nitrogen fixing ability. Most of the leguminous species are Nitrogen Fixing Tree species (NFTs).
- No allelopathic effect.
- Deciduous nature, leaf falling during crop season like *Melia azedarach*.
- Should not act as an alternative host for common insect pest and pathogen in crop field.
- Should yield nutritive and palatable fodder.
- Absence of competition at root zone level for moisture and nutrients.

Source: Huxley and Westley (1989), Mishra et al (2009), Chhetri (2010), Devendra and Leng (2011), Dawson et al (2014), Gunasekaran et al (2014), Soni et al (2016)

Tree Spacing

Tree spacing varied for different tree species, and also varied due to availability of moisture and nutrients. Spacing for different tree species as given in Table 2. However, for the purpose of fodder production, spacing may reduce up to 1m×1m for many species. Paired row of fodder tree/shrubs at wider spacing may provide a chance to grow crop after companion establishment of fodder tree/shrubs. The grass/legumes may be allowed to grow in the area between two paired rows. Competition between tree/shrub and grass should be minimum in order to reduce competition between main and companion crops, tree/shrub should grow on bunds and grass/legumes are allowed to grow in the field. Raj et al (2016) reported 60cm x 60 cm spacing for *Leucaena leucocephala*, *Calliandra calothyrsus*, and *Morus indica* to develop three tire silvo-pasture model in humid tropics.

MANAGEMENT

Age at First Harvest

Growth rate of fodder trees varied from species to species. So, the first harvest should be delayed until the fodder tree/shrub is ready to sprout and withstand pruning practices. Generally, first pruning may carried out at the 12-16 months age. Under the adverse conditions, growth rate may be slow and the first cutting/pruning may be delayed up to one or two months depending upon situations. If the growth rate is fast, then the first pruning may be earlier. The goal is to allow fodder tree species to establish deep roots and developed thick trunk. Papanastasis et al (1998) reported that the cutting or grazing should not be start earlier than the third year since establishment for attaining an optimum combination of quantity and quality of biomass in deciduous fodder tree and shrub plantations.

Grazing

Animals graze on the plants, usually herbaceous species. The grazing systems in which trees are present and play an interactive role with animal production by providing shade to animals, made available fodder or other products can be considered as silvo-pastoral systems. This system saves labour cost, time and effort, whereas it damages the existing plants and causes wastage of fodder due to trampling. The key to direct grazing is subdivision of the fodder plot into paddocks and adopt deferred grazing system. Generally, grazing periods are 1-2 weeks, followed by recuperation periods of 3-6 week. These practices minimize the losses and provide better conditions to silvo-pastoral systems.

Cut-and-Carry (Protein Bank)

High nutrient value fodder tree/shrub species are grown in block configurations or along the boundaries or other designated places; the foliage is lopped periodically and for stall feeding. It reduces fodder wastages however, the labour cost may be increases. An important management factors to be consider under *cut-and-carry* system are cutting height, cutting frequency, and management in lean period. (Raj et al 2016). At the end of lean period, fodder from trees and shrubs may be provided to the animals as a supplemental feed through a *cut-and-carry* operation, when other resources are deficient in nutrients (Papanastasis et al 1998).

Cutting Height

The number of research studies has been carried out in different parts of and results of their studies revealed that the cutting height of 50-150 cm is optimum, however it depends upon the tree/shrub species. Raj et al (2016) reported that the 1m height is suitable for pruning of fodder tree species such as *Leucaena leucocephala*, *Calliandra calothyrsus*, and *Morus indica* in humid tropics. Isah et al (2014) recommended 0.02m cutting height in order to improve the fodder yield of *Moringa oleifera*.

Cutting Frequency

The common recommendation of cutting frequencies as 6-18 weeks for fodder tree/shrub species. Generally, longer cutting frequencies (12-18 weeks) provide more biomass but increase the proportion of tender wood. Shorter cutting frequencies of 6-12 weeks, favoured for fodder yields and fodder quality. Pruning interval of 3 months is reported for fodder tree species of *Leucaena leucocephala*, *Calliandra calothyrsus* and *Morus indica* in humid tropics (Raj et al 2016). Isah et al (2014) recommended that the progressive increase of 2 weeks interval of cutting may improve the fodder yield of *Moringa oleifera*.

Dry-Season Management

Six to eight weeks before beginning of dry-season, trees should be cut from recommended height. The new foliage produced over the next few weeks will retain properly during the dry-season. When the dry-season is very long or the area of fodder bank is very large, the pre-dry-season harvest should be in phases. It will assure the availability of fodder throughout the dry-season. The excess quantity may be used by converting in to the silage as feed/fodder for milch animals.

NUTRIENT CONTENT AND TOXICITY

Feeding of palatable, nutritious fodder/ forage had direct impact on the animal growth rate and health, productivity as well as product quality. Poor quality feed and husbandry systems with imbalance feeding of total dry matter and nutrients intake causing inferior production (Gebremedhin 2015). Nutritive value in case of crude protein, digestible crude protein, total digestible crude nutrients, crude fibre, calcium and phosphorus content (%) of some fodder tree species are given in Table 3. Nutrient content of fodders greatly affected due to soil composition, manure and fertilizer application, irrigation, growth stage, frequency of cutting and variety of feed resources (Upreti and Shrestha 2006). Ramana et al (2008) reported that the medium to high CP (13.1–22.8%) and optimum NDF (29.4–32.7%) found in foliage of *Leucaena leucocephala* and *Dichrostachys cinerea*, whereas low to medium CP (4.1 -14 4%) recorded in forage obtained from understorey vegetation (perennial and annual grasses as well as forbs).

Toxicity

Subabul leaves and seeds both contain a toxic and non-protein substance known as mimosine (Meena et al 2013). It is an amino acid creates toxicity, including hyper active, goiter development and stillbirths in pregnant animals. Alopacia in sheeps, goats and in cattle are common. Therefore, to minimize the toxic effect, the 30% of the recommended level of subabul forage may be used for sheep, goat and cattle (Babu et al 2012).

Table 3: *Nutritive values of fodder trees cultivated under silvo-pastoral system.*

Sl. No	*Tree Species*	*Crude protein (%)*	*Digestible crude protein (%)*	*Total digestible nutrients (%)*	*Crude fibre (%)*	*Ca (%)*	*P (%)*
1	Sirish	14.9-29.2	11.6	49.3	25.3-37.5	1.1-2.7	0.1-0.3
2	Neem	12.4-18.3	8.4-9.3	42.8-53.3	11.4-23.1	0.9-4.0	0.1-0.3
3	Kachnar	10.7-15.9	5.0-9.2	47.9-55.5	20.7-33.0	1.4-4.1	0.2-0.4
4	Shisham	2.7-24.1	3.7-9.1	20.9-52.2	12.5-32	2-2.3	0.2
5	Bamboo (*D. strictus*)	14.2-15.1	9.3	48.9	15.6-23.5	1.1-1.6	0.2-0.3
6	Jamun	8.8-10.2	0.1	43.8	19.8	1.3	0.1-0.2
7	Dhamni (*Grewia tilifolia*)	13.2	-	-	-	1.5	0.1
8	Anjan (Hardwickia)	9.0	-	-	30.4	2.3-3.3	0.1
9	Subabul	15.2-27.6	12.6-16.4	57.1-70.2	10.2-17.2	2.7-3.1	0.2

10	Mulberry	15-27.6	10.7	59.6	9.1-15.3	2.4-4.7	0.1-0.2
11	Khejri	13.9-15.3	-	-	17.5-22.1	1.9-3.6	0.2-0.5
12	Jharber (*Z. numularia*)	11.5	5.5	51.1	33.8	1.9	0.3
13	Imli (*Tamarindus indica*)	11.2-15.4	-	-	14.2-22	1.7-3.2	0.1-0.6
14	Agast	25-30	-	75	18.4	1.48	0.34
15	Gliricidia	14.7	-	-	19.9	1.58	0.29
16	Acacia	15.1	-	-	22.6	1.21	0.06
17	Pigeon pea	22.8	-	-	20.1	0.37	0.17
18	Tamarind	14.0			21.0	2.81	0.20

Source: Dwivedi 1992, Devendra 1992

FODDER PRODUCTION

The comparative study at NRCAF, Jhansi revealed that on an average 4.55 t from pasture + 0.51 t from tree leaves and pods produced from silvi-pastoral system which is just double the yield than natural grassland. These results indicate the possibilities to get more biomass by establishing the silvi-pasture, which is producing less than 2 t ha^{-1} year forage through natural vegetation. The forage production from one tier (*Cenchrus* pasture alone), two tier (*Cenchrus* + *Ailanthus excelsa*) and three tier (*Cenchrus* + *A. excelsa* + *Dichrostachys cinerea*) systems were compared with natural pasture at CSWRI, Avikanagar, Rajasthan under semi-arid condition. Results revealed that three tier silvi-pasture system provided maximum forage production (2.78 t ha^{-1} dry forage from pasture + 0.95 t ha^{-1} green tree leaves).

Roy et al (1983) studied 8-9 years old silvi-pasture system in semi-arid conditions and reported that the mean green and dry leaf fodder of 9.63 and 5.28 t ha^{-1}, respectively with annual lopping at 1/3 intensity in *Albizia procera*, whereas *A. lebbeck* provide of 6.24 and 2.78 t ha^{-1} green and dry leaf fodder, respectively in case of through biannual lopping at 2/3 intensity. *Bahunia purpurea* yielded green leaf fodder of 10.02 kg tree^{-1} year at Jhansi. Top feed production (kg tree^{-1}) from six fodder trees of Bundelkhand region showed the maximum dry leaf fodder of 11.38 in *A. procera* followed by *A. amara* (11.20), *A. lebbeck* (4.21), *Hardwickia binata* (3.67) and *D. cinerea* (2.76) and a minimum of 0.51 in *A. tortilis*, respectively. In *S. grandiflora* and *S. sesban*, the dry leaf fodder yield of 0.3 kg tree^{-1} was recorded after 3.5 years of establishment when grown at the density of 5000 ha^{-1} (Gupta et al 1983).

A three tier silvi-pastoral system developed at NRCAF, Jhansi with woody components of *Dichrostachys cinerea* + *A. amara* + *L. leucocephala* showed top feed (dry leaf + pod) production of 0.04, 0.26, 0.48, 1.15, 0.53, 0.55 and 0.54 t ha^{-1} during 1, 2, 3, 4, 5, 6 and 7 year, respectively, when pruned up to 50 per cent height of the trees from ground level. The average leaf fodder yield in young trees of *A. excelsa*, *P. cineraria*, *A. indica* and *Dichrostachys cinerea* were 21.13, 25.70, 12.65 and 0.67 kg/tree, respectively. The corresponding yields of older trees were 46.14, 57.81, 46.45 and 1.12 kg tree^{-1}, respectively. Raj et al (2016) conducted two year field study involving nine modules They observed that the highest dry matter (31.49 t ha^{-1}) and crude protein yield (4.75 t ha^{-1}) was obtained from two-tier Hybrid Napier + trees (mulberry + calliandra) system, which is a better option for producing quality forage at small holder stall –fed dairy farms under humid tropical climate.

CONCLUSION

India is sustainably nourishing 18 per cent livestock population of the world with a limited grazing sources. The gap between demand and supply of dry forage is regularly increasing, whereas for green forage gap is decreasing with slow rate. Now a days continuous grazing is hazardous to sustainable fodder production system. Hence, the nutritious fodder production through silvo-pastoral system is one of the options for sustainable fodder production. Cut-and-carry module may help in minimizing problems arises in *Rabi* season due to open grazing. Introduction of grass/ forage crop components on natural pasture or degraded lands to provide nutritious green fodder round the year for livestock. This system also provides an opportunity to generate employment to the rural people by involving in activities of animal husbandry i.e. collection, processing, and marketing of value added products get from trees and grasses. An appropriate management practices in silvo-pastoral systems will increase production of forage to full fill demand of fodder.

REFERENCES

Anonymous 2002 *Forage Plant Protection*, Indian Grassland and Fodder Research Institute, Jhansi (UP), India 23

Anonymous 2004 *Annual Report: Grassland and Silvi-pasture management*. Indian Grassland and Fodder Research Institute, Jhansi (UP), India-47-60

Anonymous 2013 *IGFRI: Vision 2050* Indian Grassland and Fodder Research Institute (Indian Council of Agricultural Research), Jhansi (UP) 25

Babu BJ, PVVS Siva Rao, Rao CC, Prasad EV and Murthy TGK 2012 Fodder revolution in East Godavari district: an initiative by CTRI – KVK, New Image Graphics, Vijayawada 27

Bijalwan A, Dobriyal MJR, Thakur TK, Verma P and Singh S 2017 Scaling-up of neem (*Azadirachta indica* A Juss) cultivation in agroforestry for entrepreneurship and economic strengthening of rural community of India, *International Journal of Current Research in Biosciences and Plant Biology* **4**(1) 113-118

Bhatt RK, G Suresh, and R Newaj 2003 Biophysical aspects of tree crop interactions in silvopastoral system in *Agroforestry Potential and Opportunities* (eds) PS Pathak and R Newaj Agrobios (India) and Indian Society of Agroforestry 54-64

Cajas-Giron Y and F Sinclair 2001 Characterization of multi-strata silvo-pastoral systems on seasonally dry pastures in the Caribbean region of Colombia, *Agroforestry System* **53** 215-225

Cameron DM, Gutteridge RC and Rance SJ 1991 Sustaining multiple production systems: Forest and fodder trees in multiple use systems in the tropics, *Tropical Grassland* **25** 165-172

Chhetri RB 2010 Some fodder yielding trees of Meghalaya, Northeast India, *Indian Journal of Traditional Knowledge* **9**(4) 786-790

Choocharoen C, Neef A, Preechapanya P and Hoffmann V 2014 Agrosilvopastoral systems in Northern Thailand and Northern Laos: minority peoples' knowledge versus government policy, *Land* **3** 414-436

Datt C, Datta M and Singh NP 2008 Assessment of fodder quality of leaves of multipurpose trees in subtropical humid climate of India, *Journal of Forestry Research* **19**(3) 209-214

Dawson IK, Carsan S, Franzel S, Kindt R, van Breugel P, Graudal L, Lilleso J-PB, Orwa C, and Jamnadass R 2014 Agroforestry, livestock, fodder production and climate change adaptation and mitigation in East Africa: issues and options, ICRAF Working 178 Nairobi. 22

Devendra C 1992 Nutritional potential of fodder trees and shrubs as protein sources in ruminant nutrition, *Legume trees and other fodder trees as protein sources for livestock* 95-108

Devendra C 2012 Agroforestry and silvo-pastoral systems potential to enhance food security and environmental sustainability in South East Asia, *Open Access Scientific Reports* **1** (2) 163-170

Devendra C 2014 Perspectives on the potential of silvo-pastoral systems, *Agrotechnology* **3**(1) 1-8

Devendra C, Thomas D, Jabbar MA and H Kudo 1997 Improvement of livestock production in crop-animal systems in the rainfed agro-ecological zones of South East Asia, Nairobi, Kenya 107

Devendra C and Leng RA 2011 Feed resources for animals in Asia: issues, strategies for use, intensification and integration for increased productivity, *Asian-Australasian Journal of Animal Science* **24**(3) 303 – 321

Dwivedi AP 1992 *Agroforestry Principles and Practices*, Oxford & Ibh Publishing Company. New Delhi 365

Feldhake C 2001 Microclimate of a natural pasture under planted *Robinia pseudoacacia* in central Appalachia, West Virginia, *Agroforestry Systems* **53** 297-303

Feldhake C 2002 Forage frost protection potential of conifer silvopastures. *Agricultural and Forest Meteorology* **112**(2) 123-130

Gebremedhin WK 2015 Nutritional benefit and economic value of feeding hydroponically grown maize and barley fodder for Konkan Kanyal goats, *IOSR Journal of Agriculture and Veterinary Science* **8**(7) 24-30

Gomase P, Priti G, Shaikh A, Sindha S and Khan MS 2012 *Sesbania sesban* Linn: a review on its ethnobotany, phytochemical and pharmacological profile, *Asian Journal of Biomedical and Pharmaceutical Science* **2**(12) 11-14

Gunasekaran S, Viswanathan K and C Bandeswaran 2014 Selectivity and palatability of tree fodders in sheep and goat fed by cafeteria method, *International Journal of Science, Environment* **3**(5) 1767-1771

Gupta RD, Arora S and V Sharma 2009 Rainfed agricultural production through watershed management approach in Siwalik region of Jammu, *Journal of Soil and Water Conservation* **8**(1) 19-24

Gupta SK, Pathak PS and Roy RD 1983 Biomass production of *Sesbania sesban* Pers. on different habitats, *Myforest* **19**(3) 145-155

Huxley PA and Westley SB 1989 *Multipurpose Trees: Selection and Testing for Agroforestry*, International Council for Research in Agroforestry, English Press. Nairobi, Kenya 109

Isah AD, Bello AG and Zarumaye SA 2014 Effects of cutting heights and interval of cutting on the yield of *Moringa oleifera* (horse raddish), *International Journal of Development and Sustainability* **3**(5) 1147-1151

Islam MM, Anjum S, Modi RJ and Wadhwani KN 2016 Scenario of livestock and poultry in India and their contribution to national economy, *International Journal of Science, Environment* **5**(3) 956 – 965

Kapur R 2016 Natural resources and environmental issues, *Journal of Ecosystem & Ecography* **6**(2) 1-4

Khanna LS 2013 *Silviculture of Useful Trees* Goyal Enterprises, Delhi 405

Mariswamy K, Venkayala J, Kammardi S and MY Earagariyanna 2017 Economic and environmental impact of legume fodders on livestock production, *International Journal of Livestock Research* **7**(4) 49-58

Mathukia RK, Sagarka BK and DM Panara 2016 Fodder production through agroforestry: a boon for profitable dairy farming, *Innovare Journal of Agriculture Science* **4**(2) 13-19

Meena DVN, Ariharan VN and Nagendra PP 2013 Nutritive value and potential uses of *Leucaena leucocephala* as biofuel – a mini review, *Research Journal of Pharmaceutical, Biological and Chemical Sciences* **4**(1) 515-521

Meena MS and Singh KM 2014 Fodder Production scenario and strategies for revitalizing fodder production technologies, Munich Personal RePEc Archive. ZPD, Jodhpur, India, RAU, Pusa, India 14

Menezes R, Salcedo I and Elliott E 2002 Microclimate and nutrient dynamics in a silvopastoral system of semiarid northeastern Brazil, *Agroforestry Systems* **56**(1) 27-38

Mishra A, Nautiyal S and Nautiyal DP 2009 Growth characteristics of some indigenous fuelwood and fodder tree species of sub-tropical Garhwal Himalayas, *Indian Forester* **135**(3) 373-379

Mishra AK, Tiwari HS and Bhatt RK 2010 Growth, biomass production and photosynthesis of *Cenchrus ciliaris* L. under *Acacia tortilis* (Forssk) Hayne based silvopastoral systems in semi-arid tropics, *Journal of Environmental Biology* **31**(6) 987-993

Misri B 1986 Forage production in the Kashmir Himalayas in: *Forage Production in India* (eds) P Singh RMSI, IGFRI, Jhansi (India) 32-38

Misri B 1995 Range and forest grazing in the Himalaya, in *Workshop Proceedings* (eds) Singh. *Temperate Asia Pasture and Fodder Sub-Regional Working Group* Kathmandu 28-33

MoA and FW 2016 *Annual Report* Department of Agriculture, Cooperation & Farmers' Welfare, Ministry of Agriculture & Farmers' Welfare, Government of India, Krishi Bhawan, New Delhi 158

Nair P K R 1984 Tropical agroforestry systems and practices in *Tropical Resource Ecology and Development,* (Eds) Furtado JI and Ruddle K, John Wiley, Chichester, England 39

Nair PKR 1993 *An Introduction to Agroforestry*, Kluwer Academic Publishers, Netherlands 491

Nhan NTH 1998 Utilization of some forages as a protein source for growing goats by smallholder farmers, *Livestock Research for Rural Development* **10**(3):http://www.lrrd.org/lrrd10/3/nhan2.htm

Newaj R, Bhargava MK, Shanker AK, Yadav RS, Ajit and Rai P 2005 Resource capture and tree-crop interaction in *Albizia procera*-based agroforestry system, *Archives of Agronomy and Soil Science* **51**(1) 51-68

Panda SC 2013 *Cropping and Farming Systems*, Agrobios (India), Jodhpur 413

Pandey A, Singh R, Radhamani J and Bhandari DC 2010 Exploring the potential of *Ziziphus nummularia* (Burm. f.) Wight et Arn from drier regions of India, *Genetic Resources and Crop Evolution* **57**(6) 929-936

Pandey KC and Roy AK 2011 *Forage Crops Varieties*, IGFRI Jhansi (India) 84

Papanastasis VP, Platis PD and Dini-Papanastasi O 1998 Effects of age and frequency of cutting on productivity of Mediterranean deciduous fodder tree and shrub plantations *Forest Ecology and Management* 110 283-292

Patidar M, Patel AK, Misra AK, Sirohi AS, Kumawat RN, Meghwal PR and Roy MM 2014 *Improving Livelihood of Farmers through Livestock Interventions in Nagaur District of Rajasthan*, Central Arid Zone Research Institute. Jodhpur 56

Prinsen JH 1986 Potential of *Albizia lebbeck* (mimosaceae) as a tropical fodder tree: a review of literature, *Tropical Grasslands* **20**(2) 78-83

Puri S 2007 *Agroforestry Systems and Practices*, New India Publishing, New Delhi 657

Raj AK, Kunhamu TK, Jamaludheen V and Kiroshima S 2016 Forage yield and nutritive value of intensive silvo-pasture systems under cut and carry scheme in humid tropics of Kerala, India. *Indian Journal of Agroforestry* **18**(1) 47-52

Ramana DBV, Raia P and Singh S 2008 Nutritional evaluation of silvo-pastoral system in lambs and kids, *Indian Journal of Animal Nutrition* **25**(3) 239- 243

Reddy PP 2016 *Sustainable Intensification of Crop Production*, Springer Nature, Singapore 399

Roy MM, Deb Roy R and Pathak PS 1987 Dichrostachys cinerea a potential fodder shrub for wasteland plantation, IGFRI, Jhansi Silver Jubilee Publication No. 6

Sarvade S 2014 Agroforestry refuge for biodiversity conservation. *International Journal of Innovative Research in Science & Engineering* **2**(5) 424-429

Sarvade S 2014 *Sesbania grandiflora* (L) Poiret: a potential agroforestry tree species. *Popular Kheti* **2**(3) 204-207

Sarvade S, D S Gautam, D Kathal, and P Tiwari 2017 Waterlogged wasteland treatment through agro-forestry: A review. *Journal of Applied and Natural Science* **9** (1) 44-50

Sarvade S, R Singh, V Ghumare, D S Kachawaya, and B Khachi. 2014 Agroforestry: an approach for food security. *Indian jornal of Eco*logy **41**(1) 95-98

Shukla NP, A Rekib and D Burman 2002 Productivity of Sesbania alley cropping system with sequentially cropped berseem and maize at different moisture regimes and fertilizer, *Tropical Agricultural Research and Extension.* **5**(1&2) 35-42

Singh KA 2008 Resource management perspective for forage production and agroforestry system development in eastern Himalayan region: a review, *Indian Journal of Agronomy* **53**(4) 255-266

Singh AP, Kumar N and Sushma 2012 Evaluation of *Leucaena* varieties for fodder and fuelwood production in semi-arid regions, *Indian Journal of Agroforestry* **14**(2) 47-52

Singh A, Satanker N, Kushwaha M, Disoriya R and Gupta AK 2013 Ethno-botany and uses of non-graminaceous forage species of Chitrakoot region of Madhya Pradesh, *Indian Journal of Natural Products and Resources* **4**(4) 425-431

Soni ML, Subbulakshmi V, Yadava ND, Tewari JC and Dagar JC 2016 Silvopastoral agroforestry systems: lifeline for dry regions in *Agroforestry Research Developments* (eds) JC Dagar and JC Tewari, Nova Science Publishers, Inc. New York 245-305

Sushant 2013 Impact of climate change in eastern Madhya Pradesh, India, *Tropical Conservation Science* **6**(3) 338-364

Tewari JC, Sharma AK, Narain P and Singh R 2007 Restorative forestry and agroforestry in hot arid region of India: a review, *Journal of Tropical Forestry* **23**(I&II) 1-16

Trivedi BK 2002 Grasses 'and Legumes for Tropical Pastures, Mini Printers, Jhansi (UP) 47

Upreti CR and Shrestha BK 2006 Nutrient contents of feeds and fodder in Nepal, Nirav Printing and General Order Suppliers, Gwarko, Nepal 163

Vaghela PO, Garasiya VR, Ansodriya VV and Madariya RB 2014 Different grasses and their management, *Rashtriya Krishi* **9**(1) 49-50

Varsha KM, Raj AK, Kurien, Bastin B, Kunhamu TK and Pradeep KP 2017 High density silvo-pasture systems for quality forage production and carbon sequestration in humid tropics of Southern India, *Agroforestry Systems* 1–14 doi:10.1007/s10457-016-0059-0

Verma S 2016 A review on *Ziziphus Nummularia*: valuable medicinal plant of desert, *World Journal of Pharmacy and Pharmaceutical Sciences* **5**(3) 539-542

Yadav RP, Sharma P, Arya SL and P Panwar 2014 *Acacia nilotica*-based silvipastoral systems for resource conservation and improved productivity from degraded lands of the Lower Himalayas, *Agroforestry System* **88**(5) 851–863

CHAPTER 31

Growth Attributes, Litter Decomposition of Certain *Populus deltoides* Bartr. Ex. Marsh. (Poplar) Clones and Yield Performance of Paddy and Wheat in Partially Reclaimed Sodic Soils

AK Singh Parihar*, SK Verma and OP Rao

N.D. University of Agriculture & Technology, Kumarganj, Faizabad (U.P.), India

**Corresponding author's Email:opraonduat@yahoo.com*

ABSTRACT

A field experiment was conducted under 8 years old Poplar clones viz., G-48, Udai, L-52, Pant-5, L-49 and L-188 at Main Experiment Station, Forestry, N.D.U.A.T., Kumarganj, Faizabad, planted at 5 m x 4m spacing. The maximum growth increment data in case tree height was recorded 0.85 m yr^{-1} for L-188. The maximum diameter at breast height (dbh) recorded 1.16 cm yr^{-1} for L-188. The yield performance of paddy (Sarjoo-52) variety was recorded 1.25 ha^{-1} followed by 1.10 t ha^{-1} for NDR-359 variety. In case of wheat yield as evident from that under control conditions (without tree cover) PBW-502 showed the largest grain yield (2.47 t ha^{-1}) followed by NW-1070 (2.35 t ha^{-1}). Under litter fall studies there was gradual increase in litter fall from August to October. However, there was sudden rise in leaf litter fall in November leading to peak leaf fall in December (45.2-46.7 per cent). Month June indicated complete leaf litter fall. The period November-January accounted for about 91.1-96.1 per cent (L-49 vs. Pant-5) of total annual leaf fall. The leaf litter decomposition showed that in annual cycle there was 82-87.8 per cent dry weight loss in litter decomposition for G-48 vs. L-49.

Keywords: Agroforestry, Litter decomposition, Poplar, Sodic soil

INTRODUCTION

Soil sodicity poses a serious threat for sustaining agricultural production. It reduces the value and productivity of about 3.77 million ha area in the Indo-gangetic plains. Sodic soil (also called alkali) contain sufficient amount of exchangeable sodium (ESP >15) to cause soil dispersion and increase the pH (>8.5), thereby adversely affecting both the physical and nutritional properties of the soil with consequent reduction in crop growth. Out of 7.0 million ha of salt affected soils in India, nearly 3.8 million ha is occupied by sodic land and are primarily spread in the Indo-gangetic alluvial plains in the states of Uttar Pradesh, Punjab, Haryana and parts of Bihar and Rajasthan. There soils are formed under the influence of carbonate and bi-carbonates of Na, which upon hydrolysis produces alkalinity. These soils can be reclaimed by adopting effective reclamation technology involving rice-wheat based cropping system and integrated use of suitable amendments preferably gypsum, organic and green manure which help in maximizing and sustaining yields, improving soil health and fertilizer use efficiency (Swarup 1998). However, nutrient management and organic matter improvement of these soils are extremely important to yields to match the potential yields obtained under partially reclaimed conditions. The human population in India is expected to reach over 1400 million by 2025. By the year 2025, the demand will reached to 320 million tones food to human population. To meet the challenge of more food for rapidly increasing population, it has to come from either the same or even shrinking cultivated area, which would be an uphill task. Therefore, the only option available with us is to put more area under cultivation by reclaiming the salt affected, degraded and barren cultivable waste lands. With rising global population and ever increasing pressure on agricultural land for industrial and other infrastructural projects, waste land reclamation has become a priority to ensure sustainable food production and environmental conservation.

Agroforestry is a land use option that increase livelihood security and reduces vulnerability to climate and environmental change (Dhyani et al 2009). In addition it is a natural resource conserving system and comparatively less resource depleting system than conventional agriculture. Thus poplar based agroforestry system can play a significant role in this proposal. Poplar (*Populus deltoides*) had gained considerable importance in agroforestry plantation, mainly due to its deciduous nature, fast growing habit and high industrial requirement (Chandra et al 2001). Poplars are multi purpose tree species, and their wood in much demand for pulp and paper, plywood, matchwood, packing peaces and light constructional timber all over the world. It can be economically harvested in 6 to 8 years. *Populus deltoides* has been successfully incorporated into agroforestry systems on account of its deciduous nature, which permits enough sunlight for cultivation of agricultural crops. Poplar known for fast growth, easy vegetative propagation, capability to enrich surface soil by adding leaf litter and the ability to provide substantial production (22-25 m^3 ha^{-1} yr^{-1} in block plantation and 2-3 m^3 ha^{-1} yr^{-1} is boundary plantation) on a short rotation of 6-8 years (Anonymous 2012). Poplar species in India are winter deciduous and remain leafless for 3-4 months, thereby favoring inter cultivations on crops like paddy-wheat crop rotation. Well managed clonal Eucalyptus and poplar plantation on good sites can produce high quality wood at rates exceeding 50 m^3 ha^{-1} yr^{-1} compared to less than 1 m^3 ha^{-1} yr^{-1} growth rate for forest in India (Lal 2008). Poplar plantation intercropped with agriculture crops has better growth than those without intercrops and the growth of trees is variable under different intercrops. The poplar based agroforestry system is a resource conserving system compared with the existing land management system involving few crops like rice, wheat, sugarcane, cotton, which are extremely resource-exhaustive, be it in terms of natural, financial or human resources. The most common crop rotation (rice-wheat) in the irrigated agro-eco system in Punjab, Haryana, Uttar Pradesh and adjoining states is over exploiting the water resources and water table. Genetic variation in poplar to salt tolerance has not been extensively studied in India. Therefore, such types of studies should under be taken to identify tolerant clones for saline and sodic environments. Complete reclamation of this soil is quite costly, particularly in absence of suitable drainage arrangements. The management practices like choice of plants, method and time of planting, irrigation and other cultural practices could be tailored as to give a better performance on this problematic soil. Keeping in view the utilization of partially reclaimed sodic soil and get to more income of the poplar based agroforestry systems can do under in this proposal.

METHODOLOGY

The present investigation was carried out at the experimental area of the Department of Agroforestry, Narendra Deva University of Agriculture & Technology, Kumarganj, Faizabad (U.P.) during the year 2015-16. The area under crops (paddy and wheat) were used 20x5 m to grow in between 6 poplar clones (G-48, Udai, L-52, Pant-5, L-49, L-188 with control) trees of 8 years of age, planted at 5x4 m spacing. In the first three to four years, three rice varieties were the crop cycle followed in the tree alleys. Three varieties of paddy viz., Sarjoo-52, BPT-5204 and NDR-359 and wheat varieties NW-1014, NW-2036, NW-1076 and PBW-502 were grown at in the month of July and the last week of November with recommended spacing in four replications. The experiment was laid out in RBD design with sowing times as main plots. The data were analysed in computer lab in the Department of Statistics in the College of Agriculture, Narendra Deva University of Agriculture & Technology, Kumarganj, Faizabad (U.P.). To compare the results, control plots were maintained for each treatment in an open area near the agroforestry plots.

Climate and Soil Conditions

The climate of the region is tropical which is governed by South-West monsoon. The mean maximum temperature ranged from 17.1°C (January) to 41.6 °C (May) and the mean monthly minimum temperature from 6.2 °C (December) to 25.9 °C (July) during the year 2015. Total annual rainfall during the year was 542.90 mm, which is very low in category from the average rainfall of 10 years. The rainy season accounted for about 53.5 per cent of the annual rainfall. The maximum rainfall was received in the month of July (198.4 mm), accounting alone for about 36.54 per cent of the total annual rainfall. During 2015, the area indicated about 45.8 per cent less rainfall from the annual rainfall (average of 10 normal years) which is 1000 mm.

The humidity percentage ranges from 40.3 per cent (April) to 82.1 per cent in the month of (January). Due to less rainfall during 2015 crops like paddy, wheat, pulses and mustard yields were affected as well as plant growth parameters were also affected. Soil had slightly lower moistures (12.86 per cent) but greater bulk density (0.99 g cm^{-3}) than the soil of agroforestry system (0.96 to 0.97 g cm^{-3}). The soil of the experimental site was clay loam with following measure characteristics are giving below:

Soil Properties	*Under poplar*	*In open*
pH (1:2.5)	8.70	8.25
EC (dsm^{-1})	2.14	2.16
OC (%)	0.28	0.25
Available N (kg ha^{-1})	146.30	140.10
Available P (kg ha^{-1})	15.30	13.70
Available K (kg ha^{-1})	294.40	288.60

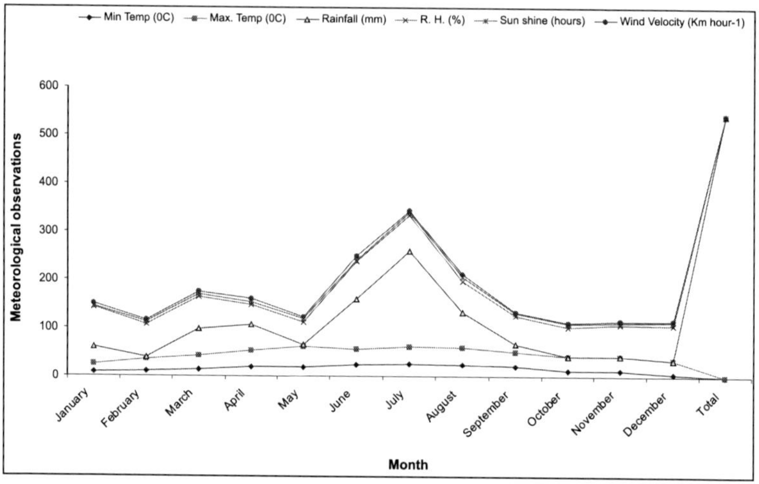

Figure 1: *Certain climatic parameters for the year 2015 recorded at Kumarganj, Faizabad*

In height and diameter of the trees was measured with the help of Ravi's multimeter and measuring tape, respectively and calculate the increment of height and dbh from March, 2015. The leaf fall and emergence pattern of the poplar trees was observed at weekly interval from March, 2015 to February, 2016. The various growth and yield parameters recorded for paddy and wheat varieties at a time of maturity.

RESULTS AND DISCUSSION

Performance of various poplar clones (G-48, Udai, L-52, Pant-5, L-49 and L-188) of willow was measured in terms of increment height and increment dbh at breast height (Table 1). Differences among the clones were found to be significant for both parameters. Maximum increment height (0.85 m yr^{-1}) was observed in clone L-188 which was significantly superior to all other clones. The minimum increment height (0.42 m yr^{-1}) was recorded in clones Pant-5. Maximum increment dbh (1.16 cm yr^{-1}) was recorded in clones L-188 which was significantly higher than all other treatments, while the minimum increment dbh (0.69 cm yr^{-1}) was recorded in clone L-52. Differences in growth parameters were due to inter-specific variation.

Table 1: *Performance of certain clones of Populus deltoides and yield potential of paddy and wheat variety under these clones*

Clone	Increment tree height (m yr^{-1})	Increment in Dbh (cm yr^{-1})	Paddy yield (q ha^{-1})			Wheat yield (q ha^{-1})			
			Sarjoo-52	BPT-5204	NDR-359	NW-1014	NW-2036	NW-1076	PBW-502
G-48 (Check)	0.80	0.89	10.2	8.4	12.1	18.7	18.1	17.4	20.2
Udai	0.67	0.91	11.4	10.9	10.6	17.2	16.1	19.6	17.8
L-52	0.69	0.69	9.7	10.2	10.1	20.4	20.2	15.8	19.3
Pant-5	0.42	0.94	11.3	9.7	12.3	19.6	16.2	18.3	21.7
L-49	0.78	1.10	12.5	8.4	9.6	16.7	16.8	20.7	17.2
L-188	0.85	1.16	10.8	9.1	8.8	22.2	19.6	17.3	22.3
Control	-	-	14.6	12.3	13.4	24.1	22.7	23.5	24.7
Average	0.70	0.98	11.5 (21.23%)	10.0 (18.70%)	11.0 (17.91%)	19.8 (17.84%)	18.5 (18.50%)	18.9 (19.57%)	20.4 (17.41%)
CD %	0.190	0.312	0.55	0.61	0.54	0.62	0.66	0.42	0.43

*Values in parentheses are grain yield reduction from open area

*Age 11 yr, Soil pH 8.7, ESP 40.1, EC 2.0 dSm^{-1}

Height, L-52-20.72 m, L-188-14.78 m, Pant-5-13.07 m, L-49-18.01 m, G-48-16.33 m and Udai-12.88 m

The grain yield of paddy varieties (Sarjoo-52, BPT-5204 and NDR-359) was significantly affected by different clones of willow. Maximum grain yield (14.60 q ha^{-1}) of paddy Sarjoo-52 recorded under control plots followed by L-49+paddy Sarjoo-52 (12.50 q ha^{-1}) which was superior over other treatments. Minimum yield of paddy was recorded in paddy variety BPT-5204 + G-48 and L-49 (8.4 q ha^{-1}). The reduction yield of paddy might be due to cultivation of partially reclaimed sodic soil and with dense leave during rice cultivation which creates shade over rice crops. The grain yield reduction was found in variety Sarjoo-52 (21.23), BPT-5204 (18.70 per cent) and NDR-359 (17.9 per cent) in compared to control. Similar results under inter crops with salix also reported by Agrawal et al (2010). Among the wheat varieties (NW-1014, NW-2036, NW-1076 and PBW-502) the maximum grain yield (20.70 q ha^{-1}) of wheat recorded under control plots in wheat variety PBW-502 followed by L-188+PBW-502 (22.3 q ha^{-1}) which was superior over other treatment and varieties under poplar based agri-silvi system maximum grain yield were recorded in PBW-502 (22.3 q ha^{-1}) followed by NW-1014 (22.2 q ha^{-1}).

Leaf Litter Fall Pattern in *Populus Deltoides* Clones Under Agroforestry Systems

Monthly leaf litter fall patter in *Populus deltoides* clones (L-49, L-52, G-48, Udai, Pant-5 and L-188) during March 2015 to February 2016 in depicted in Fig. 1. A perusal of data indicated that litter fall occurred in *Populus deltoides* clones from August to January month. A major peak of litter fall in month of November to January and January was due to complete leaf fall and initiation new leaves in the month of April. There was a gradual increase in leaf litter fall from August to October. However, there was sudden in rise in leaf litter fall in November lead in to peak leaf litter fall in December (45.2-46.7 per cent). Month of January indicated complete leaf litter fall. The period November to January accounted for about 91.1-96.1 per cent (L-49 vs Pant-5) of the total annual leaf litter fall. Similar observations were also reported by Ankurdeep Preety et al (2016).

Monthly variation in leaf litter fall in P. deltoides clones under AF system

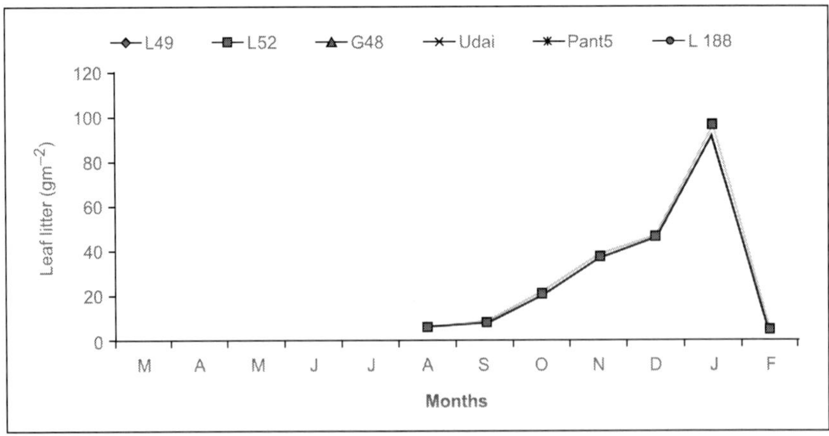

Annual leaf litter fall (gm-2 yr-1) *Range: 91.1 (L-49)-96.1 (Pant-5)*

Figure 2: *Monthly pattern of litter fall of P. deltoides during March 2015 to February 2016*

Monthly Changes in Dry Weight of *Populus deltoides* Leaf Litter During Decomposition Under Agri-Silvi System in the Year 2015

Leaf litter decomposition is a fundamental process of ecosystem functioning, which is closely linked to the nutrient recycling on plant growth. Leaf litter fall and decomposition pattern was studied in 8 years old poplar (*Populus deltoides*) planted as a field plantation in rice-wheat cropping system under the agroforestry system. It is evident from the Table-2 that highest percentage of decomposition (87.8) was recorded in poplar L-49 clone followed by Pant-5 (85.0) and L-188 (84.8). The minimum percentage of decomposition (82.0) noticed in G-48 clone. It is also apparent from the table that December month showed highest percentage of decomposition in all the poplar clones taken, while lowest percentage of decomposition recorded in the month of January in all the poplar clones. It is generally observed that average percentage of decomposition noticed in the month of June 2015 to September 2015. It is also interesting to mention that rate of decomposition started higher from the month August 2015 an onward December 2015. The present findings in the conformity with Kaushal et al (2005), who studied decomposition patterns and nutrient dynamics in leaf litter of *Populus deltoides*. They concluded that soil moisture and fertility are most significant variables which influence the rate of decomposition. Similarly, Boerner (1984) reported that mass loss from litter bags was significantly higher in more fertile soil. He further suggested that litter fall, nutrient transfers and decomposition rate were under the control of soil nutrient level. Adedeji (1986) reported that addition of urea and NPK mixture to air dried litter increased decomposition rate.

Table 2: *Monthly changes in dry weight of P. deltoides leaf litter during decomposition under agrisilvicultural system*

Month	*Leaf litter dry weight (g)*					
	L-188	*Udai*	*G-48*	*L-49*	*L-52*	*Pant-5*
Jan 15	4.40 (12.0)	4.36 (12.8)	4.38 (12.4)	4.35 (13.0)	4.36 (12.8)	4.34 (13.2)
Feb 15	4.35 (13.0)	4.20 (16.0)	4.10 (18.0)	4.10 (18.0)	4.12 (17.6)	3.95 (21.0)
Mar 15	4.22 (15.6)	3.92 (21.6)	3.91 (21.8)	3.80 (24.0)	3.90 (22.0)	3.65 (27.0)
April 15	3.85 (23.0)	3.78 (24.4)	3.60 (28.0)	3.60 (28.0)	3.61 27.8)	3.25 (35.0)
May 15	3.60 (28.0)	3.60 (28.0)	3.45 (31.0)	3.42 (31.6)	3.55 (29.0)	3.10 (38.0)
June 15	3.15 (37.0)	3.10 (38.0)	2.85 (43.0)	2.60 (48.0)	2.95 (41.0)	2.80 (44.0)
July 15	1.75 (65.0)	2.40 (52.0)	2.20 (56.0)	2.25 (55.0)	2.20 (56.0)	2.10 (58.0)
Aug 15	1.40 (72.0)	1.60 (68.0)	1.40 (72.0)	1.40 (72.0)	1.60 (68.0)	1.60 (68.0)
Sept. 15	1.26 (74.8)	1.10 (78.0)	1.20 (76.0)	0.85 (83.0)	1.25 (75.0)	1.25 (75.0)
Oct. 15	1.05 (79.0)	0.90 (82.0)	1.15 (77.0)	0.68 (86.4)	1.16 (76.8)	0.90 (82.0)
Nov. 15	0.80 (84.0)	0.81 (83.8)	1.11 (77.8)	0.65 (87.0)	0.90 (82.0)	0.85 (83.0)
Dec. 15	0.76 (84.8)	0.78 (84.4)	0.90 (82.0)	0.61 (87.8)	0.85 (83.0)	0.75 (85.0)
Annual weight loss (%)	84.8	84.4	82.0	87.8	83.0	85.0
CD (5%)	0.048	0.047	0.045	0.271	0.046	0.184

* Value in parentheses leaf weight per cent loss are based on initial weight (5g) of leaf litter in a bag

Nutrients Return Through Litter Fall in the Plantations of Different Poplar Clones

Weights of nutrient returned through annual litter fall from different poplar clones are given in Table-3. Poplar clones Pant-5 returned the highest amount of N (2.15 g m^2 per yr) through leaf litter followed by poplar L-49 (2.10 g m^2 per yr) Udai (2.05 g m^2 per yr), L-188 (2.03 g m^2 per yr), L-52 (1.96 g m^2 per yr) and G-48 (1.88 g m^2 per yr). Pant-5 also showed the highest amount of P return (0.46 g m^2 per yr) through leaf litter. However, annual return of P through leaf litter fall in all the species was for less than those of other major nutrients (N, K, Ca and Mg). The amount of K return through leaf fall was highest (1.15 g m^2 per yr) in the Pant-5 poplar clones. Calcium and magnesium return through leaf fall were highest for poplar Pant-5 and lowest for Udai clones. Among the micro-nutrient in leaf litter Fe (0.10 g m^2 per yr) and Cu (0.005 g m^2 per yr) were highest in Pant-5. The leaf fall of woody and miscellaneous litters had annual returns of (0.30-0.50 g m^2 per yr) and (0.02-0.04 g m^2 per yr), respectively. Similar to leaf litter fall, the woody and miscellaneous litter fall also reflected same patterns of nutrient returns, although their amounts were considerably lower than those in leaf litter.

The leaf fall of the six poplar clones studied did not show seasonal equanimity. However, for all the six clones, the rate of leaf fall per day basis was greatest in winter followed by summer and rainy season. A similar pattern was also observed for total above ground poplar clones litter fall under all plantations. Poplar is a deciduous species and the leaf fall was concentrated only for a few months. Although the litter showed monthly and round the year leaf fall, higher values were recorded during November to May than in rainy season. The amount of N, P, K, Ca and Mg returned through litter fall in poplar Pant-5 plantations were greater than in other poplar clones. The order of relative abundance of major nutrients (N, P, K) in the litter fall in the present study was N>K>P which agreed with the findings of Joshi et al (1997) and Rana et al (2007) for certain plantations. In the present plantations, woody and miscellaneous litter fall jointly for (13.67-20.0 per cent), foliage accounted for most of the N return to litter fall (79.0-80.0 per cent). Similar observations were also made for other plantation (Lodhiyal et al 1995). A comparison of species of the present study indicated that Pant-5 poplar clones had greater nutrient return in nutrient partially reclaimed sodic soil conditions; this may be due to its better growth on poor soils (Panda and Mohanty, 1998). The highest nutrient return through litter fall by this species contributed a sizeable improvement in soil conditions and this confirms with results by Rana et al (2003) for certain plantations.

Table 3: *Amount of nutrient return through litter fall in plantation of different poplar clones*

Poplar clones and litter decomposition	*Nutrient return (gm m^{-2} yr^{-1})*								
	N	*P*	*K*	*Ca*	*Mg*	*Cu*	*Zn*	*Mn*	*Fe*
L-49									
Leaf	2.10	0.40	1.05	2.15	1.40	0.003	0.03	0.06	0.08
Woody	0.45	0.23	0.32	1.15	0.70	0.002	0.013	0.05	0.07
Miscellaneous	0.03	0.02	0.03	0.05	0.02	0.0001	0.0005	0.002	0.003
Total	2.58	0.65	1.40	3.35	2.12	0.0051	0.0435	0.112	0.153

L-52									
Leaf	1.96	0.36	1.00	2.05	1.34	0.002	0.02	0.04	0.06
Woody	0.38	0.22	0.29	1.12	0.65	0.015	0.010	0.03	0.05
Miscellaneous	0.02	0.01	0.02	0.04	0.01	0.0001	0.0004	0.001	0.015
Total	2.36	0.59	1.31	3.21	2.00	0.0171	0.0304	0.071	0.125
G-48									
Leaf	1.88	0.35	1.05	2.01	1.33	0.018	0.02	0.03	0.05
Woody	0.30	0.20	0.30	1.10	0.61	0.013	0.011	0.02	0.03
Miscellaneous	0.02	0.01	0.02	0.03	0.01	0.0001	0.0003	0.001	0.010
Total	2.34	0.56	1.37	3.14	1.95	0.0311	0.0313	0.051	0.090
Udai									
Leaf	2.05	0.38	1.06	2.00	1.34	0.019	0.03	0.04	0.06
Woody	0.40	0.21	0.32	1.11	0.63	0.015	0.014	0.03	0.04
Miscellaneous	0.03	0.01	0.03	0.03	0.01	0.0001	0.0003	0.001	0.012
Total	2.48	0.60	1.41	3.14	1.98	0.0341	0.0443	0.171	0.112
Pant-5									
Leaf	2.15	0.46	1.15	2.24	1.51	0.005	0.05	0.07	0.10
Woody	0.50	0.25	0.38	1.19	0.77	0.003	0.016	0.06	0.08
Miscellaneous	0.04	0.03	0.04	0.06	0.03	0.0001	0.0006	0.003	0.004
Total	2.69	0.74	1.57	3.49	2.31	0.0081	0.0666	0.133	0.184
L-188									
Leaf	2.03	0.39	1.05	2.05	1.35	0.002	0.03	0.05	0.07
Woody	0.39	0.23	0.30	1.12	0.65	0.018	0.015	0.03	0.05
Miscellaneous	0.02	0.01	0.02	0.03	0.01	0.0001	0.0002	0.001	0.010
Total	2.44	0.63	1.37	3.20	2.01	0.0381	0.0452	0.081	0.130

CONCLUSION

Maximum increment height was recorded in clone Pant-5. The maximum increment in dbh was recorded in clone L-188, which was significantly higher than other treatment. The grain yield reduction was found in variety Sarjoo-52 (21.23 per cent), BPT-5204 (18.70 per cent) and NDR-359 (17.9 per cent) as compared to control. Among the wheat varieties (NW-1014, NW-2036, NW-1076 and PBW-502) the maximum grain yield (20.70 q ha^{-1}) of wheat recorded under control plots in wheat variety PBW-502 followed by L-188+PBW-502 (22.3 q ha^{-1}) which was superior over other treatment and varieties under poplar based agri-silvi system maximum grain yield were recorded in PBW-502 (22.3 q ha^{-1}) followed by NW-1014 (22.2 q ha^{-1}). It is generally observed that average percentage of decomposition noticed in the month of June 2015 to September 2015. It is also interesting to mention that rate of decomposition started higher from the month of August December. A comparison of species of the present study indicated that Pant-5 poplar clones had greater nutrient return in nutrient partially reclaimed sodic soil conditions; this may be due to its better growth on poor soils.

ACKNOWLEDGEMENT

The authors would like to acknowledge the financial support from the Indian Council of Agriculture Research (ICAR) through All India Coordinated Research Project on Agroforestry, (AICRP) on Agroforestry project for this research work.

REFERENCES

Adedeji FO 1986 Effect of fertilizer on microbial decomposition of leaf litter of a regenerating bush fallow in sub-humid tropical Nigeria, *Agriculture Ecology of System Environment* **18** 155-166

Agrawal VK, Shroti SK, and Chhumkar DS 2010 Willow (*Salix alba*) in Agroforestry, its effects on some agricultural crops, Proceedings of the National Symposium on Food Security in Context of Changing Climate. C.S.A. University of Agriculture and Technology, Kanpur, U.P., October, 30 November, **1** 86

Ankurdeep Preety RIS, Gill Baljit Singh, and Dhatt AS 2016 Optimization of Potato Varieties and their sowing time for cultivation under poplar (*Populus deltoides* Bortr. Ex. Marsh) based agri-silvi systems. *Indian Journal of Agroforestry,* **18** (1)70-76

Anonymous 2012 Country report-poplars and willows in India. National Poplar Commission of India, ICFRE, Dehradun, 1-64

Boerner RE 1984 Nutrient fluxes in litter fall and decomposition in four forests along a gradient of soil fertility in Southern Ohio, *Canada journal of Forest Research* **4** 794-802

Chandra JP 2001 Scope of poplar cultivation, *Indian Forester* **127** 51-60

Dhyani SK, Newaj R, and Sharma AR 2009 Agroforestry its relation with Agronomy, Challenges and Opportunities, *Indian Journal Forestry* **54** 249-66

Joshi CS, Rao PB, and Singh RP 1997 Comparative study of litter fall and nutrient return in some plantations of Central Himalaya, *Proceedings of Indian National Science Academy* **63** 617-624

Kaushal R, Verma KS, and Singh KN 2005 Seasonal dynamics and role of substrate quality in leaf litter decomposition of different agroforestry tree species, *Indian Journal of Agroforestry* **7** (1) 44-50

Lal P 2008 Clonal Eucalyptus plantations in India, *Indian Forester* **134** 1561-1570

Lodhiyal LS, Singh RP, and Singh SP 1995 Structure and function of an age series of poplar plantations in Central Himalaya. II Nutrient dynamics, *Annals of Botany* **76** 201-210

Panda T and Mohanty RB 1998 Litter production by *Casuarina equisetifolia* L. in coastal sandy belt of Orissa, *Tropical Ecology* **39** 149-150

Rana BS, Saxena AK, Rao OP, and Singh BP 2007 Nutrient return to the soil through litter fall under certain tree plantations on sodic wastelands in Northern India, *Journal of Tropical Forest Science* **19**(3) 141-149

Rana BS, Parihar AKS, and Singh BP 2003 Periodic change in crop yield and certain soil characters in *C. equisetifolia* based agri-silvicultural system developed on sodic wasteland 97-101 in Chauhan SK, Gill SS, Sharma SC and Chauhan R (Eds.) *Agroforestry in 21st Century.* Agrotech Publishing Academy, Udaipur

Swarup A 1998 Soil fertility problems and their management, in *Agricultural Salinity Management in India* (eds) NK Tyagi and PS Minhas, Central Soil Salinity Research Institute, Karnal 145-156

CHAPTER 32

Ecologically Based Pest Management for Sustainable Agroecosystem

Madhulika Srivastava, VeereshKumar, Usha[1], Anil Kumar, Inder Dev, Asha Ram, Naresh Kumar, Dhiraj Kumar, Lal Chand and OP Chaturvedi*

ICAR-Central Agroforestry Research Institute, Jhansi (U.P.), India

[1]RLB Central Agricultural University, Jhansi (U.P.), India

**Corresponding author's Email: madhulika.csir@gmail.com*

ABSTRACT

Manipulation of natural enemy behaviour to improve biological control of crop pests is a pleasing concept, but research on how best to achieve this in agroecosystem is a beginning. Habitat of natural enemies need to be fitted to the region, crop, and management system i.e. being used to ensure the greatest potential for benefits and to minimize undesirable effects on crop yield. HIPVs, combined with other semiochemicals such as volatiles produced from natural enemies themselves i.e. pheromones or the herbivores i.e. kairomones can be used to attract the natural enemies. The costs of any strategy for manipulation of natural enemy behaviour in agroecosystems have not been estimated. The adoption of natural enemies will depend on their costs in respect to the currently available pest management practices. Despite the costs of habitat establishment of HIPV deployment, the increasing costs of pesticides and consumer interest about pesticide residues on fresh farm products, biological control is becoming a better option. Habitat diversification or the deployment of semiochemical lures to manipulating the behaviour of natural enemies in agroecosystem, but it is possible to further increase the efficacy of natural enemies by combining more than one strategy to manipulate their behaviour in field crops.

Keywords: Agroecosystem, Agroforestry, Ecological engineering, Integrated pest management, Push pull technology

INTRODUCTION

Since the beginnings of agriculture about 10,000 years ago, farmers have had to compete with harmful organisms i.e. the pest. Pests adversely affect crop productivity and the stability of production. In India, the annual losses amount to ₹ 45,000 crore (Singh 2004).

Crop losses due to pests are higher in India than in other parts of the world (APO, 1993). Herbivorous insects are predicted to outbreak on plants (Atwal and Singh 1990; Oerke et al 1994). Insects are undoubtedly the most adaptable form of life as their total number far exceeds that of any other animal category. Without preventive protection with pesticides, natural enemies, host plant resistance and other non-chemical control, 70% of crops could have been lost due to pests (Folnovic 2015). Estimated percentage losses to different crops in India has been summarized by different workers (Singh 2004).

MANAGEMENT OF ENVIRONMENT TO FAVOUR PEST CONTROL

Managing the environment is important to increase the efficacy of natural enemies, which will be dependent upon certain factors viz; varietal development, cropping systems, tillage practices and chemical inputs. Environmental management for biological control is concerned with the functional environment i.e. the physical and biotic elements that directly or indirectly affect survival, migration, reproduction, feeding and the behaviour associated with these life processes. Recent trends in agriculture have been towards decreasing environmental diversity, increasing pesticide input and fertilizer, increasing mechanization and decreasing genetic diversity. Because of these factors an environment is created that checks the pest population regulation by natural enemies. Several research has been conducted to document the importance of manipulating environmental properties of crop fields to make them more favourable to natural enemies. (Price and Wald bauer 1982; Risch et al 1983; Herzog and Funderburk 1985). Most efforts have been directed for analysing the effect of reduced cultivated land and vegetational diversification of agroecosystems. Other types of cultural manipulations such as strip- harvesting, trap cropping, use of nests or artificial shelters, use of food sprays (Hagen 1986) and kairomones (Lewis et al 1976; Nordlund et al 1981; Srivastava 2008) that enhance the activity of specific natural enemies can be developed. Multiple cropping system enhances reproduction, survival and efficacy of natural enemies (Bach 1980 Risch 1981; Andow 1983; Letourneau and Altieri 1983; Altieri and Liebman 1986). The pest populations can be controlled directly through cultural control methods, that modify the habitat, the main thrust is conservation (maintenance of natural enemy abundance and diversity) and enhancement (increased immigration, tenure time, longevity, fertility and efficacy) strategies that can be used to manipulate natural enemies in agroecosystems. For increasing beneficial insects in our ecosystem/agroforestry systems following should be done:

- Enhance habitat suitability for immigration and host finding.
- Render alternate prey/hosts when pests are deficient.
- Provide supplementary food (food sprays, nectar and pollen) for predators/ parasitoids.
- Maintenance of noneconomic levels of the pest of alternate hosts over long periods to ensure continued survival of natural enemies.
- Provide refugia for mating or overwintering (Powell 1986; Gurr 2009).

MANAGEMENT OF VEGETATION

Rapid colonization and growth of herbivore is favoured by monoculture. These factors can be eliminated by giving a continuous vegetation, polyculture and associated food and shelter, so that natural enemies could be increased. There are numerous behavioural and physiological effects of plants on various natural enemies. Certain plants or chemicals released by the herbivore host plants attract the natural enemies (Altieri et al 1981; Letourneau and Altieri 1983; Boethel and Eikenbary 1986; Letourneau 1987). Therefore, natural enemies are sometimes more abundant in the presence of certain plants. (Nordlund et al 1985; Johnson & Hara 1987). Biocontrol agents can also be enhanced by the presence of non-crop-plants. (Barney et al 1984; Norris 1986). Natural enemies are more abundant in rapidly colonizing, fast growing plants, such as alternate prey or hosts, pollen or nectar and microhabitats, which are not available in weed free monoculture. (Van Emden 1965; Doutt and Nokata 1973). Carabids, syrphids and coccinellids are abundant in weed diversified system. Presence of specific weeds has enhanced the biological control of particular pests (Powell et al 1986). There are certain parasitoids, which can be used as biological control agents (Table 1).

Table 1: *List of some classical biological control in India*

S.No	Pest	Parasitoid
1	*Opuntia sp*	*Dactylopius ceylonicus*
2	*Salvinia molesta* (Water fern)	Weevil (*Cyrtobagous salviniae*)
3	*Icerya purchasi* (Cottony cushion scale)	*Rodolia cardinalis* (Coccinellid beetle)
4	*Nipaecoccus viridis* (Mealy bugs)	*Cryptolaemus montrouzieri*
5	*Quadraspidiotus perniciosus* (Apple pest San Jose scale)	*Encarsia perniciosi*
6	*Eriosoma lanigerum*	*Aphelinus mali* (Aphelinid parasitoids)
7	*Parthenium hysterophorus*	*Zygogramma bicolorata*
8	*Lantana camara*	*Ophiomyia lantanae*
9	*Pyrilla perpusilla*	*Epiricania melanoleuca*
10	*Scirpophaga excerptalis*	Isotima javensis

Source: Singh 2004

Increasing of Parasitoid Fitness and Efficacy by Adding Floral Diversity

Flowering field margins adjacent to crop fields may provide necessary resources for natural enemies of crop pests during periods when crop flowers are not present, thus maintaining high populations of insect predators and parasitoids, which are supported by a provision of nutrients throughout the season (Sotherton, 1984; Ahern and Brewer 2002; Buchi 2002; Sanchez et al 2003; Wanner et al 2006). Adjacent to field, flowering plant strips help support beneficial insect biodiversity in agricultural landscapes (Baggen and Gurr, 1998; Dufour, 2000; Carreck and Williams, 2002; Fiedler and Landis, 2007; Tuell et al 2008; Landis et al 2000; Isaacs et al 2009; Hogg et al 2011). Flower resource provide as an additional food for predators and parasitic Hymenoptera, potentially enhancing their fitness and efficacy. Nectar, pollen or both are required by many adult parasitoids (Powell 1986; Jervis et al 1992; 1996). Nectar rich in carbohydrate provides energy and pollen, may provide nutrients for egg production in some species (Jervis et al 1996) (Fig. 1).

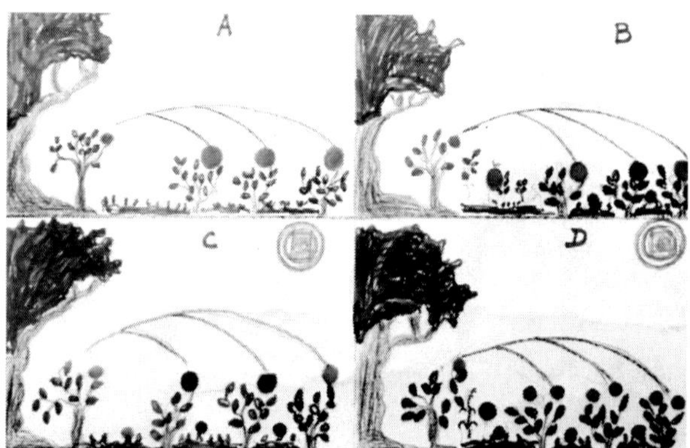

Figure 1: *Strategies for manipulating natural enemies in agroecosystems for enhanced insect pest control*

A. Conventional method with no manipulation of habitat leads to high pest numbers and few natural enemies entering the crop from the surrounding landscapes. Natural enemy populations and pest abundance are shown by circles.

B. The addition of inter-cropping, cover crops, or supplemental food sources to an agroecosystem may lead to an increase in natural enemy abundance and a potential decrease in insect pest abundance within field settings, but depends on the presence of insect pests to attract natural enemies into the crop

C. The addition of semiochemical-based lures, such as herbivore –induced plant volatiles (HIPVs) may attract natural enemies into the crop to enhance biological control, but does not provide resources to directly enhance natural enemy abundance.

D. Combining habitat manipulation and HIPV strategies in an agroecosystem may increase natural enemy abundance within field settings as well as directly attract natural enemies into the crop to enhance biological control.

As the lack of suitable floral resources in many agroecosystems may be a limiting factor in the effectiveness of natural enemies as biological control agents in some areas. So, the provision of these resources by planting flowers in an agroecosystem could therefore result in an increase in biological control of some pests. Adjacent to crop fields, placement of flowering resources, can minimize exposure to insecticides and side by side also provide the resources for natural enemies that can then re-invade the crop fields. For example, flowering borders adjacent to blueberry fields have enabled natural enemy's populations inside the crop to rebound more rapidly after insecticide applications (Walton and Isaacs 2011). We can observe an increase in natural enemies, particularly parasitoids have been observed in apple orchards adjacent to flowering borders compared to orchards without flowering borders. The same increase in parasitoid numbers coincided with a doubling of parasitism of spotted tentiform leafminer (*Phyllonorycter blancardella* Fabr.) in apple trees adjacent to flowering borders, compared with orchards without flowering borders.

It has been observed that the presence of flowers in an agroecosystem can increase the abundance of natural enemies, including parasitoids. The production of lettuce increased due to presence of a range of flower species (Chaney, 1998). More No of parasitoids were reported from a flowering buckwheat (*Fagopyrum esculentum* Moench) cover crop than from a grass and clover sward. In an apple orchard and in vineyard, presence of parasitoids increased due to the presence of buckwheat flowers (Berndt et al 2002). Gurr et al 2009 observed that due to parasitoids aggregate at or near the flowers following is improved:

(a) Parasitoid fitness is improved
(b) Parasitism rate is increased
(c) Host population is reduced
(d) Host population is reduced below an economic threshold

Improvement in Fitness

In the presence of flowers, parasitoid efficacy is increased by improving its fitness. In the field, this may be observed as a long - term increase in abundance, or as an increase in parasitism rate. Aspects of parasitoid fitness, such as longevity, fecundity and off-spring sex ratio are influenced by the presence of floral resources. Longevity of *Trichogramma carverae* Oatman and Pinto doubled, when the parasitoids were provided with honey (Gurr and Nicol 2000). The *D. tasmanica* longevity was six times greater, when these parasitoids had access to alyssum flowers (*Lobularia maritima* L.) than when they were fed water only (Berndt and Wratten 2001). The potential fecundity of *Diaeretiella rapae* (Mc Intosh) was greater, when they were given pollen or no food. The number of F_1 cocoons produced over a lifetime by *D. tasmanica* was more than seven times greater for parasitoid with access to alyssum flowers than for those fed water only. (Berndt, unpublished data). The sex ratio of parasitoid offspring can also affect fitness. For example *D. tasmanica* produced a greater proportion of female offspring, when flowers were available, than when no flowers were present in both field and laboratory (Berndt et al 2002).

Increase in Parasitism Rate

An increase in the rate of parasitism has been observed, due to presence of flowers. For example, coriander (*Coriandrum sativum* L.) and faba beans (*Vicia faba* L.) planted next to potato plots increased the parasitism rate of potato moth (*Phthorimaea operculella* Zeller) by *Copidosoma koehleri* Blanchand (Baggen and Gurr 1998). Greater parasitism rate of leafrollers (*Epiphyas postvittana*) (walker) in an apple orchard having more buckwheat and faba beans were also observed.

Reduction in Host Population

Due to presence of flowers in an agroecosystem, pest population have been found to be reduced. For example, in an eggplant (*Solanum melongena* L.) fields reduced the survivorship of Colorado potato beetle (*Leptinotarsa decemlineata* Say) larvae (Patt et al 1997). Fewer aphids in cabbage plots where there was a border of phacelia (*Phacelia tonacetifolia* Bentham), than in those without it (White et al 1995).

Pest Population Below Economic Threshold

Flowers may be added to an agroecosystem to enhance the effectiveness of classically or argumentatively released natural enemies (Gurr and Wratten 1999) or their use can be incorporated with the use of selective or reduced rate of insecticides (Van Emden 1988; Poehling 1989).

Push–Pull Strategy

A combination of stimuli are used to manipulate the behaviour of insect pests or natural enemies and to alter and change their distribution and abundance in agroecosystems, during "Push-pull strategy", (Miller and Cowles 1990; Khan et al 1997). The push-pull approach works by repelling or deterring the pest insects (push) away from the main crop with the help of deterring chemical stimuli. On the other hand, highly appealing stimuli are used to attract the pests (pull) from the main crop to other areas such as trap crops, where the pests aggregate and are easier to control (Khan et al 1997). Each individual component (the push and the pull) of the approach may be effective at reducing pest numbers below economic thresholds, combining the push and pull components increases the efficacy of such a strategy. However, since the push and pull components are commonly nontoxic, this strategy is compatible with supporting insect natural enemies. (Khan and Pickett 2004). Push –pull strategies are one of the most successful complexes, developed recently in Africa for controlling stem borers in cereal crops (Khan et al 2001). Many cereal crops are destroyed due to presence of different stem borers. Stem borers are controlled by different chemical pesticides, but this is not an economical or safe approach for different resource –poor, small –scale farms, where these pests are common. Keeping these facts in mind a push-pull strategy was developed using technologies appropriate and economical for such farmers (Khan and Pickett 2004). In this strategy are included plants that involves a combined use of intercropping and trap crops and uses plants that are also locally – appropriate and that can be used in their agricultural system. The best advantage of this strategy is that it does not use any chemical deterrents or toxins, but instead uses plant species that repel the pest away from the main crop, while also attracting insect natural enemies to the fields. The repelled stem borers are attracted to nearby trap plant, where the moths lay their eggs, but their larvae are unable to develop thus reducing the number of trapped pest insects (Khan and Pickett 2004). In Africa, where the push-pull strategy have been experienced, lowers the pest abundance and overall enhancement of beneficial insect abundance. This push-pull approach can be used for overall enhancement of beneficial insect abundance (Khan et al 2001). So, we can say this push-pull approach can be applicable to a much wider range of agricultural pest problems in a variety of crops if the appropriate components can be developed and implemented (Fig. 2).

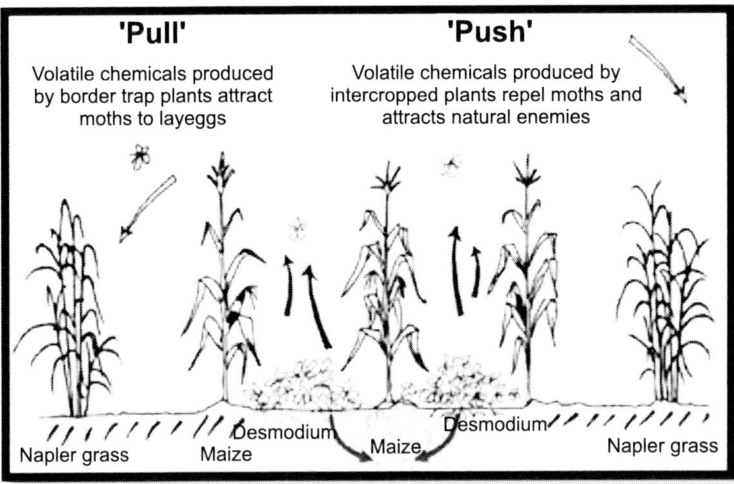

Figure 2: *Push pull technology for pest control*

Source: *Adapted from Khan et al 1997*

Intercropping, Polyculture and Cover Crops

Habitat manipulation for the beneficial insects depends upon their ability to use or exploit one or more of the plant components of the agroecosystem (Altieri and Nicholls 2004). Monoculture are an example of agroecosystems with low diversity and may be more susceptible to pest or disease outbreak (Theunissen 1994; Altieri and Nicholls 2004). Intercropping, which is the cultivation of two of more species, within the same field is a common method to increase beneficial insect diversity within agroecosystem (Vanderneer 1989). Flowering species such as clover can also provide a favourable habitat for a variety of beneficial insects that may not otherwise survive in a single crop environment and hence intercropping may provide natural pest management by increasing the abundance and diversity of insect natural enemies in agroecosystem (Thaurissan 1994). Diverse systems, stimulate complex food webs that involve more interaction among vegetation, pests and natural enemies, providing resources for a diverse group of organisms and allowing for alternative resources and food sources. Thus, polycultures and natural ecosystems with higher diversity tend to be more stable and less subject to fluctuations in pest and disease populations. Pear orchards with aromatic plants, significantly reduced pest abundance and increased the ratio of natural enemies to pest's abundance, when compared to orchards with only natural grass or clean tillage. They also found higher abundance of natural enemies and reduced numbers of major pests in intercropped orchards.

Hence, intercropping the pear orchard with aromatic plants led to improved insect pest management by enhancing the activity of the insect natural enemy community (Altieri and Nicholls 2004).

Cover crops are a major means of conserving beneficial insects and stabilizing their populations is to meet the ecological requirements of these insects within or near the cropping environment (Landis et al 2000).To be effective, many natural enemies and pollinators need access to alternate hosts, overwintering habitats, constant food supply and appropriate microclimates. These requirements can be fulfilled via the inclusion of a different assemblage of flowering plants within agricultural landscapes. Cover crops can be planted in both ways, either in rotation with annual crops or within perennial crops. These plants have many different advantages *viz.,* used to reduce soil erosion ,add or retain soil nutrients produce organic matter, reduce soil compaction and also aid in pest control (Bugg 1991; Bugg and Waddington 1994). They are used in sustainable and organic agricultural systems to enhance soil health and crop nutrition. Buckwheat (*Fagopyrum esculentum* Moench) a flowering species or clovers (*Trifolium spp*) have been promoted as cover crops to provide flowering resources for insects, when the crop is not in bloom. Cover crops having flowers is seen as one component of an overall 'farmscaping" approach that aims to make farmland suitable for a range of beneficial organisms throughout the growing season. In fact, cover crops provide shelter and resources for natural enemies, enhancing their populations and hence biological control of insect pests (Bugg and Waddington 1994). Plants should be screened for their attractiveness to not only target biological control agent, but also to other potential competitors for floral resources (Hogg et al 2011).

Pollen and Nectar Resources

Sufficient flower abundance and appropriate vegetation structure are required to support diverse populations of insects (Zurbrugg and Frank 2006) and therefore manipulation of structurally resource-poor habitats through the addition of flowering plants and grasses can increase beneficial insect populations in agricultural landscapes (Long et al 1998; Kells et al 2001; Rebek et al 2005). Many natural enemies, require access to alternate hosts, overwintering habitats a constant food supply and appropriate microclimates in order to survive (Johnson and Triplehorn 2005). Among all, most of the predators and parasitoids are omnivores and require non –prey food, such as pollen and nectar as part of their diet. Natural enemies from a broad range of orders including Hymenoptera, Diptera, Coleoptera, Heteroptera, Neuroptera,Araneae and Acari have been observed to require and /or benefit from access to flowering resources (Wacher et al 2005). Additionally, the need to select plants that are beneficial to the natural enemies without providing resources to pest insect has driven the careful evaluation of pest and natural enemy life history traits on candidate floral resources. Lacy phacelia (*Phacelia tanacetifolia* Benth) and Nasturtium (*Tropaeolum majus* L) provided resources for natural enemies, without enhancing moth pest performance.

Alternative Prey and Shelters

Most natural enemies benefit from having access to alternate hosts/prey. Banker plant systems are used to increase populations of natural enemies in greenhouse and field settings on ornamental and food crops (Van Driesche et al 2008). It generally consist of a non-crop plant that is deliberately infested with a non pest herbivore. Non-pest herbivores serve as alternative hosts for a desired parasitoid or predator of the target crop pest. A banker plant system target crop pest. A banker plant system is generally based around on the use of surrogate food, such as pollen for generalist natural enemies (Huang et al 2011). As a form of conservation biological control, banker plant systems provide alternative food or hosts for natural enemies so they can live and reproduce for long periods even when no pests are present in the crop (Frank, 2010). Banker plants can conserve released natural enemies to provide sustainable, long term suppression of crop-pests.

Natural enemies of insect pests require shelter from environmental hazards and absence of shelter during periods of heat, cold, rain, or pesticide application may be highly detrimental to their survival. Availability of suitable habitats can promote foraging, resting, overwintering, or nesting of natural enemies. Physical environmental conditions profoundly affect natural enemy activity during the growing season. For example, excessive wind is thought to limit foraging by adult hoverflies. In the windy areas, hedgerows, windbreaks or shelter-belts can reduce soil erosion, and it can improve photosynthetic and water-use efficiency by crop-plants. It can lead to locally elevated temperatures in sheltered areas. Hedgerows and windbreaks often contain flowering plants used by many natural enemies, the effect of shelter and of flowers may be difficult to separate. (Beane and Bugg 1998). In the crop fields various natural enemies are often observed for overwintering and resting aggregations. Typical sites for such aggregations vary among species and include herbaceous and woody plants as well as human-made structures (Beane &Bugg, 1998). Shelters are also necessary for the survival of predatory species. Predatory mites use a protected location as 'domatium" which is naturally –occuring shelter. (Karban et al 1995; Loughner et al 2008).

Effect of Landscape on Natural Enemies

Research on landscape effect on biological control of insect pests has developed rapidly over the past 20 years. (Turner et al 2001). For the aerial imaginary of remotely sensed data of the landscape surrounding crop fields and combining this with measurements of pest natural enemy interactions in crop fields has provided new situation. (Bianchi et al 2006). High landscape diversity including flowering plants, near crop fields also increases natural enemy population in crop fields. Aphid population growth was slower in organic than conventional farms. The agricultural habitats can be manipulated to provide resources for natural enemies. It is necessary that natural enemies are more required, where number of pests are more in number.

ROLE OF SEMIOCHEMICALS IN AGROECOSYSTEM

The term "semiochemical" (semeion = sign or signal in Greek) is used to describe a chemical or a mixture of chemicals that can act as messengers in interactions among organisms (Nordlund and Lewis 1976; Dicke and Sabelis 1988; Vet and Dicke1992). It contains intraspecific communications i.e. Pheromones-which contains chemicals that mediate interactions among individuals of the same species and interspecific communications (allelochemicals) i.e. allomones that mediate interactions among individuals of the different species. Among allelochemicals, compounds can be classified as allomones, if their production benefit the emitter, kairomones if their production benefit the receiver or synomones if their production benefit both the emitter and receiver (Dicke and Sabelis 1992; Vet et al 1995). Natural enemies are influenced by plants emitting behaviour modifying volatile organic compounds. Plants, damaged by herbivores often produce a blend of volatiles (HIPVs) (Mumm and Dicke 2010). These HIPVs consist of a mixture of the so called green –leaf volatiles (C_6 aldehydes, alcohols and acetates), terpenes (monoterpenes, sesquiterpenes, homoterpenes) and aromatic compounds, among others (Pochersky et al 2006).This release of HIPVs can signal the presence of potential prey or hosts and, therefore can be exploited by natural enemies to locate the prey organism (Sabelis et al 1999; Verkerk, 2004). Vinson (1976) described five steps for natural enemy's selection and are; (1) Host habitat location, (2) Host location, (3) Host acceptance, (4) Host suitability and (5) Host regulation. Chemical cues are particularly important for host-habitat location. In order to locate the host habitat, natural enemies use long-distance cues from plants. These HIPVs begins mainly from plants damaged by the natural enemies (Vet et al 1995). Once the host-habitat is located, natural enemies utilizes compounds produced by the host or prey (kairomones) such as volatiles emitted from body scales, honey dew or herbivore's frass.

GENETIC ENGINEERING

Plants when attacked by herbivores, can be selected for enhanced HIPV emissions. During genetic engineering programme, two approaches can be taken:

(a) Selective breeding for the selection of plants for enhancing the foraging efficiency of natural enemies.

(b) Transgenic plants, where specific genes are incorporated to prime plants for an enhanced HIPV response.

SELECTIVE BREEDING

Crops with enhanced volatile emissions are produced by plant breeding (Nordlund et al 1981, 1988), but selective breeding for high HIPV production in plants has not been discovered yet. Selection of plants that are more attractive to natural enemies may thus help biological control. For example Glanded cotton (*Gossypium hirsutum* L.) produced more natural enemies than non glanded cotton. Plants having greater induced volatile responses (HIPVs) are selected. For instance, HIPV emissions increased 8- fold in maize cultivars. (Gouinguene et al 2001; Degen et al 2004). Among below ground interactions, E-β-caryophyllene is a volatile induced from maize roots by herbivory that attracts entomopathogenic nematodes (Rasmann et al 2005).

Transgenic Plants

The transgenic plants can be used increase HIPVs (Degenhardt et al 2003; Aharoni et al 2005 2006; Turling and Ton 2006; Dudareva and Pichersky 2008; Kos et al 2009). We have to manipulate plant defence pathways. Impaired genes in mutant plants are especially involved in defence pathways (Van Poecke and Dicke 2002; Van Poecke and Dicke 2003; Shiojiri et al 2006). Transgenic plants with modified production of HIPVs will be more useful for manipulation of natural enemy behaviour. We have to modify 'terpenoid biosynthesis" because of dominance of terpenes in the HIPV blends of plants (Aharoni et al 2005; 2006). Different studies were conducted to enhance the efficacy of natural enemies under all the conditions. Transgenic plants with modified production of

HIPVs will be more useful for manipulation of natural enemy behaviour. We have to modify "terpenoid biosysthesis" because of dominance of terpenes in the HIPV blends of plants. (Aharoni et al 2005; 2006). Sometimes, pheromones can be combined with HIPVs to increase natural enemy attraction in agroecosystems. For example, attraction of the lacewing species to HIPVs in apple orchards and MeSA and iridodial, i.e. a male produced aggregation pheromone, a combination of these two was a stronger attractant than each compound alone.

USE OF TRICHOGRAMMA IN AGROECOSYSTEM

The biocontrol agents like egg parasite *Trichogramma* species have been used successfully against many lepidopterous pests. The tests of three exotic *Trichogramma* species namely *T. evanescence*, *T. brasiliensis* and *T. pickel* against *E. mechaeralis* eggs in Karnataka have given encouraging results in field conditions. *Lobesia* sp, which eats the terminal buds on new shoots is controlled by the release of parasitoids such as *Trichogramma dendrolimi* during egg production. (CABI Encyclopedia of Forest Trees). In China, eri and oak silk worm eggs are used for the mass production of Trichogramma (Wang et al 2014). In India, rice meal moth, *Corcyra cephalonica* (Stainton) is used as the laboratory host of *Trichogramma*. Eggs of *C. cephalonica* were found to be superior to other hosts of *Trichogramma* in the quality of mass reared parasitoids. In China, the idea of augmenting field populations of natural enemies is seen as just one of many techniques, which are used together for control. For example, *Polistes* augmentation is compatible with *Trichogramma* releases, intercropping and other techniques. Various attractants have been used for natural enemies in host parasitoids interaction. (Table 2).

Table 2: *Attractants used for natural enemies in host parasitoid Interaction*

S.No.	Name of attractant	Effect	References
1	Terpenoid biosynthesis	Dominance of terpenes in the HIPV	Aharoni et al 2005, 2006
2	Linalool / nerolidol synthase gene	Emission of the HIPV nerolidol and attraction of the predatory mite *P.persimilis*	Kapper et al 2005
3	Sesquiterpene synthase gene which can induce herbivory	(E) β-farnescene (E)-α-bergamotene and other herbivory induced sesquiterpenes	Schnee et al 2006
4	Extract from the host (moth) scales or synthetic kairomones (Tricosane) onto soyabean plants	To enhance the efficiency of natural enemies	Lewis et al 1975a 1975b
5	Volatiles from the ovipositor gland of female *H.zea* (Contains the moth sex pheromone)	Help in Trichogramma spp host search behaviour	Lewis et al 1982
6	Kairomone o-caffeoyltyrosine	*Aphytis melinus* Debac: A parasitoid of the California red scale (*Aonidiella aurantii* Maskell)	Hare et al 1997

Agroforestry-A Tool for Increased Natural Enemy

Agroforestry is the intentional combining of agricultural and forestry elements to create productive and sustainable farms, ranches and woodlands by reducing threats and enhancing resilience. In recent years, agroforestry has developed as an autonomous science that can help farmers to increase their profitability and sustainability of land. Trees in a variety of agroforestry systems enhance nutrient cycling, improve soil conservation and soil faunal activities and contribute to food security (Landis et al 2000; Schoeneberger et al 2012). Agroforestry plantings provide critical habitat to the native pollinators and natural enemies to the crop pests. Greater natural enemy presence has been observed in the more diverse and shaded agroforestry systems. Due to combination of number of factors i.e. trees, crops, and environment in agroforestry systems, there is an integrated approach combining host –plant resistance to pests, exploiting alternative tree species, measures that prevent pest build up, but favours natural enemies so biological control is enhanced for pest management. For avoiding pests, species substitution is feasible, if trees are grown.

There is a huge pressure on agriculture to reduce trust on chemicals and have more and more focus on sustainable methods of production (Brown and Glenn 1999). Integrated pest management (IPM) (Fig. 3) has shown immense potential for reducing the dependence of crop protection on chemical control methods (Pretty et al 1998, Atanassov et al 2002). Pest management during IPM requires a coordinated approach, integrating different tactics including cultural, biological and chemical control (Dent 1991). Plants are not capable of running away from their enemies i.e. the herbivores that may hurt them. However, plants can depend on the natural enemies of insect herbivore for protection, under certain circumstances. These natural enemies include predators and parasitoids. Practical methods have been developed and evaluated to conserve and augment natural enemies of different agricultural pests to protect plants from damage caused by insect herbivores. For attracting more and more natural enemies in an ecosystem, it is essential to improve the suitability of the crop landscape for natural enemies by manipulating the resources available for these insects and use of semiochemicals to attract predators and parasitoids.

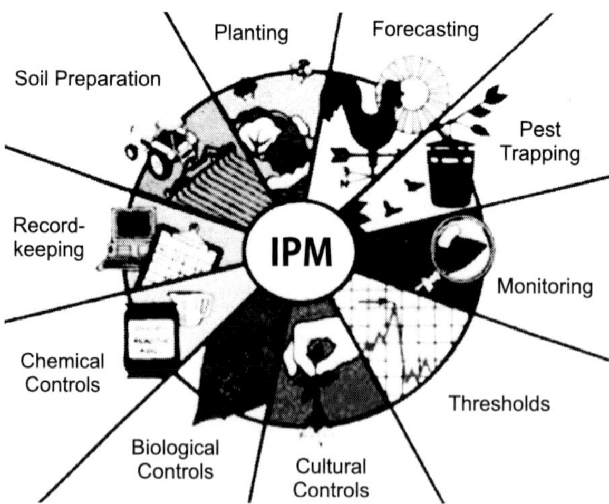

Figure 3: *Integrated pest management (modified from https://www.linkedin.com/pulse/integrated-pest-management-dr-tohid-nooralvandi)*

The manipulation of the behaviour of natural enemies, through vegetational diversification of crop habitats, and use of semiochemicals to enhance biological control in agroecosystems and combination of these to improve crop protection can be used in agroforestry systems. This system due to the use of litters and organic matters, non-crop habitat can be increased which will support natural enemies populations. To increase the population of natural enemies the non –crop habitat early in the season can be developed before the arrival of pests (Fig. 4).

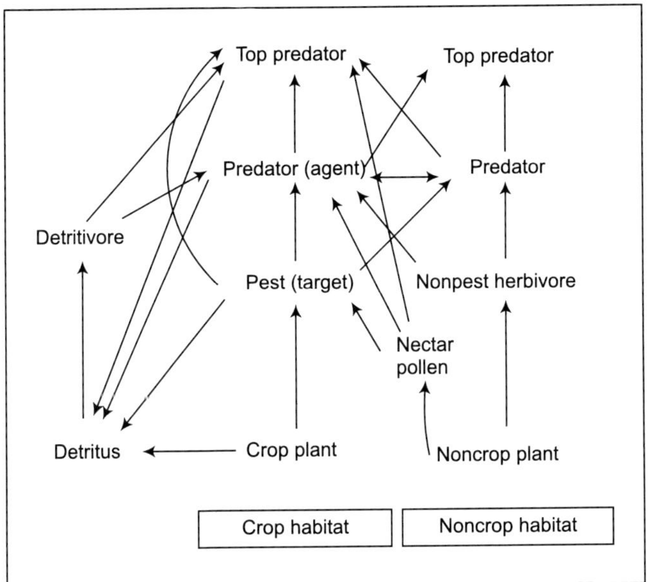

Figure 4: *Potential importance of non-crop habitat and detritus (Adapted from Gurr 2009)*

CONCLUSION

Plants are not capable of running away from their enemies, i.e. the herbivores that may eat them. IPM appears to be the only viable option to manage future insect pest problems for sustainable agriculture. To avoid use of pesticides, having enormous ill effects, we have to strengthen our natural enemies via different tactics for a sustainable Agroecosystem. Besides this certain reduced risk insecticides e.g. Indoxacarb, Emmamectin benzoate, Coregen and many others should be used to avoid the ill effects of pesticides like problems of insecticide resistance, insect resurgence, secondary pest outbreak etc.

REFERENCES

Aharoni A, Jongsma MA and Bouwmeester H J 2005 Volatile science? Metabolic science? Metabolic engineering of terpenoids in plants, *Trends in Plant Science* 594-602

Aharoni A, Jongsma MA, Kim TY, Ri MB, Giri AP, Verstappen FWA, Schwab and W Bouwmeester HJ 2006 Metabolic engineering of terpenoid biosynthesis in plants, *Phytochemistry Reviews* 49-58

Ahern RG and Brewer M J 2002 Effect of different wheat production systems on the presence of two parasitoids (Hymenoptera: Aphelinidae, Braconidae) of the Russian wheat aphid in the North American Great Plains, *Agriculture, Ecosystems, and Environment* 201-210

Altieri MA, Lewis W J, Nordlund DA, Gueldner RC and Todd JW 1981 Chemical interactions between plants and Trichogramma wasps in Georgia soybean fields, *Environmental protection and ecology* **3** 259-263

Altieri MA and Liebman M 1986 Insect, weed, and disease management in multiple cropping. In Multiple Cropping Systems (eds) CA Francis, New York, Macmillan, 183-218

Altieri MA and Nicholls CI 2004 (2nd ed) *Biodiversity and Pest Management in Agroecosystems,* Food Products Press, New York

Andow DA 1983 Plant diversity and insect populations: Interactions among beans, weeds and insects, Ph.D. dissertation, Cornell University, Ithaca, New York 201

APO 1993 Pest Control in Asia and the Pacific. Report of a seminar, September 24-October 4, 1991, Asian Productivity Organization, Tokyo, Japan

Atanassov A, Shearer PW, Hamilton G,Polk D 2002 Development and implementation of a reduced risk peach arthropod management program in New Jersey, *Horticulture Entomology* **95** 803-812

Atwal AS and Singh B1990 Pest population and Assessment of crop losses. Indian Council of Agricultural Research, New Delhi, India, 131

Bach CE 1980 Effects of plant density and diversity in the population dynamics of a specialist herbivore, the striped cucumber beetle, *Acalymma vittata* (Fabr), *Ecology* **61** 1515-1530

Baggen LR and Gurr GM 1998 The influence of food on *Copidosoma koehleri* (Hymenoptera : Encyrtidae), and the use of flowering plants as a habitat management tool to enhance biological control of potato moth *Phthorimaea operculella* (Lepidoptera: Gelechiidae). Biological Control,9-17

Barney RJ, Lamp WO, Ambrust EJ and Kapusta G 1984 Insect predator community and its response to weed management in spring-planted alfalfa, *Environmental Protection and Ecology* **6** 23-33

Beane KA and Bugg RL 1998 Natural and artificial shelter to enhance arthropod biological control agents, In Enhancing Biological control: Habitat Management to Promote Natural Enemies of Agricultural Pests, (eds) CH Pickett and RL Bugg, University of California, 239-254

Berndt LA and Wratten SD 2001 Biodiversity in the vineyard-worth the bother? *Winepress* **101** 27-30

Berndt LA, Wratten SD and Hassan P 2002 Effects of buckwheat flowers on leaf roller (lepidoptera: Tortricidae) parasitoids in a New Zealand vineyard, *Agricultural and Forest Entomology* **4** 39-45

Bianchi FJJA, Booij CJH and Tscharntke T 2006 Sustainable pest regulation in agricultural landscape composition, biodiversity and natural pest control, *In* Proceedings of the Royal Society 1715-1727

Boethel DJ and Eikenbary RD (eds) 1986 *Interactions of plant resistance and parasitoids and predators of insects*. Chichester, UK, Wiley & Sons 224

Brown MW, Glenn DM 1999 Ground cover plants and selective insecticides as pest management tools on apple orchards, *Economic Entomology* **92** 899-905

Buchi R 2002 Mortality of pollen beetle (*Meligethes spp*) larvae due to predators and parasitoids in rape fields and the effect of conservation strip, *Agriculture Ecosystems and Environment* 255-263

Bugg RL 1991 Cover crops and control of arthropod pests of agriculture, Proceedings of cover crops for clean Water, Jackson 25

Bugg RL and Waddington C1994 Using cover crops to manage arthropod pests of orchards: a review, *Agriculture Ecosystems and Environment* 11-28

Carreck N L &Williams I H 2002 Food for insect pollinators on farmland; insect visits to flowers of annual seed mixtures, *Journal of Insect Conservation* 13-23

Chaney WE 1998 Biological control of aphids in lettuce using in –field insectaries, In Enhancing Biological Control: Habitat Management to Promote Natural Enemies of Agricultural Pests, (eds) Pickett CH and RL73-83 Bugg University of California Press, Berkeley, California U.S.A.

Degen T, Dillmann C, Marion-Poll F & Turlings TCJ 2004 High genetic variability of herbivore-induced volatile emission within a broad range of maize inbred lines, *Plant Physiology* 1928-1938

Degenhardt J, Gershenzon J, Baldwin IT and Kessler A 2003 Attracting friends to feast on foes: engineering terpene emission to make crop plants more attractive to herbivore enemies, *Current Opinion in Biotechnology* 169-176

Dent D 1991 Insect pest management. Wallingford (UK): CAB International 604

Dhaliwal GS, Dilawari VKand Smith CM1999 Host plant defence against insects. In Environmental stress in crop plants (eds) GS Dhaliwal and R Arora, Commonwealth Publishers, New Delhi 172-210.

Dicke M and Sabelis M W 1992 Costs and benefits of chemical information conveyance: Proximate and ultimate factors, In: Insect chemical ecology, an evolutionary approach, (eds) BD Roitberg and B Isman, 122-155

Dicke M and Sabelis MW 1988 How plants obtain predatory mites as bodyguards, *Netherlands journal of Zoology,* 148-165

Doutt RLK and Nokata J 1973 The Tubus leafhopper and its egg parasitoid: an endemic biotic useful in grape pest management, *Environment Entomology* **2** 381-86

Dudareva N and Pichersky E 2008 Metabolic engineering of plant volatiles. Current Opinion in Biotechnology, 181-189

Dufour R 2000 Farmscaping to enhance biological control, In NCAT Sustainable Agriculture Project, *Ecology* **68** 1616–1622

Fielder AK and Landis DA 2007 Attractiveness of Michigan native plants to arthropod natural enemies and herbivores. *Environmental Entomology* 751-765

Folnovic Tanja 2015 Yield losses Due to pests, Pests and Diseases 2015

Frank SD 2010 Biological control of arthropod pests using banker plant systems: Past progress and future directions, *Biological Control* 8-16

Gouiguene S, Degen T and Turlings TCJ 2001Genotypic variation in induced odour emissions among maize cultivars and wild relatives, *Chemoecology* 9-16

Gurr GM 2009 Prospects for ecological engineering for planthoppers and other arthropod pests in rice. In Planthoppers: new threats to the sustainability of intensive rice production system in Asia (eds) KL Heong and Hardy B, Los Banos (Philippines), International Rice Research Institute 371-388

Gurr GM and H I Nicol 2000 Effect of food on longevity of adults of Trichogramma carverae Oatman and Pinto and Trichogramma brasscicae Bezdenko (Hymenoptera: Trichogrammatidae), *Australian Journal of Entomology* **39** 185-187

Gurr GM and Wratten SD1999 Integrated biological control: a proposal for enhancing success in biological control, *Pest Management* **45**(2)81-84

Hagen KS 1986 Ecosystem Analysis: Plant cultivars (HPR), Entomophagous species and food supplements. In "Interactions of Plant Resistance and Parasitoids and Predators of Insects, (eds) DJ Boethel and RD Eikenbary, John Wiley and Sons, New York 153-197

Hare JD, Morgan DJW and Nguyun T 1997 Increased parasitization of California red scale in the field after exposing its parasitoid, *Aphytis melinus*, to a synthetic kairomone, *Entomologia Experimentalis et Applicata* **82** 73-81

Herzog DC and Funderburk JE 1985 Plant resistance and cultural practice interactions with biological control, 67 -88 In Biological control in agricultural IPM systems, (eds) DC Herzog and MA Hoy, Academic Press, Orlando, Florida.

Hogg BN, Bugg RL and Daane KM 2011 Attractiveness of common insectary and harvestable floral resources to beneficial insects. Biological Control 76-84, https://attra.ncat.org/attra-pub/summaries/summary.php?pub =145

Huang N, Enkegaard A, Osborne LS, Ramakers PMJ, Messelink GJ, Pijnakker J and Murphy G 2011 The banker plant method in biological control, *Critical Reviews in Plant Sciences* 259-278

Isaacs R, Tuell J, Fiedler A, Gardiner M and Landis D 2009 Maximizing arthropod-mediated ecosystem services in agricultural landscapes: the role of native plants. *Frontiers in Ecology and the Environment* 196-203

Jervis MA, Kidd NAC and Heimpel GE 1996 Parasitoid adult feeding behaviour and biocontrol-a review, *Biocontrol News and Information* **17**(1)11

Jervis MA, Kidd NAC and Walton M 1992 A review of methods for determining dietary range in adult parasitoids. Entomophaga **37** 565-574

Johnson MW and Hara AH 1987 Influence of host crop of parasitoids (Hymenoptera) of Liriomyza spp. (Diptera:Agromyzidae), *Environmental Entomology* **16** 339-44

Johnson NF and Triplehorn CA 2005 Borror and DeLong's Introduction to the Study of Insects (7th edition), Brooks Cole, Belmont.

Kappers IF, Aharoni A, van Herpen TW, Luckerhoff LL, Dicke M and Bouwmeester HJ 2005 Genetic engineering of terpenoid metabolism attracts bodyguards to Arabidopsis, *Science* **309** 2070–2072

Karban R, English-Loeb G, Walker MA and Thaler J 1995 Abundance of phytoseiid mites on *Vitis* speices: effects of leaf hairs, domatia, prey abundance and plant phylogeny. *Experimental and Applied Acarology* 189-197

Kells A, Holland J and Goulson D 2001 The value of uncropped field margins for foraging bumblebees, *Insect Conservation* 283-291

Khan AG, Azim A, Nadeem MA, Khan MA 1997 Effect of growing fattening diets on the growth performance of intensified Afghani lambs, *Small Ruminate Research* **25**(1) 39-42

Khan Z R, Pickett JA, Wadhams L and Muyekho F 2001 Habitat management for the control of cereal stem borers in maize in Kenya, *Insect Science and its Application* 375-380

Khan ZR and Pickett JA 2004 The "push-pull" strategy for stem borer management a case study in exploiting biodiversity and chemical ecology, In Ecological engineering for pest management advances in habitat manipulations for arthropods, (eds) GM Gurr, SD Wratten and MA Altieri, CSIRO and CABI and publishing, UK 155-164

Kos M, van Loon JJA, Dicke M and Vet LEM 2009 Transgenic plants as vital components of integrated pest management, *Trends in Biotechnology* 621-627

Landis DA, Wratten SD and Gurr GM 2000 Habitat management to conserve natural enemies of arthropod pests in agriculture, *Annual Review of Entomology* 175-201

Landis DA, Wratten SD, Gurr GM 2000 Habitat management to conserve natural enemies of arthropods pests in agriculture, *Annual review of entomology, annual reviews.org*

Letourmeau DK and Altieri MA 1983 Abundance patterns of a predator. *Orius tristicolor* (Hemiptera:Anthocoridae) and its prey, *Frankiniella occidentalis* (Thusanoptera :Thripidae):Habitat attraction in polyculture versus monocultures, *Environment Entomology* **12** 1464-1469

Letourneau DK1987 The enemies hypothesis: Tritrophic Interactions and vegetational Diversity in tropical agroecosystems DOI: 102307/1939853

Lewis WJ, Jones R, Nordlund DA and Gross HR Jr 1975 Kairomones and their use for management for entomophagous insects. II Mechanism causing increase in rate of parasitization by Trichogramma spp., *Journal of Chemical Ecology* **1** 349-360

Lewis WJ, Jones RL, Nordlund DA and Sparks AN 1975 Kairomones and their use for management of entomophagous insects. I Evaluation for increasing rates of parasitization by Trichogramma spp. in the field. *Journal of Chemical Ecology* 343-347

Lewis WJ, Nordlund DA, Gueldner RC, Teal PEA and Tumlinson JN 1982 Kairomones and their use for management of entomophagous insects. XII Kairomonal activity for Trichogramma spp of abdominal tips, excretion and a synthetic sex pheromone blend of *H zea* (Boddie) moths, *Journal of Chemical Ecology* 8 1323-1331

Lewis WJRL, Jones DA, Nordlund HR, Gross Jr 1976 Kairomones and their role in pest management. 7-14. *In*: Proceedings of North-eastern Forest Instt. Work Conf. Gen. Tech. Rept. NE-27.USDA

Li LY 1994 Worldwide use of Trichogramma for biological control on different crops: A survey In: Wajnberg E, Hassan S (Eds.). Biological control with Egg Parasitoids.CAB International, Oxon 37-51

Long RF, Corbett A, Lamb C, Reberg-Horton C, Chandler J & Stimmann M 1998 Beneficial insects move from flowering plants to nearby crops, *California Agriculture* 23-26

Loughner R, Goldman K, Loeb G and Nyrop J 2008 Influence of leaf trichomes on predatory mite (*Typhlodromus pyri*) abundance in grape varieties, *Experimental and Applied Acarology* 111-122

Miller JR and Cowles RS 1990 Stimulo-deterrent diversion: a concept and its possible application to onion maggot control, *Chemical Ecology* 3197-3212

Mumm R and Dicke M 2010 Variation in natural plant products and the attraction of bodyguards involved in indirect plant defence, *Canadian Journal of Zoology* 628-667

Nordlund DA and Lewis WJ 1976 Terminology of chemical releasing stimuli in intraspecific and interspecific interactions. *Journal of Chemical Ecology* 211-220

Nordlund DA, Chalfant RB and Lewis WJ 1985 Response of Trichogramma pretiosum females to volatile synomones from tomato plants, *Entomology Science* **20** 372-376

Nordlund DA, Lewis W J and Altieri MA 1988 Influences of plant produced allelochemicals on the host-prey selection behaviour of entomophagous insects, In Novel Aspects of Insect-Plant Interactions (eds) Barbosa P and Letourneau New York 65-90

Nordlund DA, Lewis WJ, Gross JR Jr and Beevers M 1981 Kairomones and their use for management of entomophagous insects. XII. The stimulatory effects of host eggs and the importance of host egg density to the effective use of kairomones for Trichogramma pretiosum Riley, *Chemical Ecology* **7** 909-917

Norris RF 1986 Weeds and integrated pest management systems, *Horticulture Science* 21 402-10

Oerke EC, Dehne HW, Schonbeck F and Weber A 1994 Crop production and Crop Protection Elsevier Science BV, Amsterdam

Orke EC, Dehne HW, Schonbeck F and Weber A 1994 Crop production and Crop protection, Elsevier Science BV, Amsterdam

Patt JM, Hamilton GC, and Lashomb JH 1997 Impact of strip-insectary intercropping with flowers on conservation biological control of the Colorado potato beetle. *Advances in Horticultural Science* **11** 175-181

Pichersky E and Gershenzon J 2002 The information and function of plant volatiles perfumes for pollinator attraction and defence. *Current Opinion in Plant Biology* 237-243

Pichersky E, Noel JP and Dudareva N 2006 Biosynthesis of plant volatiles: nature's diversity and ingenuity, *Science* 808-811

Poehling HM 1989 Selective practices for insecticides and agricultural crops In Effects of pesticides on Non –target Invertebrates (eds) P Jepson, Intercept, Wimborne, UK 151-175

Powell W 1986 Enhancing parasitoid activity in crops. In: Insect parasitoids 13th symposium of the Royal Entomological Society of London, (eds.) J Waage & D Greathead, Academic Press, San Diego CA 319-340

Powell W, Dean GJ and Wilding N 1986 The influence of weeds on aphid-specific natural enemies in winter wheat. *Crop Protestation* **4** 182-89

Pretty J, Vorley W and Keeny D1998 Pesticides in world agriculture: causes, consequences and alternative courses. *In* Bugs in the system: redesigning the pesticide industry for sustainable agriculture Vorley W, Keeney D, editors. London, UK Earthscan 17-49

Price PW, Waldbauer GP 1982 Ecological aspects of pest management. In "Introduction to Insect Pest management" Metcalf RL & W H Luckmann, (eds), Wiley, New York 33-68

Rasmann S, Kollner TG, Degenhardt J ,Hiltpold I,Toepfer S ,Kuhlmann U, Gershenzon J and Turlings TCJ 2005 Recruitment of entomopathogenic nematodes by insect-damaged maize roots, *Nature* 732-737

Rebek EJ, Sadof CS &Hanks L M 2005 Manipulating the abundance of natural enemies in ornamental landscapes with floral resource plants, *Biological Control* 203-216

Risch SJ 1981 Insect herbivore abundance in tropical monocultures and polycultures. An experimental test of two hypotheses, *Ecology* **62** 1325-1340

Risch SJ, Andow D and Altieri MA 1983 Agroecosystem diversity and pest control: data, tentative conclusions, and new research directions, *Environmental Entomology 12* 625-629

Sabelis MW , Janssen A ,Pallini A, Venzon M Bruin j Drukker B & Scutareanu P1999 Behavioural responses of predatory and herbivorous arthropods to induced plant volatiles:from evolutionary ecology to agricultural applications,in:Induced Bent, APS PressSt Paul,MN269-296

Sanchez JA, Gillespir DR and Mc Gregor RR 2003 The effects of mullein plants (*Verbascum thapsus*) on the population dynamics of Dicyphus Hesperus (Heteroptera: Miridae) in tomato greenhouses, *Biological Control* 313-319

Schnee C, Köllner TG, Held M, Turlings TCJ, Gershenzon J, Degenhardt J 2006 The products of a single maize sesquiterpene synthase form a volatile defense signal that attracts natural enemies of maize herbivores, *Proceedings National Academy of Science* USA **103** 1129–1134

Schoeneberger M, Bentrup G, Gooijer H de 2012 Branching out agroforestry as a climate mitigation and adaptation tool for agriculture, *Soil,* Jswconline.org

Shiojiri K, Kishimoto K, Ozawa R, Kugimiya S,Urashimo S,Arimura G, Horiuchi J, Nishioka T, Matsui K and Takabayashi J 2006 Changing green leaf volatile biosysthesis in plants, An approach for improving plant resistance against both herbivores and pathogens, In Proceedings of the National Academy of Sciences of the USA 16672-16676

Singh 2004/2 Some success stories in classical Biological control of Agricultural Pests in India. Asia –Pacific Association of Agricultural Research Institution (APAARI), FAO Regional office for Asia and the Pacific Bangkok, Thailand

Singh DK 2004 Insecticidal methods of pest control. Applied Entomology, Department of Zoology, University of Delhi, Delhi-110007

Sotherton NW1984 The distribution and abundance of predatory arthropods overwintering in farmland, Annals of Applied Biology 49-54

Srivastava M, Paul AVN, Singh AK and Dureja 2008. Response of egg parasitoid *Trichogramma chilonis* Ishii (Hymenoptera:Trichogrammatidae) Kairomones from three host insects, *Journal of Biological Control* **22** (2) 333-340

Theunissen J 1994 Intercropping in field vegetable crops Pest management by agroecosystem diversification-an overview, *Pesticide Science* 65-68

Tuell JK, Fiedler AK, Landis DA and Issacs R 2008 Visitation by wild and managed bees (Hymenoptera;Apoidea) to Eastern U.S. native plants for use in conservation programs, *Environmental Entomology* 707-718

Turling TCJ and Ton J 2006 Exploiting scents of distress: the prospect of manipulating herbivore-induced plant odours to enhance the control of agricultural pests, *Current Opinion in Plant Biology* 421-427

Turner MG, Gardner RH and O'Neill RV 2001 Landscape ecology in theory and practice Pattern and process, Springer-Verlag, New York

Van Driesche RG , Lyon S, Sanderson J P, Bennett KC, Stanek EJIII and Zhang R 2008 Greenhouse trials of *Aphidius colemani* (Hymenoptera; braconida) banker plants for control of aphids (Hemiptera:Aphididae) in greenhouse spring floral crops, *Florida Entomologist* 583-591

Van Emden HF 1988 The potential for managing indigenous natural enemies of aphids on field crops. Philisophical Transactions of the Royal Society of London (B) **318** 183-201

Van Emden HF 1965 The role of uncultivated land in the biology of crop pests and beneficial insects, *Science Horticulture* **17** 121-36

Van Poecke RMP and Dicke M 2002 Induced parasitoid attraction by *Arabidopsis thaliana*: involvement of the octadecanoidd and the salicylic acid pathway, *Experimental Botany* 1793-1799

Van Poecke RMP and Dicke M 2003 Signal transduction downstream of salicylic and jasmonic acid in herbivory-induced parasitoid attraction by Arabidopsis is independent of JARI and NPRI, *Plant Cell and Environment* 1541-1548

Vanderneer JH 1989 The *ecology of intercropping*. Cambridge University Press, ISBN 9780521346894 Cambridge.

Verkerk RHJ 2004 Manipulation of tritrophic interactions for IPM, *In*: Integrated Pest Management Potential, Constraints and Challenges (eds.) O Koul, GS Dhaliwal and GW Cuperus, CABI, Oxfordshire UK 55-71

Vet LEM, Lewis WJ and Carde RT1995 Parasitoid foraging and learning, In Chemical ecology of insects (eds) RT Carde and WJ Bell 65-101, Chapman and Hall, ISBN 9780412039515, London.

Vet LEM, Wackers FL and Dicke M 1992 Ecology of infochemical use by natural enemies in a tritrophic context, *Annual Review of Entomology* 141-172

Vinson SB 1976. Host selection by insect parasitoids, *Annual Review of Entomology* **121** 109-133

Wackers FL, Rijn PCJ and Bruin J 2005 Plant-provided food for carnivorous insects: a protective mutualism and its applications, Cambrdge University Press

Walton N &Isaacs R 2011 Evaluation of flowering plant strips for support of beneficial insects in blueberry, *Environmental Entomology* 697-705

Wang Z, Lai lle K, Zhang F, Lu X, Babendreier D 2014 Mass rearing and release of Trichogramma for biological control of insect pests of corn in China, *Biological Control* **68** 136-144

Wanner H, Gu H, Gunther D, Hein S and Dorn S 2006 Tracing spatial distribution of parasitism in fields with flowering plant strips using stable isotope marking, *Biological Control* 240-247

White AJ, Wratten SD, Berry NA and Weigmann U 1995 William Olkowski, Alternative to toxic pesticides: Learning from China. Technical Director Bio Integral Resource Centre (BIRC). P.O. Box 7414, Berkeley, CA-94706, 415/524-2567.

Zurbrugg C and Frank T 2006 Factors influencing bug diversity (Insecta: Heteroptera) in semi-natural habitats, *Biodiversity and Conservation* 275-294

CHAPTER 33

Clonal Propagation of *Eucalyptus* Using Mini Cuttings

K Srinivas* and DSN Raju

Srikiran Biotech,Gopalpuram, Khammam. -507163, Telangana, India

**Corresponding author's Email: srinivas_koka@hotmail.com*

ABSTRACT

The study was aimed to evaluate the technical efficiency of mini-cuttings technique for vegetative propagation of *Eucalyptus*. The mini-cuttings were prepared with about 6 to 8 cm in length and two leaves reduced by about 50% in the upper third, being remained an area of, approximately 78 cm^2 (10 cm diameter). The mini-cuttings were placed in tubes of 93 cm^3, with substrate formed with fine vermiculite and carbonized rice hulls (1:1 v/v) and rooted in acclimatized greenhouse. After 30 days we evaluated the percentage of rooted mini-cuttings, radical vigor (number and length of roots / mini-cutting), callus formation, emission of new shoots and maintenance of the original leaves. The mini-stumps showed 100% survival after five collects and an average production of 450-500 mini-cuttings m^{-2} $month^{-2}$ and rooting ranged from 70 to 90%. Mini-cuttings technique is efficient to propagate adult propagules of the species and there are no anatomical barriers for rooting. It is now possible to determine the optimum time for permanence of vegetative propagules (mini-cuttings) inside a greenhouse for rooting, and this value can be used to optimize the structure of the nursery.

Keywords: Mini cuttings, Mini-stumps, Rejuvenation

INTRODUCTION

Vegetative propagation or cloning consist of multiplying asexually plant parts (cells, tissues, organs) to generate new plants genetically identical to the mother tree. In several hardwood species, most notably in Eucalypts, the popular stem-cutting method poses limitations in rooting behaviour, such as rapid loss of rooting competence, poor rooting quality and thereby hindering field deployment. In this context, mini-cuttings technique can be classified as an evolution of the cutting process, since it tends to reduce the area required for mother plants conduction by adopting clonal mini-garden, reducing also the rooting and acclimatization period and, above all, making unnecessary in some cases the application of plant growth regulators to induce rooting (Ferreira et al 2004, Wendling and Xavier 2005, Wendling et al 2010). Mini-cutting is a propagation technique that has largely been used in recent years for the commercial cloning of *Eucalyptus* in Brazil. The mini-cutting technique can be divided into the following stages: (i) production of mini-cuttings in clonal mini-garden, (ii) induction of adventitious rooting in the greenhouse under intermittent mist and with a controlled temperature, (iii) acclimation in a shade house and (iv) growth and hardening in outdoor conditions (Brondani et al 2010a).Rooting processes for the production of vegetative propagules of *Eucalyptus* such in mini-cutting techniques have been used for the propagation of selected clones (superior trees), which allows considerable gains for commercial production due to higher rooting rates and a reduced time for mini-cutting formation (Stape et al 2001, Titon et al 2002). This mass propagation technique originated as a response to the limitations of the stem-cutting technique used to obtain rejuvenated material from open field operational limitations as well as the cost. The present study was conducted to evaluate the efficiency of the mini cutting in propagation of eucalyptus.

METHODOLOGY

The sprouts used to obtain the mini-cuttings originated from clones of *Eucalyptus camaldulensis* mini-stumps (Fig. 1), grown in a semi-hydroponic system using elevated cement troughs, containing sand and good drainage system. The mini-stumps were produced by conventional cuttings from trees selected on the field at one year of age (Brondani et al 2010b), which were planted in troughs with a spacing of 10 cm x 10 cm. At 45 days after planting, pruning of the mini-stumps was conducted in the apical sprouts at 7 cm above the root collar, keeping a pair of leaves per mini-stump to induce new axial sprouts (Fig. 2). The first collection of sprouts was conducted 30 days after the mini-stumps were pruned, forming a clonal mini-garden. The shoots of *Eucalyptus* clones were collected. The sprouts were stored in Ice boxes containing water to prevent the loss of cellular turgor in the tissues and reduce oxidation. After the sprouts were collected, shoot tips of the mini-cuttings were retained, and 3 to 4 pair of leaves was maintained with a reduction of 50 per cent of the total leaf area. Each mini-cutting was 8 cm (± 1 cm) in total length. These mini stumps are subject to reiterated successive cut-back of the main axis to shorten them and to maintain them near the root system. The basal region of each mini-cutting was immersed in a solution containing Fungicide Bavistin 2% (Carbendezim a.i 20%) for 10 seconds. The mini-cuttings were planted in conical plastic tube containers (93 cm^3) with 2 cm of the basal region inserted in the substrate. The mixture substrate was prepared with decomposed coir pith and vermiculite (1:1, $v\ v^{-1}$). The main feature of the technique is the use of juvenile plants or plants rejuvenated *in vitro*, as source of vegetative propagules. Shoot apices are used as mini-cuttings, which are placed to root in a greenhouse equipped with temperature and humidity control. The actual size of micro-cuttings is about 7 to 8 cm with two to three leaf-pairs. Presence of the shoot apex is important for quality of the root system, because its presence induces taproot-like system. The mini-stumps left after mini-cutting harvest, sprout rapidly producing new mini-propagules, which can be harvested for use within a period of 10 days.

Figure 1: *Mini-garden under protective cover*

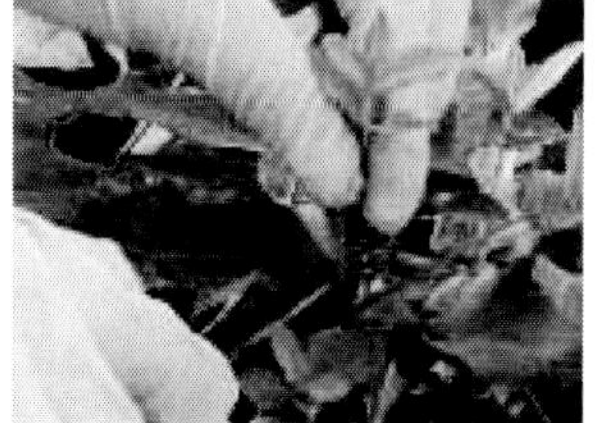

Figure 2: *Collection of apical shoots*

In the mini-garden system nutrients are supplied in ideal concentrations to promote rooting. The mini-stumps receive nutrient solution daily, prepared from a commercial fertilizer mixture. The mother plants should be well nourished with macronutrients like phosphorus, potassium, calcium and magnesium and should be moderately deficient in nitrogen. The effects of micro-nutrients on rooting are variable. Manganese, due to its inhibitory effect on rooting (possibly related to the promotion of auxin oxidative enzymes) should be used at minimal concentrations, but boron and zinc are essential for rooting (Anderson 1986).The mini-cuttings cultivated in well-balanced, nutritive solutions root better, as a result of improved nutritional status. The mini-cuttings were kept in a greenhouse with an intermittent mist system 10 - 12 daily sprays for a duration of 1 min with an interval of 15-30 min. from 10 am to 4 pm. The relative humidity at above 80 per cent and temperature maintained at 35 °C which are automatically controlled by a humidistat and timer, respectively. At the end of rooting stage, plantlets are transferred from mist chamber to 50 per cent shade for 4 days and changing of watering duration, from 1 min all hours to 15 seconds all 2 hr. After weaning, cases containing young rooted cuttings are placed in acclimation conditions during approximately 45 days, in the total absence of shade and a watering of 15mn once to twice per day according to the sunning. During this phase, the fertilization is reduced to the utilization of granular NPK 13-13-21 once by week.

RESULTS AND DISCUSSION

Rooting of the Mini-Cuttings

The quality of the resulting plants is improved: rooted cuttings with terminal bud produced from intensively managed container-grown stock plants give rise to straight, single-stemmed and true-to-type clonal offspring. This contrasts with field system derived nodal cuttings which, once rooted, needed to be first sorted out, then hand pruned to keep only the most promising axillary shoot. Morphological comparison between different material (stool plant cuttings and rooted cuttings) according to their origin field-grown stock plants or intensively managed container-grown:

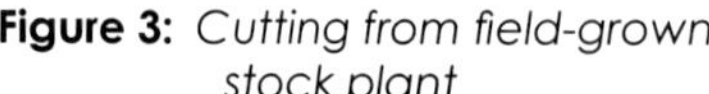

Figure 3: *Cutting from field-grown stock plant*

Figure 4: *Cutting from intensively managed sand bed grow*

Traditional cuttings issued from field-grow stool plant (Fig. 3) are more lignified stem portion of 5 to 8 cm length, taken in the median part of the stem, with 2 halved and opposed leaves, and two buds in dormant condition. Cuttings issued from intensively managed container-grown (Fig. 4) are entire short orthotropic shoots newly formed with 5 to 7 cm of long, with active shoot apex and with an average number of three entire leaf pairs. These entire shoots are harvested to its basis on the remaining fragment of the main axis of the stool plant after cut-back.

Figure 5: *Rooted Cutting from field-grown* **Figure 6:** *Rooted cutting from sand bed*

The field-grown stock plants rooted cuttings (Fig. 4 a, b) comes from improvement of one of two lateral buds of lignified stem portion. Root's system developed by this cuttings is less dense (Fig. 5) with 1 to 3 main roots and very few secondary roots.

Propagules Productivity

Certain practices for managing mini-stumps in sand beds produce vegetative propagules similar to the macro-cutting system. The sprouts are allowed to grow to bigger sizes, and after collection each sprout is divided into three mini-cuttings (basal, middle and apical). This system produces higher number of propagules m^{-2} but it loses the best characteristic of the mini-cuttings provided by the presence of the shoot apex, which is the production of a high quality root system (a taproot-like system, as discussed earlier).

One question that arises is the influence of plant spacing in mini-hedges on the rooting of mini-cuttings and mini-stump productivity. Trials at our nursery showed that 5 × 5 cm spacing between plants does not interfere significantly in the rooting capacity of the mini-cuttings, but wider spacing allows for higher survival rates of mini stumps hence a spacing of 10 × 10 cm was followed. Compared to the field orchards, the capacity of producing vegetative propagules in the sand bed is highly expressive, consequently, more cuttings can be produced $year^{-1}$ $unit^{-1}$ constructed area of mini hedge, besides they are ease of handle, manage and they have a lower cost per plant.

Table 1: *Comparison of the cuttings production between intensively managed sand bed-grown and the field-grown stock plants in 100 m^2 area*

Stages	***Number of cutting obtained***	
	Using field-grown stock plants spaced at 1mt x1mt. 100 plants population @30 cuttings / month	***Using intensively managed sand bed -grown at 10cm X10cm 10,000 plants population @ 5 cuttings/ month***
After 6 months	9,000	1,50,000
After 9 months	18,000	3,00,000
After 1 year	27,000	4,50,000
After 2 year	54,000	9,00,000

Replacing the traditional field-grown stock plants by intensively managed container-grow ones resulted in easier and more economical maintenance, while increasing the number and improving the quality of the cuttings, having productivity 15 folds higher than the traditional system (Table 1).

Use of the Mini-Cutting System for Cloning Other Woody Species

The applicability of the mini-cutting technique has also been tested on other broadleaf and coniferous woody species. In general, the methods used are similar to those used with *Eucalyptus*, with minor adaptations. The results have shown that this technique can be fully applied to species like *Casuarina, Melia dubia, Acacia mangium and Dalbergia sissoo* and many other woody and non-woody species. The best results were obtained with juvenile plants. Thus, the mini-cutting method is feasible for use in family multiplication in these species. Use of mini-cuttings can be a possible alternative for production of family forests by multiplication of full-sib or tested half sib families. The advantages of using intensively managed container-grown stock plants are becoming more and more obvious for many eucalyptus species or inter-specific hybrids (Wendling and Xavier 2003, Titon 2006). It can reasonably be assumed that the technique detailed in this paper for successfully mass propagating by rooted cuttings selected genotypes in our conditions can also benefit other eucalypt species in different environments. It is indeed much cheaper and easier to monitor and manage than the open field CMA which is currently more and more adopted for mass producing eucalypts and other forest tree species by rooted cuttings in many parts of the world (Jimenez 2005). The financial aspects, the overall efficiency, including the degree of sophistication and relevant skillfulness requirements, as well as the final destination of the planting material constitute indeed determining issues in choosing between nursery and *in vitro* clonal propagation system, and being aware that only tissue-cultured clones comply with international phytosanitary dispatching restrictions.

All these crucial issues applied to the selected clones mass produced for local utilization warrant the new nursery technique developed from intensively managed and container-grown stock plants. The rationale for promoting mass clonal propagation by rooted cuttings of selected tree genotypes by intensive stock plant management has been argued and demonstrated for a long time on different arborescent species (Franclet 1977). The advantages of preferring container-grown stock plants to field-established ones for achieving this goal can be summed up as follows:

- Better control and therefore optimization of stock plant environment with regard to light, watering, soil fertility, wind, etc.
- Proximity to the misting area resulting in shorter time and distance for collecting the cuttings that can be set sooner with less stress under proper rooting conditions;
- Easier maintenance – weeding, feeding, pest and disease treatments– and control;
- More adapted to intensive management– frequent pruning and pinching operations;
- Higher density of stock plants per area;
- Greater overall efficiency and cost effectiveness.

CONCLUSION

There are many possibilities for improvements in the current clonal propagation methods used for hardwoods. An understanding of adventitious rooting will become increasingly detailed through experimental approaches that range from simply revisiting previous practices, with the various phases of the rooting process and their specific features in mind, that lead to the expression of a new root system.Intensive management technique in sand-grown for propagating *Eucalyptus* is an easy technique to popularize. Its exponential increase of cuttings production increases its interest, in particular for the developing countries which often do not have means to acquire expensive technologies. This study emphasizes the importance of suitable nursery techniques for improving the ability for true-to-type cloning by rooted cuttings of *Eucalyptus* clones. The physiological changes induced by such practices, which are responsible for this greater responsiveness of the plant material selected to be mass clonally propagated are under investigations at a more basic level. However, the practical benefits are from now on obvious, first for clonal selection, then for the large-scale utilization of the clones selected.

REFERENCES

Anderson AS 1986 New root formation in plants and cuttings, M.B. Jackson (Ed.),. Martinus Nijhof, Dordrecht, 223

Brondani GE, Grossi F, Wendling I, Dutra LF, Araujo MA 2010a IBA application for rooting of *Eucalyptus benthamii* Maiden & Cambage x *Eucalyptus dunnii* Maiden minicuttings, *Acta Scientiarum Agronomy* **32** (4) 667-674

Brondani GE, Wendling I, Grossi F, Dutra LF, Araujo MA, 2010b *Eucalyptus benthamii* x *Eucalyptus dunnii* minicutting technique: minicutting survival and rooting in relation to collection and seasons, *Ciência Florestal*, **20** (3) 453-465

Franclet A1977 Manipulation des pieds-mères et amélioration de la qualité des boutures. Afocel, Études et Recherches **12** 20

Ferreira EM, Alfenas AC, Mafia RG. Letite HG, Sartorio RC, Penchel Filho RM 2004 Determination of the optimum time for rooting of mini-cuttings of *Eucalyptus* spp. Clones, *Revista Árvore*, **28** (2)183-187

Jimenez CM 2005 Minijardines clonales hidroponicas: la experiencia con Eucalyptus de Smurfit Carton de Colombia. Oral presentation given during the workshop on "Avances en Silvicultural Clonal". GenFores, Costa Rica, 26-29 July 2005.

Stape JL, Goncalves JL M, Goncalves AN 2001 Relationships between nursery practices and field performance for *Eucalyptus* plantations in Brazil, *New Forests*, **22 (1-2)**19-41

Titon M, Xavier A and Otoni WC 2002 Rooting dynamics of microcuttings and minicuttings of *Eucalyptus grandis* clones, *Revista Árvore*, **26** (6) 665-673

Titon NM, Xavier A and Otoni WC 2006 Clonal propagation of *Eucalyptus grandis* using the mini-cutting and micro-cutting techniques, *Scientia Forestalis* **71** 109-117

Wendling I and Xavier A 2003 Miniestaquia seriada no rejuvenescimento de clones de *Eucalyptus*, *Pesq. Agropec.Bras, Brasileira*, **38** (4) 475-480

Wendling I and Xavier A 2005 Indolbutiric acid and serial minicutting technique on rooting and vigor of *Eucalyptus grandis* clone minicuttings, *Revista Árvore*, **29** (6) 921-930

Wendling I, Brondani GE, Dutra LF and Hansel F 2010 A Mini-cuttings technique: a new *ex vitro* method for clonal propagation of sweetgum, *New Forests*, **39** (3) 343-353

CHAPTER 34

Genetic Diversity Studies on *Calophyllum Inophyllum* (Undi) Progenies Through Inter Simple Sequence Repeat Markers

B Palanikumaran*, KT Parthiban, PR Renganayaki and C Cinthia Fernandaz

Forest College and Research Institute, Mettupalayam - 641301, Tamil Nadu, India

**Corresponding author's Email: kumaranbass@gmail.com*

ABSTRACT

The present study was carried out at Forest College and Research Institute, Mettupalayam with the objective of studying the genetic divergence among 30 undi progenies using 10 primers of inter simple sequence repeat (ISSR) markers. Among them, 2 ISSR markers were found to be polymorphic and the number of bands ranged from eight to eleven. The highest polymorphic information content (PIC) value 0.58 was observed in primer UBC823 and average PIC value was 0.45. Cluster analysis using NTSYS generated dendrogram divided all the 30 progenies into ten distinct groups. The progenies A4 and A7, A5 and A8 were closest at the similarity coefficient of 1.000.

Keywords: Diversity, Inter simple sequence repeats, Markers, Primers

INTRODUCTION

Selection is the most important activity in all breeding programmes (Zobel and Talbert 1984). Tree yield, i.e., tree volume is a complex and highly variable character which is influenced by many component characters (Deepak Pandey and Hooda 1997). In the integrated structure of the plant, the overall correlation observed between two variables is a function of a series of direct and indirect relationships between different variables which could be affected through path coefficient studies (Wright, 1921). However, research pertaining to correlation and path analysis in Undi is very scant variable (Divakara et al 2010, Pavithra et al 2013) and thus needs intensive research. The nature and degree of divergence in the base population is a pre-requisite for any improvement and conservation strategies (Gradual et al 1999). The genetic diversity has been assessed traditionally either by provenance testing using Mahalanob is D^2 statistics and Canonical vector analysis or electrophoretic analysis of enzymes. Study of metric characters in field trials was earlier the dominating technique and it is still today the most robust and valid way of assessing genetic variation. Information from field experiments is very valuable, because assessed adaptive genetic variation is still the best for breeding and conservation activities (Eriksson 1995). Recently, DNA based molecular markers have increasingly been applied in tree improvement programmes for quantification of genetic diversity. These markers have little been affected by environment and developmental stages (Haines and Martin 1995). Markers also have important immediate application in supportive research for tropical hardwoods and non-industrial species mainly for quantification of genetic diversity. There are several methods based on DNA markers viz., RFLP (Restriction Fragment Length Polymorphism), RAPD (Randomly Amplified Polymorphic DNA), SSR (Simple Sequence Repeats), ISSR (Inter Simple Sequence Repeats), AFLP (Amplified Fragment Length Polymorphism) etc. Among these techniques, ISSR marker is very significant because this marker in its simplest form, no prior genomic knowledge is required to design arbitrary sequence oligonucleotide primers (Gonzalez et al 2005). But the use of molecular markers for genetic diversity studies in Undi dismally modest thus underscores the research needs.

METHODOLOGY

The study was carried out at Forest College and Research Institute, Mettupalayam with the objective of studying the genetic divergence among 30 Undi progenies using 10 primers of Inter Simple Sequence Repeats (ISSR) markers. It is necessary to check the quality and concentration of DNA for carrying out PCR reaction. Genomic DNA was quantified using the spectrophotometer in the concentration of nucleic acids are usually determined by measuring the sample OD (optical density) at 260 nm against blank. The ration A260/A280 is used to estimate the purity of nucleic acid. Pure DNA should have a ratio approximately 1.8. The quality of DNA was also checked using 0.8 per cent agarose gel electrophoresis. The DNA was then diluted to appropriate concentration for use in molecular analysis (Table 1). Extracted and quantified genomic DNA has been subjected to Polymerase chain reaction (PCR) using a thermal cycler (Eppendorf master cycler personal, Germany) for performing ISSR profiling (Williams et al 1990) to identify polymorphism from the collected germplasm. This includes random decamers as its primers and the list of primers used for the study has been provided in the Table 2. This will help in identification of polymorphic markers towards identification and characterization of the germplasm and the varieties within the species. DNA from the collected 30 genotypes from each species was amplified using a set of 15 arbitrary oligonucleotide decamer primers. Amplification reactions were in volumes of 25.0 μL each containing 13.0 μL of sterilized double distilled water, 2.5 μL of 10 X buffer A, 1.0 μL of dNTP (2.5 mM each), 1.0 μL of primer (1 μM) and 0.3 μL (~1.0U) of Taq polymerase in 4.0 μL of the template DNA. Amplification were performed using 0.2 ml PCR tubes (Axygen Inc., USA) in an eppendorf master cycler personal programmed for 40 cycles consisting of 30 seconds in 94°C ; one minute in 36° C and 2.5 minutes in 72 °C denaturing, annealing and extension respectively. These cycles were preceded by a step for an initial denaturation of five minutes at 94 °C and the cycles followed by a final extension of ten minutes at 72 °C. Then it is programmed to maintain at 4 °C till we use it for electrophoresis. The steps involved in PCR amplification of ISSR reaction are presented (Table 3).

Table 1: *Reaction mixtures for ISSR amplification*

S.No.	Reaction mix	Quantity
1.	Template DNA	4.0 μl
2.	dNTP's	1.0 μl
3.	Assay buffer 10x	2.0 μl
4.	$MgCl_2$	4.0 μl
5.	Primer	1.0 μl
6.	Taq DNA polymerase	0.3 μl
7.	Sterile milli – Q water	13.0 μl

Table 2: *List of primers used for ISSR amplification*

S.No	Primers	Primer sequence 5'-3'
1.	UBC808	AGAGAGAGAGAGAGAGC
2.	UBC809	AGAGAGAGAGAGAGAGG
3.	UBC810	GAGAGAGAGAGAGAGAT
4.	UBC811	GAGAGAGAGAGAGAGAC
5.	UBC823	TCTCTCTCTCTCTCTCC
6.	UBC834	AGAGAGAGAGAGAGAGYT
7.	UBC835	AGAGAGAGAGAGAGAGYC
8.	UBC844	CTCTCTCTCTCTCTCTRC
8.	UBC848	CACACACACACACACARG
9.	UBC881	GGGTGGGGTGGGGTG
10.	UBC890	VHVGTGTGTGTGTGTGT

Amplified products along with the 6X tracking dye were resolved electrophoretically based on their molecular weight by running the products on a 1.2 per cent agarose gel matrix using a submarine electrophoresis with 1 x TAE buffer (Sambrook and Russell 2001) at 5 V cm^{-1}. After the run, the gel was stained using ethidium bromide (Sharp et al 1973) solution for five to ten minutes and was destained using dH_2O for two to three minutes. Ethidium bromide fluoresces orange upon UV illumination (310-320 nm). The gel image has been documented using a Herolab documentation unit (Herolab 442 K, E.A.S.Y., Germany). Documented gel was used to score polymorphism by the presence or absence of the bands. Also, monomorphic bands were scored to calculate the percentage of polymorphism. Clear and unambiguous bands of ISSR markers were scored based on the presence or absence of the corresponding band among the genotypes. The scores '1' and '0' indicates the presence and absence of bands respectively. In case of binary coding, a data matrix computing of '1' and '0' was formed and was subjected to cluster analysis. Sequential Agglomerative Hierarchial Non-overlapping (SAHN) clustering was performed on similarity matrix using dice coefficient for the binary data (obtained through scoring) utilizing the Unweighted Pair Group Method with Arithmetic Averages (UPGMA) method (Table 4). This was done by using NTSYS pc version 2.02 (Rohlf

2002). The confidence limits for the groupings by dendrograms were computed by using WINBOOT, a program for performing bootstrap analysis of binary data to determine the confidence limits of UPGMA based dendrograms (Yap and Nelson 1996). Bootstraping involves repeated sampling with replacement of the characters in matrix of operational taxonomic unit (OUT) x characters to create numerous bootstrap matrices of the same size as the original matrix.

Table 3: *Random ISSR primers used for DNA amplification in Calophyllum inophyllum*

Sl.No	*Primers*	*Sequence 5'-3'*	*Total number of bands*	*PIC values*
1	UBC823	TCTCTCTCTCTCTCTCC	11	0.58
2	UBC834	AGAGAGAGAGAGAGAGYT	08	0.33
	Mean			0.45

Computation of Similarity Indices

From the binary code matrices obtained from the ISSR markers, the Jaccard's similarity indices were computed for the 30 genotypes of Undi using the SIMQUAL programme of NTSYS pc version 2.02 as follows, Jaccard (1908) similarity index

$$SI\ ij = a/(n-d)$$

Where, SI ij = Similarity between i^{th} and j^{th} samples

a = Number of electromorphs/ traits commonly shared

d = Number of electromorphs/ traits absent in i^{th} and j^{th} samples but present in other samples

n = (b + c) i.e. the number of unmatches

b = Number of electromorphs / traits present in i^{th} samples but absent in j^{th} samples

C = Number of electromorphs / traits present in j^{th} samples but absent in i^{th} samples.

Polymorphism Information Content

Polymorphism Information Content (PIC) or expected heterozygosity scores for each ISSR marker was calculated based on the formula given by Anderson et al (1993)

$$PIC = 1 - \sum_{j=1}^{n} Pij2$$

Where, Pij is the frequency of jth allele for the marker i and the summation extends over n alleles.

RESULTS AND DISCUSSION

Assessment of genetic diversity is a prerequisite for efficient conservation and utilization of genetic resources. During the past two decades, several high-throughput PCR-based technologies such as Inter-Simple Sequence Repeats (ISSR) and Amplified Fragment

Length Polymorphisms (AFLP) have been developed to assay genetic polymorphism at the DNA level (Shyamsundar et al 2014). ISSR has been used for diversity assessment by many researchers in tropical and temperate trees. Extensive studies have already been reported in many tropical trees *viz.*, *Pongamia pinnata* (Shyamsundar et al 2014); *Casuarinas* (Yasodha and Kathirvel 2004) ISSR techniques was used to study the genetic structure and relationship between species and populations. Similarly, DNA based ISSR techniques have been successfully utilized in many genera and species of forest trees (Mohapatra and Singhal 2000) for diversity study. In the present study, 19 reproducible bands were obtained from 2 primers (Fig. 1). The primers produced 100 per cent polymorphism (UBC823 and UBC834). The highest PIC value was observed with the primer UBC823 and the lowest PIC value was observed with primer UBC834. The present investigation revealed that Undi accessions used for diversity assessment through ISSR, showed moderate level of genetic variations. Polymorphism in several tree species were documented by several authors *viz.*, *Jatropha* spp. (Kumar et al 2013); and *Tamarindus indica* (Nandini et al 2011).

In Undi, the average linkage between the genotypes was used for constructing a phylogenetic tree. The association amongst different genotypes was presented in the form of dendrogram (Fig. 2). The progenies A4 and A7, A5 and A8, were closest at the similarity coefficient 1.000. The progenies A1 and A2, A2 and A19, A2 and A20, A2 and A25, A5 and A15, A8 and A15, A12 and A28 were the farthest or most diverse and dissimilar to all others. Similar grouping pattern within species and related genotypes are evidenced from the earlier reports in *Eucalyptus tereticornis* (Tewari et al 2013).The results revealed that ISSR is an efficient technique to characterize the Undi genotypes and classify different genotypes based on the ISSR markers generated. It was also indicated that ISSR analysis has determined the genetic relationships and estimated the genetic diversity among the genotypes of Undi. The results of the present study can be used as a stepping stone for evolving well defined approach based on evaluation and characterization of genetic variation in Undi genotypes which can be further used for the improvement of these two species for various traits through different breeding methods.

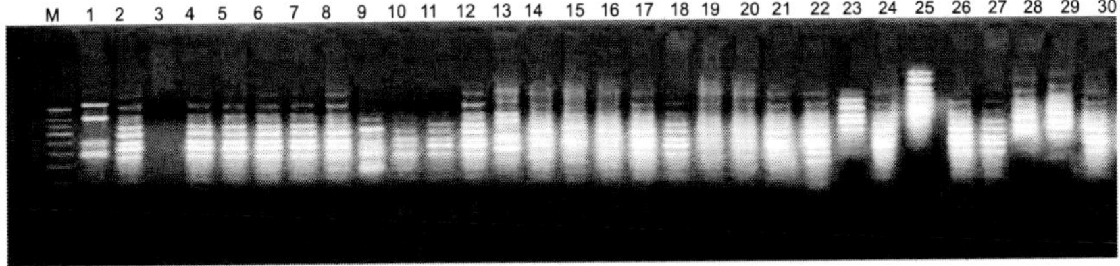

UBC834 (AGAGAGAGAGAGAGAGYT)

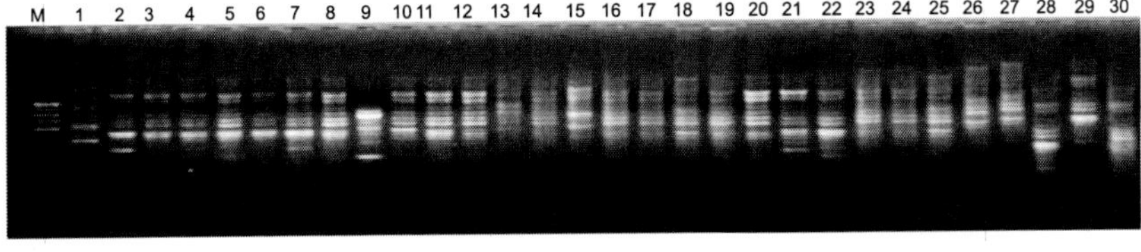

Figure 1: *ISSR profile of Calophyllum inophyllum*

Table 4: *ISSR similarity matrix of Calophyllum inophyllum progenies*

Seed Source	1	2	3	4	5	6	7	8	9	10	11	12	13	14	15	16	17	18	19	20	21	22	23	24	25	26	27	28	29	30
1	1.0000																													
2	0.3157	1.0000																												
3	0.5789	0.5263	1.0000																											
4	0.4210	0.8947	0.6315	1.0000																										
5	0.3684	0.8421	0.5789	0.9473	1.0000																									
6	0.4736	0.8421	0.4736	0.8421	0.7894	1.0000																								
7	0.4210	0.8947	0.6315	1.0000	0.9473	0.8421	1.0000																							
8	0.3684	0.8421	0.5789	0.9473	1.0000	0.7894	0.9473	1.0000																						
9	0.4736	0.6315	0.5789	0.5263	0.4736	0.5789	0.5263	0.4736	1.0000																					
10	0.5789	0.6315	0.7894	0.7368	0.6842	0.5789	0.7368	0.6842	0.7894	1.0000																				
11	0.4736	0.6315	0.6842	0.7368	0.7894	0.5789	0.7368	0.7894	0.6842	0.8947	1.0000																			
12	0.4210	0.6842	0.4210	0.7894	0.8421	0.6315	0.7894	0.8421	0.4210	0.6315	0.7368	1.0000																		
13	0.6315	0.4736	0.6315	0.4736	0.4210	0.5263	0.4736	0.4210	0.7368	0.7368	0.6315	0.5789	1.0000																	
14	0.6315	0.5789	0.7368	0.6842	0.6315	0.6315	0.6842	0.6315	0.6315	0.7368	0.6315	0.5789	0.7894	1.0000																
15	0.5263	0.3684	0.7368	0.3684	0.3157	0.4210	0.3684	0.3157	0.7368	0.6315	0.5263	0.3684	0.7894	0.6842	1.0000															
16	0.6842	0.5263	0.7894	0.6315	0.5789	0.5789	0.6315	0.5789	0.6842	0.7894	0.6842	0.5263	0.8421	0.9473	0.7368	1.0000														
17	0.5263	0.6842	0.6315	0.7894	0.7368	0.7368	0.7894	0.7368	0.5263	0.6315	0.6315	0.6842	0.6842	0.8947	0.5789	0.8421	1.0000													
18	0.6315	0.5789	0.6315	0.6842	0.6315	0.6315	0.6842	0.6315	0.5263	0.6315	0.6315	0.5789	0.6842	0.8947	0.5789	0.8421	0.8947	1.0000												
19	0.6842	0.3157	0.7894	0.4210	0.4736	0.3684	0.4210	0.4736	0.4736	0.5789	0.5789	0.4210	0.6315	0.7368	0.7368	0.7894	0.6315	0.7368	1.0000											
20	0.5789	0.3157	0.6842	0.4210	0.4736	0.3684	0.4210	0.4736	0.3684	0.5789	0.5789	0.5263	0.5263	0.6315	0.6315	0.5789	0.5263	0.6315	0.7894	1.0000										
21	0.6315	0.4736	0.7368	0.5789	0.5263	0.5263	0.5789	0.5263	0.3157	0.5263	0.5263	0.4736	0.4736	0.6842	0.5789	0.6315	0.6842	0.7894	0.7368	0.8421	1.0000									
22	0.6842	0.5263	0.6842	0.6315	0.5789	0.5789	0.6315	0.5789	0.4736	0.5789	0.5789	0.5263	0.5263	0.7368	0.6315	0.6842	0.7368	0.8421	0.7894	0.6842	0.8421	1.0000								
23	0.4736	0.5263	0.6842	0.6315	0.6842	0.5789	0.6315	0.6842	0.3684	0.4736	0.4736	0.6315	0.5263	0.7368	0.6315	0.6842	0.7368	0.6315	0.7894	0.6842	0.6315	0.6842	1.0000							
24	0.6315	0.4736	0.8421	0.5789	0.5263	0.5263	0.5789	0.5263	0.5263	0.6315	0.6315	0.4736	0.6842	0.7894	0.7894	0.8421	0.7894	0.7894	0.8421	0.6315	0.7894	0.8421	0.7368	1.0000						
25	0.7894	0.3157	0.5789	0.4210	0.3684	0.3684	0.4210	0.3684	0.3684	0.4736	0.3684	0.4210	0.6315	0.6315	0.5263	0.6842	0.6315	0.6315	0.6842	0.4736	0.5263	0.5789	0.5789	0.6315	1.0000					
26	0.5263	0.5789	0.6315	0.6842	0.6315	0.6315	0.6842	0.6315	0.6315	0.7368	0.7368	0.6842	0.7894	0.8947	0.6842	0.8421	0.8947	0.8947	0.6315	0.6315	0.6842	0.7368	0.6315	0.7894	0.5263	1.0000				
27	0.4736	0.5263	0.6842	0.6315	0.5789	0.5789	0.6315	0.5789	0.5789	0.6842	0.6842	0.6315	0.7368	0.8421	0.7368	0.7894	0.8421	0.8421	0.6842	0.6842	0.7368	0.7894	0.6842	0.8421	0.4736	0.9473	1.0000			
28	0.4736	0.5263	0.6842	0.4210	0.3684	0.4736	0.4210	0.3684	0.6842	0.4736	0.3684	0.3157	0.6315	0.7368	0.8421	0.6842	0.6315	0.6315	0.6842	0.5789	0.6315	0.6842	0.6842	0.7368	0.4736	0.6315	0.6842	1.0000		
29	0.5789	0.4210	0.7894	0.5263	0.4736	0.4736	0.5263	0.4736	0.5789	0.6842	0.5789	0.5263	0.7368	0.8421	0.8421	0.7894	0.7368	0.7368	0.7894	0.7894	0.7368	0.7894	0.7894	0.8421	0.5789	0.8421	0.8947	0.7894	1.0000	
30	0.5263	0.5789	0.6315	0.5789	0.5263	0.6315	0.5789	0.5263	0.7368	0.7368	0.7368	0.5789	0.8947	0.7894	0.7894	0.8421	0.7894	0.7894	0.6315	0.5263	0.5789	0.6315	0.5263	0.7894	0.5263	0.8947	0.8421	0.6315	0.7368	1.0000

1	FCRICI 1	11	FCRICI 11	21	FCRICI 21
2	FCRICI 2	12	FCRICI 12	22	FCRICI 22
3	FCRICI 3	13	FCRICI 13	23	FCRICI 23
4	FCRICI 4	14	FCRICI 14	24	FCRICI 24
5	FCRICI 5	15	FCRICI 15	25	FCRICI 25
6	FCRICI 6	16	FCRICI 16	26	FCRICI 26
7	FCRICI 7	17	FCRICI 17	27	FCRICI 27
8	FCRICI 8	18	FCRICI 18	28	FCRICI 28
9	FCRICI 9	19	FCRICI 19	29	FCRICI 29
10	FCRICI 10	20	FCRICI 20	30	FCRICI 30

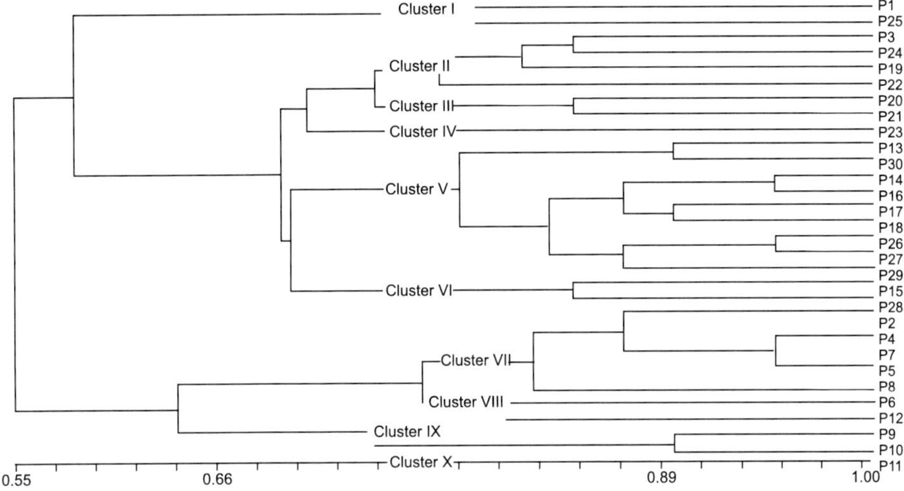

Figure 2: *Dendrogram of Calophyllum inophyllum progenies based on Jaccard's similarity coefficient*

REFERENCES

Anderson JA, Churchill GA, Autrique JE, Tanksley SD, and Sorrells ME 1993 Optimizing parental selection for genetic linkage maps, *Genome* **36** 181-186

Basha SD, George Francis HPS, Makkar K, Becker M, and Sujatha M 2009 A comparative study of biochemical traits and molecular markers for assessment of genetic relationships between *Jatropha curcas* L. germplasm from different countries, *Plant Science* **176** 812-823

Deepak Pandy and Hooda MS 1997 Genetic variability and correlation studies in seed traits of mesquite *Prosopis juliflora*, *Indian Journal of Forestry* **24**(2) 162-165

Divakara BN, Alur AS, and Tripathi S 2010 Genetic variability and relationship of pod and seed traits in *Pongamia pinnata* (L.) Pierre a potential agroforestry tree, *International Journal of Plant Production* **4**(2) 129-141

Elavazhagan T, Ramakrishnan M, Jayakumar S, Chitravadivu C, and Balakrishnan V 2009 DNA finger printing analysis in *Casuarina equisetifolia* by using RAPD markers, *Botany Research International* **2**(4) 244-247

Eriksson G, Baradat PH, Adams WT, and Muller Starch G 1995 SPB Academic Publishing, Amsterdam, the Netherlands. 349-358

Gonzalez A, Wong A, Delgado-Salinas A, Papa R, and Grepts P 2005 Assessment of inter simple sequence repeat markers to differentiate sympatric wild and domesticated populations of common bean, *Crop Science* **45**(2) 606-615

Gradual L, Kjaer ED, Suangtho P, and Kaosaard A 1999 Conservation of genetic resources of teak in Thailand. Technical Note No. 52. Danida Forest Seed Centre, Denmark.

Haines RJ and Martin BE 1995 Biotechnology and the sustainable production of tropical timber, ITTO Pre Project Report PPR **42 /97(F)**: 168

Jaccard P 1908 Nouvelles recherches sur la distribution florale, Bulletin de la Société Vaudoise des Sciences Naturelles **44:** 223-270

Jubera BS, Janagoudar DP, Biradar RL, Ravikumar R, Koti V, and Patil SJ 2009 Genetic diversity analysis of elite *Jatropha curcas* (L.) genotypes using Randomly Amplified Polymorphic DNA markers, *Karnataka Journal of Agricultural Sciences* **22** (2) 293-295

Kumar S, Virendra Kumar, Manoj Kumar Sharma, Narendra Kumar, Sengar, and Nidhi Jaiswal 2013

Assessment of genetic diversity in Jatropha (*Jatropha curcas* L.) by using RAPD markers, *International Journal of Biotechnology and Bioengineering Research* **4**(4) 383-394

Li F, Gan S, Weng Q, Zhao X, Huang S, Li M, Chen S, Wang Q, and Shi F 2008 RAPD and morphological diversity among four populations of the tropical tree species *Paramichelia baillonii* (Pierre) Hu in China, *Forest Ecology and Management* **255** 1793-1801

Mohapatra and Singhal 2000 Role of molecular markers in biodiversity conservations, *Ann. For.,* **8**(1) 1-7

Nandini R, Singh TR, and Dhanapal GN 2011 Morphometric and molecular diversity studies in *Tamarindus indica* population, *Plant Molecular Biology and Biotechnology* **2**(2) 26-33.

Pavithra HM, Gowda B, Prasanna KT, and Shivanna MB 2013 Pod and seed traits in candidate plus trees of *Pongamia pinnata* from southern peninsular India in relation to provenance variation and genetic variability, *Korean Society of Crop Science* **012** 0052-8.

Rohlf FJ 2002 NTSYS-pc: Numerical taxonomy system ver. 2.1. Exeter Publishing Ltd., Setauket, New York.

Sharp PA, Sugden B, and Sambrook J 1973 Detection of two restriction endonuclease activities

Shyamsundar S, Aadil K, Negi MS, and Tripathi 2014 Efficiency of two dominant marker systems, ISSR and TE-AFLP for assessment of genetic diversity in biodiesel species *Pongamia pinnata*, *Current Science* **106** 1576-1580

Tiwari S, Rakhi Tiwari, Richa Srivastava, Pratibha Kumari, and Sujeet Kr Singh 2013 Molecular genetic analysis of *Eucalyptus tereticornis* by using RAPD markers, *European Journal of Experimental Biology* **3**(6) 103-110

Williams JGK, Kubelik AR, Livak KJ, Rafalski JA, and Tingey SV 1990 DNA polymorphisms amplified by arbitrary primers are useful as genetic markers, *Nucleic Acids Research* **18** 6531-6535

Wright S 1921 Correlation and causation, *Journal of Agricultural Research* **20** 557-585

Yap IV and RJ Nelson 1996 WINBOOT: a program for performing bootstrap analysis for binary data to determine the confidence limits of UPGMA-based dendrograms. IRRI Disc, *Ser. No. 14. International Rice Research Institute,* Manila, Philippines

Yasodha R, and Kathirvel M 2004 Genetic analyses of Casuarinas using ISSR and FISSR markers, *Genetica* **122** (2) 161-172

Zabeau M and Vos P 1993 Selective restriction fragment amplification: a general method for DNA Fingerprinting. *European Patent.*

Zobel BJ and Talbert J 1984 Applied tree improvement, John *Wiley & Co.,* New York, 503 p.

CHAPTER 35

Standardization of Growing Medium and Container Type for Quality Stock Production of *Emblica Officinalis* Gaertn.

Raveena Negi*, Manisha Thakur and G S Shamet

Dr. Y.S. Parmar University of Horticulture and Forestry, Nauni, Solan, (H.P.), India

**Corresponding author's Email:negiraveena1991@gmail.com*

ABSTRACT

The containerized production system has advantage over the bare root seedlings. The modern container systems currently being used are root trainers which avoid root spiraling, encourage lateral root developments and to limit the egress of roots from the container. Besides container types, the growing medium also assumes great significance as far as production of quality stock is concerned. The aonla (*Emblica officinalis* Gaertn.) is one of the most important fruit crop of India. Its fruit have its medicinal properties and also used in tannin and dyeing industries. Keeping in view the importance of species, studies were conducted to examine the best growing medium and container type for quality stock production. The growing medium (M_3) consisting of soil + sand + mushroom compost (1:1:2) in combination with polybag (C_2) having size 9x4.5 inches with sixteen bottom perforations (i.e. M_3C_2), registered significantly maximum germination (76.00 %) and growth parameters viz., height growth (51.89 cm), collar diameter (3.9 mm) and root length (34.41 cm), dry root weight (3.19 g), dry shoot weight (4.23 g) and stock quality index (0.51). This may be attributed to larger container size and perforations with good amount of growing mixture, good porosity, good moisture retention-etc., which helped in faster and better growth of the seedling.

Keywords: *Emblica officinalis*, Growing medium, Root trainers

INTRODUCTION

There is a great pressure on the forest resource these days to meet the requirement of timber and other forest products in the country. To reduce this pressure, new approaches like plantation forestry, social forestry and other afforestation practices have been mooted as alternative strategies to prevent large scale decline in this valuable resource. It has been observed that nearly 50% failure in the field establishment was due to poor quality seedlings and choice of species. The ultimate success of the various plantation programmes, therefore, depends upon the right choice of species and quality of planting stock. The nursery stocks are usually raised as bare root or the container seedling stock. Bare root seedlings are, however, not produced in large quantity due to transpiration stress, packaging problems and greater planting shock which often results in high mortality rates in the field. The containerized production system which evolved during the last 50 years varies from simple tar paper pots of 1930s (Strachan1974), to wide variety of rigid-walled containers in use these days. The most common containers which are, however, extensively used in India are the polybags. These are easily available, cheap and can be used for raising large quantities of seedlings. The main problem with these, however, is root coiling, as the bottom is closed and hence hamper development of quality stock production (Gera et al 1998). The containerized production system has advantage over the bare root seedlings. The modern container systems currently being used are root trainers which avoid root spiraling provide air pruning, encourage lateral root developments and to limit the progress of roots from the container. Besides container type, the growing medium also assumes great significance as far as production of quality stock is concerned. The growing medium should be fertile, light, well drained and aerated. The commonly used growing media besides soil contain compost, animal manure and sawdust, etc. The container size also has a control on quality of raised stock. The aonla (*Emblica officinalis* Gaertn.) belonging to family Euphorbic is one of the most important wild fruit crops of India. The species is indigenous to South-East Asia, particularly central and southern India, where it is preferred for cultivation since time immemorial. The species is quite hardy and found mostly in dry deciduous forests in the country. The fruit is highly nutritious and regarded as the richest source of vitamin C (400-1300 mg/100 g). Medicinally, it is an important constituent of *chyavanprash* and *triphala*-the ayurvedic tonic. The species is also used to cure chronical infections of ear, nose and throat and widely used for gastrointestinal disorders and stomach ailments. It is also used in tannin and dyeing industries in the country.

METHODOLOGY

The experiment was conducted in polyhouse of the Department of Silviculture and Agroforestry, Dr. Y. S. Parmar University of Horticulture and Forestry, Solan, Himachal Pradesh from March 2013 to April 2014. Seeds were raised in 4 types of containers (Table 1) i.e. 9″ × 4.5″ polybag with 8 side perforations (C_1), 9″ × 4.5″ with 8 side and 8 bottom perforations (C_2), 7″ × 4.5″ with 8 side perforations (C_3), 7″ × 4.5″ with 8 side and 8 bottom perforations (C_4), Root trainers 300cc (C_5) and growing media soil, sand and FYM (1:1:1) (M_1), soil, sand and mushroom compost (1:1:1) (M_2), soil, sand and mushroom compost (1:1:2) (M_3), soil, sand and mushroom compost (1:1:3) (M_4), Soil, sand and vermicompost (1:1:1) (M_5) and soil, sand and compost (1:1:2) (M_6). The experiment was in Randomized Block Design (factorial), no. of treatment combinations were 30 and replicated thrice (Plate 1).

Plate 1: *Emblica officinalis seedlings raised in the containers*

Growing Medium Preparation

Mushroom compost, vermicompost, FYM, sand and soil were sieved with a fine mesh (1cm^2) to remove the undecomposed part, stones and other foreign material and then the media for respective treatments were prepared by mixing different ingredients in required proportions.

Table 1: *Main features of containers*

Containers	*Volume (cc)*	*Length (cm)*	*Top diameter (cm)*	*Growing density m^{-2}*
C_1	1130	22.50	8.00	169
C_2	1130	22.50	8.00	169
C_3	955.17	19.00	8.00	150
C_4	955.17	19.00	8.00	150
C_5	300	15.00	7.00	148

After sowing, weeding was done regularly as and when required and seedlings irrigated twice a day during summer months.

RESULTS AND DISCUSSION

Seedlings raised in polybag of size 9″ × 4.5" having eight side + eight bottom perforations, filled with medium consisting of soil + sand + mushroom compost (1:1:2) (M_3C_2) resulted in significantly maximum germination of 76.00 per cent. This was however, closely followed by seedlings raised in M_2C_1 and M_3C_4 each giving a value of 74.67 per cent. This was at par with the seedlings raised in M_4C_2 (71.00%), M_6C_4 (68.33%), M_3C_5 (64.00%) and M_3C_3 and M_5C_1. The significantly least value of 21.33 percent was obtained for M_2C_5 consisting of root trainer (300cc) filled with soil + sand + vermicompost (1:1:1). The growing medium (M_3) consisting of soil + sand + mushroom compost (1:1:2) in polybag C_2 attained significantly maximum height growth (51.89 cm) of the seedlings.

It was closely followed by M_3C_3 combination treatment giving the value of 49.03 cm. The treatment combination M_1C_5 produced the significantly smallest seedling height (27.82 cm) growth in the species. Similarly, polybag of size 9 x 4.5" having eight side + eight bottom perforations (C_2) filled with growing media of soil + sand + mushroom compost (1:1:2) (M_3) mixture, resulted in significantly highest collar diameter of 3.90 mm, being superior to all other combinations in this regard. The treatment combination M_1C_5 resulted in significantly least diameter growth of 2.45 mm in the seedlings. The interaction of container type and growing medium was found to exert significant influence on root length of *Emblica officinalis* seedlings (Table 2). The significantly maximum root length (34.41cm) was observed in M_3C_2 treatment combination, being superior compared to other treatment combinations. The significantly minimum root length (21.08 cm) was recorded in M_1C_5 treatment combination.

Table 2: *Effect of growing media and container type on germination and other growth parameters of Emblica officinalis Gaertn. seedlings*

Treatment	*Germination (%)*	*Height growth (cm)*	*Collar diameter (mm)*	*Root length (cm)*	*Dry root weight (g)*	*Dry shoot weight (g)*	*Root-shoot ratio*	*Stock quality index*
M_1C_1*	48.33 (44.02)	35.14	2.69	26.58	1.77	2.00	0.89	0.26
M_1C_2	61.00 (51.66)	36.48	3.50	26.06	2.63	2.94	0.89	0.47
M_1C_3	43.33 (41.02)	36.47	2.89	28.87	1.66	2.72	0.61	0.32
M_1C_4	42.67 (40.63)	33.01	3.19	23.26	1.71	2.02	0.85	0.33
M_1C_5	37.33 (37.28)	27.82	2.45	21.08	1.70	2.75	0.62	0.38
M_2C_1	74.67 (59.86)	38.79	3.50	28.47	2.19	3.00	0.73	0.42
M_2C_2	66.67 (55.73)	34.4	2.84	27.99	2.29	2.88	0.79	0.4
M_2C_3	54.00 (47.29)	45.07	3.45	30.82	1.94	3.32	0.55	0.31
M_2C_4	52.33 (46.33)	46.06	3.30	28.56	2.36	3.50	0.67	0.38
M_2C_5	21.33 (27.08)	37.57	3.46	28.13	2.13	2.54	0.83	0.37
M_3C_1	63.00 (52.55)	44.44	3.73	30.95	3.15	3.66	0.85	0.48
M_3C_2	76.00 (60.69)	51.89	3.90	34.41	3.19	4.23	0.76	0.51
M_3C_3	63.67 (53.22)	49.03	3.39	30.68	2.96	3.25	0.91	0.4
M_3C_4	74.67 (59.84)	45.97	3.32	30.89	2.98	3.38	0.88	0.42
M_3C_5	64.00 (53.43)	38.42	3.22	31.27	2.38	3.43	0.67	0.42
M_4C_1	57.00 (49.01)	45.89	3.05	30.13	2.51	3.21	0.74	0.33
M_4C_2	71.00 (57.43)	47.25	3.15	27.88	2.92	3.50	0.83	0.39
M_4C_3	52.00 (45.99)	44.43	2.58	28.94	2.59	2.90	0.89	0.3
M_4C_4	42.67 (40.54)	40.12	3.12	28.88	3.09	3.26	0.95	0.46
M_4C_5	43.00 (40.96)	36.99	3.10	21.93	1.91	2.41	0.67	0.23
M_5C_1	63.67 (53.22)	33.74	2.90	26.99	1.84	2.26	0.81	0.32
M_5C_2	62.67 (52.4)	42.55	3.63	31.30	2.67	3.39	0.78	0.45
M_5C_3	46.67 (42.96)	30.70	2.69	30.19	2.64	2.99	0.89	0.44

M_5C_4	48.00 (43.84)	38.37	3.16	31.52	1.96	2.99	0.65	0.37
M_5C_5	41.33 (39.91)	36.64	3.73	25.22	2.01	2.41	0.82	0.38
M_6C_1	48.00 (43.71)	40.14	3.24	24.35	2.70	3.28	0.78	0.41
M_6C_2	57.33 (49.2)	44.07	3.50	30.88	2.52	3.20	0.79	0.41
M_6C_3	48.67 (44.2)	33.18	3.45	25.74	2.04	2.49	0.82	0.41
M_6C_4	68.33 (55.8)	35.82	3.37	29.97	1.92	2.35	0.81	0.33
M_6C_5	60.00 (50.83)	30.54	2.79	27.12	2.03	2.33	0.71	0.35
SE±	6.37	3.94	0.38	2.97	0.35	0.35	0.90	0.07
$CD_{0.05}$	12.76	7.88	0.76	5.94	0.71	0.71	0.19	0.14

Figures in the parenthesis are arc sin transformed values

* *M_1C_1 (medium consisting of soil, sand and FYM 1:1:1 ratio in poly bag of size 9″ × 4.5″with 8 perforations); M_1C_3 (consisting of soil, sand and FYM 1:1:1 ratio in poly bag of size 7″ × 4.5″with 8 bottom perforations); M_1C_5 (medium consisting of soil, sand and FYM 1:1:1 ratio in root trainers 300cc); M_2C_3 (medium consisting of soil, sand and mushroom compost 1:1:1 ratio in poly bag of size 7″ × 4.5″ with 8 perforations); M_2C_5 (medium consisting of soil, sand and mushroom compost 1:1:1 ratio in root trainers 300cc); M_3C_2 (medium consisting of soil, sand and mushroom compost 1:1:2 ratio in poly bag of size 9″ × 4.5″with 16 perforations); M_4C_4 (medium consisting of soil, sand and mushroom compost 1:1:3 ratio in poly bag of size 7″ × 4.5″with 16 bottom perforations).*

Moreover, treatment M_3C_2 recorded significantly maximum dry root weight of 3.19 g in the seedlings. The significantly least root weight (1.66 g) was recorded in seedling raised with M_1C_3 treatment combination. For dry shoot weight seedlings raised in polybag (C_2) containing growing medium M_3 recorded significantly maximum value 4.23 g in the seedlings. This was however, followed by M_3C_1 treatment combination giving values of 3.66 g. The significantly minimum shoot weight (2.0 g) was observed in M_1C_1 treatment combination. On the other hand significantly maximum root-shoot ratio (0.95) was registered by M_4C_4 treatment combination. However, significantly minimum value was recorded in M_2C_3 treatment combination giving value of 0.55.

It is apparent from the data in Table 2 that significantly maximum stock quality index of 0.51 was recorded in M_3C_2 treatment combination and significantly minimum stock quality index of 0.26 was recorded in M_1C_1 treatment combination. The results might be attributed to larger container size (1131cc) and side + bottom perforations with good amount of growing mixture, good porosity, good moisture retention and low plant density, which helped in faster and better growth of the seedling. These results are in agreement with the findings of Chand (2008) who reported that growing media and container type interacted positively and performed better in the form of growth and biomass of *Terminalia chebula, Terminalia bellirica and Emblica officinalis* seedlings. Earlier, Magotra (2005) had also found that interaction between growing media and container type significantly affected the growth and biomass in *Dendrocalamus strictus and Dendrocalamus hamiltonii.* Similar results were recorded by Thakur et al., (2000) in *Terminalia chebula* and Kumar et al., (2005) in *Pinus roxburghii* where they found that size of container significantly affected plant growth under each growing media. Similarly, the dry weight of stock was significantly affected by interactions between bag size and potting mixtures (Mishra and Jaiswal 1993). The reason for varied seedling growth in container and media interactions might be attributed to the variation in container size and nature of potting mixture which control the moisture retention capacity, aeration and nutrient availability to the seedlings (Jabula and Sargenta 1977). The growing medium in combination with container size has a significant role in carrying capacity of potting mixture, favorable pH, EC and availability of nutrients. The larger volume of container size and low growing density allowed the seedlings to grow vigorously and healthy compared to smaller size container. The favorable results might also be due to larger container size supporting seedling for longer periods by providing nutrients and moisture as these are the basic requirements for establishment and faster growth as suggested by Gupta et al., (1992). Many other researchers (Bahuguna and Lal (1990) in *Acacia nilotica*; Rathore et al., (2004) in *Casuarina equisetifolia*; Al- Menaie et al., (2010) in *Cassia nodosa* and *Cassia fistula* and Lambert et al (2010) in *Taxodium distichum*, *Acer rubrum* and *Pinus palustris*) have also observed similar influence of growing media x container type interaction on seedling growth and development

CONCLUSION

The combination M_3C_2 i.e. polybag (C_2) of size 9" × 4.5" with eight side + eight bottom perforations filled with soil, sand and mushroom compost in the ratio of 1:1:2 (M_3) gave significantly maximum germination per cent, growth and biomass. Hence, this combination of container type and growing media may be used for quality seedling production of *Emblica officinalis*.

REFERENCES

Al-Menaie H S, Al-Ragom O, Al-Shatti A, Mathew M and Suresh N 2010 Evaluating the growth performance of *Cassia nodosa* and *Cassia fistula* L. seedlings using different potting mixtures, *Academic Journal of Plant Sciences* **3**(1) 33-36

Bahuguna V K and Lal P 1990 To study the effect of environment and different soil mixture on germination of *Acacia nilotica* seed at nursery stage, *Indian Journal of Forestry* **116** (6) 474-478

Chand T 2008 Effect of nursery management practices on stock quality and outplanting performance of some medicinal trees in mid hills of H.P. PhD. Thesis, Dr Y S Parmar University of Horticulture and Forestry, Nauni, Solan, H.P. India 225

Gera MN, Gera SS and Bhandari AS 1998 Improved seedling quality of polybag plants, use of mounted angle-iron beds, *Indian Journal of Forester* **124**(2) 116-122

Gupta GN, Rajawat MS and Singh G 1992 Potting mixtures for nursery plants of *Acacia* in arid region, *Van Vigyan* **30**(3) 175-178

Jabula N Q and Sargento 1977 Growth and survival of Kaaton Bangkal (*Anthocephalus cadamba*) seedlings in different size pots and in various potting media, *Phillipp Lumberm* **17**(9) 48-53

Kumar S, Dhiman RC and Nayital RK 2005 Morphology and biomass parameters of *Pinus roxburghii* Sargent seedlings as affected by container type and growing medium, *Indian Forester* **131**(1) 37-44

Lambert BB, Harper SJ and Robinson DS 2010 Effect of container size at time of planting on tree growth rates for baldcypress (*Taxodium disticjum* Rich), red maple (*Acer rubrum* L) and longleaf pine (*Pinus palustris* Mill.), *Arboriculture and Urban Forestry* **36**(3) 93-98

Magotra K 2005 Studies on container type, growing media and phosphorus doses for raising quality nursery stock of *Dendrocalamus strictus* and *D. hamiltonii*. MSc.Thesis, Dr Y S Parmar University of Horticulture and Forestry, Nauni, Solan, H.P. India 99

Misra KK and Jaiswal HR 1993 Effect of size of polythene bags and potting mixtures on the survival and growth of silver oak (*Grevilla robusta* Parker) seedlings, *Indian Forester* **119**(11) 941-943

Rathore TS, Annapurna D, Joshi G and Shrivastava A 2004 Studies on potting mixture and size of containers on the quality of seedling production in *Casuarina equisetifolia*, *Indian Forester* **130**(3) 323-332

Strachan MD 1974 Tar paper containers. Containerized Forest Tree Seedling Symposium, (August 26-29) Great Plant Agricultural Council 209-210

Thakur S, Badiyala SD and Sharma NK 2000 Relative growth performance of harar (*Terminalia chebula* Retz.) seedlings in different soil mixtures and polybag size, *Journal of Non Timber Forest Products* **7**(3/4)199-201

CHAPTER 36

Effect of Anthracnose Disease on Photosynthetic Traits in Groundnut (*Arachis Hypogaea*)

Ankur Jha*, Mayank Chaturvedi[1], Badre Alam[1] and Ram Newaj[1]

Krishi Vigyan Kendra, Auraiya -206244 U.P., India

[1]ICAR-Central Agroforestry Research Institute, Jhansi -284003 U.P., India

**Corresponding author's Email: jhaankur111@gmail.com*

ABSTRACT

Colletotrichum is one of the most important plant pathogen, causing economically important disease anthracnose in a wide range of hosts including cereals, legumes, vegetables, perennial crops and tree fruits. Anthracnose disease (caused by *Colletotrichum dematium*) of groundnut (*Arachis hypogaea*) has become a serious problem in recent years in central India. The disease is characterized by leaf spotting resulting in 'shot hole' symptoms and finally defoliation, which affects the yield greatly. Infection of leaves directly hampered the photosynthetic machinery and intercellular mechanisms of the plant. Therefore, it is necessary to understand the severity of the disease and factors affecting under diseased conditions, which will help in devising suitable and effective management practices for proper yield. In India, research work on important aspects of the anthracnose disease of groundnut has not been done systematically. Information on crop photosynthetic traits under diseased conditions and related strategies to overcome yield loss is lacking. The purpose of this research was to understand the effects of anthracnose disease on some important photosynthetic traits such as rate of CO_2 assimilation, thylakoid electron transport, water use efficiency and stomatal conductance, etc.

Keywords: Anthracnose, CO_2 assimilation, Stomatal conductance, Thylakoid electron transport rate

INTRODUCTION

Groundnut (*Arachis hypogaea*) is one of the most important oil yielding crop as well as used as dry fruits, grown mainly in the Asian countries. India is the largest producer of groundnut in the world. The average yield of groundnut is approximately 745 kg ha^{-1}

(Jha et al 2012). One of the important factors contributing to low yield is disease attack. More than 55 pathogens including viruses, fungus and nematodes have been reported to affect groundnut production. Anthracnose disease is a pre harvest disease caused by the fungus *Colletotrichum dematium* which develops under humid condition. The disease has been reported from all major bean crops growing regions of India in mild to severe form and in tropical and subtropical areas (Jha et al 2012); it causes considerable damage by reducing seed quality and yield. The disease causes qualitative as well as quantitative losses. The disease process that is a result of the interaction between a pathogen and its host leads to changes in several physiological processes of the host plant, including dark respiration, photosynthesis, translocation of water and nutrients, transpiration, and regulation of growth (Lucas 1998). Fungal infection generally reduces the photosynthetic rate by decreasing functional leaf area and reducing the photosynthetic efficiency of the remaining green leaf area (Bastiaans 1991, Raggi 1978). Dark respiration of leaves usually rises after infection in consequence of increased metabolic activity of cells of the diseased leaf (Bastiaans 1993, Owera et al 1981). The transpiration rate can go up or down during the infection process, depending on the pathosystem (Bassanezi et al 2001). Water loss from an infected leaf can increase due to destruction of part of the leaf cuticle, increased permeability of leaf cell membranes, or inhibition of stomatal closure. Reduction of transpiration may result from stomatal closure, reduction of air space by hyphae or hypertrophy of mesophyll cells, obstruction of conducting tissue and stomata, and defoliation. Although it is well known that plant pathogens alter the physiological processes of their hosts, comparative studies involving biotrophic, hemibiotrophic and necrotrophic pathogens and their changes in the host physiological processes are not very common and are made at a specific stage of disease development (Bassanezi et al 2001). The relative photosynthetic rate per unit of wheat leaf area were smaller for rust and powdery mildew, caused biotrophic pathogens, than for septoria leaf blotch, caused by a necrotrophic pathogen. The greatest impact on photosynthesis of the necrotrophic pathogens could be explained by the secretion of enzymes or phytotoxic compounds that diffuse to uncolonized portions of the leaf (Bassanezi et al 2001). Necrotrophic pathogens that cause vein infection can induce a systemic stomatal closure in green areas of the leaf (Bassanezi et al 2002), reducing the photosynthetic rate of the whole leaf considerably even when disease severity is low. For biotrophic pathogens, however, the greatest decline in photosynthesis occurs in fungal invaded regions of the leaf or in the areas adjacent to the fungal mycelium (Shtienberg 1992). The reduction in the rate of transpiration of plants infected by necrotrophic pathogens was proportional to the diseased area (Bassanezi et al 2001). The destruction of cells was directly related to the presence of the fungus. Plants with rust, however, had a decrease in transpiration that was proportionally smaller than the corresponding reduction of healthy leaf area due to disease. At low levels of rust infection, the transpiration rate exceeded that of healthy leaves. This has been attributed to the epidermal rupture after pustule formation. The methods commonly used for in vivo measurements of photosynthesis are gas exchange and chlorophyll fluorescence emission (Martin 1986). Variables usually obtained by gas exchange systems are CO_2 assimilation, transpiration rate, intercellular concentration of CO_2, and stomatal conductance. Measurements of chlorophyll fluorescence provide information about photosystem II (PSII) activity and changes in photosynthetic metabolism of diseased leaves (McGrath and Pannypacker 1990, Mayer et al 2001, Scholes and Rolfe 1995). The aim of this study was to compare the alterations in photosynthetic mechanism especially in CO_2 assimilation, thylakoid electron transport, water use efficiency, transpiration and other related photosynthetic traits in leaves infected with *Colletotrichum dematium* pathogens.

METHODOLOGY

Plants Inoculation

Groundnut (*Arachis hypogaea*) plants were grown at ICAR-Central Agroforestry Research Institute Jhansi, India (25° 27′ N latitude and 78° 35′ E longitude, 271 m above MSL) during *kharif* (rainy) season in 2012-13 from seed in pots in natural environment. Soil used in the experiment was black having a mean pH 7.02. Air temperature was 32 ± 4 °C. Irradiance (PAR) at the top of the plants was 1250– 1400 μ mol m^{-2} s^{-1} during the 14 h photoperiod. All measurements were performed on the central leaflet of the first trifoliate leaves. Two-week-old seedlings were spray-inoculated with spore suspensions of inoculum of *Colletotrichum dematium* according to standard refereed techniques. After inoculation, plants were covered with transparent plastic bags for 24 h. First symptoms became visible 2 days after inoculation, and then the disease developed over a period of 20 days. Spraying with water and covering with bags for 24 h after spraying did not significantly affect photosynthesis in control plants. The crop was grown taking all the standard agronomic practices.

Assessment of Gas Exchange

The assessments of net photosynthetic rate, dark respiration, stomatal conductance, transpiration rate and intercellular concentration of CO_2 were made with a Portable Photosynthesis System (LI-6400XT; Licor, USA) on an area of 2 cm^2. The conditions during the assessments were: leaf temperature at 25-30 °C; CO_2 concentration in the air coming into the ranged from 380-390; relative humidity in the ranged from 30 to 50 per cent. The central leaflet was kept under light until the stabilization of P_N, E, gs, and Ci values and then the central leaflet was kept in the dark for 15 min. to measure dark respiration. Assessment of variables of chlorophyll fluorescence the minimal fluorescence (Fo), maximal fluorescence (Fm) and optimal quantum yield of PSII (Fv per Fm) were assessed in the leaflets of non-inoculated plants, and in two regions of inoculated leaves: lesioned and apparently healthy areas. These fluorescence variables were assessed only on plants kept at room temperature (25-30 °C) in the experiment. The Fo was assessed after the emission of a modulated red light beam of 600 Hz and very low intensity. The Fm was obtained with a saturating light pulse for 0.6 s. The value of Fv per Fm was calculated using Fv per Fm = (Fm-Fo) per Fm. The efficiency of PSII photochemistry was calculated as: ΦPSII = (Fm'-Ft) per Fm'. Since ΦPSII is the quantum yield of PSII photochemistry, it can be used to determine linear electron transport rate (ETR) as described by Genty et al (1989): ETR = ΦPSIIxPPFDx0.5x0.84. NPQ is linearly related to heat dissipation and varies on a scale from 0 until infinity even if in a typical plants value ranges between 0.5 and 3.5 at light saturation level, calculated as: NPQ = (Fm-Fm') per Fm'.

Assessment of Disease Severity

The trial consisted of two treatments viz., *Colletotrichum dematium* inoculated groundnut plants (infected) and un-inoculated or healthy plants (control), and each treatment was replicated six times (total 12 pots). Seedlings were harvested after 12 weeks growth (one per replicate), and observations on shoot length (cm) and fresh weight (gm per plant) were taken by standard methods. The area assessed for gas exchange on the central leaflet was marked so that the next assessments could be made in the same area. The necrotic area as well as the yellowish halo and the soaked, faded or dried areas that surrounded the lesions were all included in the assessment of disease severity. The disease severity of the leaf region where the variables of chlorophyll fluorescence were assessed was estimated by counting the number of lesions and multiplying this number by a visual estimate of average lesion size with the aid of standard area diagrams, then dividing by assessed leaf area. To study the anthracnose disease severity index in Groundnut (*Arachis hypogaea*) plants, similar set of experiments were carried out with aforesaid treatments (i.e. four) in two successive years i.e. 2012 and 2013. All the treatments were replicated three times where one plant was maintained per pot, and observations were taken every 10 days interval after inoculation up to complete wilting of the seedlings. The severity of disease on individual plants were rated (infected stages) on a scale from 0 to 4 point scale; where 0 = 0 per cent (no wilting), 1 = 1–30 per cent, 2 = 31–60 per cent, 3 = 61–90 per cent and 4 = 91 per cent -100 per cent wilting or dead plant (Shukla et al 2014). The disease severity index at the time of harvesting is being presented here.

Chlorophyll Estimation

Six leaves from each treatment were collected for chlorophyll estimation. Chlorophyll (Chl) was extracted from the leaves with acetone and DMSO (Dimethyl sulphoxide) solvents (Arnon 1949). Data from the two repetitions of experiments for disease were analyzed together. Data were analyzed through standard deviation (SD) and standard error (SE) calculation followed by t-test through SYSTET-11 software. All the graphs were generated through Microsoft Excel-97-2003. In the present experiment it has been observed through P_{Nmax} versus PPFD curve that the rate of CO_2 assimilation (P_{Nmax}) reduced under anthracnose disease infection (Table 1). The value of P_{Nmax} under non-infected condition was approximately two fold than the infected condition. In non-diseased (healthy) groundnut plants the rate of CO_2 assimilation was 24.37 μ mol m^2 s^{-1}, whereas it reduced upto 13.70 μ mol per m^2 s^{-1} under disease infection. Similar to CO_2 assimilation thylakoid electron transport rate (ETR) was also reduced under infected condition. Under infected condition the ETR was 88.40 μ mol m^{-2} s^{-1}, whereas under non-infected condition the ETR was 119.08 μ mol m^{-2} s^{-1}. Net photochemical quenching (NPQ) was also reduced from 3.24 to 2.79 under infection. Water use efficiency reduced under infection significantly from 14.78 to 8.87. The efficiency of PSII photochemistry (ΦPSII) was also reduced under infection. Minimal fluorescence (Fo) increased under infected condition from 461.32 to 543.44, whereas maximal fluorescence (Fm) decreased in groundnut leaves from 2311.55 to 1982.83 under infection. Optimal quantum yield of PSII (Fv per Fm) considerably reduced from 0.80 to 0.72 under infected condition. Stomatal conductance and rate of transpiration reduced under infection. First symptoms became visible 2 days after inoculation, and then the disease developed over a period of 20 days. The severity of disease was gradually increased after inoculation. After 20 days of inoculation the disease severity index was about 22 per cent, whereas after 40 days of inoculation disease severity index reached up to 54 per cent, while after 60 days of inoculation disease severity index was 73 per cent (Fig. 1). Chlorophyll a, chlorophyll b and total chlorophyll decreased under infection significantly (Fig. 2).

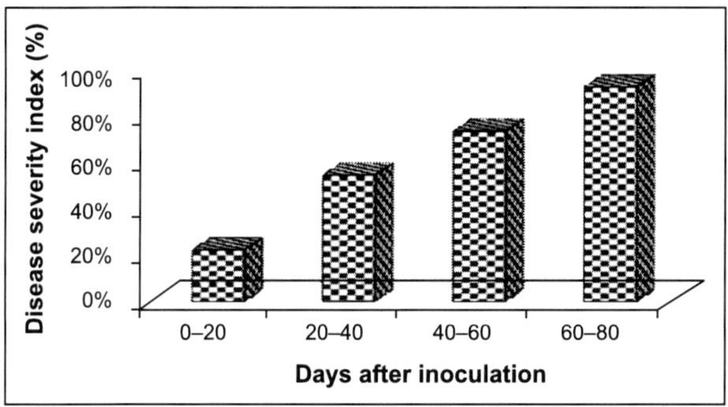

Figure 1: *Disease Severity Index (DSI) of groundnut plants infected by Colletotrichum dematium.*

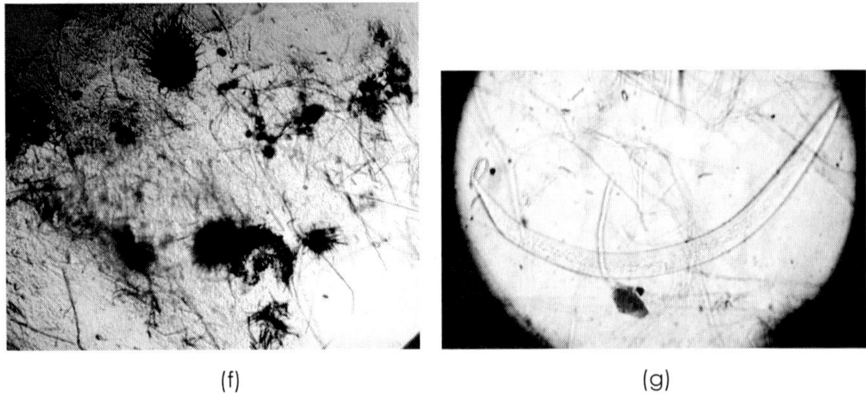

(f) (g)

a. Control or non-infected groundnut plants b. Infected plants of groundnut, c. Comparison in non-infected and infected plant leaves, d. Close view of infected leaves e. Conidia and setae from infected leaves, f. Acervuli fruiting body, g. Falcate shape conidia

Table 1: *Comparative photosynthetic traits of groundnut indicating limitations of CO_2 assimilatory functions under three different stages of Anthracnose disease infection*

Treatment	$P_{N\,max}$	gs	Fo	Fm	Fv/Fm	ΦPSII	NPQ	ETR	E	WUE
Non infected Plants	24.371 ±0.518	0.055 ±0.003	461.325 ±5.393	2311.553 ±3.563	0.800 ±0.002	0.136 ±0.001	2.791 ±0.007	119.082 ±1.292	1.652 ±0.033	14.787 ±0.525
Infected Plants (stage-1)	19.584 ±0.521	0.048 ±0.000	500.584 ±1.914	2099.215 ±4.936	0.772 ±0.003	0.121 ±0.002	3.014 ±0.024	101.226 ±0.878	1.624 ±0.005	11.558 ±0.333
Infected Plants (stage-2)	13.705 ±0.783	0.040 ±0.001	543.442 ±3.709	1982.837 ±3.283	0.725 ±0.001	0.101 ±0.001	3.241 ±0.008	88.400 ±1.388	1.595 ±0.103	8.871 ±1.061
Infected Plants (stage-3)	8.861 ±0.514	0.032 ±0.000	575.891 ±1.047	1754.081 ±14.080	0.701 ±0.006	0.093 ±0.002	3.326 ±0.010	63.558 ±1.513	1.238 ±0.009	6.881 ±0.195
LSD (at 0.05)	1.604	0.012	9.291	21.120	0.0120	0.003	0.038	3.475	0.147	1.677

$P_{N\,max}$= rate of CO_2 assimilation at maximum PPFD (2000 μmol $m^{-2}s^{-1}$) (μmol $m^{-2}s^{-1}$); gs= stomatal conductance (m mol $m^{-2}s^{-1}$); Fo= minimal fluorescence; Fm= maximal fluorescence; Fv/Fm= dark adapted value of variable fluorescence/maximal fluorescence; Φ_{PSII}= effective PSII quantum yield; NPQ= net photochemical quenching; ETR = rate of thylakoid electron transport (μ mol $m^{-2}s^{-1}$); E= rate of transpiration (m mol) ; WUE = water use efficiency (μ mol CO_2 $mmol^{-1}$ water) . Mean ± S.E, (n = 6)

With the development of integrated pest management work, knowledge of the effect of pests and disease on the host physiological processes is must, because there are several microorganisms which act as photosynthetic rate reducer (Boote et al 1983). Since photosynthesis is the biological process that leads to yield accumulation and because the water status of the crop markedly affects photosynthesis, carbon metabolism and water relations are the critical areas of concern in understanding the effects of disease on plants (Lakso et al 1982). Cells of leaves penetrated by the fungal structure or killed by the secretion of some toxic biochemical lose their chlorophyll and probably their capacity to photosynthesize and transpire. In present case study the destruction of cells is directly related to the presence of the fungus, and uncolonized portion of the leaves remain normally active. The effect of infection on the physiological processes completely accounted for by the loss of leaf area to the pathogen. In present study the rate of CO_2 assimilation (P_{Nmax}) observed in P_N versus PPFD curve under diseased and healthy (non-infected) groundnut plants clearly demonstrated the status of photosynthetic efficiency of groundnut (Table 1). It has been observed through P_N versus PPFD curve that the reduction in P_{Nmax} under infection was mainly due to decrease in photosynthetic leaf area under diseased condition. A similar study in maize and banana showed a net decrease in photosynthetic efficiency due to fungal infection (Rodrigues et al 2000). Low absorption of photosynthetic active radiation diminished thylakoid electron transport rate, which ultimately decreased net rate of CO_2 assimilation (Schnabel et al 1998). It is clear that *Colletotrichum dematium* infection poses a great limitation to ETR. ETR is extremely essential for photosynthetic reactions. This ultimately led to reduced photosynthetic process due to lack of sufficient photosynthetic electrons under infected conditions (Alam et al 2012). This has been supported by relatively enhanced net photochemical quenching in infected plants. These facts are well corroborating with some earlier reports. The dark adapted Fv per Fm values around 0.80 in non-infected plants indicate that the plants were in good condition especially free from any cellular damage. Under infected condition decrease in Φ_{PSII} and increase in NPQ were related with the potential efficiency of PSII photochemistry. Further, results indicated that gs gradually decreased under infected condition (Table 1). Reduction in gs under infection in addition to the decrease in PSII activity especially P_{Nmax} and ETR clearly demonstrated the reduced functioning of photosynthetic machinery under infection. The water use efficiency (WUE) decreased under infected condition. Water loss from an infected leaf can increase due to destruction of part of the leaf cuticle, increased permeability of leaf cell membranes or inhibition of stomatal closure. Reduction of transpiration may result from stomatal closure, reduction of air space by hyphae or hypertrophy of mesophyll cells, obstruction of conducting tissue and stomata and defoliation etc.

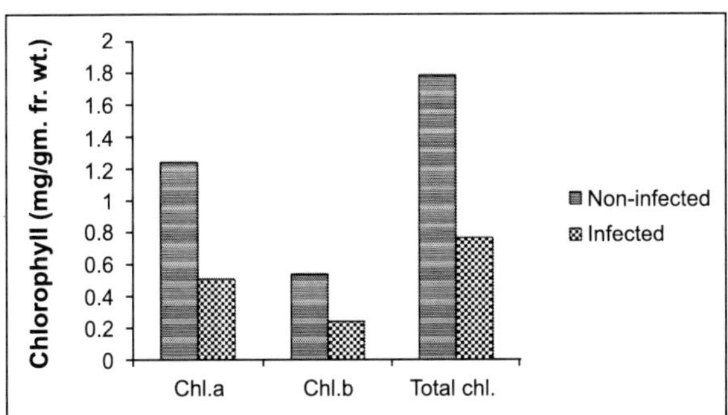

Figure 2: *Pigment profile of infected and non-infected leaves of groundnut*

Under infected condition leaves of groundnut plants could not use CO_2 assimilation as proficiently as the plants did in case of non-infected condition. This indicated that anthracnose disease infection has conspicuously affected the whole photosynthetic processes at cellular and functional level and finally affected yield and productivity of crop as it has also been noted in other fungal diseases (Bassanezi et al 2001, Bassanezi et al 2002, Shtienberg 1992). Under infected condition chlorophyll a, chlorophyll b and total chlorophyll decreased significantly because of reduction in green leaf area under infection (Fig. 2).

CONCLUSION

Under anthracnose disease infection, production of infected plants diminished due to loss in photosynthetic pigments of leaves. Loss in pigments like chlorophyll a and chlorophyll b and green photosynthetic surface of leaves ultimately reduced the rate of CO_2 assimilation and electron transport rate as well.

ACKNOWLEDGEMENT

Authors are grateful to Dr. S.K. Dhyani, Ex Director, Central Agroforestry Research Institute, Jhansi (U.P) and Dr. Ram Newaj, Principal Scientist, Central Agroforestry Research Institute, Jhansi (U.P) for providing necessary research facilities for this study.

REFERENCES

Alam B, Newaj R, Ram M, and Dhyani SK 2012 An adaptive mechanism in mustard (*Brassica juncea*) for abiotic Stress Tolerance, some preliminary results, *Agroforestry News Letter* **24** (1) 3

Arnon DI 1949 Copper enzymes isolated chloroplast polyphenoloxidase in *Beta vulgaris*, *Plant Physiology* **24** 1-5

Bassanezi RB, Amorim L, Filho AB, and Berger RD 2002 Gas exchange and emission of chlorophyll fluorescence during the monocycle of rust, angular leaf spot and anthracnose on bean leaves as a function of their trophic characteristics, *Journal of Phytopathology* **150** 37–47

Bassanezi RB, Amorim L, Filho AB, Hau B, and Berger RD 2001 Accounting for photosynthetic efficiency of bean leaves with rust, angular leaf spot and anthracnose to assess crop damage, *Plant Pathology* **50** 443–452

Bastiaans L 1991 Ratio between virtual and visual lesion size as a measure to describe reduction in leaf photosynthesis of rice due to leaf blast, *Phytopathology* **81** 611-615

Bastiaans L 1993 Effects of leaf blast on photosynthesis of rice. 1. Leaf photosynthesis, *Netherlands Journal of Plant Pathology* **99** 197-203

Boote KJ, Jones JW, Mishoe JW, and Berger RD 1983 Coupling pests to crop growth simulators to predict yield reductions, *Journal of Phytopathology* **73** 1581-1587

Genty B, Briantais JM, and Baker NR 1989 The relationship between the quantum yield of photosynthetic electron transport and quenching of chlorophyll fluorescence, *Biochim Biophys Acta* **990** 87–92

Jha A, Tiwari S, Zacharia S, and Simon S 2012 First report of anthracnose disease on groundnut caused by *Colletotrichum dematium* from Allahabad Uttar Pradesh, in India, *International Journal of Agriculture Science*, **8**(2) 465-457

Lakso AN, Pratt C, Pearson RC, Pool RM, Seem RC, and Welser MJ 1982 Photosynthesis,

Transpiration and water use efficiency of mature grape leaves infected with Uncinula necator (Powdery mildew), *Journal of Phytopathology* **72** 232-236

Lucas JA 1998 Plant Pathology and Plant Pathogens. Blackwell Science, Oxford, UK

Martin PJ 1986 Gaseous exchange studies of barley leaves infected with *Rhynchosporium secalis* (Oudem) J J Davis, *Physiology and Molecular Plant Pathology* **28** 3-14

McGrath MT and Pennypacker SP 1990 Alteration of physiological processes in wheat leaves caused by stem rust and leaf rust, *Journal of Phytopathology* **80** 677-686

Meyer S, Saccardy-Adji K, Rizza F, and Genty B 2001 Inhibition of photosynthesis by *Colletotrichum lindemuthianum* in bean leaves determined by chlorophyll fluorescence imaging, *Plant, Cell and Environment* **24** 947-956

Owera SAP, Farrar JF, and Whitbread R 1981 Growth and photosynthesis in barley infected with brown rust, *Physiology of Plant Pathology* **18** 79-90

Raggi V 1978 The CO_2 compensation point, photosynthesis and respiration in rust infected bean leaves, *Physiology of Plant Pathology* **13** 135-139

Rodrigues CPLS, Azevedo JL, Pereira JO, Carneiro ML, Vieira CA, and Labate 2000 Symptomless infection of banana and maize by endophytic fungi impairs photosynthetic efficiency, *New Phytology* **147** 609-615

Schnabel G, Strittmatter G, and Noga G 1998 Changes in photosynthetic electron transport in potato cultivars with different resistance after infection with *Phytophthora infestans*, *Journal of Phytopathology* **146** 205-210

Scholes JD and Rolfe SA 1995 How do biotrophic pathogens affect the photosynthetic metabolism of their hosts? *Aspects of Applied Biology* **42** 91-99

Shtienberg D 1992 Effects of foliar diseases on gas exchange processes: a comparative study, *Journal of Phytopathology* **82** 760-765

Shukla A, Dehariya K, Vyas B, and Jha A 2014 Interactions between arbuscular mycorrhizae and *Fusarium oxysporum* f. sp. *ciceris*: effects on fungal development, seedling growth and wilt disease suppression in *Cicer arietinum* L, *Archives of Phytopathology and Plant Protection* DOI: 10.1080/03235408.2014.884831

CHAPTER 37

Silvipasture Model of Agroforestry in Augmenting Fodder Production for Sustainable Livestock Farming

S Gunasekaran*, C Valli and C Bandeswaran

Institute of Animal Nutrition, Tamil Nadu Veterinary and Animal Sciences University, Kattupakkam - 603 203, Tamil Nadu, India

**Corresponding author's Email: gunaj2@gmail.com*

ABSTRACT

The critical role of trees and shrubs in livestock production has been well recognized in India. The present system of feeding livestock by landless / marginal farmers is to allow them to graze / browse on the available vegetations viz. grasses, legumes, tree leaves, shrubs and agricultural by products. The available feed and fodder in the country are not enough to sustain the existing population of livestock. Agroforestry can play an important role to overcome the fodder shortage. One of the common agroforestry models is silvipasture which can produce forage for animals on a sustainable basis by optimizing land productivity, conserving soils and nutrients. Valuable data on tree fodder yield, nutritional value of tree fodder and productivity of animals maintained in the silvipasture model of agroforestry has been documented in this chapter. In silvipasture, it is possible to get about 18-20 of green fodder per hectare by integrating *Cenchrus ciliaris* with *Stylosanthus hamata* together with fodder trees. The shrubs like *Leucaena leucocephala*, *Gliricidia*, etc could be harvested for leaf fodder for 6-7 times annually. The total tannin ranged between 0.81 to 4.02 per cent, which was below 5 per cent. The growth rate of kids was not affected when 500g of tree leaves (*Leucaena leucocephala per Inga dulce*) were replaced with 100g concentrate feed in kid ration and there is reduction of feed cost. *Gliricidia sepium* and *Leucaena leucocephala* can replace upto 15 per cent and 20 per cent of green fodder respectively in lactating crossbred cows. Animal integration studies, tree fodder products and livelihood improvement were also explained.

Keywords: Agroforestry, Silvipasture model, Sustainable livestock farming

INTRODUCTION

Livestock rearing is one of the major occupations in India and is making significant contribution to the country's GDP. The present system of feeding livestock by landless

/ marginal farmers is to allow them to graze / browse on the available vegetations viz. grasses, legumes, tree leaves, shrubs and agricultural by products. The available feed and fodder in the country are not enough to sustain the existing population of livestock. To overcome this shortage, growing food and fodder crops on the same unit of land in rainfed situations and integrating trees and grasses with crop farming on marginal and sub marginal land with improved technology deserve high priority. It is in this scenario that to alleviate the existing fodder shortage, to create fodder banks for the future, to overcome problems associated with deforestation and mitigate climate change the concept of integrating fodder trees in agricultural land, wasteland, community land and pastures is gaining momentum. Agroforestry will play very effective roles in the utilisation of the natural resources in a most effective manner for sustainable crop production and socio - economic upliftment of livestock farmers.

Silvipasture Model of Agroforestry

Important agroforestry models are agrisilviculture, silvihorticulture, silvipasture, hortipasture and agrisilvipasture. One of the common agroforestry models is silvipasture defined as growing ideal / suitable combination of grasses, legumes and preferably fodder trees for producing forage for animals on a sustainable basis by optimizing land productivity, conserving soils and nutrients. In degraded soils where cropping is not possible, the silvipasture system serves the purpose of forage production with system integration. Silvipasture model in degraded waste lands has been recognized as low input technology, in this system fodder trees are grown in degraded wastelands / barren wastelands and they serve as fodder bank for livestock (Eg. *Leucaena leucocephala* + *Stylosanthus*). Another silvipasture model is three tier agroforestry model where trees are grown with grasses and legumes in three tiers so that year round fodder is ensured. (Eg. *Gliricidia sepium* + fodder sorghum + *Stylosanthus* sp). The practice of growing trees in field boundaries is known as boundary plantations where the fodder from trees is utilized for feeding livestock.

Silvipasture system combines livestock and trees / grasses that offer two main advantages for the animals. First, trees modify microclimatic conditions including temperature, water vapour content or partial pressure, and wind speed, which can have beneficial effects on pasture growth and animal welfare. Second, trees also provide alternative feed resources during periods of low forage availability.

Tree Fodders

Depending on rainfall and soil fertility, fodder trees like *Acacia nilotica, Acacia leucocephala, Albizzia lebbeck, Leucaena leucocephala, Lannea coromandalica* could be integrated in silvipasture. These trees / shrubs grow well even under drought conditions and produce fodder in two years. About 10 t of leaf fodder can be obtained annually from trees raised in one hectare. The shrubs like *Leucaena leucocephala*, *Gliricidia,* etc could be harvested for leaf fodder for 6-7 times annually. As leaves and pods of Acacia trees contain about 11-15 per cent of crude protein, such fodders are very good for livestock, especially sheep and goat (Table 1).

Table 1: *Common fodder trees of Tamil Nadu with their biomass yield and nutritive value*

S.No.	*Botanical name*	*Biomass yield (t DM ha^{-1})*	*Crude protein (%)*	*Crude fibre (%)*	*TDN (%)*
1.	*Acacia nilotica*	0.5 - 2	12	28	45
2.	*Acacia leucophloea*	0.5 - 2	14	25	45
3.	*Acacia planifrons*	0.5 - 2	9.6	22	45
4.	*Ailanthes exelsa*	3 - 4.5	6.0	23	63
5.	Albizia lebbek	1 - 2	16	18	50
6.	*Albizia amera*	2 - 3	15	20	50
7.	*Azardirachta indica*	-	6	27	55
8.	*Bahunia sp*	1.5 – 2.5	17	22	55
9.	*Dalbergia sisso*	-	17	25	50
10	*Erythrina indica*	1.0 – 2.0	26	12	60
11.	*Ficus bengalensis*	-	12	27	52
12	*Hardwickia binata*	-	10	26	50
13.	*Gliricidia sepium*	2-15	22	25	65
14.	*Inga dulci*	2 - 4	20	25	55
15.	Milligtonia hortensis	-	6	20	62
16.	*Leucaena leucocephala*	2-20	21	24	68
17.	*Lannea coromandalica*	-	4	18	65
18.	*Sesbania sesban*	2 -10	24	28	65
19.	*Sesbania grandiflora*	2 -10	22	27	65
20.	*Thespesia populnea*	-	11	20	60

The nutritive values of tree fodders are influenced by various factors. Studies have revealed that per cent crude protein of Gliricidia was highest when it was pruned once in 45 days. Tannin is the common anti nutritional factor present in tree leaves. A total of ten tree species were tested for their total tannin in their leaves. The total tannin ranged between 0.81 to 4.02 per cent, which was below 5 per cent (maximum tolerance level).

Pasture Grasses

Cenchrus ciliaris grass can also be grown in rainfed condition between the fodder trees. They can be raised with seeds or slips. As seeds fall down and grow on their own, no reseeding is required. Initially, about 7-10 kg of seeds is required per hectare. This grass yields forage of about 15-20 t ha^{-1}. As it contains 9 per cent crude protein and sufficient amounts of calcium, it is a good source of green fodder for livestock. Leguminous plants like *Stylosanthus* can also be integrated along with fodder trees. About 20-25 kg of seeds is needed per hectare. About 10-20 t of forage per hectare can be obtained in dry lands. This fodder contains 18-20 per cent of crude protein. In silvipasture, it is possible to get about 18-20 MT of green fodder per hectare by integrating *Cenchrus* with *Stylosanthus* together with fodder trees, with this about 12-16 sheep / goat can be raised annually which will fetch ₹ 36,000 income per hectare to the farmers.

Animal Integration Studies

Browse from trees and shrubs plays an important role in feeding ruminants in many parts of the World, particularly in the tropics, and there has been considerable research into the nutritional potential and limitations of many tropical fodder species (Devendra 1992).

Small Ruminant Studies

Feeding mixture of tree leaves containing equal proportion of *Albizia lebbeck, Ficus bengalensis* and *Leuceana leucocephala* along with green grass at 1:1 ratio significantly improves the feed efficiency by 27 per cent in sheep. Mixed silvipasture (*Leucaena leucocephala, Gliricidia sepium, Azadirachta indica, Stylosanthus*) system was able to support about 53g of daily weight gain in sheep when compared to 35g daily weight gain in natural grazing land without supplementation. *Leuceana leucocephala* leaves and grass were fed at 50 per cent level each, the sheep gained 49.1 g body weight per day against 41.5g when fed with grass alone.

In neem with bajra agri-silvi model, sheep fed with 60 per cent roughage as bajra straw and the rest as combinations of *Azadirachta indica* and *Gliricidia sepium* leaves recorded a average daily gain of 58.0 g which is about 80 per cent as that of sheep fed with conventional ration. The feed efficiency was 8.08 kg. Sole feeding of *Leucaena leucocephala* has no ill effect on growth, reproduction and pregnancy in sheep. The average daily body weight gain of 100g, conception rate of 90 per cent and birth weight of 2.3 kg can be achieved. The growth rate of kids were not affected when 500g of tree leaves (*Leucaena leucocephala / Inga dulce*) were replaced with 100g concentrate feed in kid ration and there is reduction of feed cost by Rs.1.70 per day. Five hundred grams of fresh *Moringa oliefera* leaves can be fed as an alternative feed supplement without deleterious effects in goats

Palatability Studies

Palatability trials were conducted with different tree fodder in sheep and goats. Study in goats revealed that tree leaves were 40-50 per cent more palatable than grass. Among

Leucaena leucocephala, Gliricidia sepium, Ficus bengalensis, Lannea coromandelica, Inga dulce and *Albizia lebbek*, leaves of *Leucaena. leucocephala* and *Inga dulce* were highly palatable. *Albizia lebbek* tree leaves were least palatable to goats. In another trial to assess palatability of *Leucaena leucocephala*, *Inga dulce*, *Albizzia lebbeck* and *Gliricidia sepium* in sheep and goats, *Leucaena leucocephala* and *Inga dulce* ranked first and second, respectively in both sheep and goats. In goats, *Gliricidia sepium* ranked third followed by *Albizzia lebbeck*. However, in sheep it was *vice - versa.*

To improve the palatability of *Azadirachta indica* and *Gliricidia sepium*, a study was carried out by sprinkling of salt solution / molasses or wilting for 8 hours and it was inferred that wilting for eight hours greatly enhances palatability.

Tree Leaf Meal in Concentrate Mixture of Animals

The tree fodder lopped is dried, ground and incorporated in concentrate mixture at varying levels in replacement of conventional feed ingredients. *Leucaena leucocephala* + *Gliricidia sepium* leaf meal (1:1) based concentrate feed as protein source partially (30 per cent) replaced the soybean meal, deoiled rice bran, sunflower oil cake mixture in conventional goat concentrate feed and reduced the cost of feed by Rs.4.86 per kg. Feed cost to the extent of 17 per cent was reduced compared to the conventional concentrate feed.

Products Prepared from Tree Fodder

Complete feed block, extruded feed and tree leaf meal can be prepared from tree fodder harvested from agroforestry models.

Feed block

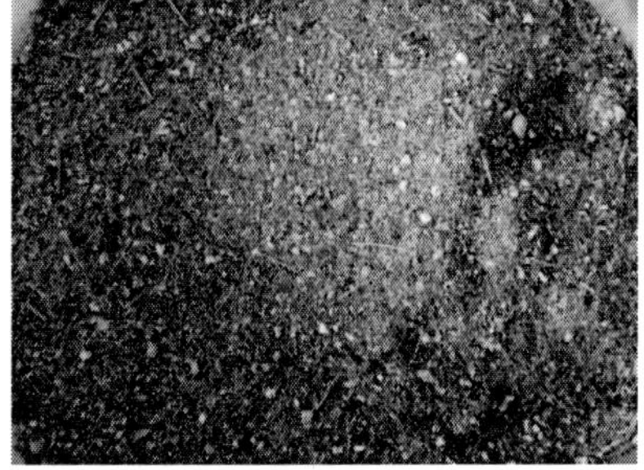

Tree leaf meal

Large Ruminants Studies

A tree leaf meal based concentrate feed (*Gliricidia sepium* and *Leucaena leucocephala* leaves, 30 per cent inclusion at 1:1 ratio) fed to buffalo calves (8-9 months old) to evaluate its potential against the conventional fed concentrate feed revealed that growth is similar to conventionally fed concentrate feed in a 90 days study. The feed cost to the extent of 25 per cent was reduced compared to conventional concentrate feed. Fresh *Azadirachta indica* leaves can contribute to the extent of 15 per cent of the roughage component in the ration of lactating dairy cows thereby reducing the fodder cost by 15 per cent without affecting their milk yield or composition. *Leucaena leucocephala* can replace upto 20 per cent of green fodder (Dry Matter basis) in lactating crossbred dairy cows thereby reducing the fodder cost by 20 per cent without affecting its milk yield or composition. *Gliricidia sepium* can replace upto 15 per cent of green fodder in lactating crossbred cows thereby reducing the fodder cost by 15 per cent without affecting the milk yield and its composition and could sustain the milk production during fodder scarcity.

Level of Feeding of Fodders from Agroforestry Models

Leuceana leucocephala	*Bajra napier hybrid grass*	*Cenchrus ciliaris*
Large ruminants: 5 - 8 kg *Small ruminants: 2 - 3 kg*	*Large ruminants: 20 - 25 kg* *Small ruminants: 3 - 4 kg*	*Large ruminants: 20 - 25 kg* *Small ruminants: 3 – 4 kg*
Inga dulce	*Fodder sorghum*	*Desmanthus virgatus*
Small ruminants: 2 - 3 kg	*Large ruminants: 20 - 25 kg* *Small ruminants: 3 – 4 kg* Anti nutritional factor: *Cyanic acid (young leaves)*	*Large ruminants: 4 - 5 kg* *Small ruminants: 0.5 – 1 kg*

Gliricidia sepium	*Guinea grass*	*Stylosanthus spp.*
 *Large ruminants: 3 - 5 kg* *Small ruminants: 1- 2 kg* *Wilting in shade for 5 hours is essential for feeding the animals.*	 *Large ruminants: 20 - 25 kg* *Small ruminants: 3 - 4 kg*	 *Large ruminants: 4 - 5 kg* *Small ruminants: 0.5 – 1 kg*

Ecological Benefits

Alternate land use systems such as agroforestry, agro-horticultural, agro-pastoral, and agro silvipasture are more effective for soil organic matter restoration (Manna et al 2003). Tree -based agro ecosystems have more closed nutrient cycles that help conserve soil productivity. Planting and pruning N-fixing legumes is a feasible way to add nitrogen to the systems. There is robust evidence that agroforestry systems have potential for improving water use efficiency by reducing the unproductive components of the water balance (Turner et al 2002). Trees and shrubs in agroforestry models play a vital role in maintaining an ecological balance and improving the livelihood of people in the arid regions. These prevent soil erosion supply, forage for livestock and act as source of fuel wood and timber.

Livelihood Improvement

The livelihoods improvement through natural resource management seeks to understand individual or household strategies through which they make long term progress towards a better quality of life (Campbell et al 2003, Pretty et al 2005). The adverse impact of climate change may be more severely felt by poor people who are more vulnerable than rich. Appropriate policy responses combining the agro ecosystems as key assets can strengthen adaptation and help build the resilience of communities and households to local and global change. It has been shown in different studies that the multiple use silvipastoral system is more economically attractive in addition to multiple ecological benefits. *Leucaena leucocephala*, *Gliricidia sepium* and *Cenchrus ciliaris* silvipasture in dry land yielded 10.47 t dry fodder biomass/ha. When 30 lambs were fed with the fodder harvested from silvipasture for 9 months, the growth rate of lambs was increased by 68 per cent. By integrating the livestock with agroforestry the farmers can improve their livelihood by increase in their revenue.

CONCLUSION

Silvipastroal systems are important source of fodder and may overcome the fodder shortage in the country. Silvipastroal systems gives about 18-20 tonne of green fodder per hectare through integrating different grasses (*Cenchrus ciliaris* with *Stylosanthus hamata* etc.) together with fodder trees. Tree leaf fodder are rich in nutrients and small and large ruminants are benefitted when these fodder are given to them in different proportion based upon palatability. Besides fodder, silvipastoral systems provide several benefits such as soil restoration, nitrogen fixation, livelihood improvement etc.

REFERENCES

Campbell BM and Sayer JA 2003 Integrated Natural Resource Management: Linking Productivity, the Environment and Development, CABI Publishing, Wallingford, UK, 315 pp.

Devendra C, 1992 Nutritional potential of fodder trees and shrubs as protein sources in ruminant nutrition. In: Speedy, A., Pugliese, P.L. (Eds.), Legume Trees and Other Fodder Trees as Protein Sources for Livestock. FAO, Rome, 1992

Manna MC, Ghosh PK and Acharya CL, 2003 Sustainable crop production through management of soil organic carbon in semiarid and tropical India. *Journal of Sustainable Agriculture*, **21**, 87-116

Pretty JN, Morison JIL, and Hine RE, 2005 Reducing food poverty by increasing agricultural sustainability in developing countries. *Agricultural Ecosystem Environment,* **95**, 217-234

Turner NC, and Ward PR, 2002 The role of agroforestry and perennial pasture in mitigating water logging and secondary salinity: summary. *Agricultural. Water Manag*ement, **53**, 271-275

CHAPTER 38

Policy Linkages for Scaling-up of Agroforestry in India

Mahendra Singh*, KB Sridhar, Dhiraj Kumar, Vikas Kumar[1], SK Singh[2], RP Dwivedi, Inder Dev, RK Tewari and OP Chaturvedi

ICAR-Central Agroforestry Research Institute, Jhansi-284003 (U.P.) India

[1]ICAR-Indian Grassland and Fodder Research Institute, Jhansi-284003 (U.P.) India

[2]Bundelkhand University, Jhansi (U.P) India

Corresponding author's Email: Mahendra.singh8@icar.gov.in

ABSTRACT

This article is an attempt to assess the understanding of policy linkages between agroforestry and other related sectors of the economy. The scaling-up of any technology in agriculture sector as elsewhere is often undermined by conflict among multiple stakeholders. Research institutions have developed various financially viable agro-specific agroforestry models for entire country. However, due to lack of coordination among various stakeholders, the adoption level so far has not yet achieved as desired. The study confirms the importance of policy interactions between agroforestry and other related sectors, particularly within the broader framework of macroeconomic policies. However, the information available on inter-sectoral policy linkages related to agroforestry forestry is fragmentary and does not lend itself to systematic and comparative analysis. A Significant research effort is thus indispensable both for generating information and for developing an appropriate analytical framework.

Keywords: Agroforestry, Financial viability, Policy-linkages, Scaling-up

INTRODUCTION

It is well documented that a combination of various policies influence the scaling-up of agroforestry research in any country. Policy framework at multiple scales play seminal role in the multi-scaled nature of agroforestry. Socio- ecological system, for example open the opportunities for markets in ecosystem services, price policy for agroforestry products, trade policy, credit policy, subsidy policy, energy policy and forest policy are substantially effects the development of agroforestry. Research in this area is needed to address the effects of these policies on agroforestry and other sectors to evaluate how these policy measures can reduce negative inter-sectoral effects and enhance positive inter-sectoral effects. Research on institutional implications of inter-sectoral linkages, such as the impact of restructuring institutional responsibilities to increase collaboration and coordination between institutions are imperative for sustainable development of agroforestry.

Scaling-up means expanding, replicating, adapting and sustaining successful projects, programs, or policies over time in a geographic space for greater benefits and impacts. Scaling up can occur horizontally, vertically and diagonally. Scaling can occur directly, in which an organization's, initiative is directly responsible for change, and indirectly, in which these actors influence others to implement key changes. Development practitioners have suggested useful insights on scaling-up such as identification of good practice indicators, stages, conditions and pathways. The conditions for scaling-up of agroforestry systems depend on resources, partnerships, local context and knowledge management. The lack of market intelligence and market failure for raw wood were recognized as major determinants for scaling –up agroforestry systems. Price volatility of raw wood is a most important factors for variability in area under agroforestry. Price-discovery mechanism of raw wood is in favour of wood-based industries. Distortions in wood market hinder the farmers enthuse to adopt agroforestry in their fields.

To promote agroforestry, various major policy initiatives, including the National Agriculture Policy 2000; Planning Commission Task Force on Greening India 2001; National Bamboo Mission 2002; National Policy for Farmers; 2007; Green India Mission 2010 and finally; a dedicated National Agroforestry Policy was approved by Government of India in 2014. This policy has recommended for setting up of a Mission or Board to address development of agroforestry sector in an organised manner ((MoA 2014). The Sub-Mission on Agroforestry (SMAF) under National Mission on Sustainable Agriculture (NMSA) is an initiative to this end. The NMSA, one of the eight mission under National Action Plan on Climate Change (NAPCC) seeks to address issues related to climate change through adaptation and mitigation strategies. The aim of the sub-mission is to expand the tree coverage on farmland in complementary with agricultural crops (MoA & FW 2016).

Farmers allocate their resources in production of various commodities on the basis of signals they receive from markets. The allocation of area under crops influenced by the perceived relative profitability of different crops. The growing of tree on farms for market seemed to farmers in many regions of India a more profitable option than field crops. However, in spite of relative higher profitability to farmers and several concerted efforts made by government, the adoption of agroforestry has not yet reached as expected.

Agroforestry in Different Agro-Climatic Zones

The total area under agroforestry in nine agro-climatic zones is 13.62 million ha. In terms of the area, maximum was in Upper-Gangetic Plain (15.47%), followed by West-Coast Plains and Hill regions (14.18%) among nine agro-climatic zones in India (Table1).

Table 1: *Agro-climatic zones wise estimated area under agroforestry, India, 2016-17*

Agro-climatic Zone	*Geographical area (million ha)*	*Agroforestry area (million ha)*	*Agroforestry area of geographical area (%)*
Lower- Gangetic Plains Region	6.73	0.80	11.91

Middle- Gangetic Plains Region	16.57	1.30	7.87
Upper- Gangetic Plains Region	14.44	2.23	15.47
Trans- Gangetic Plains Region	11.60	1.14	9.85
Central Plateau and Hill Region	37.84	1.93	5.09
Western Plateau and Hill Region	32.74	1.56	4.75
West Coast Plains and Hill Region	11.68	1.66	14.18
Gujarat Plains and Hill Region	18.98	2.57	13.56
Western Dry Region	17.87	0.43	2.41
Total/Average	168.46	13.62	8.09

Source: ICAR (2017) Annual Report (2016-17), Department of Agricultural Research and Education, Ministry of Agriculture and farmers Welfare, Government of India

Financial Viability of Agroforestry Models for Various Agro-Climatic Zones

The financial viability of prominent crop rotation i.e. rice-wheat in the state of Haryana is presented in table 2, which clearly indicates that this rotation is not able to support farmers to lend a dignified living. Therefore, diversification of this rotation is required through agroforestry which is much profitable and the farmers toward climate smart agriculture.

Table 2: *Economics of rice-wheat rotation in Haryana, 2013-14 (₹/ha)*

Particular	*Crop*	
	Paddy	*Wheat*
Cost of production (₹/q)	1389	1045
Average yield (q/ha)	47.6	45.3
MSP (₹/q)	1345	1400
Total cost of cultivation (₹/ha)	66133	47292
Total income of cultivation (₹/ha)	64022	63364
Net income of cultivation (₹/ha)	-2111	16072
Annual Net income of cultivation (₹/ha)	13961	

Source: Authors' estimate on is based on the data of Commission on Agricultural Costs and Prices, MoA & FW, Govt. of India.

The financial viability of agroforestry models for specific agro-climatic regions is presented in Table 3. It is verified that net present worth (NPW) of the models are profitable and if we convert with price index, than the value would change many times. All the financial indicators show sufficient return on investment in agroforestry.

Table 3: *Financial viability of prominent agroforestry systems in various Agro-climatic zones, India (Year: 2001)*

S. No	Agro-climatic Zone	Situation (Rainfed/ Irrigated / Non arable land)	Agroforestry system	Duration (year)	Tree (Common name)	Tree (Scientific name)	Crop/ Grass	NPV (₹/ha)	B/C ratio	FIRR (%)	AEV (₹/ha)	LEV (₹/ha)
1.	Western Himalayan	Rainfed	Agrihorticulture	15	Peach	*Prunus persica*	Maize, Soyabean	12,039	1.21	30	2,059	13,726
2.	Eastern Haimalayan	Irrigated	Agrisilviculture	8	Kadamba	*Anthocephalus cadamba*	Paddy	24,348	1.56	32	5,426	36,173
3.	Lower Gangetic Plains	irrigated	Agrisilviculture	10	Eucalyptus	*Eucalyptus hybrid*	Paddy, Wheat	114,532	2.33	33	22,821	152,138
4.	Lower Gangetic Plains	Rainfed	Agrisilviculture	15	Australian Babool	*Acacia auriculiformis*	Jute	4,558	1.17	30	779	5,197
5.	Middle Gangetic Plains	Rainfed	Agrisilviculture	20	Sheesham	*Dalbergia sissoo*	Seasamum	9,030	1.63	36	1,443	9,618
6.	Middle Gangetic Plains	Non- arable land	Silvipasture	10	Bamboo	*Bambusa vulgaris*	Marvel grass	18,907	1.76	56	3,767	25,115
7.	Upper Gangetic Plains	Irrigated	Agrisilviculture	7	Poplar	*Populus deltoides*	Paddy, Wheat	190,241	3.31	69	45,726	304,843
8.	Trans Gangetic Plains	Rainfed	Agrisilviculture	20	Babul	*Acacia nilotica*	Bajra, Guar	27,788	2.61	39	4,439	29,596
9.	Eastern Plateau and Hill	Irrigated	Agrisilviculture	12	Gumhar	*Gmelina arborea*	Paddy, Linseed	50,857	1.74	34	9,382	62,548
10.	Eastern Plateau and Hill	Non- arable land	Agrisilviculture	15	Anjan	*Hardwickia binata*	Dinanath grass	68,035	2.13	48	11,635	77,568
11.	Central Plateau and Hill	Rainfed	Agrihorticulture	8	Aonla	*Emblica officinalis*	Groundnut, Gram	7,409	1.09	28	1,651	11,007
12.	Western Plateau and Hill	Non- arable land	Silvipasture	16	Siris	*Albizia amara*		43,768	1.69	44	7,351	49,005
13.	Southern Plateau and Hill	Irrigated	Agrisilviculture	15	Kapok	*Ceiba pentendra*	Groundnut	46,746	1.43	29	7,994	53,296
14.	Southern Plateau and Hill	Rainfed	Agrisilviculture	10	Imli	*Tamarindus indica*	Chilli	29,101	2.03	33	5,799	38,657
15.	West Coast and Ghat	Irrigated	Agrisilviculture	7	Beefwood	*Casurina equisetifolia*	Paddy	33,907	1.39	26	8,150	54,332
16.	West Coast and Ghat	Irrigated	Agrihorticulture	8	Coconut	*Cocos nucifera*	Elephant Foot, Ginger	17,650	1.27	33	3,933	26,222
17.	West Coast and Ghat	Rainfed	Agrisilviculture	20	Teak	*Tectona grandis*	Sweet Potato	210,246	2.00	31	33,589	223,928
18.	Gujarat Plains and Hill	Rainfed	Agrisilviculture	20	Sheesham	*Dalbergia Sissoo*	Castor	8,037	2.14	31	1,284	8,560
19.	Western Dry	Rainfed	Agrisilviculture	20	Khejri	*Prosopis cineraria*	Bajra	54,409	2.53	37	8,692	57,950
20.	Western Dry	Non- arable land	Silvipasture	10	Ber	*Ziziphus nummularia*	Cenchrus ciliaris	19,810	1.53	34	3,947	26,315
21.	The Isalands	Rainfed	Agrisilviculture	6	Coconut	*Cocos nucifera*	Paddy	6,325	1.17	32	1,671	11,143

Source: Planning Commission (2001) Report of the Task Force on Greening India for Livelihood Security and Sustainable Development, Government of India, New Delhi

Note:

(i) **NPV:** Net present value is determined by discounting all revenues and costs to the present;

(ii) **B/C ratio:** benefit/cost ratio is calculated by dividing the sum of discounted revenues by the sum of discounted costs;

(iii) **FIRR:** Financial internal rate of return is a discount rate at which an investment's NPV equal zero;

(iv) **AEV:** Annual equivalent value is an indicator that expresses NPV in annual equivalents distributed equally over the years of the lifespan of the investment; and

(v) **LEV:** Land expectation value is interpreted as the maximum amount of money a landowner can pay for the land and still earn the minimum acceptable rate of return of an agroforestry investments.

Price Crash of Poplar Since 2012-13

The price slump in short duration wood species like poplar and eucalyptus impacted farmers' income and diverted the investment in new plantations. The trend in nominal price of 'Over' category of Poplar shows continuous upward movement during 2005-06 to 2012-13 and sharp declining afterward. The 'Under' and 'Sokta' category of poplar followed the same trend except declining trend in 2006-07 to 2007-08. The highest price of all category of polar recorded in the year 2012-13. The nominal price of the poplar decreased to one third during last three years. The highest price of all category of poplar recorded in the year 2012-13 (Fig. 1).

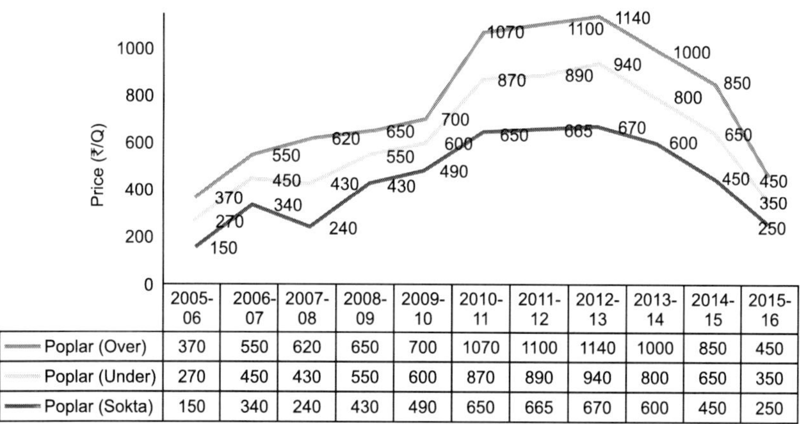

	2005-06	2006-07	2007-08	2008-09	2009-10	2010-11	2011-12	2012-13	2013-14	2014-15	2015-16
Poplar (Over)	370	550	620	650	700	1070	1100	1140	1000	850	450
Poplar (Under)	270	450	430	550	600	870	890	940	800	650	350
Poplar (Sokta)	150	340	240	430	490	650	665	670	600	450	250

Figure 1: *Trend in nominal price of poplar 2005-06 to 2015-16*

Import of Wood and Wood-Based Products, India

Despite the sharpened price crash of wood in India since 2013-14, the quantity of net import of wood and wood-based products is increasing. This is an example of policy fatigue towards agroforestry farmers and discourage them for new plantations (Table 4).

Table 4: *Status of Import and export of wood based products, India, 2015-16 (Value in ₹ Crore)*

HS code	Commodity	Import	Share (%)	Export	Share (%)	Net import	Custom duty in 2005 (%)	Custom duty in 20015 (%)
44	Wood and articles of wood; wood charcoal.	15502	34.6	2912	23.4	12590	12.5	5
45	Cork and articles of cork.	35	0.1	13	0.1	22	12.5	5
46	Manufactures of straw, of esparto or of other plaiting materials; basketware and wickerwork.	14	0.0	49	0.4	-35	12.5	5
47	Pulp of wood or of other fibrous cellulosic material; waste and scrap of paper or paperboard.	10699	23.9	74	0.6	10625	12.5	5
48	Paper and paperboard; articles of paper pulp, of paper or of paperboard.	16205	36.2	7494	60.2	8711	6.5	10
49	Printed bookds, newspapers, pictures and other products of the printing industry; manuscripts, typescripts and plans.	2288	5.1	1898	15.3	390	6.5	10
	Total	44743	100	12440	100	32303		

Source: Authors estimates based on data from office of the Directorate General of Foreign Trade (DGFT), Ministry of Commerce and Industry, Govt. of India

Perceptions of Various Stakeholders in Agroforestry Sector for Price Crash of Poplar Since 2012-13

The perceptions of various stakeholders in agroforestry sector for price crash of poplar since 2012-13 is presented in table 5.

Table 5: *Perceptions of various stakeholders in agroforestry sector for price crash of poplar since 2012-13*

Stakeholder	Reasons narrated for price crash of timber logs
Agroforestry Farmers	Price decided by plywood industry's representative resulted distress sale by farmers, collusion of commission agent and plywood industry representative
Forest officials	Normal price discovery due to market demand and supply mechanism
Plywood industry managers	High supply of raw wood and low demand of finished goods due to slump in real estate market, substitution of wood products by plastics, cheap import of plywood, higher cost of inputs like adhesive and industry running under capacity due to low demand
Paper industry managers	Slow demand of paper products in market and availability of cheap raw material from plywood industry (Waste material), Only very small quantity of Sokta quality of Eucalyptus and Poplar purchase as raw material
Commission agents	Farmers are solely responsible due to higher production at once

Furniture manufacturers	Cheaper import of timber logs and unprocessed plyboard especially from China
Agroforestry researchers	Cyclic fluctuations, Normal phenomena on the basis of past price trend
Policy analyst and agricultural economists	Restriction of new entrant in plywood industry, Oligospony market, probability of cartel in plywood industry, Plywood Industry making super-normal profit, lack of government control on plywood industry, market policy failure, low custom duty on import of wood and wood products, lack of minimum support price on timber logs, term of trade not in in favour of agroforestry farmers

Inter-Sectoral Policy Linkages Related to Agroforestry Sector in India

The present study is based on various policy documents and their interaction with agroforestry sector. Secondary data published from various ministry and primary data collected from Yamuna Nagar timber market in Haryana were used to test the hypothesis of the study. The inter-sectoral policy linkages related to agroforestry sector are presented in following policy framework (Table 6):

Table 6: *Inter-sectoral policy linkages related to agroforestry sector in India*

Policy area	Policy instrument	linkage	Agroforestry related effect
Macroeconomic policies	Fiscal policies-taxation	Taxation on agroforestry products	Capacities for reinvestment by farmers and wood-based industries
	Monetary and credit policies	Credit allocation and interest rate	Investment capacities in resource management
	Foreign investment regulations	Infrastructure development	Reduction or increase in agroforestry based industries
	Trade policies: WTO commitments	To liberalize trade in industrial products include reductions in tariffs and removal of quantitative restrictions	Incentive or constraint to production and export
	Trade policies: exchange rate	Valuation of imported and exported commodities	Costly or cheaply import or export
	Import control	Custom duties and fumigation of imported products, which reduced imports of equipment and efficient technologies	Positive and negative impact on farmers and wood based industries respectively.
	Export promotion	Export subsidies	Overutilization of selected species

Structural adjustment	Public investment and privatization	Investment through Agroforestry Mission and expansion of role of private sector	Social welfare and efficiency in production, marketing and trade
	Pricing policies for inputs and outputs	Correcting market distortions	Valuation of outputs
Land distribution and reservation	Legislation on land ownership	Land titling and tenure	Security of rights and exploitation of resources
	Legislation on right for production, transportation and trade	Adoption of agroforestry systems and Investment	Production of Selective commercial tree species, forest reserves and encroachment of forest resources
Agriculture	Food policy	Food production versus wood production	Intensification of agriculture through agroforestry and complementarity between these two sector
	Pricing policy	Agricultural pricing versus wood pricing	Allocation of scarce resources due to price risk aversion
	Agricultural research policy	The investment on agricultural research and economic rates of return to agriculture / agroforestry research	Share of agricultural research to total agricultural gross Domestic Products (GDP) and share of agroforestry in agricultural research
Energy	Energy self-reliance	Household and industries consumption and fuel substitution	Fuelwood demand and pollution
Environment	India's Intended Nationally Determined Contribution (INDC) is to create an additional carbon sink of 2.5 to 3 billion tonnes of CO2 equivalent through additional forest and tree cover by 2030	Investment in plantation	Agroforestry is viable option for area expansion under green cover
Poverty	Income and employment generation policy	Resource poor marginal and small farmers and disguised unemployment in rural areas	Agroforestry sector generates income and employment and reduce disguised unemployment in rural areas

CONCLUSION

The study verified the impact of various inter-sectoral policy such as taxation on agroforestry products discourage the investment made by both farmers and wood-based industries. While, credit facilities at cheaper interest rate, infrastructure development positively influenced the agroforestry sector. Term of trade between agroforestry and other sector with price volatility of outputs are key issues in term of policy perspective. Inter-sectoral policy linkages are necessary condition for scaling-up agroforestry in India. The study confirms the importance of policy interactions between agroforestry and other related sectors, particularly within the broader framework of macro-economic policies. However, the information available on inter-sectoral policy linkages related to agroforestry is fragmentary and does not lend itself to systematic and comparative analysis. A significant research effort is thus indispensable both for generating information and for developing an appropriate analytical framework. The gaps in understanding of how inter-sectoral linkages operate should be filled as soon as possible by using General Equilibrium Model. A challenging and urgent task lies ahead, calling for the mobilization of efforts and resources.

REFERENCES

IFPRI 1992 Priorities for forestry and agroforestry policy research: report of an international workshop. In: H Gregersen, P Oram and J Spears. International Food Policy Research Institute, Washington, D.C.

ICAR 2017 Annual Report (2016-17), Department of Agricultural Research and Education, Ministry of Agriculture and farmers Welfare, Government of India

Planning Commission 2001 Report of the Task Force on Greening India for Livelihood Security and Sustainable Development, Government of India, New Delhi.

CACP 2015 Commission on Agricultural Costs and Prices, Mo A & FW, Govt. of India.

Directorate General of Foreign Trade (DGFT), Ministry of Commerce and Industry, Govt. of India.